The Greek Alphabet

Alpha	A	α	Iota	I	ι	Rho	P	ρ
Beta	B	β	Kappa	K	κ	Sigma	Σ	σ
Gamma	Γ	γ	Lambda	Λ	λ	Tau	T	τ
Delta	Δ	δ	Mu	M	μ	Upsilon	Υ	υ
Epsilon	E	ϵ	Nu	N	ν	Phi	Φ	ϕ
Zeta	Z	ζ	Xi	Ξ	ξ	Chi	X	χ
Eta	H	η	Omicron	O	o	Psi	Ψ	ψ
Theta	Θ	θ	Pi	Π	π	Omega	Ω	ω

Abbreviations for Units

A	ampere		L	litre
Å	angstrom (10^{-10} m)		m	metre
atm	atmosphere		MeV	megaelectronvolts
C	coulomb		Mm	megametre (10^6 m)
°C	degree Celsius		mi	mile
cal	calorie		min	minute
cm	centimetre		mm	millimetre
dyn	dyne		ms	millisecond
eV	electronvolt		N	newton
°F	degree Fahrenheit		nm	nanometre (10^{-9} m)
fm	femtometre, fermi (10^{-15} m)		pt	pint
ft	foot		qt	quart
Gm	gigametre (10^9 m)		rev	revolution
G	gauss		s	second
g	gram		T	tesla
H	henry		u	unified mass unit
h	hour		V	volt
Hz	hertz		W	watt
in	inch		Wb	weber
J	joule		y	year
K	kelvin		yd	yard
kg	kilogram		μm	micrometre (10^{-6} m)
km	kilometre		μs	microsecond
keV	kiloelectronvolts		μC	microcoulomb
lb	pound		Ω	ohm

PHYSICS

SECOND EDITION

VOLUME 2 # Physics

PAUL A. TIPLER

Oakland University

Rochester, Michigan

WORTH PUBLISHERS, INC.

Physics Second Edition, Volume 2

Paul A. Tipler

Copyright © 1976, 1982 by Worth Publishers, Inc.

All rights reserved

Printed in the United States of America

Library of Congress Catalog Card No. 81-70206

ISBN: 0-87901-183-1

Second Printing, July 1982

Designed by Malcolm Grear Designers

Composition by Progressive Typographers, Inc.

Illustrated by Felix Cooper and John Kramer

Cover by Harold E. Edgerton, MIT, and J. Kim Vandiver, MIT: Vortex formed at the tip of a fan blade by heated air above an alcohol lamp (stroboscopic color schlieren photograph at $\frac{1}{3}$ μs exposure)

Worth Publishers, Inc.

444 Park Avenue South

New York, NY 10016

For my parents, Perry and Ella Tipler

Preface

Preface

This second edition of *Physics* reflects the many helpful responses and suggestions from reviewers and users of the first edition. Like the first edition, it is a textbook for the standard two-semester or three-semester elementary physics course for engineering and science majors. It is assumed that the students have taken calculus, or are concurrently taking it. In the second edition, the order of topics has been changed to correspond more closely to the structure of most courses in elementary physics. In particular, two chapters on mechanical waves and sound (Chapters 14 and 15) now come right after mechanics so they can be taught in the first semester; and electricity and magnetism is now discussed before optics. The overall organization is now mechanics (Chapters 1–13), mechanical waves and sound (Chapters 14 and 15), thermodynamics (Chapters 16–19), electricity and magnetism (Chapters 20–31), optics (Chapters 32–34), and modern physics (Chapters 35–37). For those who prefer (as I do) to teach all waves together, Chapters 14, 15, 32, 33, and 34 can be taught easily as a unit, either before or after electricity and magnetism.

Many chapters, particularly in the mechanics portion of the book, have been rewritten, reordered, and combined for greater clarity and ease in reading and teaching. Some examples are

1. Chapter 2 ("Motion in One Dimension") has been reorganized and rewritten so that it is easier for the student who is just beginning calculus. Calculus topics such as the initial-value problem and integration are now in optional sections at the end of the chapter.

2. Chapters 4 and 5 ("Newton's Laws" and "Applications of Newton's Laws") have been rearranged so that the techniques of problem solving with examples of constant forces in one dimension appear in Chapter 4, after a statement of Newton's laws and a discussion of force and mass. The material on the basic interactions (formerly in Chapter 6) is now at the beginning of Chapter 5, before the applications of Newton's laws. In that chapter there is also a new section on torque and the static equi-

librium of rigid bodies. This early introduction of the concept of torque allows students to work problems in statics immediately after studying Newton's laws, and it also makes the study of rotational dynamics in Chapters 9 and 10 easier.

3. The chapters on work and energy have been rewritten and rearranged so that the somewhat formal and mathematical material formerly in Chapter 8 has become a more qualitative discussion that appears after energy has been defined in Chapter 6.

4. All the material on the center of mass is now presented together in Chapter 8, "Systems of Particles," which also includes collisions. Collisions are first discussed in a general reference frame, with the treatment of the center-of-mass reference frame relegated to an optional section.

5. The discussion of damped and driven oscillators (formerly in Chapter 15) has been shortened and simplified and placed at the end of the chapter on simple harmonic motion (Chapter 11).

6. The material on Gauss' law is now consolidated in Chapter 21, which also includes a discussion of conductors in electrostatic equilibrium.

There are three new chapters and several new sections. Chapter 13, "Solids and Fluids," contains material on elasticity and on the statics and dynamics of fluids. Chapter 18, "Thermal Properties and Processes," contains new material on the van der Waals gas, phase diagrams, and heat transfer. Chapter 37, "Nuclear Physics," is an elementary introduction to the properties of nuclei, radioactivity (including carbon dating), and nuclear reactions (including both fission and fusion). Among the new sections are "Scientific Notation and Significant Figures," in Chapter 1; "Systems with Varying Mass: Rocket Motion," in Chapter 8; and "General Relativity," in Chapter 35.

There are over 2,000 exercises and problems in the new edition. The number of problems has been increased by about 33%. Most of the new problems are fairly easy; they will provide a smooth transition in level between the exercises and the more challenging problems. However, a few of the new problems should challenge even the best students.

SI units are now used almost exclusively throughout the text, with all examples, exercises, and problems given in SI units. The only exceptions are in Chapter 4, where there are a few exercises on the conversion of force units (pounds to newtons, etc.), and in Chapter 18, where R factors for insulation materials are used with U.S. customary units.

Three new essays have been written for this edition, "Solar Energy" by Laurent Hodges, "Exponential Growth" by Albert Bartlett, and "James Clerk Maxwell" by C. W. F. Everitt.

A feature new to this edition is the short list of learning objectives at the beginning of each chapter, to help students focus on the important information to be presented, to give them an idea of the depth of understanding that will be required, and to serve as an aid for later review. Individual instructors may wish to add or subtract from these lists according to their preferences.

Some features of the first edition that have been retained are:

1. The book is written at a level at which it can be read and enjoyed by the less well-prepared students, and yet be challenging to the better-prepared students. To make the book appeal to a wide range of student abilities, many new concepts are introduced in an intuitive, nonmathe-

matical way, often proceeding from a specific example to a more general relation. Typically, the general or formal derivation follows in an optional paragraph at the end of the section or chapter.

2. Questions are presented at the ends of various sections within each chapter. Some of these questions are routine and can be easily answered from the material in the preceding section; others are open-ended and can serve as a basis for classroom discussion.

3. A review section consisting of a checklist of words and phrases and a set of true-false questions appears at the end of each chapter. Each term in the checklist is accompanied by the page number where it is defined, and answers to all the true-false questions are given at the back of the book.

4. There is an extensive set of relatively easy exercises followed by a set of more difficult problems at the end of each chapter. The exercises are organized by sections within the chapter. Answers to all the odd-numbered exercises and problems are given at the back of the book. (Answers to the even-numbered exercises and problems, and some worked-out solutions, are in the Instructor's Manual.)

5. Fourteen essays from the first edition have been retained. Of the seventeen essays now in the book, four are biographical (on Isaac Newton, Benjamin Franklin, James Maxwell, and Albert Einstein) and the rest are on applications (e.g., xerography, transistors, radar astronomy) or topics of current interest (e.g., thermal pollution, exponential growth, black holes, sonic booms). These brief essays are included for the enjoyment and enlightenment of students and instructors. There are no exercises or problems connected with the essays.

6. The book's readability is enhanced by its attractive physical layout: its large pages, wide margins, two-color line drawings, many photographs, margin comments highlighting important concepts and equations, ruled color lines setting off important equations, and optional material in the right-hand column bounded on each side by colored rules.

Acknowledgments

Many people have contributed to one or both editions of this book. I would like to thank everyone who used the first edition and sent me comments and corrections, and all those who tried out early drafts of the book. Some of the people who were particularly helpful with the first edition were James Gerhart, Jerry Griggs, John McKinley, Granvil Kyker, and Stanley Williams. I would like to thank Oakland University for granting me a leave during which much of the work on the new edition was done, and the physics department at the University of California at Berkeley for its kind hospitality during my stay there. Peter Matthews of the University of British Columbia contributed most of the interesting new problems, and Mark Scheuern of Oakland University was a great help in the preparation of the answer section. Ron Brown of California Polytechnic State University deserves special thanks for his review of the first edition, the entire revised manuscript, and the galleys of the second edition. He made many important and useful suggestions for improvement.

Others who reviewed various parts of the first edition to offer suggestions for the second edition, or who reviewed the manuscript for the second edition, were: Zvi Bar-Yam, Southeastern Massachusetts University; Henry Bass, University of Mississippi; Ernst Bleuler, Pennsylvania State University, University Park; Sumner Davis, University of California, Berkeley; Paul Doherty, Oakland University; Lowell Eliason, California State University, Long Beach; David A. Glocker, Rochester Institute of Technology; Richard Haracz, Drexel University; Richard T. Kouzes, Princeton University; Andrew C. Kowalik, Nassau Community College; Alan J. Lazarus, Massachusetts Institute of Technology; Sherman Lowell, Washington State University; Robert Luke, Boise State University; W. Anthony Mann, Tufts University; Paul F. Nichols, San Diego State University; Pierre Piroue, Princeton University; L. Arthur Read, Wilfrid Laurier University; Charles Rogers, Southwestern Oklahoma University; John Stevens, Sacramento State University; Jacques Templin, San Diego State University; T. A. Wiggins, Pennsylvania State University, University Park; Robert Williamson, Oakland University; John Van Zytveld, Calvin College.

I was fortunate in having two excellent typists for this edition, Julie Thompson and Marshall Tuttle. As usual, my wife Sue has given me continual support throughout this long project. Finally, I would like to thank everyone at Worth Publishers for their help and encouragement, and particularly June Fox, for her tireless efforts in coordinating this project.

Paul A. Tipler
Rochester, Michigan

February 1982

Contents

Contents in Brief

VOLUME 2

VOLUME 1   Contents

VOLUME 2   Contents

PHYSICS

CHAPTER 20   The Electric Field

Objectives   After studying this chapter you should:

1. Be able to state Coulomb's law and use it to find the force exerted by one point charge on another.

2. Know the value of the Coulomb constant in SI units.

3. Know the magnitude of the electronic charge e in coulombs.

4. Be able to use Coulomb's law to calculate the electric field for both discrete and continuous charge systems.

5. Be able to draw the lines of force for simple charge systems and to obtain information about the direction and strength of an electric field from such a diagram.

6. Know that a spherically symmetric shell charge distribution produces zero electric field inside the shell and produces a field outside the shell the same as that of a point charge at the center of the shell.

7. Be able to state the difference between a polar and nonpolar molecule and describe the behavior of each in a uniform electric field and in a nonuniform electric field.

Bruce Roberts/Rapho Guillumette

A lightning discharge is a spectacular effect resulting from electromagnetic forces.

The electrostatic force between two point charges was examined briefly in Chapter 5 in our discussion of forces in nature. We now begin a more detailed study of electricity. After a short history of the development of the concept of electric charge, we again state Coulomb's law for the force between two charges and define the electric field. We then discuss some general properties of electric fields, show how to calculate them from various charge distributions, and look at the behavior of point charges and electric dipoles in electric fields.

20-1   Electric Charge

Observations of electrical attraction can be traced back to the ancient Greeks.* The Greek philosopher Thales of Miletus (640–546 B.C.) observed that when amber is rubbed, it attracts small objects such as straw or feathers. (This attraction was often confused with the magnetic attraction of lodestone for iron.) Little more of electrical phenomena was understood until the sixteenth century, when the English physician William Gilbert (1540–1603) studied electrical and magnetic phenomena systematically. Gilbert showed that many substances besides amber acquire an attractive property when rubbed. He was one of the first to understand clearly the distinction between this attraction and the magnetic attraction, and he introduced the terms electric force, electric attraction, and magnetic pole. (The word electric comes from the Greek *elektron*, meaning amber, and the word magnetic comes from *Magnesia*, the country where magnetic iron ore was found.) Gilbert is perhaps best known for his discovery that the behavior of a compass needle results because the earth itself is a large magnet with poles at the north and the south; the orientation of a compass needle near the earth is merely an example of the general phenomenon of the attraction or repulsion of the poles of two magnets. Gilbert apparently failed to observe electric repulsion.

Around 1729 Stephen Gray, an Englishman, discovered that electric attraction and repulsion can be transferred from one body to another if the bodies are connected by certain substances, particularly metals. This discovery was of great importance, since previously experimenters could electrify an object only by rubbing it. The discovery of electric conduction also implies that electricity has an existence of its own and is not merely a property somehow brought out in a body by rubbing. The existence of two kinds of electricity was suggested by Charles François Du Fay (1698–1739), who described an experiment in which a gold leaf is attracted by a piece of glass rod previously rubbed. When the leaf is touched by the glass, it acquires the "electric virtue" and then repels the glass. "It is certain that bodies which have become electric by contact are repelled by those which have rendered them electric; but are they repelled likewise by other electrified bodies of all kinds?"** Du Fay answers this question by noting that the gold leaf, which is repelled by the glass rod, is attracted by an electrified piece of amber or resin. If the gold leaf is electrified by touching it to an electrified piece of amber, it is repelled by the amber but attracted by the electrified glass. He gave the two kinds of electricity the names vitreous and resinous and postulated the existence of two fluids which become separated by friction*** and are neutralized when they combine.

In 1747 the great American statesman and scientist Benjamin Franklin proposed a one-fluid model of electricity, described in the following experiment. If person *A* standing on wax (to insulate him from

Portrait of Charles Augustin de Coulomb.

---

* Much of this discussion follows Sir Edmund Whittaker's excellent *A History of Theories of Aether and Electricity*, Nelson, London, 1953; Torchbook edition, Harper, New York, 1960.

** *Memoir de l'Académie* (1733), as quoted by Whittaker.

*** We now know that the transfer of charge from one substance to another is not associated with friction but with the close contact of the substances achieved by rubbing them together.

A 1745 woodcut showing electrical effects being transmitted through two persons before attracting feathers or bits of paper at table on left. (*Courtesy of the Deutsches Museum, Munich.*)

the ground) rubs a glass tube with a piece of silk cloth, and another person B, also standing on wax, touches the glass tube, both A and B become electrified. They can each give a spark to person C standing on the ground. However, if A and B touch each other before touching C, the electricity of A and B is neutralized. Franklin proposed that every body has a "normal" amount of electricity. When a body is rubbed against another, some of the electricity is transferred from one body to the other; thus one has an excess and the other an equal deficiency. The excess and deficiency can be described with plus and minus signs: one body is plus and the other minus. An important feature of Franklin's model is the implication of conservation of electricity, now known as the *law of conservation of charge.* The electric charge is not created by the rubbing; it is merely transferred. Since Franklin chose to call Du Fay's vitreous kind of electricity positive, Du Fay's resinous electricity was merely the lack of vitreous electricity, or negative. A glass rod when rubbed acquires an excess of electricity and is positive, whereas an amber rod loses electricity when rubbed and becomes negative.

Conservation of charge

Franklin's choice was unfortunate because we now know that it is electrons that are transferred in the rubbing process, and according to Franklin's convention the electrons have a negative charge. Thus when we say that a (positive) charge is transferred from body A to body B, it is really electrons that are transferred from body B to body A. When glass is rubbed with silk, electrons are transferred from the glass to the silk, leaving the glass positive and the silk negative; when amber is rubbed with fur, electrons are transferred from the fur to the amber.

In the course of his experiments, Franklin noticed that small cork balls inside a metal cup seemed to be completely unaffected by the electricity of the cup. He asked his friend Joseph Priestley (1733–1804) to check this fact, and Priestley began experiments which showed that there is no electricity on the inside surface of a hollow metal vessel (except near the opening). From this result Priestley correctly deduced that the force between two charges varies as the inverse square of the distance between them, just like the gravitational force between two masses.

The inverse-square force law for electricity was confirmed by experiments of Charles Coulomb (1736–1806) using a torsion balance of his own invention. [The torsion balance was independently invented by the English scientist John Michell (1724–1793), who used it to show that the force between two magnetic poles also varies as the inverse square of the distance between the poles.] Coulomb's experimental apparatus was essentially the same as that described (Chapter 12) for the Cavendish experiment, with the masses replaced by small charged balls. For the magnitudes of charges easily transferred by rubbing, the gravitational attraction of the balls is completely negligible compared with the electric attraction or repulsion. Coulomb also used his torsion balance to confirm that the force between two magnetic poles varies as the inverse square of the distance.

The nineteenth century saw rapid growth in the understanding of electricity and magnetism, culminating with the great experiments of Michael Faraday (1791–1867) and the mathematical theory of James Clerk Maxwell (1831–1879). The experiments at the beginning of the twentieth century should be mentioned in this brief historical discussion of electric charge. In 1897 the English physicist J. J. Thomson showed that all materials contain particles which have the same charge-to-mass ratio. We now know that these particles, called *electrons*, are a fundamental part of the makeup of all atoms. In 1909 the American physicist Robert Millikan discovered that electric charge always occurs in integral amounts of a fundamental unit; i.e., the charge is *quantized*. Any charge can be written $q = Ne$, where N is an integer and e is the magnitude of the fundamental unit. The electron has charge $-e$, and the proton $+e$. The electron and the proton are very different particles. For example, the mass of the proton is nearly 2000 times that of the electron, and protons exert strong nuclear forces on neutrons, protons, pi mesons, etc., whereas electrons do not participate in the strong nuclear interaction. Yet the magnitude of the charge of the electron is exactly equal to that of the proton. Most elementary particles have either no charge (e.g., the neutron), a charge of $+e$, or a charge of $-e$. Particles made up of combinations of elementary particles may have a charge of $2e$, $3e$, etc. For example, the alpha particle, which is the nucleus of the helium atom, consisting of two protons and two neutrons bound tightly together, has a charge of $+2e$. An atom with Z protons in its nucleus (Z is the *atomic number*) has Z electrons outside the nucleus and is electrically neutral.

In this brief historical survey we have mentioned three important properties of electric charge:

1. Charge is conserved.

2. Charge is quantized.

3. The force between two point charges varies as the inverse square of the distance between the charges.

Coulomb's torsion balance. (*Courtesy of the Smithsonian Institution.*)

Charge quantization

Questions

1. Compare the properties of electric charge with those of gravitational mass. Discuss similarities and differences.

2. How might the world be different if the charge of the proton were slightly greater in magnitude than that of the electron?

20-2 Coulomb's Law

The experiments of Coulomb and others on the forces exerted by one point charge on another are summarized in *Coulomb's law:*

The force exerted by one point charge on another is along the line joining the charges. It is repulsive if the charges have the same sign and attractive if the charges have opposite signs. The force varies inversely as the square of the distance separating the charges and is proportional to the magnitude of each charge.

Coulomb's law can be written as a simple vector equation. Let q_1 and q_2 be two point charges separated by a distance r_{12}, which is the magnitude of the vector $\mathbf{r}_{12}$ pointing from charge q_1 to charge q_2 (Figure 20-1). The force exerted by charge q_1 on charge q_2 is then

$$\mathbf{F}_{12} = \frac{kq_1 q_2}{r_{12}^2} \, \hat{\mathbf{r}}_{12} \qquad\qquad 20\text{-}1 \qquad \textit{Coulomb's law}$$

where k is a constant and $\hat{\mathbf{r}}_{12} = \mathbf{r}_{12}/r_{12}$ is the unit vector pointing from q_1 toward q_2. The force $\mathbf{F}_{21}$ exerted by q_2 on q_1 is the negative of $\mathbf{F}_{12}$ by Newton's third law. That is, $\mathbf{F}_{21}$ is equal in magnitude to $\mathbf{F}_{12}$ but opposite in direction.

Equation 20-1 includes the result that like charges repel and unlike charges attract. If both q_1 and q_2 are positive or both negative, the force $\mathbf{F}_{12}$ on q_2 is in the direction $\hat{\mathbf{r}}_{12}$ away from charge q_1 and the force $\mathbf{F}_{21}$ on q_1 is directed away from q_2. That is, if both charges have the same sign, the force is repulsive. Similarly, if one charge is positive and the other negative, the product $q_1 q_2$ is negative and the forces are attractive.

The constant k depends on the choice of units for charge. In SI units the unit of charge is the *coulomb* (C), which is defined as the amount of charge which flows past a point in a wire in one second when the current in the wire is one ampere. The ampere (A) is defined in terms of a magnetic force measurement which we shall describe later. In this system of units, the quantities F_{12}, q_1, q_2, and r_{12} in Equation 20-1 are all defined independently, and the constant k is determined by experiment. The measured value of k is

$$k = 8.99 \times 10^9 \text{ N·m}^2/\text{C}^2 \qquad\qquad 20\text{-}2$$

In most calculations, it is convenient to use the approximation

$$k \approx 9 \times 10^9 \text{ N·m}^2/\text{C}^2 \qquad\qquad 20\text{-}3 \qquad \textit{The Coulomb constant}$$

With his torsion balance Coulomb was able to show that the exponent of r_{12} in the force equation is 2 with an uncertainty of a few percent. Although this was the most accurate measurement then, experiments similar to Priestley's can now be done with much greater precision, and we now know that the exponent is 2 with an experimental uncertainty of about 2 parts in 10^9.

In a system of charges, each charge exerts a force on each other charge given by Equation 20-1. The resultant force on any charge is the vector sum of the individual forces exerted on that charge by all the other charges in the system.

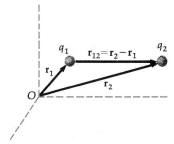

Figure 20-1
Charge q_1 at position $\mathbf{r}_1$ and charge q_2 at $\mathbf{r}_2$ relative to origin O. The force exerted by q_1 on q_2 is in the direction of the vector $\mathbf{r}_{12} = \mathbf{r}_2 - \mathbf{r}_1$ if both charges have the same sign, and in the opposite direction if the charges have opposite signs.

Example 20-1 Three positive point charges lie on the x axis; $q_1 = 25 \ \mu\text{C}$ is at the origin, $q_2 = 10 \ \mu\text{C}$ is at $x = 2$ m, and $q_3 = 20 \ \mu\text{C}$ is at $x = 3$ m (Figure 20-2). Find the resultant force on q_3.

The force on q_3 due to q_2, which is 1 m away, is in the positive x direction and has the magnitude

$$F_{23} = \frac{kq_2q_3}{r_{23}^2} = \frac{(9 \times 10^9 \text{ N·m}^2/\text{C}^2)(10 \times 10^{-6} \text{ C})(20 \times 10^{-6} \text{ C})}{(1 \text{ m})^2}$$
$$= 1.8 \text{ N}$$

The force on q_3 due to q_1, which is 3 m away, is also in the positive direction. Its magnitude is

$$F_{13} = \frac{kq_1q_3}{r_{13}^2} = \frac{(9 \times 10^9 \text{ N·m}^2/\text{C}^2)(25 \times 10^{-6} \text{ C})(20 \times 10^{-6} \text{ C})}{(3 \text{ m})^2}$$
$$= 0.50 \text{ N}$$

Since both forces are in the positive x direction, the resultant force is also in that direction and is just the sum of the magnitudes:

$$F = F_{23} + F_{13} = 2.3 \text{ N}$$

Note that the charge q_2, which is between q_1 and q_3, has no effect on the force F_{13} exerted by q_1 on q_3, just as the charge q_1 has no effect on the force exerted by q_2 on q_3.

Figure 20-2
Three point charges on the x axis for Example 20-1.

Example 20-2 Charge $q_1 = +25 \ \mu\text{C}$ is at the origin, charge $q_2 = -10 \ \mu\text{C}$ is on the x axis at $x = 2$ m, and charge $q_3 = +20 \ \mu\text{C}$ is at the point $x = 2$ m, $y = 2$ m. Find the resultant force on q_3.

Since q_2 and q_3 have opposite signs, the force exerted by q_2 on q_3 is attractive in the negative y direction, as shown in Figure 20-3. Its magnitude is

$$F_{23} = \frac{(9 \times 10^9)(10 \times 10^{-6})(20 \times 10^{-6})}{2^2} = 0.45 \text{ N}$$

The distance between q_1 and q_3 is $2\sqrt{2}$ m. The force exerted by q_1 on q_3 is directed along the line from q_1 to q_3 and has the magnitude

$$F_{13} = \frac{(9 \times 10^9)(25 \times 10^{-6})(20 \times 10^{-6})}{(2\sqrt{2})^2} = 0.56 \text{ N}$$

The resultant force is the vector sum of these two forces. Since F_{13} makes an angle of 45° with the x and y axes, its x and y components are equal to each other and to $F_{13}/\sqrt{2} = 0.40$ N. The x and y components of the resultant force are therefore

$$F_x = F_{23x} + F_{13x} = 0 + 0.40 = 0.40 \text{ N}$$

$$F_y = F_{23y} + F_{13y} = -0.45 + 0.40 = -0.05 \text{ N}$$

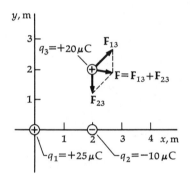

Figure 20-3
Force diagram for Example 20-2. The resultant force on charge q_3 is the vector sum of the forces $\mathbf{F}_{13}$ due to q_1 and $\mathbf{F}_{23}$ due to q_2.

The fundamental unit of electric charge e is related to the coulomb by

$$e = 1.60 \times 10^{-19} \text{ C} \qquad \text{20-4}$$

The charge of an electron

Typical laboratory charges of 10 to 100 nC, which can be produced by rubbing, involve the transfer of many electrons. For example, the number of electrons in 10 nC is

$$10 \times 10^{-9} \text{ C} \frac{1 \text{ electron}}{1.60 \times 10^{-19} \text{ C}} = 6.25 \times 10^{10} \text{ electrons}$$

Such charges do not reveal that electric charge is quantized. A million electrons could be added to, or subtracted from, this charge without detection by ordinary instruments.

Example 20-3 In the hydrogen atom the electron is separated from the proton by a distance of about 5.3×10^{-11} m on the average. What is the electrostatic force exerted by the proton on the electron?

Since the proton charge is $+e$ and the electron charge $-e$, the force is attractive and has the magnitude

$$\frac{ke^2}{r^2} = \frac{(9 \times 10^9)(1.6 \times 10^{-19})^2}{(5.3 \times 10^{-11})^2} = 8.2 \times 10^{-8} \text{ N}$$

This is much greater than the negligible gravitational force between the particles, which is

$$F_G = \frac{Gm_e m_p}{r^2} = \frac{(6.67 \times 10^{-11})(9.1 \times 10^{-31})(1.67 \times 10^{-27})}{(5.3 \times 10^{-11})^2}$$

$$= 3.6 \times 10^{-47} \text{ N}$$

Question

3. If the sign convention for charge were changed so that the electron charge was positive and the proton charge negative, would Coulomb's law be written the same or differently?

20-3 The Electric Field

In Chapter 12 we defined the gravitational field due to a set of masses to be the gravitational force exerted by these masses on a test mass m_0 divided by m_0. The electric field due to a set of charges is defined analogously. If we have a set of point charges q_i at various points in space and place a test charge q_0 at some point P, the force on it will be the vector sum of the forces exerted by the individual charges. Since each of these forces is proportional to the charge q_0, the resultant force is proportional to q_0. The electric field $\mathbf{E}$ at point P is defined to be this force divided by q_0*:

$$\mathbf{E} = \frac{\mathbf{F}}{q_0} \qquad \qquad \text{20-5} \qquad \textit{Electric field defined}$$

The SI unit of electric field is the newton per coulomb (N/C).

Example 20-4 When a 5-nC test charge is placed at a point, it experiences a force of 2×10^{-4} N in the x direction. What is the electric field $\mathbf{E}$ at that point?

From the definition, the electric field is

$$\mathbf{E} = \frac{2 \times 10^{-4} \mathbf{i} \text{ N}}{5 \times 10^{-9} \text{ C}} = 4 \times 10^4 \mathbf{i} \text{ N/C}$$

where $\mathbf{i}$ is the unit vector in the x direction.

---

* In this discussion we assume that the presence of the test charge q_0 does not change the original distribution of the other charges. This is true in practice if there are no conductors present or if q_0 is so small that its influence on the original charge distribution is negligible.

We think of the electric field as a condition in space set up by the system of point charges. This condition is described by the vector **E**. By moving the test charge q_0 from point to point we can find the electric field vector **E** at any point (except one occupied by a charge q_i). The electric field **E** is thus a vector function of position. The force exerted on a test charge q_0 at any point is related to the electric field at that point by

$$\mathbf{F} = q_0\mathbf{E} \qquad\qquad\qquad 20\text{-}6$$

Example 20-5 What is the force on an electron placed at the point in Example 20-4 where the electric field is $4 \times 10^4\mathbf{i}$ N/C?

Since the charge of the electron is $-e = -1.6 \times 10^{-19}$ C, the force is

$$\mathbf{F} = (-1.6 \times 10^{-19}\ \text{C})(4 \times 10^4\mathbf{i}\ \text{N/C}) = -6.4 \times 10^{-15}\mathbf{i}\ \text{N}$$

Although it is customary to think of the test charge q_0 as positive, we see from Example 20-5 that the sign of q_0 has no effect on the definition of **E**. If q_0 is positive, the force on q_0 is in the direction of the field **E**; if q_0 is negative, the force is in the direction opposite to **E**. The ratio $\mathbf{F}/q_0$ is the same in either case.

Like the gravitational field, the electric field is more than a calculation device. This concept enables us to avoid the problem of action at a distance if we concede that the field is not propagated instantaneously. We thus think of the force exerted on charge q_0 at point P as being exerted *by the field at point P* rather than by the charges, which are some distance away. Of course the field at point P is produced by the other charges, but not instantaneously.

Consider a single point charge q_1 at the origin. If we place a test charge q_0 at some point given by the position vector **r**, the force on this charge will be

$$\mathbf{F} = \frac{kq_1q_0}{r^2}\,\hat{\mathbf{r}}$$

The electric field at point **r** due to the charge q_1 is thus

$$\mathbf{E} = \frac{kq_1}{r^2}\,\hat{\mathbf{r}} \qquad\qquad\qquad 20\text{-}7$$

Electric field of a point charge

If charge q_1 is suddenly moved at time $t = 0$, the change in field at **r** is not instantaneous. The change in the field is propagated with the speed of light c, and so the test charge at **r** will not react to the change in position of the charge q_1 until a later time $t = r/c$, the time required for propagation of the change in the field. In this chapter we shall be concerned only with static electric fields.

The electric field due to a system of point charges can be found from Coulomb's law. Let $\mathbf{r}_{i0}$ be the vector from the ith charge q_i to point P. The force on a test charge q_0 at P is then

$$\mathbf{F} = \frac{kq_1q_0}{r_{10}^2}\,\hat{\mathbf{r}}_{10} + \frac{kq_2q_0}{r_{20}^2}\,\hat{\mathbf{r}}_{20} + \cdots + \frac{kq_iq_0}{r_{i0}^2}\,\hat{\mathbf{r}}_{i0} + \cdots$$

$$= q_0 \sum_i \frac{kq_i}{r_{i0}^2}\,\hat{\mathbf{r}}_{i0}$$

where $\hat{\mathbf{r}}_{i0} = \mathbf{r}_{i0}/r_{i0}$ is the unit vector pointing away from q_i toward point P. The electric field at point P due to the charges q_i (but excluding the test charge q_0) is then

$$\mathbf{E} = \frac{\mathbf{F}}{q_0} = \sum \frac{kq_i}{r_{i0}^2}\,\hat{\mathbf{r}}_{i0} \qquad\qquad\qquad 20\text{-}8$$

Example 20-6 A positive charge q_i is at the origin, and a second positive charge q_2 is on the x axis at $x = a$. Find the electric field at points on the x axis (Figure 20-4).

In the region $x > a$ the electric field due to each charge is along the x axis in the positive x direction. The distance r_{10} to the charge q_1 is x, and the distance r_{20} to q_2 is $x - a$. The electric field is then

$$\mathbf{E} = \frac{kq_1}{x^2}\,\mathbf{i} + \frac{kq_2}{(x-a)^2}\,\mathbf{i} \qquad x > a$$

where $\mathbf{i}$ is the unit vector in the x direction. In the region $0 < x < a$, the field due to q_1 is in the positive direction, but the field due to q_2 is in the negative x direction. The distance to q_1 is again x, and that to q_2 is $a - x$. Since these distances are squared in calculating the field, we can write either $(a - x)^2$ or $(x - a)^2$. The electric field between the charges is thus

$$\mathbf{E} = \frac{kq_1}{x^2}\,\mathbf{i} - \frac{kq_2}{(x-a)^2}\,\mathbf{i} \qquad 0 < x < a$$

To the left of the origin, the field due to each charge is in the negative x direction. The distance to q_1 is $-x$, and that to q_2 is $-x + a$ (since x is a negative number). Then for negative x we have

$$\mathbf{E} = -\frac{kq_1}{x^2}\,\mathbf{i} - \frac{kq_2}{(x-a)^2}\,\mathbf{i} \qquad x < 0$$

Figure 20-4
Point charges on x axis for Example 20-6.

Example 20-7 Find the electric field at points on the y axis for the charges in Example 20-6.

The fields due to each charge at a point on the positive y axis are shown in Figure 20-5. The field due to q_1 has magnitude $E_1 = kq_1/y^2$ and is in the $+y$ direction. The field due to q_2 has magnitude $E_2 = kq_2/(y^2 + a^2)$ and is in a direction making an angle θ with the y axis. The resultant electric field is the vector sum of these fields, $\mathbf{E} = \mathbf{E}_1 + \mathbf{E}_2$. The y component of the resultant electric field is

$$E_y = \frac{kq_1}{y^2} + \frac{kq_2}{y^2 + a^2}\cos\theta = \frac{kq_1}{y^2} + \frac{kq_2 y}{(y^2 + a^2)^{3/2}}$$

using $\cos\theta = y/\sqrt{y^2 + a^2}$. The x component of the resultant electric field is

$$E_x = -\frac{kq_2}{y^2 + a^2}\sin\theta = -\frac{kq_2 a}{(y^2 + a^2)^{3/2}}$$

using $\sin\theta = a/\sqrt{y^2 + a^2}$.

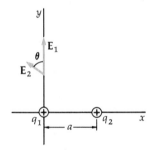

Figure 20-5
Example 20-7. On the y axis, the electric field $\mathbf{E}_1$ due to charge q_1 is along the y axis, and the field $\mathbf{E}_2$ due to charge q_2 makes an angle θ with the y axis. The resultant electric field is the vector sum $\mathbf{E} = \mathbf{E}_1 + \mathbf{E}_2$.

Example 20-8 A point charge $+q$ is at the origin and an equal but opposite charge $-q$ is on the x axis a small distance away at $x = -a$ (Figure 20-6). Find the electric field at points on the x axis far from the origin.

At great distances from the origin, the electric field due to the positive charge is nearly canceled by that due to the negative charge. At points on the x axis, $E_y = E_z = 0$ and E_x is given by

$$E_x = \frac{kq}{x^2} + \frac{k(-q)}{(x+a)^2} = kq\left[\frac{1}{x^2} - \frac{1}{(x+a)^2}\right]$$

Putting the terms in brackets over a common denominator, we get

$$\frac{1}{x^2} - \frac{1}{(x+a)^2} = \frac{(x+a)^2 - x^2}{x^2(x+a)^2} = \frac{2ax + a^2}{x^2(x+a)^2}$$

$$= \frac{2ax(1 + a/2x)}{x^4(1 + a/x)^2}$$

Figure 20-6
Equal and opposite charges separated by a distance a for Example 20-8. Such a charge system is called an electric dipole.

For $x \gg a$, we can neglect $a/2x$ compared with 1 in the numerator and a/x compared with 1 in the denominator. We then have

$$E_x \approx \frac{2kqa}{x^3}$$

A system of equal and opposite charges separated by a small distance is called an *electric dipole*. A vector from the negative charge to the positive charge having the magnitude qa is called the dipole moment vector **p**. In terms of the dipole moment, the electric field on the axis of the dipole at a great distance is given by

$$E_x = \frac{2kqa}{x^3} = \frac{2kp}{x^3} \qquad\qquad 20\text{-}9$$

The electric field far from a dipole is proportional to the dipole moment and decreases as the cube of the distance.

20-4 Lines of Force

It is convenient to picture the electric field by drawing *lines of force* to indicate the direction of the field at any point, just as we did for the gravitational field in Chapter 12.* The field vector **E** is tangent to the line at each point and indicates the direction of the electric force experienced by a positive test charge placed at that point. Because there are an infinite number of points in space through which to draw a line, we choose only a few representative lines to be drawn. Since we could choose any point to indicate the direction of the field by a line, the lines need not be continuous, but it is both customary and useful to draw continuous lines which begin at positive charges and end at negative charges. At any point near a positive charge, a positive test charge is repelled from the charge; the field lines therefore diverge from a point occupied by a positive charge. Similarly, the lines converge toward a point occupied by a negative charge. (This differs somewhat from the gravitational field lines, which end at point masses but have no points from which they diverge because the gravitational field is always attractive.)

Figure 20-7 shows the lines of force, or electric field lines, of a single positive point charge. Consider a spherical surface of radius r with its center at the charge. For a fixed number of lines emerging from the charge, the number of lines per unit area on the sphere is inversely proportional to the area of the sphere, $4\pi r^2$. Thus the density of lines decreases with distance as $1/r^2$, just as the magnitude of the electric field decreases. If we adopt the convention of drawing a fixed number of lines from a point charge, the number being proportional to the charge strength, and if we draw the lines symmetrically about the point charge, the field strength is indicated by the density of the lines.

The electric field near a negative point charge points inward toward the charge, so that the lines of force point toward the negative charge. If we have more than one point charge, we can indicate the electric field strength by choosing the number of lines diverging from the positive charges or converging toward the negative charges to be proportional to the charge. Figure 20-8 shows the lines of force for two equal positive point charges separated by a distance a.

* Much of the discussion in Chapter 12 of the gravitational field lines applies also to electric field lines. We repeat that discussion here in case Chapter 12 was omitted in your study of mechanics.

(a)

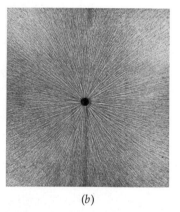

(b)

Figure 20-7
(a) Electric field lines, or lines of force, of a single charge. (b) Bits of thread suspended in oil. The electric field of the charged object in the center induces opposite charges on the ends of each bit of thread, causing the thread to align itself parallel to the field. (*Courtesy of Harold M. Waage, Princeton University.*)

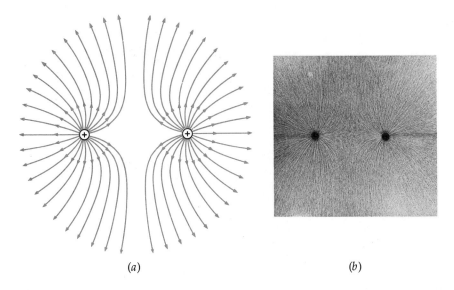

(a)   (b)

Figure 20-8
(a) Lines of force due to two positive point charges. (b) Electric lines of force of two equal charges of the same sign, shown by bits of thread suspended in oil. (*Courtesy of Harold M. Waage, Princeton University.*)

We construct this pattern without calculating the field at each point. At points near one of the charges, the field is approximately due to that charge alone because of the inverse-square-distance dependence of the field.

Thus on a sphere of very small radius ($r \ll a$) about one of the charges, the field lines are radial and equally spaced. Since the charges are equal, we draw an equal number of lines from each charge. At very large distances from the charges, the field is approximately the same as that due to a point charge of magnitude $2q$. So on a sphere of radius $r \gg a$, the lines are approximately equally spaced. (In our two-dimensional drawings, these spheres are replaced by circles.) We can see by merely looking at the figure that the electric field in the space between the charges is weak because there are few lines in this region compared with the region just to the right or left of the charges, where the lines are more closely spaced. This information can of course also be obtained by direct calculation of the field at points in these regions.

We can apply the above reasoning to drawing the lines of force for any system of point charges. Very near each charge, the field lines are equally spaced and leave or enter the charge depending on the sign of the charge. Very far from all the charges, the detailed structure of the system cannot be important, and the field lines are the same as those of a single point charge equal to the net charge of the system. For future reference, we summarize our rules for drawing lines of electric force.

1. The number of lines leaving a positive point charge or entering a negative charge is proportional to the charge.

Rules for drawing lines of force

2. The lines are drawn symmetrically leaving or entering a point charge.

3. Lines begin or end only on charges.

4. The density of lines (number per unit area perpendicular to the lines) is proportional to the magnitude of the field.

5. No two field lines can cross.

Rule 5 follows from the fact that **E** has a unique direction at any point in space (except at a point occupied by a point charge). If two lines crossed, two directions would be indicated for **E** at the point of inter-section.

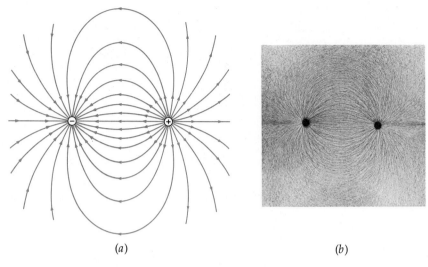

(a)   (b)

Figure 20-9
Example 20-8. (*a*) Lines of
force for an electric dipole. (*b*)
Electric lines of force of two
equal but opposite charges,
shown by bits of thread suspended in oil. (*Courtesy of
Harold M. Waage, Princeton
University.*)

Figure 20-9 shows the lines of force for an electric dipole. Very near
the positive charge the lines are radially outward. Very near the negative charge the lines are radially inward. Since the charges have equal
magnitude, the number of lines that begin at the positive charge equals
the number that end at the negative charge. In this case the field is
strongest in the region between the charges, as indicated by the high
density of field lines.

Example 20-9 Sketch the lines of force for a negative charge $-q$ a distance a from a positive charge $+2q$.

Since the positive charge is twice the magnitude of the negative
charge, twice as many lines leave the positive charge as enter the negative charge. At great distances from the charges the system looks like a
single charge $+q$. Half the lines beginning on the positive charge leave
the system. On a sphere of radius r, where r is much larger than the separation of the charges a, these lines are approximately symmetrically
spaced and point radially outward, the same as the lines from a single
positive point charge $+q$. The other half leaving the positive charge
$+2q$ enter the negative charge. Field lines are shown in Figure 20-10.

It is important to realize that the *convention indicating the electric field
strength by the lines of force works only because the electric field varies
inversely as the square of the distance from a point charge*. Since the gravi-

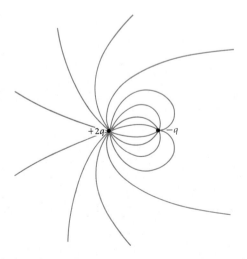

Figure 20-10
Example 20-9. Lines of force
for a point charge $+2q$ and a
second point charge $-q$. At
great distances from the
charges the lines are the same
as for a single charge $+q$.

tational field of a point mass also varies inversely as the square of the distance, the concept of lines of force is also useful in picturing the gravitational field. In that case, however, the lines of gravitational field enter point masses because the field points toward the mass and there are no points in space from which the gravitational field lines diverge; i.e., the gravitational force is always attractive, whereas the electric force can be either repulsive or attractive, depending on whether the charges have like or unlike signs.

As an illustration of the power of the use of lines of force, we now consider the electric field due to a spherically symmetric shell of charge of radius R and negligible thickness. Let the total charge on the shell be Q. (Such a charge distribution can be realized in practice by placing the charge on a spherical conductor. As we shall see in the next chapter, the charge will distribute itself uniformly over the surface of such a conductor.) We choose the origin to be at the center of the shell. Because of the symmetry of the spherical shell, the only possible direction for the electric field lines is along radii, either toward the origin or away from it. Let us assume the charge to be positive. At great distances compared with the shell radius R, the field must look like that of a point charge Q. Thus the lines are radially outward and equally spaced far from the shell. The lines of force for this system are shown in Figure 20-11. Outside the shell the lines are exactly the same as those for a point charge. Thus the electric field produced outside the shell is in the radial direction and has the magnitude

$$E = \frac{kQ}{r^2} \qquad r > R \qquad\qquad 20\text{-}10a$$

The lines must begin at the charge on the shell. There can be no lines inside the shell because if there were, they would diverge from, or converge to, the center of the shell at the origin. But there is no charge at the origin, so there can be no lines converging or diverging from that point. We therefore have the remarkable result that a spherically symmetric shell of charge produces no electric field inside it:

$$E = 0 \qquad r < R \qquad\qquad 20\text{-}10b$$

If we place a test charge q_0 anywhere inside the shell, it will experience no electric attraction or repulsion by the shell. This is easily under-

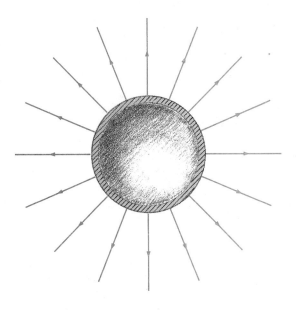

Figure 20-11
Lines of force for a spherical-shell charge distribution. Outside the shell the lines are the same as from a point charge at the center of the shell. Inside the shell there are no lines, indicating that the electric field is zero there.

stood for a test charge at the center of the shell because of the symmetry, but it is also true at every other point inside the shell. This remarkable result is a consequence of the inverse-square nature of the electric force. We shall derive it directly from Coulomb's law in the next section, but the derivation is very difficult because of the difficult spherical geometry needed in setting up the problem. In Chapter 21 we shall extend the idea of lines of force and develop a mathematical equation known as Gauss' law which can be used to calculate the electric field due to highly symmetric charge distributions such as the spherical shell of charge.

20-5 Calculation of the Electric Field for Continuous Charge Distributions

The electric field produced by a given charge distribution can be calculated in a straightforward way from Coulomb's law. For a set of point charges, the field is obtained from Equation 20-8. We used this to calculate the field for two positive point charges and for an electric dipole in Examples 20-6 to 20-8. In this section we consider charge distributions consisting of many charges so close together that the charge can be considered to be continuously distributed over a surface or through a volume, even though electric charge is a discrete quantity microscopically. The use of a continuous charge density to describe a distribution of a large number of discrete charges is similar to the use of a continuous mass density to describe air, which in reality consists of a large number of discrete molecules. In either case, it is usually easy to find a volume element $\Delta \mathcal{V}$ large enough to contain many individual charges or molecules (billions) and yet small enough to ensure that replacing $\Delta \mathcal{V}$ by a differential $d\mathcal{V}$ and using calculus will introduce negligible error. If charge ΔQ is distributed throughout a volume $\Delta \mathcal{V}$, the *charge density* ρ is defined by

$$\rho = \frac{\Delta Q}{\Delta \mathcal{V}}$$ 20-11   *Volume charge density*

Often charge is distributed in a thin layer on the surface of an object. (We shall show later that a static charge on a conductor always resides on the surface of the conductor.) In these cases, it is convenient to define a *surface density* σ. Let t be the thickness of the layer of charge. Then in a volume element of area ΔA the charge is

$$\Delta Q = \rho t \,\Delta A = \sigma \,\Delta A$$

where σ is the charge per unit area:

$$\sigma = \frac{\Delta Q}{\Delta A} = \rho t$$ 20-12   *Surface charge density*

Similarly, if the charge is along a line, e.g., on a string of cross-sectional area A, we choose a volume element of length ΔL, $\Delta \mathcal{V} = A \,\Delta L$, and define the charge per unit length λ by

$$\Delta Q = \rho A \,\Delta L = \lambda \,\Delta L$$

or

$$\lambda = \frac{\Delta Q}{\Delta L} = \rho A$$ 20-13   *Linear charge density*

Example 20-10 Calculate the electric field on the axis of a uniform ring of charge.

Let Q be the total positive charge, which is distributed uniformly on a ring of radius a. We wish to find the field at a point on the axis of the ring a distance x from the center of the ring.

Figure 20-12 shows the part of the field $\Delta \mathbf{E}$ due to a portion of the charge ΔQ. This field has a component ΔE_x along the axis of the ring and a component $\Delta E_\perp$ perpendicular to the axis. From the symmetry of the figure we see that the resultant field due to the entire ring must lie along the axis of the ring; i.e., the perpendicular components will sum to zero. In particular, the perpendicular component shown will be canceled by that due to another portion of the charge on the ring directly opposite the one shown.

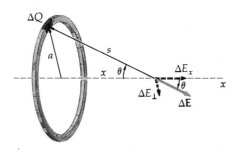

Figure 20-12
Charged ring of radius a. The electric field on the x axis is in the x direction.

The axial component due to the part of the charge shown is

$$\Delta E_x = \frac{k\,\Delta Q}{s^2} \cos \theta = \frac{k\,\Delta Q}{s^2} \frac{x}{s} = \frac{k\,\Delta Q\ x}{(x^2 + a^2)^{3/2}}$$

where

$$s^2 = x^2 + a^2 \qquad \text{and} \qquad \cos \theta = \frac{x}{s} = \frac{x}{\sqrt{x^2 + a^2}}$$

Since the distance from the field point at which we are calculating the field to all parts of the charge is the same, and since the angle θ is the same for all parts of the charge, the field due to the entire ring of charge is

$$E_x = \sum \frac{k\,\Delta Q\ x}{(x^2 + a^2)^{3/2}} = \frac{kx}{(x^2 + a^2)^{3/2}} \sum \Delta Q = \frac{kQx}{(x^2 + a^2)^{3/2}} \qquad \text{20-14}$$

Example 20-11 Calculate the electric field on the perpendicular bisector of a uniform line charge.

Let λ be the linear density. At a point on a perpendicular bisector the field $\mathbf{E}$ will not have a component parallel to the line of charge.

Figure 20-13 shows the geometry of the problem. We have chosen a coordinate system such that the origin is at the center of the line charge, the charge is on the x axis, and our field point is on the y axis at P. The magnitude of the field produced by an element of charge $\Delta Q = \lambda\,\Delta x$ is

$$|\Delta \mathbf{E}| = \frac{k\lambda\,\Delta x}{s^2}$$

The perpendicular component (in this case, the y component) is

$$\Delta E_y = \frac{k\lambda\,\Delta x}{s^2} \cos \theta = \frac{k\lambda\,\Delta x}{s^2} \frac{y}{s} \qquad \text{20-15}$$

The total field E_y is computed by summing over all Δx of the line charge.

Figure 20-13
Line charge. The electric field on the y axis which bisects the charge is in the y direction.

For our case, we can sum (integrate) from $x = 0$ to $x = L/2$ and multiply by 2 because, by symmetry, each half of the line charge gives an equal contribution. The integration is somewhat simplified if we change from the variable x to θ and integrate from $\theta = 0$ to θ_0, defined by

$$\tan \theta_0 = \frac{\frac{1}{2}L}{y}$$

Equation 20-15 in terms of the differentials dE_y and dx is

$$dE_y = \frac{ky\lambda}{s^3} \, dx$$

The distance x is related to the angle θ by

$$x = y \tan \theta$$

Then

$$dx = y \sec^2 \theta \, d\theta = y \left(\frac{s}{y}\right)^2 d\theta$$

and

$$dE_y = \frac{ky\lambda}{s^3} \frac{s^2}{y} \, d\theta = \frac{k\lambda}{y} \frac{y}{s} \, d\theta = \frac{k\lambda}{y} \cos \theta \, d\theta$$

The total y component of the field is twice the integral of this from $\theta = 0$ to $\theta = \theta_0$:

$$E_y = 2 \int_{\theta=0}^{\theta=\theta_0} dE_y = \frac{2k\lambda}{y} \int_0^{\theta_0} \cos d\theta = \frac{2k\lambda}{y} \sin \theta_0 \qquad \text{20-16}$$

where

$$\sin \theta_0 = \frac{\frac{1}{2}L}{\sqrt{(\frac{1}{2}L)^2 + y^2}}$$

The field produced by a line charge of infinite length is obtained from this result by setting $\theta_0 = 90°$. Then

$$E_y = \frac{2k\lambda}{y} \qquad \text{20-17}$$

Example 20-12 Calculate the electric field on the axis of a uniformly charged disk.

Let the disk have a radius R and contain a uniform charge per unit area σ. The electric field on the axis of the disk will be parallel to the axis. We can calculate this field by treating the disk as a set of concentric ring charges and using our result from Example 20-10. Consider a ring of radius r and width dr, as in Figure 20-14. The area of this ring is $2\pi r \, dr$, and its charge is $dq = 2\pi\sigma r \, dr$. The field produced by this ring is given by Equation 20-14, replacing Q by $2\pi\sigma r \, dr$ and a by r. Thus

$$dE_x = \frac{kx2\pi\sigma r \, dr}{(x^2 + r^2)^{3/2}}$$

The total field produced by the disk is found by integrating from $r = 0$ to $r = R$:

$$E_x = kx\pi\sigma \int_0^R (x^2 + r^2)^{-3/2} 2r \, dr$$

$$= kx\pi\sigma \left[\frac{(x^2 + r^2)^{-1/2}}{-\frac{1}{2}}\right]_0^R$$

$$= -2kx\pi\sigma \left(\frac{1}{\sqrt{x^2 + R^2}} - \frac{1}{x}\right) = 2\pi k\sigma \left(1 - \frac{x}{\sqrt{x^2 + R^2}}\right) \qquad \text{20-18}$$

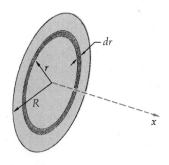

Figure 20-14
Disk carrying uniform surface charge density. The electric field at a point on the axis of the disk is along the axis. It can be calculated by finding the field due to a ring of radius r and area $2\pi r \, dr$ and summing over the rings from $r = 0$ to $r = R$, the radius of the disk.

The interesting and important result for the case of the field near an infinite plane of charge can be obtained from this result by letting either R go to infinity or x go to zero. Then

$$E_x = 2\pi k\sigma \qquad\qquad 20\text{-}19$$

Thus the field due to an infinite-plane charge distribution is uniform; i.e., the field does not depend on x.

Optional

Figure 20-15
Spherical shell of radius R carrying a uniform surface charge. The field at point P a distance r from the center of the sphere is found by first finding the field of a ring of width $R\,d\theta$ and radius $R\sin\theta$ and summing over the rings from $\theta = 0$ to $\theta = \pi$. The resulting field for $r > R$ is the same as if all the charge were at the origin. When point P is inside the shell $(r < R)$, the field due to the shell is zero.

Example 20-13 Calculate the electric field due to a spherical shell of charge.

We first consider a field point P outside the shell. The geometry is shown in Figure 20-15. By symmetry, the field must be radial. We can consider the spherical shell to be a set of ring elements and use our result found in Example 20-10. We thus choose for our charge element the strip shown, which has circumference $2\pi R \sin\theta$ and width $R\,d\theta$. The area of this strip is $dA = 2\pi R^2 \sin\theta\,d\theta$, and the charge is

$$dQ = \sigma\,dA = \sigma 2\pi R^2 \sin\theta\,d\theta$$

The radial component of the field due to this ring is

$$dE_r = \frac{k\,dQ}{s^2}\cos\alpha = \frac{k\sigma 2\pi R^2 \sin\theta\,d\theta}{s^2}\cos\alpha \qquad 20\text{-}20$$

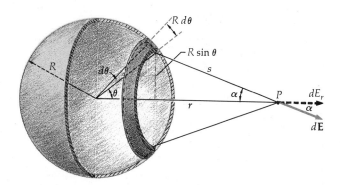

Before we integrate over the charge distribution, we must eliminate two of the three related variables s, θ, and α. It is convenient to write everything in terms of s, which varies from $s = r - R$ at $\theta = 0$ to $s = r + R$ at $\theta = 180°$. By the law of cosines, we have

$$s^2 = r^2 + R^2 - 2rR\cos\theta \qquad\qquad 20\text{-}21$$

Differentiating gives

$$2s\,ds = +2rR\sin\theta\,d\theta$$

or

$$\sin\theta\,d\theta = \frac{s\,ds}{rR} \qquad\qquad 20\text{-}22$$

An expression for $\cos\alpha$ can also be obtained from the law of cosines applied to the same triangle. We have

$$R^2 = s^2 + r^2 - 2sr\cos\alpha$$

or

$$\cos\alpha = \frac{s^2 + r^2 - R^2}{2sr} \qquad\qquad 20\text{-}23$$

Substituting these results into Equation 20-20 then gives

$$dE_r = \frac{k\sigma 2\pi R^2}{s^2}\frac{s\,ds}{rR}\frac{s^2 + r^2 - R^2}{2sr}$$

$$= \frac{k\sigma\pi R}{r^2}\left(1 + \frac{r^2 - R^2}{s^2}\right)ds \qquad\qquad 20\text{-}24$$

The field due to the entire shell of charge is found by integrating from $s = r - R$ $(\theta = 0)$ to $s = r + R$ $(\theta = 180°)$:

$$E_r = \frac{k\sigma\pi R}{r^2}\int_{r-R}^{r+R}\left(1 + \frac{r^2 - R^2}{s^2}\right)ds$$

$$= \frac{k\sigma\pi R}{r^2}\left[s - \frac{r^2 - R^2}{s}\right]_{r-R}^{r+R} \qquad\qquad 20\text{-}25$$

Substitution of the upper and lower limits yields $4R$ for the quantity in brackets. Thus

$$E_r = \frac{k\sigma 4\pi R^2}{r^2} = \frac{kQ}{r^2} \qquad\qquad 20\text{-}26$$

where $Q = 4\pi R^2\sigma$ is the total charge. The field is therefore the same as that due to a point charge Q at the origin.

We now consider a field point inside the shell. The calculation for this case is identical except that s now varies from $R - r$ to $r + R$. Thus

$$E_r = \frac{k\sigma\pi R}{r^2}\left[s - \frac{r^2 - R^2}{s}\right]_{R-r}^{r+R} \qquad\qquad 20\text{-}27$$

Substitution of these upper and lower limits yields 0. Therefore

$$E_r = 0 \qquad \text{point inside shell} \qquad\qquad 20\text{-}28$$

These are the same results as we obtained by considering the lines of force for this charge distribution.

Example 20-14 Calculate the electric field due to a solid sphere of constant charge density $\rho = Q/V$, where $V = \frac{4}{3}\pi R^3$ is the volume of the sphere of charge.

We consider the sphere to consist of a set of concentric spherical shells and use the result already obtained, that the field produced by a shell is zero inside the shell and kq/r^2 outside the shell, where q is the charge of the shell. For points outside our solid sphere of charge each shell produces a field identical to that of a point charge at the center of the sphere, and for $r > R$ we have

$$E_r = \frac{kQ}{r^2}$$

Consider now a point P inside the charge, $r < R$. In Figure 20-16 the dotted line encloses a sphere of radius r. Let q' be the portion of the charge inside this sphere. Since the charge density is uniform, q' is given by

$$q' = \rho\mathcal{V}' = \rho\frac{4}{3}\pi r^3 = \frac{Q}{\frac{4}{3}\pi R^3}\frac{4}{3}\pi r^3 = \frac{Qr^3}{R^3}$$

The field at P due to q' is the same as if q' were at the origin. The rest of the charge produces no field at point P. This part of a charge can be con-

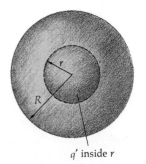

q′ inside r

Figure 20-16
Solid ball of charge. The electric field at a point $r < R$ inside the ball is kq'/r^2, where $q' = (r^3/R^3)Q$ is that part of the total charge Q which is inside the sphere of radius r.

sidered to be a set of spherical shells, and for each shell, point P is inside. The field at a distance $r < R$ is thus

$$E_r = \frac{kq'}{r^2} = \frac{kQr^3/R^3}{r^2} = \frac{kQ}{R^3}\,r \qquad\qquad 20\text{-}29$$

Figure 20-17 shows a sketch of E_r versus r. This function is sometimes used to describe the electric field of an atomic nucleus, which can be considered to be approximately a uniform sphere of charge.

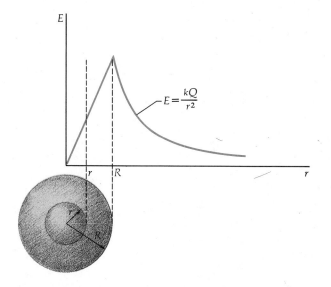

Figure 20-17
Plot of E_r versus r for a solid ball of charge of radius R. For r less than R the field increases linearly with r. Outside the charge, the electric field is kQ/r^2, the same as that due to a point charge Q at the origin.

Question

4. Explain why the electric field increases with r rather than decreasing as $1/r^2$ as one moves out from the center inside a spherical charge distribution of constant density.

20-6 Motion of Point Charges in Electric Fields

When a particle with charge q is placed in an electric field $\mathbf{E}$, it experiences a force $q\mathbf{E}$. If this is the only force on it, the particle has an acceleration $q\mathbf{E}/m$, where m is the mass of the particle.* If the electric field is known, the charge-to-mass ratio of the particle can be determined from the measured acceleration. For example, for an electric field uniform in space and constant in time, the path of the particle is a parabola, similar to that of a projectile in a uniform gravitational field. The measurement of the deflection of electrons in a uniform electric field was used by J. J. Thomson in 1897 to demonstrate the existence of electrons and to measure their charge-to-mass ratio. (Since magnetic fields were also used in this measurement, we defer discussion of the Thomson experiment until Chapter 26.) We shall give some examples of motion of electrons in

---

* We are assuming that the speed of the particle is small enough for us to use classical mechanics and neglect special relativity. This assumption is often *not* valid for the motion of electrons in electric fields.

constant electric fields. Problems of this type can be worked using the constant-acceleration formulas from Chapter 2 or the equations for projectile motion from Chapter 3.

Example 20-15 An electron is projected into a uniform electric field $E = 1000$ N/C with initial speed $v_0 = 2 \times 10^6$ m/s parallel to the field. How far does the electron travel before it is brought momentarily to rest?

Since the charge of the electron is negative, the force $-e\mathbf{E}$ is in the direction opposite the field. We thus have a constant-acceleration problem in which the acceleration is opposite to the initial velocity, and we are asked to find the distance traveled. We can use the constant-acceleration expression relating the distance to the velocity,

$$v^2 = v_0^2 + 2a(x - x_0)$$

Using $x_0 = 0$, $v = 0$, $v_0 = 2 \times 10^6$, and $a = -eE/m$, we find for the distance

$$x = \frac{mv_0^2}{2eE} = \frac{(9.11 \times 10^{-31})(2 \times 10^6)^2}{2(1.6 \times 10^{-19})(1000)} = 1.14 \times 10^{-2} \text{ m}$$

Example 20-16 An electron is projected into a uniform electric field $E = 2000$ N/C with initial velocity $v_0 = 10^6$ m/s perpendicular to the field. Compare the weight of the electron to the electric force on it. By how much is the electron deflected after it has traveled 1 cm?

The electric force on the electron is eE, and the gravitational force is mg. Their ratio is

$$\frac{F_{\text{elec}}}{F_{\text{grav}}} = \frac{eE}{mg} = \frac{(1.6 \times 10^{-19})(2000)}{(9.1 \times 10^{-31})(9.8)} = 3.6 \times 10^{13}$$

As in most common cases, the electric force is huge compared with the gravitational force, which is wholly negligible.

It takes the electron a time

$$t = \frac{x}{v_0} = \frac{10^{-2} \text{ m}}{10^6 \text{ m/s}} = 10^{-8} \text{ s}$$

to travel the distance of 1 cm perpendicular to the field. In this time it is deflected antiparallel to the field a distance given by

$$y = \tfrac{1}{2}at^2 = \frac{1}{2}\frac{eE}{m}t^2$$

Substituting the known values of e/m, E, and t gives

$$y = 1.76 \times 10^{-2} \text{ m} = 1.76 \text{ cm}$$

Questions

5. The direction of the force on a positive charge in an electric field is, by definition, in the direction of the field line passing through the position of the charge. Must the acceleration be in this direction? The velocity? Explain.

6. A positive charge is released from rest in an electric field. It starts out in the direction of a field line. Will it continue to move along the field line?

20-7 Electric Dipoles in Electric Fields

Although atoms and molecules are electrically neutral, they are affected by electric fields because they contain positive and negative charges. In some molecules the center of positive charge does not coincide with the center of negative charge. These *polar molecules* are said to have a permanent electric dipole moment. When such a molecule is placed in a uniform electric field, there is no net force on it but there is a torque, which tends to rotate the molecule. In a nonuniform electric field the molecule experiences a net force because the field at the center of the positive charge is different from that at the center of the negative charge. An example of a polar molecule is NaCl, which is essentially a positive sodium ion of charge $+e$ combined with a negative chlorine ion of charge $-e$.

Polar molecules

Atoms and molecules for which the centers of positive and negative charge coincide are also affected by an electric field. Because the electric force on the positive charge is in the direction opposite that on the negative charge, the electric field tends to separate, or polarize, these charges. These systems then have an induced dipole moment when they are in an electric field, and they also experience a net force in a nonuniform field. The force produced by a nonuniform electric field on an electrically neutral charge system is responsible for the familiar attraction of a charged comb for uncharged bits of paper.

Figure 20-18 shows a schematic picture of a polar molecule with the centers of positive and negative charge indicated. (The center of charge is defined analogously to the center of mass, with the mass replaced by the charge.) The behavior of such a molecule can be described by a vector **p**, called the *dipole moment*. Let the total positive charge of the molecule be q and the total negative charge $-q$. The dipole moment of the molecule is defined to be the product of the charge q and the displacement vector **L** pointing from the center of negative charge to the center of positive charge:

$$\mathbf{p} = q\mathbf{L} \qquad\qquad 20\text{-}30$$

Figure 20-18
Schematic diagram of a polar molecule. The centers of positive and negative charge are separated by a distance L. The molecule behaves like an electric dipole of moment $\mathbf{p} = q\mathbf{L}$, where q is the magnitude of the total positive or negative charge.

Dipole moment defined

The diameter of an atom or molecule is of the order of 10^{-10} m = 0.1 nm. A convenient unit for electric dipole moments of atoms and molecules is the fundamental electronic charge e times the distance 1 nm. For example, the dipole moment of NaCl in these units has a magnitude of about $0.2e$ nm.

We often simplify the description of the behavior of a polar molecule in an electric field by replacing the complicated charge distribution of the molecule with a simple electric dipole consisting of two charges q and $-q$ separated by a distance L and having the same dipole moment as the molecule.

Figure 20-19 shows a simple electric dipole whose dipole moment makes an angle θ with an external uniform electric field **E**. The forces acting on the dipole,

$$\mathbf{F_1} = q\mathbf{E} \qquad \mathbf{F_2} = -q\mathbf{E}$$

are opposite in direction and equal in magnitude since the field is uniform. The net force on the dipole is thus zero. However, the two forces produce a torque, which tends to rotate the dipole so that it points in the direction of the field. We found in Chapter 5 that the torque produced by two equal and opposite forces, called a couple, is the same about any point in space. From Figure 20-19 we see that the torque about the negative charge has the magnitude $F_1 L \sin\theta = qEL \sin\theta =$

Figure 20-19
Electric dipole in a uniform electric field. The net force on the dipole is zero, but there is a net torque $\mathbf{L} \times \mathbf{F_1} = q\mathbf{L} \times \mathbf{E} = \mathbf{p} \times \mathbf{E}$ which tends to align the dipole in the direction of the field.

$pE \sin \theta$. The direction of the torque is into the paper and such that it rotates the dipole moment **p** into the direction of the electric field **E**. This torque can be conveniently written as the cross product of the dipole moment **p** and the electric field **E**:

$$\tau = \mathbf{p} \times \mathbf{E} \qquad\qquad 20\text{-}31$$

Torque on a dipole

If an electric dipole is placed so that it makes an angle θ with a uniform electric field, it will experience a torque of magnitude $pE \sin \theta$ tending to rotate it toward its equilibrium position $\theta = 0$. If the dipole is free, the torque will cause it to rotate so that the vector **p** oscillates about the equilibrium direction $\theta = 0$ until its energy is dissipated. We can calculate the potential energy of a dipole in a uniform field by computing the work we must do to increase the angle θ. If we apply a torque of magnitude $pE \sin \theta$ and rotate the dipole through an angle $d\theta$, the work we do equals the increase in the potential energy:

$$dU = pE \sin \theta \, d\theta$$

Integrating gives

$$U = -pE \cos \theta + U_0$$

Potential energy of a dipole in an electric field

It is customary to choose the potential energy to be zero when the dipole is perpendicular to the electric field, i.e., when $\theta = 90°$. Then $U_0 = 0$, and the potential energy of the dipole is

$$U = -pE \cos \theta = -\mathbf{p} \cdot \mathbf{E} \qquad\qquad 20\text{-}32$$

If the electric dipole is placed in a nonuniform electric field, as in Figure 20-20, there is a net force acting on the dipole in addition to the torque tending to align the dipole with the field. The net force depends on the orientation of the dipole in the field, the dipole moment, and how rapidly the field varies in space. For the situation pictured in Figure 20-20 the net force on the dipole has a component downward and one in the direction of increasing field.

Most molecules and all atoms are nonpolar; i.e., the centers of positive and negative charge coincide, and there is no permanent electric

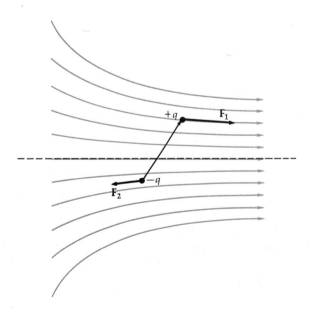

Figure 20-20
For a general orientation of an electric dipole in a nonuniform electric field there is both a net torque tending to align the dipole with the field and a net force on the dipole. Here, the net force has a component parallel to **E** and a small component downward. Its magnitude depends on the dipole moment **p**, its orientation, and on how rapidly the field varies in space.

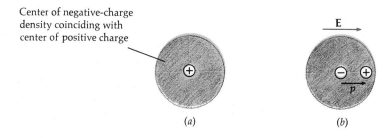

Center of negative-charge
density coinciding with
center of positive charge

(a)   (b)

Figure 20-21
Schematic diagram of a non-polar molecule. (*a*) In the absence of an external electric field the center of the positive charge coincides with that of the negative charge, and there is no dipole moment. (*b*) In the presence of an external electric field the centers of positive and negative charge are displaced, producing a dipole moment parallel to the external field. The magnitude of the dipole moment depends on the magnitude of the electric field **E**.

dipole moment. However, an external electric field produces a separation of the positive and negative charge distribution, thus inducing a dipole moment (Figure 20-21). Since the positive charge is displaced in the direction of the field and the negative charge is displaced in the opposite direction, the induced dipole moment is always in the direction of the external electric field. Thus no torque is exerted because the angle between the dipole moment **p** and the field **E** is zero. However, if the electric field is not uniform, there will be an external force acting on the dipole. Figure 20-22 shows a nonpolar molecule in an external electric field of a positive point charge Q. The induced dipole moment is parallel to **E** in the radial direction from the point charge. Since the field is stronger at the negative charge nearer the point charge, the force on the dipole is toward the point charge and the dipole is attracted toward the point charge. If the point charge were negative, the induced dipole would be in the opposite direction and the dipole would again be attracted to the point charge.

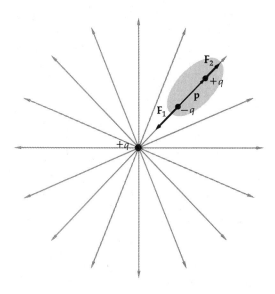

Figure 20-22
Nonpolar molecule in the electric field of a positive point charge. The induced dipole moment is parallel to the field **E**. Since the point charge is closer to the center of negative charge than to the center of positive charge, the dipole is attracted to the point charge.

Question

7. A small, light conducting ball with no net electric charge is suspended from a thread. When a positive charge is brought near the ball, the ball is attracted toward the charge. How does this come about? If the ball is an insulator instead of a conductor, it still is attracted. How? How would the situation have been different if the charge brought near were negative instead of positive?

Benjamin Franklin (1706–1790)

I. Bernard Cohen
Harvard University

When Franklin's contemporaries wanted to express their admiration for his scientific achievements, they could think only of comparing him to Newton. Joseph Priestley wrote that Franklin's book on electricity "bid fare to be handed down to posterity as expressive of the true philosophy of electricity; just as the Newtonian philosophy is of the true system of nature in general."

In order to appreciate Franklin's contribution, we must remember that electricity is a young branch of physics. Whereas we can trace an early history of statics and dynamics, heat and light, and even atomic theory back to the Greeks, electricity emerged as a proper subject of scientific study only in the days of Newton. How meager the information about electricity was in the early eighteenth century can be seen from the fact that the fundamental distinction between conductors and nonconductors had not yet been made, nor had it been discovered that there are two kinds of electric charge. When Franklin took up this new subject, he was almost forty, past the age when we usually think of great scientific discoveries being made. Largely self-educated, Franklin had already obtained a solid grounding in experimental physics, having studied Newton's *Opticks* and many of the primary textbooks of newtonian experimental science. His printing and newspaper business was successful enough to allow him to retire from active participation, and he eagerly seized upon the new subject and began to work at it intensively, together with a small group of coworkers.

His interest had first been sparked when he attended some public lectures on science given by a Dr. Adam Spencer, first in Boston (where Franklin was visiting his family) and then in Philadelphia (where Franklin sponsored Spencer's lectures). He purchased Spencer's apparatus in order to perform experiments himself. Soon thereafter the Library Company of Philadelphia (of which Franklin was the principal founder) received a gift of electrical apparatus, with instructions for using it, from Peter Collinson, a London merchant.

Before long Franklin and his fellow experimenters realized that they had progressed in knowledge far beyond the literature that accompanied the apparatus. Franklin periodically sent reports of his work to Collinson and to other London correspondents, which eventually were assembled into a book entitled *Experiments and Observations on Electricity, Made at Philadelphia*, first published in England in 1751.

For his pioneering research in electricity, Franklin was elected a Fellow of the Royal Society of London, with the singular distinction of being forgiven the annual dues. Shortly thereafter he was awarded the Society's Copley Gold Medal, the highest scientific honor then being awarded in England. His book on electricity was a spectacular success, going through five editions in English and being translated into French (three editions in two different translations), Italian, and German. In 1773 Franklin was elected a Foreign Associate of the French Academy of Sciences, an extraordinary honor, since, according to the terms of the Academy's foundation, there could be only eight such foreign associates at any one time. No American was similarly honored again for another century.

Franklin's discoveries included not only important new experimental evidence but a new theory of electricity which made a science of the subject. This theory was based upon the fundamental postulate that all electrostatic phenomena (charging and discharging) result from the motion or transfer of a single electrical fluid. This hypothetical fluid was made up of "particles," or atoms, of electricity that repelled one another but were attracted by the particles of "ordinary" matter. A charged body was one that had either lost or gained elec-

Title page of the first edition of Franklin's book on electricity, from the copy he presented to Harvard College. (*Courtesy of the Houghton Library, Harvard University.*)

EXPERIMENTS
AND
OBSERVATIONS
ON
ELECTRICITY,
MADE AT
Philadelphia in *America*,
BY
Mr. BENJAMIN FRANKLIN,
AND
Communicated in several Letters to Mr. P. COLLINSON,
of *London*, F. R. S.

LONDON:
Printed and sold by E. CAVE, at St. *John's* Gate. 1751.
(*Price 2s. 6d.*)

Portrait of Franklin.

The Bettmann Archive

trical fluid and accordingly was in a state that Franklin called "plus" (positive) or "minus" (negative). This postulate was closely associated with a major theoretical principle, known today as the law of conservation of charge: whatever charge is lost by one body must be gained by one or more other bodies, so that negative and positive charges always appear simultaneously or are simultaneously canceled out in equal amounts. This law implied that electrical effects were not the results of the "creation" of some mysterious entity (as had often been supposed before Franklin) but merely followed from an alteration or redistribution of the amount of electrical fluid in a body. A variety of experiments confirmed the universality of this principle of conservation of charge, which remains (along with conservation of momentum) one of the most fundamental principles of physical science.

One of Franklin's most startling experiments was the analysis of the Leyden jar, the first "condenser," or capacitor. The Leyden jar is a glass jar with a metal-foil outer coating. Inside, the jar has another metal-foil coating, or contains water or lead shot, which is in contact with a wire passing through the cork of the jar. The jar is usually charged by bringing the wire into contact with an electrostatic generator or a bit of rubbed amber, glass, or sulfur, while the outer coating is grounded. Franklin discovered that when a jar is charged, and the two conductors are then separated from the glass, neither the inner nor the outer conductor will show any sign of being charged; but when they are placed once again on the two sides of the jar, that jar will produce all the familiar phenomena of being charged. Franklin said that the whole charge "resides" in the glass, but today we refer to this phenomenon under the name of "polarization of the dielectric." Franklin also showed that such a device does not depend upon the shape of the bottle, and he invented the parallel plane capacitor.

The sentry-box experiment, from a manuscript of Franklin's with drawings. (*Courtesy of the American Academy of Arts and Sciences, Office of Charles and Ray Eames.*)

In a variety of experiments, Franklin also showed the effects of grounding and insulation. He discovered that a grounded pointed conductor can actually "draw off" the charge of a nearby charged object and that contrariwise a charged pointed conductor, however well-insulated, will "throw off" the charge through its point. This led him to study the electrical nature of the lightning discharge. Franklin's most important experiment with lightning was not made with the familiar kite but was the experiment of the sentry box. Franklin proposed that on a high building a sentry box be erected, with a long pointed conductor rising up through the roof. To determine whether the pointed conductor were charged, an experimenter would stand on an insulated stool inside the sentry box and bring up to the pointed conductor a grounded conductor set in an insulating handle. His objective was to draw a spark. If, as Franklin supposed, thunderclouds are electrically charged, then the pointed conductor would always become charged by induction when such clouds passed overhead, and a spark could be obtained. In practice, this experiment not only proved that clouds are electrically charged, so that lightning is only an ordinary electrical discharge from clouds on a large scale, but such rods also attracted a stroke of lightning and showed it to be an electrical phenomenon. Described in his book on electricity, this experiment was first performed in France. Franklin had been waiting for the completion of the spire of Christ Church in Philadelphia, where he hoped to erect a sentry box in order to perform the experiment. The kite was thought of as an alternative, an afterthought.

The lightning experiments brought fame to Franklin far and wide. Today the historical importance of these experiments is often misunderstood. In proving that lightning is an electrical discharge, Franklin showed that the electrical experiments performed in the laboratory are directly related to events in the natural world on a large scale. Thereafter, any general science of nature that did not include electricity would obviously be incomplete. Furthermore, the lightning experiments led Franklin to the invention of the lightning rod. For the first time in history, research in pure science led to a practical invention of major consequence. Bacon had indeed been correct in predicting that pure scientific knowledge would lead men to practical applications which would enable them to control their environment.

When Franklin began his research in electricity, the great French scientist Buffon (1707–1788) pointed out that electricity was not yet a science, but only a collection of bizarre phenomena subject to no single law. After Franklin produced his theory of electricity, which explained and correlated the known phenomena and also predicted verifiable new ones, it was generally agreed that electricity had indeed become a science. In fact, electricity was the first new science (or branch of science) to arise since Newton.

It is often thought that Franklin was not really a pure scientist in the ordinary sense of this expression, that he was, rather, a gadgeteer and inventor whose claim to science was aggrandized by his success as a statesman. In point of fact the opposite is true. By 1776, when Franklin was sent to France as the American representative, he had already gained an international reputation for his scientific work, a factor of the greatest importance in his diplomatic success. This was no unknown local patriot, but one of the leading figures of the scientific world. We continually pay tribute to Franklin's scientific genius whenever we use the many words he introduced into the language of electricity: plus and minus, positive and negative, electric battery, and a host of others.

When Franklin retired from business, he hoped to devote the rest of his life to the peaceful pursuit of a career in science. The demands of his community and his country, however, all too soon drew him into a life of public service. Franklin's choice between his love of science and his duty to his fellow men was expressed by him as follows: "Had Newton been Pilot but of a single common Ship, the finest of his Discoveries would scarce have excused, or atoned for his abandoning the Helm one Hour in Time of Danger; how much less if she carried the Fate of the Commonwealth."

Review

A. Define, explain, or otherwise identify:

Charge conservation, 585
Charge quantization, 586
Coulomb's law, 587
Electric field, 589

Electric dipole, 592
Lines of force, 592
Polar molecule, 603
Dipole moment, 603

B. Without looking back at the examples, draw a careful diagram and set up the integration to find (a) the electric field on the bisector of a line charge (Example 20-11) and (b) the electric field on the axis of a uniformly charged disk (Example 20-12). Then check your diagram and integral against those in the examples.

True or false:

1. The electric field of a point charge always points away from the charge.

2. The charge of the electron is the smallest charge found.

3. Electric lines of force never diverge from a point in space.

4. Electric lines of force cannot cross at a point in space.

Exercises

Section 20-1, Electric Charge, and Section 20-2, Coulomb's Law

1. Find the number of electrons in a charge of (a) 1 μC; (b) 1 pC.

2. In electrolysis a quantity of electricity called the faraday $\mathscr{F}$, equal to Avogardro's number of electron charges, will deposit 1 gram ionic mass of monovalent ions, for example, 23 g of Na and 35.5 g of Cl in the decomposition of NaCl. Calculate the number of coulombs in a faraday. (Avogadro's number is 6.02×10^{23}.)

3. Two protons in the helium nucleus are about 10^{-15} m apart. Calculate the electrostatic force exerted by one proton on the other.

4. A charge $q_1 = 4.0$ μC is at the origin, and a charge $q_2 = 6.0$ μC is on the x axis at $x = 3.0$ m. (a) Find the force on charge q_2. (b) Find the force on q_1. (c) How would your answers differ if q_2 were -6.0 μC?

5. Three point charges are on the x axis; $q_1 = -6.0$ μC is at $x = -3.0$ m, $q_2 = 4.0$ μC is at the origin, and $q_3 = -6.0$ μC is at $x = 3.0$ m. Find the force on q_1.

6. Two equal charges of 3.0 μC are on the y axis, one at the origin and the other at $y = 6$ m. A third charge $q_3 = 2$ μC is on the x axis at $x = 8$ m. Find the force on q_3.

7. Three charges are at the corners of a square of side L. The two charges at the opposite corners are positive, and the other is negative. All have the same magnitude q. Find the force exerted by these charges on a fourth charge $+q$ placed at the remaining corner.

Section 20-3, The Electric Field

8. A charge of 4.0 μC is at the origin. What is the magnitude and direction of the electric field on the x axis at (a) $x = 6$ m and (b) $x = 10$ m? (c) Sketch the function E_x versus x for both positive and negative x. (Remember that E_x is negative when $\mathbf{E}$ points in the negative x direction.)

9. Two charges each $+4$ μC are on the x axis, one at the origin and the other at $x = 8$ m. (a) Find the electric field on the x axis at $x = 10$ m and at $x = 2$ m. (b) At what point on the x axis is the electric field zero? (c) What is the direction of $\mathbf{E}$ at points on the x axis just to the right of the origin? Just to the left of the origin? (d) Sketch E_x versus x.

10. Two equal positive charges of magnitude $q_1 = q_2 = 6.0$ nC are on the y axis at points $y_1 = +3$ cm and $y_2 = -3$ cm. (a) What is the magnitude and direction

of the electric field at the point on the x axis at $x = 4$ cm? (b) What is the force exerted on a test charge $q_0 = 2$ nC placed on the axis at $x = 4$ cm?

11. Charge $q_1 = +6.0$ nC is on the y axis at $y = +3$ cm, and charge $q_2 = -6.0$ nC is on the y axis at $y = -3$ cm. (a) What is the magnitude and direction of the electric field on the x axis at $x = 4$ cm? (b) What is the force exerted on a test charge $q_0 = 2$ nC placed on the x axis at $x = 4$ cm?

12. When a test charge $q_0 = 2$ nC is placed at the origin, it experiences a force of 8.0×10^{-4} N in the positive y direction. (a) What is the electric field at the origin? (b) What would be the force on a charge -4 nC placed at the origin?

Section 20-4, Lines of Force

13. Figure 20-23 shows lines of force for a system of two point charges. (a) What are the relative magnitudes of the charges? (b) What are the signs of the charges? (c) In what regions of space is the electric field strong? In what regions is it weak?

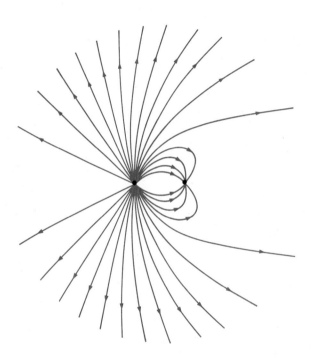

Figure 20-23
Exercise 13.

14. Two charges $+q$ and $-3q$ are separated by a small distance. Draw lines of force for this system.

15. Three equal positive point charges are situated at the corners of an equilateral triangle. Sketch the lines of force in the plane of the triangle.

Section 20-5, Calculation of the Electric Field for Continuous Charge Distributions

16. Show that the electric field E_x on the axis of a ring charge of radius a approaches that of a point charge, as expected, when the distance x from the plane of the ring is much greater than a.

17. A 0.5-μC charge is uniformly distributed on a ring of radius 3 cm. Find the electric field on the axis of the ring at (a) 1 cm, (b) 2 cm, (c) 3 cm, and (d) 600 cm from the center of the ring.

18. Show that the electric field on the perpendicular bisector of a line charge approaches that of a point charge, as expected, when the distance y from the charge is much greater than the length L.

19. A uniform line charge of density λ is on the x axis from $x = -a$ to $x = +b$. (a) Find an expression for the y component of the electric field at a point on the y axis. (b) Show that your expression is the same as Equation 20-16 when $a = b = \frac{1}{2}L$.

20. A dipole consists of a charge $+q$ on the x axis at $x = a$ and a charge $-q$ on the x axis at $x = -a$. (a) What direction is the electric field at a point on the y axis? (b) Find an expression for the electric field at a point on the y axis. (c) Show that for y much greater than a, the electric field on the y axis is given by $E_x \approx -kp/y^3$, where $p = q(2a)$ is the dipole moment.

21. A line charge has density $\lambda = 2\ \mu C/m$ and a total length of 6 m. Use approximate expressions to find the electric field on the bisector of the line at distances 2 cm and 60 m.

22. (a) Using the fact that at very great distances a disk charge should look like a point charge, write down the expected form of the electric field on the axis and very far from a uniformly charged disk of radius a and charge density σ. (b) Derive this result from Equation 20-18 by showing that $x/\sqrt{x^2 + a^2} = (1 + a^2/x^2)^{-1/2}$ and approximating this expression for a/x much less than 1, using the binomial expansion $(1 + \epsilon)^n \approx 1 + n\epsilon$ for small ϵ.

23. A disk of radius 6 cm carries a uniform charge density 30 $\mu C/m^2$. Using reasonable approximations, find the electric field on the axis of the disk at distances (a) 0.01 cm, (b) 0.02 cm, (c) 0.03 cm, and (d) 500 cm.

24. Show that the electric field just outside a spherical shell of charge is σ/ϵ_0, where σ is the charge per unit area.

25. A spherical shell of radius 10 cm carries a charge of 2 μC uniformly distributed on its surface. Find the electric field at the following distances from the center of the shell: (a) 5 cm, (b) 9.99 cm, (c) 10.01 cm, (d) 20 cm, and (e) 40 cm.

26. A sphere of radius 10 cm has a charge 2 μC uniformly distributed throughout its volume. Find the electric field at the following distances from the center of the sphere: (a) 5 cm, (b) 9.99 cm, (c) 10.01 cm, (d) 20 cm, (e) 40 cm. Compare these results with those of Exercise 25.

Section 20-6, Motion of Point Charges in Electric Fields

27. In finding the acceleration of an electron or other charged particle the ratio of the charge to mass of the particle is important. (a) Compute e/m for an electron. (b) What is the magnitude and direction of acceleration of an electron in a uniform electric field of magnitude 100 N/C? (c) Nonrelativistic mechanics can be used only if the speed of the electron is significantly less than the speed of light c. Compute the time it takes for an electron placed at rest in an electric field of magnitude 100 N/C to reach a speed $0.01c$. (d) How far does the electron travel in that time?

28. (a) Compute e/m for a proton and find its acceleration in a uniform electric field of magnitude 100 N/C. (b) Find the time it takes for a proton initially at rest in such a field to reach the speed of $0.01c$ (see Exercise 27).

29. An electron has an initial velocity 2.0 $\mu M/s$ in the x direction. It enters a uniform electric field $\mathbf{E} = 400\mathbf{j}$ N/C which is in the y direction. (a) Find the acceleration of the electron. (b) How long does it take for the electron to travel 10 cm in the x direction? (c) By how much and in what direction is the electron deflected after traveling 10 cm in the x direction?

30. Calculate the magnitude and direction of the electric field which would be needed to balance the weight of (a) an electron, (b) a proton, (c) an oil drop which has a mass of 0.2 ng and carries a charge of $+10e$, (d) a pingpong ball of mass 25 g and charge $+0.01\ \mu C$.

31. The earth has an electric field in its atmosphere which is about 150 N/C directed downward. Compare the upward electric force on an electron with the downward gravitational force.

32. An electron is projected with an initial speed of 3.0×10^5 m/s at an upward angle of 30° with the horizontal. There is a vertically upward electric field of strength 10^5 N/C. (a) What is the highest point reached by the electron? (b) How long does it take to reach this height? (c) What is the horizontal range of the electron projectile?

Section 20-7, Electric Dipoles in Electric Fields

33. Two point charges $q_1 = 2.0$ pC and $q_2 = -2.0$ pC are separated by 4 μm. What is the dipole moment of this pair of charges? Sketch the pair and indicate the direction of the dipole moment.

34. A dipole of moment $0.5e$ nm is placed in a uniform electric field of strength 4.0×10^4 N/C. What is the magnitude of the torque on the dipole when (a) the dipole is parallel to the electric field, (b) the dipole is perpendicular to the electric field, (c) the dipole makes an angle of 30° with the electric field?

Problems

1. Four charges of equal magnitude are arranged at the corners of a square of side L, as shown in Figure 20-24. (a) Find the magnitude and direction of the force exerted on the charge on the lower left corner by the other charges. (b) Show that the electric field at the midpoint of one of the sides of the square is directed along that side toward the negative charge and has magnitude E given by

$$E = k\frac{8q}{L^2}\left(1 - \frac{\sqrt{5}}{25}\right)$$

Figure 20-24
Problem 1.

2. Two charges q_1 and q_2 are placed a distance L apart. (a) What must the relative values of q_1 and q_2 be (including sign) if the electric field is to be zero at some point on the line of the charges a distance D from q_2 and $D + L$ from q_1? (b) Is there a second point anywhere on the line of the charges at which the electric field is zero?

3. Two charges q_1 and q_2 when combined give a total charge of 6 μC. When they are separated by 3 m, the force exerted by one charge on the other has the magnitude 8 mN. Find q_1 and q_2 if (a) both are positive, so that they repel each other; (b) one is positive and the other negative, so that they attract each other.

4. A positive charge Q is to be divided into two positive charges q_1 and q_2. Show that for a given separation D the force exerted by one charge on the other is greatest if $q_1 = q_2 = \frac{1}{2}Q$.

5. Two equal positive charges q are on the y axis; one is at $y = a$ and the other at $y = -a$. (a) Show that the electric field on the x axis is along the x axis with E_x given by $E_x = 2kqx(x^2 + a^2)^{-3/2}$. (b) Show that near the origin, when x is much smaller than a, E_x is approximately $2kqx/a^3$. (c) Show that for x much larger than a, E_x is approximately $2kq/x^2$. Explain why you would expect this result even before calculating it.

6. (a) Show that the electric field for the charge distribution in Problem 5 has its greatest magnitude at the points $x = a/\sqrt{2}$ and $x = -a/\sqrt{2}$ by computing dE_x/dx and setting the derivative equal to zero. (b) Sketch the function E_x versus x using the results of part (a) and parts (b) and (c) of Problem 5.

7. For the charge distribution in Problem 5 the electric field at the origin is zero. A test charge q_0 placed at the origin will therefore be in equilibrium. (a) Discuss the stability of the equilibrium for a positive test charge by considering small displacements from equilibrium along the x axis and small displacements along the y axis. (b) Repeat part (a) for a negative test charge. (c) Find the magnitude and sign of a charge q_0 which can be placed at the origin so that the net force on each of the three charges is zero. Consider what happens if any of the charges is displaced slightly from equilibrium.

8. Show that E_x on the axis of a ring charge (Equation 20-14) has its maximum and minimum values at $x = +a/\sqrt{2}$ and $x = -a/\sqrt{2}$. Sketch E_x versus x.

9. (a) Find the slope dE_x/dx at the origin for E_x due to a ring charge (Equation 20-14). (b) A bead of mass m carrying a negative charge $-q$ slides along a stretched thread which is along the axis of a ring of radius a and total charge Q. Show that if the bead is displaced slightly from the origin and released, the motion is simple harmonic. Find the frequency of the motion.

10. A disk of radius 30 cm carries a uniform charge density σ. Compare the approximation $E = \sigma/2\epsilon_0$ with the exact expression for the electric field (Equation 20-18) by computing the neglected term as a percentage of the field for distances $x = 0.1, 0.2$, and 3 cm. At what distance is the neglected term 1 percent of $\sigma/2\epsilon_0$?

11. A line charge of density λ has the shape of a square of side L which lies in the yz plane with its center at the origin. Find the electric field on the x axis at arbitrary distance x and compare your result to the field on the axis of a charged ring of approximately the same size carrying the same total charge.

12. A semi-infinite line charge of uniform density λ lies along the x axis from $x = 0$ to $x = \infty$. Find both E_x and E_y at a point on the y axis.

13. Two infinite planes are parallel to each other and separated by a distance d. Find the electric field to the left of the planes, to the right of the planes, and between the planes when (a) each plane contains a uniform charge density $\sigma = +\sigma_0$, and (b) the left plane has a uniform charge density $\sigma = +\sigma_0$ and the right plane $\sigma = -\sigma_0$. Draw the lines of force for each case.

14. A thin wire carries a uniform linear charge density λ and is bent into a circular arc which subtends an angle $2\theta_0$, as shown in Figure 20-25. Show that the electric field at the center of curvature of the arc has the magnitude $E = (2k\lambda \sin \theta_0)/R$. *Hint:* Consider an element of length $dl = R\, d\theta$ at some angle θ; find the component of the field along the line OC, and integrate from $\theta = -\theta_0$ to $\theta = +\theta_0$.

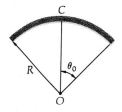

Figure 20-25
Problem 14.

15. An electric dipole consists of two charges $+q$ and $-q$ separated by a very small distance $2a$. Its center is on the x axis at $x = x_1$, and it points along the x axis toward positive x. It is in a nonuniform electric field, which is also in the x direction, given by $\mathbf{E} = Cx\mathbf{i}$, where C is a constant. (a) Find the force on the positive charge and that on the negative charge and show that the net force on the dipole is $Cp\mathbf{i}$. (b) Show that, in general, if a dipole of moment $\mathbf{p}$ lies along the x axis in an electric field in the x direction, the net force on the dipole is given approximately by $(dE_x/dx)p\, \mathbf{i}$.

16. A positive point charge $+Q$ is at the origin, and a dipole of moment $\mathbf{p}$ is a distance r away and in the radial direction, as in Figure 20-22. (a) Show that the force exerted by the electric field of the point charge on the dipole is attractive with approximate magnitude $2kQp/r^3$ (see Problem 15). (b) Consider now the dipole at the origin and a point charge Q a distance r away along the line of the dipole. From your result of part (a) and Newton's third law, show that the magnitude of the electric field of the dipole along the line of the dipole a distance r away is approximately $2kp/r^3$.

17. Two small spheres of mass 10 g are suspended from a common point by threads of length 50 cm. Find the angle made by either thread with the vertical when each sphere carries a charge of $2\mu C$ by (a) using the small-angle approximation $\sin \theta \approx \tan \theta$ and (b) using numerical methods.

18. A nonconducting thread forms a circle of radius a and lies in the xy plane with its center at the origin. It carries a nonuniform charge density $\lambda = \lambda_0 \sin \theta$, where θ is measured from $\theta = 0$ on the positive x axis. Find the electric field at the origin.

CHAPTER 21

Gauss' Law and Conductors in Electrostatic Equilibrium

Objectives   After studying this chapter, you should:

1. Be able to state Gauss' law and use it to find the electric field produced by various symmetrical charge distributions.

2. Be able to discuss the difference between conductors and insulators.

3. Be able to prove that in electrostatic equilibrium the free charge on a conductor resides on the surface of the conductor.

4. Know and be able to derive the result that the electric field just outside the surface of a conductor has the magnitude σ/ϵ_0.

5. Be able to discuss charging by induction.

Radio Times/Hulton Picture Library

Karl Friedrich Gauss, German mathematician, astronomer, and physicist (1777–1855).

In this chapter we put the idea of lines of force on a quantitative basis and develop a mathematical equation known as Gauss' law. We then show how Gauss' law can be used to find the electric field in problems with a high degree of symmetry, such as that of a uniform sphere of charge. We shall see that in such situations it is much easier to find the field using Gauss' law than it is to calculate the field directly from Coulomb's law. We then discuss electric conductors and apply Gauss' law to show that in electrostatic equilibrium the charge on a conductor resides completely on the surface.

Consider a mathematical surface enclosing the dipole shown in Figure 21-1 (the surface may be spherical or any other shape). The number of lines of force coming from the positive charge and crossing the surface going out of the enclosure depends on where the surface is drawn, but this number is exactly equal to the number of lines entering the enclosure and ending on the negative charge. If we count the number leaving as positive and the number entering as negative, the net number leaving or entering is zero. In the lines-of-force figures for other charge distributions, such as that shown in Figure 21-2, the net number of lines leaving any surface enclosing the charges is proportional to the net charge enclosed by the surface. The number is the same for all surfaces enclosing the charges. This result in its quantitative form (developed in Section 21-2) is known as Gauss' law.

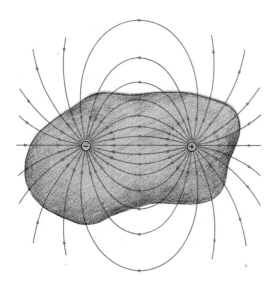

Figure 21-1
Surface enclosing an electric dipole. The number of lines leaving the surface is exactly equal to the number of lines entering the surface no matter where the surface is drawn, as long as it encloses both charges.

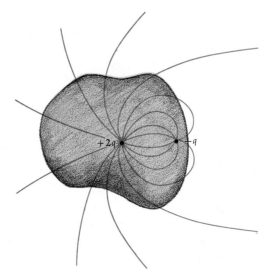

Figure 21-2
Surface enclosing charges $+2q$ and $-q$. The net number of lines leaving the surface (those leaving minus those entering) is proportional to the net charge inside the surface. In this drawing the number of lines leaving or entering a charge has been chosen to be $8q$ (16 lines leave charge $+2q$, and 8 enter $-q$). The net number of lines leaving the surface is therefore 8 no matter where the surface is drawn, as long as it encloses both charges.

21-1 Electric Flux

The mathematical quantity related to the number of lines of force through a surface area is called the *electric flux*. Figure 21-3 shows an electric field which is uniform in magnitude and direction over some region. The electric flux ϕ through a surface of area A which is perpendicular to the field is defined as the product of the field E and the area A:

$$\phi = EA \qquad\qquad 21\text{-}1$$

The units of flux are newton-metres^2 per coulomb (N·m^2/C). Since the electric field is proportional to the number of lines per unit area, the electric flux is proportional to the number of lines of force through the surface area:

$$\phi \propto N \qquad\qquad 21\text{-}2$$

The proportionality constant depends upon the choice of the number of lines leaving or entering a unit charge.

In Figure 21-4 the surface of area A_2 is not perpendicular to the electric field **E**. The number of lines that cross area A_2 is the same as the number that cross area A_1. The areas are related by

$$A_2 \cos \theta = A_1$$

where θ is the angle between **E** and the unit vector $\hat{n}$ perpendicular to surface A_2, as shown. The flux through a surface not perpendicular to **E** is defined to be

$$\phi = \mathbf{E} \cdot \hat{n} A = EA \cos \theta = E_n A \qquad\qquad 21\text{-}3$$

where $E_n = \mathbf{E} \cdot \hat{n}$ is the component of the electric-field vector perpendicular, or normal, to the surface.

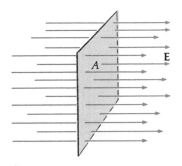

Figure 21-3
Lines of force for a uniform electric field crossing an area A perpendicular to the field. The product EA is called the electric flux.

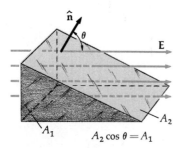

$$A_2 \cos \theta = A_1$$

Figure 21-4
Lines of force for a uniform
electric field which is perpendicular to the area A_1 but
makes an angle θ with the
unit vector $\hat{\mathbf{n}}$ normal to area
A_2. When $\mathbf{E}$ is not perpendicular to the area, the flux
through the area is $E_n A$,
where $E_n = E \cos \theta$ is the
component of $\mathbf{E}$ perpendicular to the area. The flux
through A_2 is then the same
as that through A_1.

We can generalize our definition of electric flux to curved surfaces
over which the electric field may vary in magnitude or direction or both,
by dividing the surface up into a large number of very small elements. If
each element is small enough, it can be considered to be a plane and the
variation of the electric field across the element can be neglected. Let $\hat{\mathbf{n}}_i$
be the unit vector perpendicular to such an element and ΔA_i be its area
(Figure 21-5). (If the surface is curved, the unit vectors $\hat{\mathbf{n}}_i$ will have different directions for different elements.) The flux of the electric field
through this element is

$$\Delta\phi_i = \mathbf{E} \cdot \hat{\mathbf{n}}_i \, \Delta A_i$$

The total flux through the surface is the sum of $\Delta\phi_i$ over all the elements.

Figure 21-5
When $\mathbf{E}$ varies in either magnitude or direction, the area
is divided into small area elements ΔA_i. The flux through
the area is computed by sum-
ming $E_n \, \Delta A_i$ over all the area
elements.

In the limit as the number of elements approaches infinity and the area
of each element approaches zero, this sum becomes an integral. The
general definition of electric flux is then

$$\phi = \lim_{\Delta A_i \to 0} \sum_i \mathbf{E} \cdot \hat{\mathbf{n}}_i \, \Delta A_i = \int \mathbf{E} \cdot \hat{\mathbf{n}} \, dA \qquad\qquad 21\text{-}4$$

Electric flux defined

As with a uniform electric field and a plane surface, the number of lines
of force through any surface is proportional to the flux.

We are often interested in the flux of the electric field through a
closed surface, i.e., a surface which separates space into two regions,
one inside the surface and one outside. On a closed surface the unit
normal vector $\hat{\mathbf{n}}$ is defined to be directed outward at each point. At a
point where a line of force leaves the surface, $\mathbf{E}$ is directed outward and
$\mathbf{E} \cdot \hat{\mathbf{n}}$ is positive; but at a point where a line of force enters the surface, $\mathbf{E}$
is directed inward and $\mathbf{E} \cdot \hat{\mathbf{n}}$ is negative. The total or net flux ϕ_{net}
through the closed surface is positive or negative depending on
whether $\mathbf{E}$ is predominantly outward or inward on the surface. Since
the flux through any part of the surface is proportional to the number of
lines through the surface, the net flux is proportional to the *net* number
of lines of force leaving the surface, i.e., the number of lines going out
of the surface minus the number going into the surface. The integral
over a closed surface is indicated by the symbol $\oint$. The net flux through
a closed surface is therefore written

$$\phi_{\text{net}} = \oint \mathbf{E} \cdot \hat{\mathbf{n}} \, dA = \oint E_n \, dA \qquad\qquad 21\text{-}5$$

Example 21-1 A point charge q is at the center of a spherical surface of radius R. Calculate the net flux of the electric field through this surface.

The electric field at any point on the surface is radial, with magnitude

$$E = \frac{kq}{R^2}$$

Since $\hat{\mathbf{n}}$ is also radial on the surface, $\mathbf{E} \cdot \hat{\mathbf{n}} = kq/R^2$ is constant everywhere on the surface. The net flux through the spherical surface is therefore just the product of the radial component of $\mathbf{E}$ and the total area of the spherical surface, $4\pi R^2$:

$$\phi_{\text{net}} = \oint \mathbf{E} \cdot \hat{\mathbf{n}}\, dA = \oint \frac{kq}{R^2}\, dA = \frac{kq}{R^2} \oint dA = \frac{kq}{R^2} 4\pi R^2$$

or

$$\phi_{\text{net}} = \oint \mathbf{E} \cdot \hat{\mathbf{n}}\, dA = 4\pi kq \qquad\qquad \text{21-6}$$

21-2 Gauss' Law

In Example 21-1 we calculated the flux of a point charge through a spherical surface of radius R to be $4\pi kq$, independent of R. The number of lines of force going out through a spherical surface of radius R is also proportional to the charge q. This is consistent with our previous observations that the net number of lines going out of a surface is proportional to the net charge inside the surface. This number of lines is the *same for all surfaces surrounding* the charge, independent of the shape of the surface. Since the number of lines and the flux are just proportional to each other, it follows that Equation 21-6 holds for the flux through any surface enclosing the point charge q. *The net flux through any surface surrounding a point charge q equals $4\pi kq$.*

We can extend this result to systems of more than one point charge. In Figure 21-6 the surface S encloses two point charges q_1 and q_2, and there is a third point charge q_3 outside the surface. Since the electric field at any point on the surface is the vector sum of the electric fields produced by each of the three charges, the net flux $\phi_{\text{net}} = \oint \mathbf{E} \cdot \hat{\mathbf{n}}\, dA$ through the surface is just the sum of the fluxes due to the individual charges. The flux through the surface due to the charge q_3, which is outside the surface, is zero because every line of force from q_3 that enters the surface at one point leaves the surface at some other point. The net number of lines through the surface from a charge outside the surface is zero. The flux through the surface due to charge q_1 is $4\pi kq_1$, and that due to charge q_2 is $4\pi kq_2$. The net flux through the surface equals $4\pi k(q_1 + q_2)$, which may be positive, negative, or zero depending on the signs and magnitudes of the two charges.

In general, for a system of charges, the net flux through any surface S equals $4\pi k$ times the net charge inside the surface.

$$\phi_{\text{net}} = \oint \mathbf{E} \cdot \hat{\mathbf{n}}\, dA = 4\pi kq_{\text{inside}} \qquad\qquad \text{21-7}$$

Gauss' law

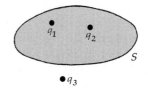

Figure 21-6
Three point charges q_1, q_2, and q_3 and a surface S which encloses q_1 and q_2.

This important result is Gauss' law. Its validity depends on the fact that the electric field due to a single point charge varies inversely with the square of the distance from the charge. It was this property of the electric field that made it possible to draw a fixed number of lines of force from a charge and have the density of lines be proportional to the field strength. It is customary to write the Coulomb constant k in terms of another constant ϵ_0, called the *permittivity of free space*:

$$k = \frac{1}{4\pi\epsilon_0} \qquad\qquad 21\text{-}8$$

With this notation Coulomb's law and Gauss' law are written

$$\mathbf{F}_{12} = \frac{1}{4\pi\epsilon_0} \frac{q_1 q_2}{r_{12}^2} \hat{\mathbf{r}}_{12} \qquad\qquad 21\text{-}9 \qquad \textit{Coulomb's law}$$

and

$$\phi_{\text{net}} = \oint \mathbf{E} \cdot \hat{\mathbf{n}}\, dA = \frac{1}{\epsilon_0} q_{\text{inside}} \qquad\qquad 21\text{-}10 \qquad \textit{Gauss' law}$$

The value of ϵ_0 in SI units is

$$\epsilon_0 = \frac{1}{4\pi k} = \frac{1}{4\pi(8.99 \times 10^9)} = 8.85 \times 10^{-12} \text{ C}^2/\text{N}\cdot\text{m}^2 \qquad 21\text{-}11 \qquad \textit{Permittivity of free space}$$

Equation 21-7 (and its equivalent, Equation 21-10) can be derived mathematically from Coulomb's law without reference to lines of force, but the derivation is rather difficult. It is given at the end of this chapter.

Gauss' law can be used to calculate the magnitude of $\mathbf{E}$ when there is a high degree of symmetry so that the direction of $\mathbf{E}$ is known everywhere and the magnitude of $\mathbf{E}$ is constant over some simple surface.

Example 21-2 Find the field of a point charge q.

We consider this simple example to illustrate the method and to show that Gauss' law is indeed equivalent to Coulomb's law. We choose our origin at the point charge and consider a spherical surface of radius r (Figure 21-7). This mathematical surface, used for calculation only, is called a *gaussian surface*. It is clear from symmetry that $\mathbf{E}$ must be radial and that its magnitude can depend only on the distance from the charge. The normal component $\mathbf{E} \cdot \hat{\mathbf{n}} = E_r$ has the same value everywhere on our spherical surface. The flux of $\mathbf{E}$ through this surface is thus

$$\phi_{\text{net}} = \oint \mathbf{E} \cdot \hat{\mathbf{n}}\, dA = \oint E_r\, dA = E_r \oint dA$$

But $\oint dA$ is just the total area of the spherical surface, $4\pi r^2$. Since the total charge inside the surface is just the point charge q, Gauss' law gives

$$E_r 4\pi r^2 = \frac{q}{\epsilon_0} \qquad \text{or} \qquad E_r = \frac{q}{4\pi\epsilon_0 r^2} \qquad\qquad 21\text{-}12$$

We have thus derived Coulomb's law from Gauss' law. Since we originally derived Gauss' law from Coulomb's law, we have now shown the two laws to be equivalent.*

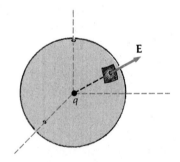

Figure 21-7
Calculation of the electric field of a point charge using Gauss' law. On a spherical surface surrounding the charge, the electric field $\mathbf{E}$ is perpendicular to the surface and constant in magnitude. The flux of the field through this surface is then $E_r 4\pi r^2$.

---

* Gauss' law and Coulomb's law are strictly equivalent only for electrostatic fields. Nonelectrostatic electric fields, produced by changing magnetic fields, are not described by Coulomb's law. Because the lines of force for these fields are closed lines, which do not diverge from, or converge to, any point in space, the net flux of nonelectrostatic fields through any closed surface is zero. Hence, Gauss' law holds for all electric fields.

Example 21-3 Find the electric field due to a uniform line charge of infinite length along the x axis.

Because of the symmetry of this situation, **E** must be perpendicular to the line and directed away from the line charge. Also, the magnitude of **E** can depend only on the perpendicular distance of the field point from the line.

Figure 21-8 shows a cylindrical can coaxial with the line charge. Let r be the radius of the cylinder, which we use for our gaussian surface. Everywhere on the cylindrical part of the surface **E** is perpendicular to the surface and constant in magnitude. On the ends of the can **E** is parallel to the surface; thus the ends make no contribution to the flux.

Figure 21-8
To calculate the electric field of an infinite line charge from Gauss' law a cylindrical surface concentric with the line charge is chosen. On the cylinder the electric field is constant in magnitude and perpendicular to the surface. The flux through the surface is $E_n 2\pi rL$, where L is the length of the cylinder.

The total charge inside the gaussian surface is λL, where λ is the charge per unit length of the line charge and L is the length of the can. Setting the flux through this surface equal to $1/\epsilon_0$ times the net charge inside, we obtain

$$\phi_{\text{net}} = \oint E_n \, dA = E_n \oint dA = \frac{\lambda L}{\epsilon_0}$$

Since the area of the cylindrical surface is $2\pi rL$,

$$E_n 2\pi rL = \frac{\lambda L}{\epsilon_0}$$

or

$$E_n = \frac{1}{2\pi\epsilon_0} \frac{\lambda L}{r} = 2k \frac{\lambda}{r} \qquad\qquad 21\text{-}13$$

in agreement with our result from Example 20-11.

Consider now a finite line charge which extends beyond the cylindrical gaussian surface shown (Figure 21-9). Since the net charge inside the surface is still λL, it might seem that we would get the same result for **E** no matter how long the line charge is. However, if the line charge is not infinite, our symmetry arguments break down. One problem is that **E** is not perpendicular to the cylindrical surface except at points equidistant from the ends of the charge. Also, E_n is not constant everywhere on the surface; that is, E_n depends not only on r but also on the distance from the center of the line charge. For this case of a finite line charge and the gaussian surface shown, Gauss' law still holds. The integral of E_n over the gaussian surface (including the faces) equals $1/\epsilon_0$ times the net charge inside the surface, λL. However, the law is *not useful* for calculating the field **E** because E_n is not constant on the gaussian surface.

This example is typical of many applications of Gauss' law. Whenever there is a high degree of symmetry (infinite line charge in this example), the field **E** is obtained from Gauss' law far more easily than by direct integration using Coulomb's law. In the absence of a high degree of symmetry, Gauss' law is of no use in obtaining **E**.

Figure 21-9
Cylindrical surface concentric with a line charge of finite length. Gauss' law is not useful for finding the electric field of a finite line charge because the electric field is not constant in magnitude on the surface and is not perpendicular to the surface.

Example 21-4 Find the electric field due to an infinite plane of uniform charge density σ.

Again, we can use symmetry to argue that the field $\mathbf{E}$ must be perpendicular to the plane, must depend only on the distance from the plane to the field point, and must have the same magnitude but opposite direction at points the same distance below the plane as above it. We choose for our gaussian surface a pillbox-shaped cylinder with its axis perpendicular to the plane and with its center on the plane (Figure 21-10). Let each end of the cylinder be parallel to the plane and have area A. In this case, $\mathbf{E}$ is parallel to the cylindrical surface, and there is no flux through this curved surface. Since the flux out of each face is $\mathbf{E} \cdot \hat{\mathbf{n}} A = E_n A$, the total flux is $2E_n A$. The net charge inside the surface is σA. From Gauss' law

$$\phi_{net} = 2E_n A = \frac{\sigma A}{\epsilon_0}$$

$$E_n = \frac{\sigma}{2\epsilon_0} = 2\pi k \sigma \qquad \qquad 21\text{-}14$$

This result agrees with that obtained from Coulomb's law in Example 20-12.

As in Example 21-3, the symmetry arguments essential for the *use* of Gauss' law depend on the fact that the plane charge distribution is infinite in extent.

Figure 21-10
Gaussian surface for the calculation of the electric field due to an infinite plane of charge. On the upper and lower faces of this pillbox surface, $\mathbf{E}$ is perpendicular to the surface and constant in magnitude. The flux through this surface is $2E_n A$, where A is the area of each face.

Example 21-5 Find the field due to a spherical shell of charge of radius R.

We choose a spherical gaussian surface of radius r, concentric with the charged shell. Whether r is greater than R or less than R, the symmetry of the problem implies that $\mathbf{E}$ is perpendicular to the gaussian surface and constant in magnitude on the surface. The flux through the gaussian surface is thus $4\pi r^2 E_r$. For a field point outside the shell $(r > R)$,

$$\phi_{net} = 4\pi r^2 E_r = \frac{Q}{\epsilon_0}$$

where Q is the total charge on the shell. Then

$$E_r = \frac{1}{4\pi\epsilon_0} \frac{Q}{r^2} = \frac{kQ}{r^2} \qquad \text{outside}$$

For a field point inside the shell $(r < R)$, the total charge inside the gaussian surface is zero, so that the flux must be zero:

$$\phi_{net} = 4\pi r^2 E_r = 0$$

and therefore

$$E_r = 0 \qquad \text{inside}$$

Since Newton's law of gravitation and Coulomb's law have the same inverse-square-distance dependence, the results of all of these calculations of the electric field using Gauss' or Coulomb's law can be applied directly to the analogous cases of calculation of the gravitational field. We need only replace the appropriate charge or charge density by the corresponding mass or mass density and change the sign, since the gravitational field of a point mass (which is always positive) points toward the mass, whereas the electric field of a (positive) point charge points away from the charge. Gauss' law for the flux of the gravitational field **g** is

$$\phi_{net} = \oint \mathbf{g} \cdot \hat{\mathbf{n}} \, dA = \oint g_n \, dA = -4\pi G m_{inside} \qquad 21\text{-}15$$

where m_{inside} is the total mass enclosed by surface S.

Questions

1. If the electric field **E** is zero everywhere on a closed surface, is the net flux through the surface necessarily zero? What then is the net charge inside the surface?

2. If the net flux through a closed surface is zero, does it follow that the electric field **E** is zero everywhere on the surface? Does it follow that the net charge inside the surface is zero?

3. Is the electric field **E** in Gauss' law the part of the electric field due to the charge inside the surface, or is it the net electric field due to all charges whether they are inside or outside the surface?

4. What information is needed in addition to the total charge inside a surface to use Gauss' law to find the electric field?

21-3 Electric Conductors

The great difference in the electrical behavior of conductors and insulators was noted even before the discovery of electric conduction. Gilbert had classified materials according to their ability to be electrified. Objects which could be electrified he called *electrics;* those which could not (metals and some other materials) he called *nonelectrics.* After Gray discovered conduction, Du Fay showed that all materials can be electrified but that care must be taken in insulating Gilbert's nonelectrics from the ground (or the experimenter) lest the charge be quickly conducted away. Using only physiological sensation for detection, Cavendish compared the conducting abilities of many substances.*

> It appears from some experiments, of which I propose shortly to lay an account before this Society, that iron wire conducts about 400 million times better than rain or distilled water—that is, the electricity meets with no more resistance in passing through a piece of iron wire 400,000,000 inches long than through a column of water only one inch long. Sea-water, or a solution of one part of salt in 30 of water, conducts 100 times, or a saturated solution of sea-salt about 720 times, better than rain-water.

* Henry Cavendish, *Philosophical Transactions of the Royal Society of London,* vol. 66, p. 196 (1776).

Because of the enormous variation in the ability to conduct electricity, it is possible and convenient to classify most materials as conductors or insulators (nonconductors). The ability of a material to conduct electricity is measured by its conductivity. The conductivity of a typical conductor is of the order of 10^{15} times that of a typical insulator, whereas within the group of conductors the conductivity varies only over several orders of magnitude. We shall give a precise definition of electric conductivity (and its reciprocal, resistivity) in Chapter 24, when we study electric currents, i.e., electric charges in motion. There we also discuss *semiconductors,* which are neither conductors nor insulators.

In this chapter we are concerned only with the behavior of conductors in electrostatic equilibrium, i.e., when all electric charges are at rest. The property of a conductor that is important in studying electrostatic fields is the availability inside the conductor of charge that is free to move about. The source of this free charge is electrons that are not bound to any atom. For example, in a single atom of copper, 29 electrons are bound to the nucleus by electrostatic attraction of the positively charged nucleus. The outermost electrons are more weakly bound than the innermost electrons because of the greater distance from the positive nucleus and because of the repulsion of the inner electrons. (This is called *screening.*) When a large number of copper atoms combine to form metallic copper, the electron binding of a single atom is changed by interaction with neighboring atoms. One or more of the outer electrons in an atom are no longer bound but are free to move throughout the whole metal, much as a gas molecule is free to move about in a box. The number of free electrons depends on the particular metal but is of the order of one per atom.

In the presence of an external electric field, the free charge in a conductor moves about the conductor until it is so distributed that it creates an electric field which cancels the external field inside the conductor. Consider a charge q inside a conductor. If there is a field **E** inside the conductor, there will be a force q**E** on this charge; and if it is free to move, i.e., if it is not bound to an atom or molecule by a stronger force, it will accelerate. Thus electrostatic equilibrium is impossible in a conductor unless the electric field is zero everywhere inside the conductor.

The electrostatic field is zero inside a conductor

Figure 21-11 shows a conducting slab placed in an external electric field $\mathbf{E}_0$. The free electrons are originally distributed uniformly throughout the slab. Since the slab is made up of neutral atoms, it is electrically neutral (assuming that no extra charge has been placed on it). If the external electric field is to the right, there will be a force on each electron $-e\mathbf{E}_0$ to the left because the electron has a negative charge, and the free electrons accordingly accelerate to the left. At the surface of the conductor, the conductor exerts forces on these electrons which balance that due to the external field, so that the electrons are bound to the conductor. (If the external field is very strong, the electrons can be stripped off from the surface. In electronics this is called *field emission.* We assume here that the external field is not strong enough to overcome the forces binding the electrons to the surface.) The result is a negative surface charge density on the left side of the slab and a positive surface charge density on the right side because of the removal of some of the free electrons from that side. Both these charge densities produce an electric field inside the slab which is opposite the external field. These two fields cancel everywhere inside the conductor, so that there is no unbalanced force on the free electrons and electrostatic equilibrium results. The behavior of the free charge in a conductor placed in an ex-

Figure 21-11
Conducting slab in an external electric field $\mathbf{E}_0$. A positive charge is induced on the right face and a negative charge on the left face, so that the resultant electric field inside the conductor is zero. The field lines then end on the left face and begin again on the right face.

ternal electric field is similar no matter what the shape of the conductor may be. When an external field is applied, the free charge quickly distributes itself until an equilibrium distribution is reached such that the net electric field is zero everywhere inside the conductor. The time to reach equilibrium depends on the conductivity. For copper and other good conductors it is less than about 10^{-16} s. For all practical purposes, electrostatic equilibrium is reached instantaneously.

Question

5. Distinguish between free charge in a conductor and net charge in a conductor.

21-4 Charge and Field at Conductor Surfaces

In this section we use Gauss' law to show that in electrostatic equilibrium (1) any net electric charge on a conductor resides on the surface of the conductor and (2) the electric field just outside the surface of a conductor is perpendicular to the surface and has the magnitude σ/ϵ_0, where σ is the local surface charge density at that point on the conductor.

To obtain the first result we consider a gaussian surface just inside the actual surface of a conductor in electrostatic equilibrium, as shown in Figure 21-12. Since the electric field is zero everywhere inside the conductor, it is zero everywhere on the gaussian surface, which is chosen to be completely within the conductor. Since $E_n = 0$ at all points on the gaussian surface, the net flux $\oint E_n \, dA$ through the surface must be zero. By Gauss' law this flux equals $1/\epsilon_0$ times the net charge inside the surface. Hence there can be no net charge inside any surface lying completely within the conductor. If there is any net charge on the conductor, it must be on the conductor surface.

Figure 21-12
The dashed line indicates a gaussian surface chosen to be just inside the actual surface of the conductor. Since the electric field is zero everywhere on the gaussian surface, Gauss' law implies that there is zero net charge inside that surface. Any charge on the conductor must therefore reside on the surface of the conductor.

To find the electric field just outside the surface of a conductor, we consider a portion of the conductor surface small enough to be considered flat, with charge density σ, which has negligible variation over the portion. We construct a cylindrical pillbox gaussian surface (Figure 21-13) with one face just outside the conductor and parallel to its surface, and the other face just inside the conductor. On the surface of the conductor, in equilibrium, the electric field must be perpendicular to the surface. If there were a tangential component of **E**, the free charge on the conductor would move until this component became zero. Since one face of the pillbox surface is just outside the conductor, we can take **E** to be perpendicular to this face. The other face of the pillbox is inside the conductor, where **E** is zero. There is no flux through the cylindrical surface of the pillbox because **E** is tangent to this surface. The flux through the pillbox is thus $E_n A$, where E_n is the field just outside the conductor surface and A is the area of the face of the pillbox. The net charge inside the gaussian surface is σA. Gauss' law gives

$$\phi_{\text{net}} = \oint E_n \, dA = E_n A = \frac{\sigma A}{\epsilon_0}$$

or

$$E_n = \frac{\sigma}{\epsilon_0} \qquad\qquad 21\text{-}16$$

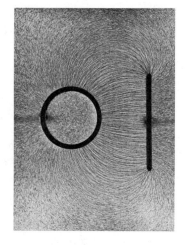

Lines of force for an oppositely charged cylinder and plate, shown by bits of fine thread suspended in oil. Note that the field lines are perpendicular to the conductors and that there are no lines inside the cylinder. (*Courtesy of Harold M. Waage, Princeton University.*)

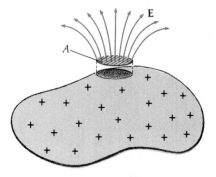

Figure 21-13
Pillbox gaussian surface with one face just outside and the other face just inside the surface of a charged conductor. The flux through the gaussian surface is $E_n A$, where E_n is the electric field at the surface of the conductor.

This result is just twice the field produced by an infinite plane of charge. We can understand this result by comparing the flux lines for an infinite plane charge with those for a conducting slab carrying the same charge density. Figure 21-14 shows a very large charged plane sheet. The flux lines are perpendicular to the sheet and point away from it on both sides. If we construct a pillbox gaussian surface with one face to the right of the sheet and the other to the left of the sheet, the net flux through this surface is $E_n 2A$, where A is the area of each face. Setting this net flux equal to $1/\epsilon_0$ times the net charge inside the pillbox A, we obtain $E_n = \frac{1}{2}(\sigma/\epsilon_0)$ for the electric field of an infinite sheet.

Figure 21-15 shows a large conducting slab. If we put a net charge on this conductor, it will be distributed equally on both faces. The conducting slab is equivalent to two charged sheets. The electric field due to each of these sheets is $\frac{1}{2}(\sigma/\epsilon_0)$. Between the sheets these fields are in opposite directions, so that their magnitudes subtract, giving zero net field inside the conductor. To the right or left of both sheets the two fields add, giving a magnitude of σ/ϵ_0 for the net electric field. The flux lines from the conducting slab are twice as dense as those for a single charged sheet if the charge *densities* are the same. (To produce equal

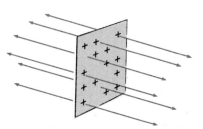

Figure 21-14
Electric field lines for a charge density σ on an infinite plane. The electric field has the magnitude $\sigma/2\epsilon_0$.

Figure 21-15
Electric field lines for a charge density σ on each face of a conducting slab. Inside the conductor, the electric field due to one face cancels that due to the other face. Outside the conductor, the field has the magnitude σ/ϵ_0 because there are two planes of charge.

charge densities on a large conducting slab and a large plane sheet, twice as much charge is needed on the slab.)

Similar but slightly more complicated reasoning can be applied to a conductor of arbitrary shape. Consider point P just outside the surface of a conductor, as shown in Figure 21-16. We can consider the charge on the surface of a conductor to consist of two parts, (1) the charge in the immediate neighborhood of point P and (2) all the rest of the charge. Since point P is just outside the surface, the charge in the immediate neighborhood looks like an infinite plane charge. It produces a field of magnitude $\frac{1}{2}(\sigma/\epsilon_0)$ at P and a field of equal magnitude just inside the conducting surface pointing away from the surface. The rest of the charge on the conductor (or elsewhere) must produce a field $\frac{1}{2}(\sigma/\epsilon_0)$ inside the conductor pointing toward the surface, so that the net field inside the conductor is zero. Whereas the field due to this second part of the charge cancels the field *inside* the conductor, it adds to that produced by the neighboring charge just outside the conductor, giving a net field $\frac{1}{2}(\sigma/\epsilon_0) + \frac{1}{2}(\sigma/\epsilon_0) = \sigma/\epsilon_0$ just outside the conductor. (The part of the field due to the distant charges has the same magnitude and direction at points just inside and just outside the surface.) This argument was first given by Laplace in about 1800.

Figure 21-16
Arbitrarily shaped conductor carrying a charge on its surface. The charge in the vicinity of point P looks like an infinite plane sheet of charge if P is very close to the conductor. This charge produces an electric field of magnitude $\sigma/2\epsilon_0$ both inside and outside the conductor, as indicated by the solid arrows. Since the resultant field inside the conductor must be zero, the rest of the charge must produce a field of equal magnitude, indicated by the dashed arrows. Inside the conductor these fields cancel, but outside at point P they add to give $E = \sigma/\epsilon_0$.

21-5 Charging by Induction

A simple and practical method of charging a conductor makes use of the free movement of charge in a conductor. In Figure 21-17 two uncharged metal spheres are in contact. When a charged rod is brought near the spheres, free electrons on one flow to the other. Suppose, for example, that the rod is positively charged. The rod attracts negatively charged electrons, and the sphere nearest the rod acquires electrons from the other, leaving the near sphere with a negative charge and the far sphere with an equal positive charge due to lack of electrons. If the spheres are separated before the rod is removed, they will have equal and opposite charges. A similar result is obtained, of course, with a negatively charged rod which drives electrons from the nearest sphere to the other. In each case, the spheres are charged without being touched by the rod, and the charge on the rod is undisturbed. This is called *electrostatic induction*.

Figure 21-17
Charging by induction. The two spherical conductors in contact become oppositely charged because the positively charged rod attracts electrons to the left side, leaving the right side positive. If the spheres are now separated with the rod in place, they retain equal and opposite charges.

A convenient large conductor is the earth itself. For most purposes we can consider the earth to be an infinitely large conductor. When a conductor is connected to the earth, it is said to be *grounded*. This is indicated by a connecting wire and parallel horizontal lines, as in Figure 21-18*b*. We can use the earth to charge a single conductor by induction.

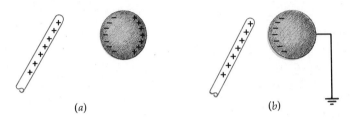

(a)   (b)

Figure 21-18
(*a*) The charge on a single conducting sphere is polarized because the electrons are attracted to the positively charged rod, leaving the opposite side positive. (*b*) When the conductor is grounded, i.e., connected to a very large conductor such as the earth, electrons from the earth neutralize the positive charge, leaving an excess negative charge. This charge remains if the ground connection is broken before the rod is removed.

In Figure 21-18*a* a positively charged rod is brought near a neutral conductor, and the conductor becomes polarized as shown. Free electrons are attracted to the side near the positive rod, leaving the other side with a positive charge. If we ground the conductor while the charged rod is still present, the conductor becomes charged oppositely to the rod because electrons from the earth travel along the connecting wire and neutralize the positive charge on the far side of the conductor. The connection to ground is broken before the rod is removed to complete the charging by induction.

Questions

6. An insulating rod is given a charge and then used to charge a set of conductors by induction. What practical limit is there on the number of times the rod can be used without recharging?

7. Can insulators as well as conductors be charged by induction?

Optional

21-6 Mathematical Derivation of Gauss' Law

Gauss' law can be derived mathematically using the concept of the solid angle. Consider an area element ΔA on a spherical surface. The solid angle $\Delta\Omega$ subtended by ΔA at the center of the sphere is defined to be

$$\Delta\Omega = \frac{\Delta A}{r^2}$$

where r is the radius of the sphere. Since ΔA and r^2 both have dimensions of length squared, the solid angle is dimensionless. The unit of solid angle is the *steradian* (sr). Since the total area of a sphere is $4\pi r^2$, the total solid angle subtended by a sphere is

$$\frac{4\pi r^2}{r^2} = 4\pi \text{ sr}$$

There is a close analogy between the solid angle and the ordinary plane angle, which is defined to be the ratio of an element of arc length of a circle, Δs, divided by the radius of the circle: $\Delta\theta = \Delta s/r$ rad. The total (plane) angle subtended by a circle is 2π rad.

In Figure 21-19 the area element ΔA is not perpendicular to the radial lines from point O. The unit vector $\hat{\mathbf{n}}$ normal to the area element makes an angle θ with the unit radial vector $\hat{\mathbf{r}}$. In this case, the solid angle subtended by ΔA is

$$\Delta\Omega = \frac{\Delta A\,\hat{\mathbf{n}}\cdot\hat{\mathbf{r}}}{r^2} = \frac{\Delta A\cos\theta}{r^2} \qquad 21\text{-}17$$

Figure 21-19
Area element ΔA whose normal $\hat{\mathbf{n}}$ is not parallel to the radial line from O to the center of the element. The solid angle subtended by this element at O is $(\Delta A\cos\theta)/r^2$.

Figure 21-20 shows a point charge q surrounded by a surface of arbitrary shape. To calculate the flux through this surface, we want to find $\mathbf{E}\cdot\hat{\mathbf{n}}\,\Delta A$ for each element of area on the surface and sum over the entire surface. The flux through the area element shown is

$$\Delta\phi = \mathbf{E}\cdot\hat{\mathbf{n}}\,\Delta A = \frac{kq}{r^2}\hat{\mathbf{r}}\cdot\hat{\mathbf{n}}\,\Delta A = kq\,\Delta\Omega$$

The solid angle $\Delta\Omega$ is the same as that subtended by the corresponding area element of a spherical surface of any radius. The sum of the flux through the entire surface is kq times the total solid angle subtended by the closed surface, which is 4π sr:

$$\phi_{\text{net}} = \oint \mathbf{E}\cdot\mathbf{n}\,dA = kq\oint d\Omega = 4\pi kq = \frac{q}{\epsilon_0}$$

Figure 21-20
Point charge q enclosed by surface S. The flux through the area element ΔA is proportional to the solid angle subtended by the area element at the charge. The net flux through the surface found by summing over all area elements is proportional to the total solid angle 4π, which is independent of the shape of the surface.

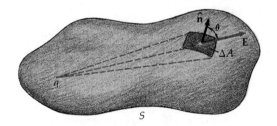

Review

A. Define, explain, or otherwise identify:

Flux, 616
Gauss' law, 617
Permittivity of free space, 618
Gaussian surface, 618

Electrostatic equilibrium, 622
Free charge in a conductor, 622
Field emission, 622
Charging by induction, 625

B. True or false:

1. If there is no charge in a region of space, the electric field on a surface surrounding the region must be zero everywhere.

2. Gauss' law holds only for symmetric charge distributions.

3. The electric field inside a uniformly charged spherical shell is zero.

4. The electric field inside a conductor is always zero.

5. The result that $\mathbf{E} = 0$ inside a conductor in equilibrium can be derived from Gauss' law.

6. A conductor has free electrons only if it has an excess negative charge.

7. If the net charge on a conductor is zero, the charge density must be zero at every point on the surface.

8. Half of the electric field at a point just outside the surface of a conductor is due to the charge on the surface in the immediate vicinity of that point.

Exercises

Section 21-1, Electric Flux, and Section 21-2, Gauss' Law

1. Consider a uniform electric field $\mathbf{E} = 2\mathbf{i}$ kN/C. (a) What is the flux of this field through a square of side 10 cm in a plane parallel to the yz plane? (b) What is the flux through the same square if the normal to its plane makes a 30° angle with the x axis?

2. What is the net flux of the uniform electric field of Exercise 1 through a cube of side 10 cm oriented so that its faces are parallel to the coordinate planes?

3. A point charge $+2\ \mu C$ is at the center of a sphere of radius 0.5 m. (a) Find the surface area of the sphere. (b) Find the magnitude of the electric field at points on the surface of the sphere. (c) What is the flux of the electric field due to the point charge through the surface of the sphere? (d) Would your answer to part (c) change if the point charge were moved so that it was inside the sphere but not at the center? (e) What is the net flux through a cube of side 1 m which circumscribes the sphere?

4. A single point charge $+2\ \mu C$ is at the origin. A spherical surface of radius 2.0 m has its center on the x axis at $x = 5$ m. (a) Sketch lines of force for the point charge. Do any lines enter the spherical surface? (b) What is the net number of lines that leave the spherical surface, counting those that enter as negative? (c) What is the net flux of the electric field due to the point charge through the spherical surface?

5. A positive point charge q is at the center of a cube of side L. A large number N of lines of force are drawn from the point charge. (a) How many of the lines pass through the surface of the cube? (b) How many lines pass through each face (assuming none are on the edges or corners)? (c) What is the net outward flux of the electric field through the cubical surface? (d) Use symmetry arguments to find the flux of the electric field through one face of the cube. (e) Which if any of your answers would change if the charge were inside the cube but not at its center?

6. Careful measurement of the electric field at the surface of a black box indicates that the net outward flux through the surface of the box is 6.0 kN·m²/C. (a) What is the net charge inside the box? (b) If the net outward flux through the surface of the box were zero, could you conclude there were no charges inside the box? Why or why not?

7. An electric field is uniform in the positive x direction for positive x, and uniform with the same magnitude but in the negative x direction for negative x. $\mathbf{E} = 200\mathbf{i}$ N/C for $x > 0$ and $\mathbf{E} = -200\mathbf{i}$ N/C for $x < 0$. A right circular cylinder of length 20 cm and radius 5 cm has its center at the origin and its axis along the x axis so that one face is at $x = +10$ cm and the other at $x = -10$ cm. (a) What is the outward flux through each face? (b) What is the flux through the side of the cylinder? (c) What is the net outward flux through the cylindrical surface? (d) What is the net charge inside the cylinder?

8. Can you use Gauss' law to calculate the electric field on the axis of an electric dipole? If so, do it. If not, explain why not.

9. A sphere of radius R has a constant volume charge density ρ. Use Gauss' law to derive expressions for the electric field a distance r from the center of the sphere for (a) $r < R$ and (b) $r > R$.

10. A thin-walled cylindrical shell of radius R and infinite length carries a uniform surface charge density σ. (a) Show that the electric field is zero for $r < R$. (b) Show that for $r > R$ the electric field has the magnitude

$$E = \frac{\sigma R}{\epsilon_0} \frac{1}{r}$$

(c) Show that your result for part (b) is the same as for an infinite line charge of the same charge per unit length.

Section 21-3, Electric Conductors

11. A penny has a mass of 3 g and is made of copper of molecular mass 63.5 g/mol. Assume one free electron for each copper atom. (a) How many free electrons are there in a penny? (b) How much free charge (in coulombs) is there?

Section 21-4, Charge and Field at Conductor Surfaces

12. A penny is in an external electric field of magnitude 1 kN/C and direction perpendicular to its faces. (a) Find the charge densities on each face of the penny. (b) If the radius of a penny is taken to be 1 cm, what is the total charge on one face?

13. A metal slab has square faces of side 10 cm. It is placed in an external electric field which is perpendicular to its faces. What is the magnitude of the electric field if the total charge on one of the faces of the slab is 1 nC?

14. A charge of 6 nC is placed uniformly on a square sheet of nonconducting material of side 20 cm in the yz plane. (a) What is the charge density σ? (b) What is the magnitude of the electric field just to the right and just to the left of the sheet? (c) The same charge is placed on a conducting square slab of side 20 cm and thickness 1 mm. What is the charge density σ? (Assume the charge distributes itself uniformly on the large square surfaces.) (d) What is the magnitude of the electric field just to the right and just to the left of each face of the slab?

15. A nonconducting spherical shell of outer radius 15 cm carries a net charge of 2 μC uniformly distributed on its surface. (a) What is the charge density σ? What is E (b) just outside the surface of the sphere and (c) just inside the shell? (d) A portion of the shell is removed, leaving a small hole. What is E just outside and just inside the shell at the hole? Assume the rest of the charge on the shell is undisturbed.

16. An irregularly shaped conductor carries a surface charge. At some point P on the surface the charge density is $\sigma = 1$ μC/m². (a) What is the electric field just outside the surface at point P? (b) The conductor is now replaced by an insulator of exactly the same shape and carrying exactly the same charge density σ at each corresponding point. What now is the electric field just outside the surface at point P? (c) If your answers for parts (a) and (b) are different, explain why. If they are the same, in what way is the insulator different from the conductor?

Section 21-5, Charging by Induction

17. Explain, giving each step, how a positively charged insulating rod can be used to give a metal sphere (a) a negative charge and (b) a positive charge. (c) Can the same rod be used to give one sphere a positive charge and another sphere a negative charge without recharging the rod?

18. Figure 21-21 shows a device called an electroscope, which consists of two metal-foil leaves attached to a conducting rod with a conducting knob on top. When uncharged, the leaves hang together vertically. When charged, the leaves repel each other. The divergence of the leaves indicates the amount of charge. (a) A positively charged nonconducting rod is brought close to the knob of the uncharged electroscope, and the leaves diverge. Explain why, with a diagram showing the charge distribution. (b) If the rod is removed, the leaves come back together. If the knob is momentarily grounded with the rod close, the leaves also come together but then diverge when the rod is removed. Explain with a diagram. What sign of net charge is now on the electroscope? (c) With the electroscope charged as in part (b), a negatively charged rod is brought close. Do the leaves come together or diverge more? Explain with a diagram.

Figure 21-21
Exercise 18.

Section 21-6, Mathematical Derivation of Gauss' Law

There are no exercises for this section.

Problems

1. A nonconducting sphere of radius R carries a volume charge density proportional to the distance from the center: $\rho = Ar$ for $r \le R$, where A is a constant; $\rho = 0$ for $r > R$. (a) Find the total charge by summing the charges in shells of thickness dr and volume $4\pi r^2\, dr$. (b) Find the electric field E_r both inside and outside the charge distribution and sketch E_r versus r.

2. Repeat Problem 1 for a sphere with charge density $\rho = B/r$ for $r \le R$ and $\rho = 0$ for $r > R$.

3. Repeat Problem 1 for a sphere with charge density $\rho = C/r^2$ for $r \le R$ and $\rho = 0$ for $r > R$.

4. Consider two infinitely long concentric cylindrical shells. The inner shell has radius R_1 and carries a uniform surface charge density σ_1 while the outer shell has radius R_2 and carries a uniform surface charge density σ_2. (a) Use Gauss' law to find the electric field in the regions $r < R_1$, $R_1 < r < R_2$, and $R_2 < r$. (b) What should the ratio σ_2/σ_1 and the relative sign be for the electric field to be zero at $r > R_2$? What then is the electric field between the shells? (c) Sketch lines of force for the situation in part (b).

5. A spherical shell of radius R_1 carries a total charge q_1 uniformly distributed on its surface. A second larger spherical shell of radius R_2 concentric with the first carries a charge q_2 uniformly distributed on its surface. (a) Use Gauss' law to find the electric field in the regions $r < R_1$, $R_1 < r < R_2$, and $R_2 < r$. (b) What should the ratio of the charges q_1/q_2 and their relative sign be for the electric field to be zero for $r > R_2$? (c) Sketch the lines of force for the situation in part (b).

6. An infinitely long cylinder of radius R has a uniform volume charge density ρ. Show that the electric field has the magnitude

$$E = \begin{cases} \dfrac{R^2\rho}{2\epsilon_0}\dfrac{1}{r} & r \ge R \\[2mm] \dfrac{\rho}{2\epsilon_0}r & r \le R \end{cases}$$

Sketch E versus r.

7. A thick nonconducting spherical shell of inner radius r_1 and outer radius r_2 has a uniform volume charge density. Find the total charge and the electric field everywhere.

8. A ring of radius R carries a uniform positive charge density λ. Figure 21-22 shows a point in the plane of the ring but not at the center. Consider two elements of the ring, of lengths s_1 and s_2 (indicated in the figure) and at distances

r_1 and r_2 from point P. (*a*) What is the ratio of the charges of these elements? Which produces the greater field at point P? (*b*) What is the direction of the field at point P due to these elements? What is the direction of the total electric field at point P? (*c*) Suppose that the electric field due to a point charge varied as $1/r$ rather than $1/r^2$. What would the electric field be at point P due to the elements shown? (*d*) How would your answers differ if point P were inside a spherical shell of charge and the elements were of *area* s_1 and s_2?

9. Use Gauss' law for the flux of the gravitational field (Equation 21-15) to derive expressions for the gravitational field both inside and outside a sphere of uniform mass density, and sketch $g(r)$ versus r.

10. A solid conductor has a cavity as shown in Figure 21-23. A small charge $+q$ is placed in the cavity. (This can be done by placing a charge on a small piece of cork and suspending the cork with an insulating string.) (*a*) Prove that there is an induced charge density on the inner surface of the cavity such that the total induced charge is $q' = -q$, independent of the location of q. (*b*) Draw lines of force for this problem.

11. A spherical conducting shell with zero net charge has inner radius a and outer radius b. A point charge q is placed at the center in the cavity. (*a*) Use Gauss' law and the properties of conductors in equilibrium to find the electric field in each of the regions $r < a$, $a < r < b$, and $b < r$. (*b*) Draw lines of force for this situation. (*c*) Describe the charge density on the outer surface ($r = b$) of the sphere. How would this charge density be affected if the point charge in the cavity were moved away from the center? Sketch lines of force for the case in which the point charge is not at the center of the cavity.

12. The electrostatic force on a charge at some point is the product of the charge and the electric field due to all other charges. Consider a small charge on the surface of a conductor, $\Delta q = \sigma \, \Delta A$. (*a*) Show that the electrostatic force on the charge is $\sigma^2 \, \Delta A / 2\epsilon_0$. (*b*) Explain why this is just half $\Delta q \, E$, where $E = \sigma/\epsilon_0$ is the electric field just outside the conductor at that point. (*c*) The force per unit area is called the electrostatic stress. Find the stress when a charge of 2 μC is placed on a conducting sphere of radius 10 cm.

13. (*a*) Show that for both the infinite plane charge and the spherical shell, the electric field is discontinuous at the surface charge by the amount σ/ϵ_0. (*b*) Prove that, in general, when there is a surface charge σ, the electric field component perpendicular to the surface is discontinuous by the amount σ/ϵ_0. Do this by constructing a gaussian pillbox with faces on each side of the surface. Use Gauss' law to find $E_2 - E_1$, where E_2 is the normal component of E on one side and E_1 is that on the other side of the surface.

14. A nonconducting sphere of radius a has a spherical cavity of radius b with center at the point $x = b$, $y = 0$, as shown in Figure 21-24. The sphere contains a uniform charge density ρ. Show that the electric field in the cavity is uniform and is given by $E_y = 0$, $E_x = \rho b/3\epsilon_0$. *Hint:* Replace the cavity by spheres of equal positive and negative charge densities.

15. A very long cylindrical shell of inner radius a and outer radius b carries a uniform charge density

$$\rho = \begin{cases} 0 & r < a \\ \rho_0 & a < r < b \\ 0 & r > b \end{cases}$$

Find the electric field everywhere.

Figure 21-22
Problem 8.

Figure 21-23
Problem 10.

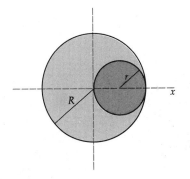

Figure 21-24
Problem 14.

CHAPTER 22   Electric Potential

Objectives   After studying this chapter you should:

1.   Be able to give definitions of electric potential difference, electric potential, and electrostatic potential energy.

2.   Be able to calculate the potential difference between two points, given the electric field in the region.

3.   Be able to give the definition of the electronvolt energy unit and the conversion factor between it and the joule.

4.   Be able to calculate the electrostatic potential energy of a system of point charges.

5.   Be able to sketch the equipotential surfaces given a pattern of electric field lines.

6.   Be able to calculate the electric potential for various charge distributions.

7.   Be able to discuss the phenomena of charge sharing, dielectric breakdown, and corona discharge.

The concept of energy, which was so important in our study of mechanics, is also important in electricity. In Chapter 6 we discussed the fact that any central force such as the electrostatic force given by Coulomb's law is conservative; i.e., the work done on a particle by this force as the particle moves from one point to another depends only on the initial and final positions and not on the path taken. There is thus a potential-energy function associated with the force. In this chapter we discuss the concept of electrostatic potential energy and a related concept of electric potential.

22-1 Potential Difference

The work we must do to move a body at constant speed from one point to another in a conservative force field equals the change in the potential energy of the body. Consider a test charge q_0 in an electrostatic field $\mathbf{E}$ produced by some system of charges. The electric force on the charge is $q_0\mathbf{E}$. If we wish to move the charge (at constant speed), we must exert a force $-q_0\mathbf{E}$ on it. If we give the charge a small displacement $d\boldsymbol{\ell}$, the work we do, $dW = -q_0\mathbf{E} \cdot d\boldsymbol{\ell}$, equals the increase in the potential energy of the test charge. The change in the potential-energy function, dU, is thus defined by

$$dU = -q_0\mathbf{E} \cdot d\boldsymbol{\ell} \qquad \text{22-1}$$

If the charge is moved from some initial point a to some final point b, the change in its potential energy is

$$U_b - U_a = \int_a^b dU = -\int_a^b q_0\mathbf{E} \cdot d\boldsymbol{\ell} \qquad \text{22-2}$$

The potential-energy change is proportional to the test charge q_0. The potential-energy change per unit charge is called the *potential difference* ΔV. For an infinitesimal displacement $d\boldsymbol{\ell}$, we obtain the potential difference from Equation 22-1 by dividing by q_0:

Potential difference defined

$$dV = \frac{dU}{q_0} = -\mathbf{E} \cdot d\boldsymbol{\ell} \qquad \text{22-3}$$

and for a finite displacement from point a to point b,

$$\Delta V = V_b - V_a = \frac{\Delta U}{q_0} = -\int_a^b \mathbf{E} \cdot d\boldsymbol{\ell} \qquad \text{22-4}$$

The potential difference $V_b - V_a$ is the work per unit charge necessary to move a test charge at constant speed from point a to point b.

Equation 22-4 defines the change in the function V, which is called the *electric potential* (or sometimes just the *potential*). Like the potential energy U, only the *change* in the electric-potential function V is significant. The value of the potential function at any point is determined by arbitrarily choosing V to be zero at some convenient point. If we choose U and V to be zero at the same point, which is customary, the electric potential at any point equals the potential energy of the charge divided by the charge q_0.

Potential

The SI unit of potential and potential difference is the joule per coulomb, called a volt (V):

$$1 \text{ V} = 1 \text{ J/C} \qquad \text{22-5}$$

Volt defined

From Equation 22-4 we note that the dimensions of potential are also those of electric field times distance. Thus the unit of electric field E, the newton per coulomb, is also equal to a volt per metre:

$$1 \text{ N/C} = 1 \text{ V/m} \qquad \text{22-6}$$

If we place a positive test charge q_0 in an electric field $\mathbf{E}$ and release the charge, it will accelerate in the direction of $\mathbf{E}$ along the field line. The kinetic energy of the charge will increase, and its potential energy will decrease. *The electric field lines therefore point in the direction of decreasing*

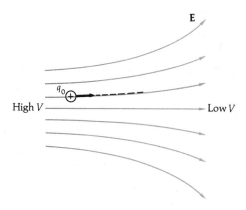

High V ———————————————————— Low V

Figure 22-1
When a positive test charge
$+q_0$ is placed in an electric
field, it accelerates in the
direction of the field. Its
kinetic energy increases, and
its potential energy decreases.
The field lines point in the
direction of decreasing
potential.

electric potential (Figure 22-1). For example, if we have a constant electric field of 10 V/m in the x direction, the potential decreases by 10 V in each metre in the x direction.

The simplest charge distribution is a single point charge Q at the origin. The electric field of such a charge is radially outward (assuming Q to be positive) and has the magnitude kQ/r^2. We can find the electric potential at any distance from this charge from the definition of potential difference (Equations 22-3 and 22-4). Consider a test charge q_0 at an initial distance r_a from the charge Q. Let us give it a small displacement in the radial direction away from the origin. Then $\mathbf{E} \cdot d\boldsymbol{\ell} = E_r \, dr$, and the change in potential energy dU is

$$dU = -q_0\mathbf{E} \cdot d\boldsymbol{\ell} = -q_0 E_r \, dr = -\frac{kq_0 Q}{r^2}\, dr \qquad \text{22-7}$$

and the change in the potential dV is

$$dV = -\mathbf{E} \cdot d\boldsymbol{\ell} = -E_r \, dr = -\frac{kQ}{r^2}\, dr \qquad \text{22-8}$$

Therefore the change in the potential from point r_a to point r_b is

$$\Delta V = V_b - V_a = -\int_a^b \frac{kQ}{r^2}\, dr = \frac{kQ}{r_b} - \frac{kQ}{r_a}$$

If we define the potential energy and the potential to be zero at an infinite distance from the point charge ($r_b = \infty$), the potential at point r_a is $V_a = kQ/r_a$. At a general distance r from the point charge Q the electric potential is

$$V(r) = \frac{kQ}{r} \qquad \text{22-9}$$

Potential of a point charge

The potential is positive or negative depending on the sign of the charge Q. The potential energy of the two-charge system is

$$U(r) = q_0 V(r) = \frac{kQq_0}{r} \qquad \text{22-10}$$

If we bring a test charge q_0 from a very large distance from the origin ($r = \infty$) to some distance r (Figure 22-2), the work we must do against the electric force equals the increase in the potential energy, which is given by Equation 22-10.

The electric potential $V = kQ/r$ is the work per unit charge that must be done to bring a positive test charge from infinity to the distance r at constant speed.

Potential is work per unit charge

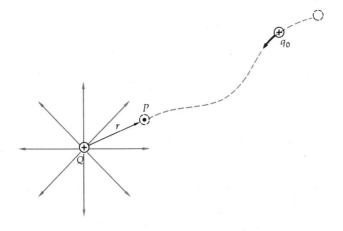

Figure 22-2
The work required to bring a test charge q_0 at constant speed from infinity to a point P that is a distance r from a charge Q at the origin is kQq_0/r. The work per unit charge is kQ/r, the electric potential at point P relative to zero potential at infinity.

Example 22-1 What is the electric potential at the distance of the first Bohr orbit, $r = 0.529 \times 10^{-10}$ m, from a proton? What is the potential energy of the electron and proton at this separation?

The charge of the proton is $q = 1.6 \times 10^{-19}$ C. Equation 22-9 gives

$$V = \frac{kq}{r} = \frac{(9 \times 10^9)(1.6 \times 10^{-19})}{0.529 \times 10^{-10}} = 27.2 \text{ V}$$

The charge of the electron is $-e = -1.6 \times 10^{-19}$ C. In SI units the potential energy of the electron a distance 0.529×10^{-10} m from a proton is

$$U = qV = (-1.6 \times 10^{-19} \text{ C})(27.2 \text{ V}) = -4.36 \times 10^{-18} \text{ J}$$

A convenient unit of energy is the *electronvolt* (eV), defined to be one volt times the magnitude of the charge of the electron. Since the electron has a charge of -1.6×10^{-19} C, and 1 C·V equals 1 J (Equation 22-5), we have

$$1 \text{ eV} = 1.6 \times 10^{-19} \text{ C·V} = 1.6 \times 10^{-19} \text{ J} \qquad 22\text{-}11 \qquad \textit{Electronvolt defined}$$

In Example 22-1 the potential energy of the electron and proton at the separation of the first Bohr orbit is 27.2 eV.

Questions

1. Explain in your own words the distinction between electric potential and electric potential energy.

2. If a charge is moved a small distance in the direction of an electric field, does its electric *potential energy* increase or decrease? Does your answer depend on the sign of the charge?

3. If a charge is moved a small distance in the direction of an electric field, does the electric *potential* increase or decrease? Does your answer depend on the sign of the charge?

4. What direction can you move relative to an electric field so that the electric potential does not change?

5. A positive charge is released from rest in an electric field. Will it move toward a region of greater or smaller electric potential?

22-2 Potential of a System of Point Charges and Electrostatic Potential Energy

If we have a system of point charges, the potential at some point (relative to zero potential at an infinite distance from the system) equals the sum of the potentials at that point due to the individual point charges. Consider, for example, two point charges q_1 and q_2 at some arbitrary positions. To bring a test charge q_0 from an infinite distance away to some position near these charges, we must do work against the resultant electric field. Since this field is just the vector sum of the individual electric fields due to each of the charges, this work is the sum of the work that must be done against the field due to each charge separately, which is $kq_1q_0/r_{10} + kq_2q_0/r_{20}$, where r_{10} is the distance from q_0 to q_1 and r_{20} is that from q_0 to q_2. The potential at some point relative to zero potential at infinity is therefore the sum of the potentials due to each charge. In general, for a system of point charges, the potential at some point P is given by

$$V = \sum_i \frac{kq_i}{r_{i0}}$$

22-12

where r_{i0} is the distance between the field point P and the ith charge q_i.

Example 22-2 Three positive 2-μC point charges are at the corners of a square of side 3 m (Figure 22-3). What is the potential V at the fourth, unoccupied corner of the square? How much work is needed to bring up a fourth positive charge of 2 μC and place it at the fourth corner of the square?

The unoccupied corner of the square is 3 m from two of the charges and $3\sqrt{2}$ m from the other charge. The potential at that point, according to Equation 22-12, is then

$$V = \frac{kq_1}{r_1} + \frac{kq_2}{r_2} + \frac{kq_3}{r_3}$$

$$= (9 \times 10^9) \left(\frac{2 \times 10^{-6}}{3} + \frac{2 \times 10^{-6}}{3} + \frac{2 \times 10^{-6}}{3\sqrt{2}} \right)$$

$$= 1.62 \times 10^4 \text{ V}$$

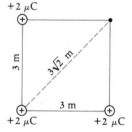

Figure 22-3
Three equal positive point charges at the corners of a square of side 3 m for Example 22-2.

Note that an alternative expression for the units of the Coulomb constant k is the volt-metre per coulomb (V·m/C).

The work needed to bring a fourth charge to the unoccupied corner of the square is the product of the charge and the potential there due to the other charges:

$$W = qV = (2 \times 10^{-6} \text{ C})(1.62 \times 10^4 \text{ V}) = 3.24 \times 10^{-2} \text{ J}$$

In the above example we calculated the work needed to bring up a fourth charge to a position near a system of three other charges. Since the electrostatic force is conservative, this work does not depend on the path taken by the charged particle but only on its position relative to the other charges. We did not calculate the work needed to assemble the original three charges. This work is also independent of the paths taken by the charges and of the order of assembly; it depends only on the initial and final positions of the charges. The work need to bring two charges Q and q_0 from an infinite separation to a final separation r is given by Equation 22-10. This work is the electrostatic potential energy of the two-charge system relative to zero potential energy when the charges are infinitely separated. In general:

The electrostatic potential energy of a system of point charges is the work needed to bring the charges from an infinite separation to their final positions.

Electrostatic potential energy

Example 22-3 Find the electrostatic potential energy of three point charges q_1, q_2, and q_3 separated by distances r_{12}, r_{23}, and r_{13} (Figure 22-4).

The work needed to bring two charges q_1 and q_2 from an infinite separation to a separation of r_{12} is kq_1q_2/r_{12}. The work needed to bring the third charge q_3 to a distance r_{13} from q_1 and r_{23} from q_2 is $W_3 = kq_1q_3/r_{13} + kq_2q_3/r_{23}$. The total work to assemble the three charges is therefore

$$W = \frac{kq_1q_2}{r_{12}} + \frac{kq_1q_3}{r_{13}} + \frac{kq_2q_3}{r_{23}}$$

This result is independent of the order in which the charges are brought to their final positions.

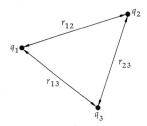

Figure 22-4
Three charges separated by arbitrary distances. The electrostatic potential energy of the system is the work required to bring the charges from an initial infinite separation to their final positions.

22-3 Electric Field and Potential: Equipotential Surfaces

The electric field lines point in the direction of decreasing potential. Consider a small displacement $\Delta\ell$ in an arbitrary electric field **E**. The work done against the electric field in giving a test charge q_0 such a displacement is $\Delta U = -q_0\mathbf{E}\cdot\Delta\ell$, and the change in potential is

$$\Delta V = -\mathbf{E}\cdot\Delta\ell \qquad\qquad 22\text{-}13$$

If the displacement $\Delta\ell$ is perpendicular to the electric field, the potential does not change. The greatest change in V occurs when the displacement $\Delta\ell$ is parallel or antiparallel to **E**. When $\Delta\ell$ is parallel to **E**, we have

$$\Delta V = -E_\ell\,\Delta\ell$$

If we divide by $\Delta\ell$ and take the limit as $\Delta\ell$ goes to zero, we have

---

$$E_\ell = -\frac{dV}{d\ell} \qquad d\ell \text{ parallel to } \mathbf{E} \qquad\qquad 22\text{-}14$$

A vector which points in the direction of the greatest change in a scalar function and which has a magnitude equal to the derivative of that function with respect to distance in that direction is called the *gradient* of the function. The electric field is the negative gradient of the potential. The field lines point in the direction of the greatest decrease in the potential function.

Figure 22-5 shows the field lines of the electric field due to a point charge Q at the origin. If we move a test charge perpendicular to these lines, no work is done and the potential does not change. A surface on which the electric potential is constant is called an *equipotential surface*.

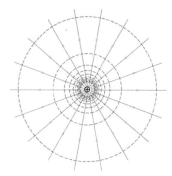

Figure 22-5
Lines of force and equipotential surfaces of a point charge. The equipotentials are spherical surfaces. The lines of force are everywhere perpendicular to the equipotentials.

For the potential $V = kQ/r$ produced by a point charge at the origin, the surfaces defined by r = constant are equipotential surfaces. These surfaces are concentric spherical surfaces. The lines of force are always perpendicular to an equipotential surface. For a point charge at the origin, the lines of force are radial lines, and the equipotential surfaces are spheres. If we have a uniform electric field in the x direction, e.g., that produced by an infinite plane charge in the yz plane, the lines of force are parallel lines in the x direction and the equipotential surfaces are planes parallel to the yz plane. Figure 22-6 shows equipotential surfaces for the field produced by an electric dipole. In a two-dimensional drawing of lines of force, the curve representing the intersection of an equipotential surface with the plane of the paper is shown by sketching a continuous curve everywhere perpendicular to the lines of force.

Since the electric field at the surface of a conductor in equilibrium is perpendicular to the surface, the conductor surface is itself an equipotential surface; since the electric field is zero inside a conductor (in electrostatic equilibrium), the potential is constant everywhere on and inside a conductor.

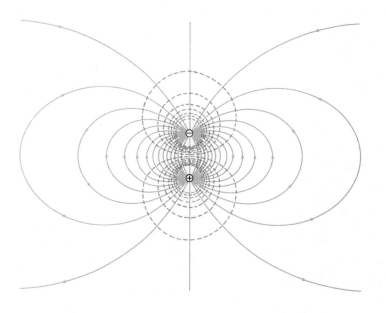

Figure 22-6
Lines of force and equipotential surfaces of a dipole. The lines of force are always perpendicular to the equipotential surfaces.

22-4 Calculation of Electric Potential

There are two ways of calculating the potential at some point due to a given charge distribution. If the electric field $\mathbf{E}$ is already known, the potential V can be found directly from its definition, Equation 22-4. Consider, for example, an infinite-plane charge distribution of surface charge density. Let the charge be in the yz plane. In Chapter 20 we found that the electric field for an infinite plane charge has magnitude $2\pi k\sigma = \sigma/2\epsilon_0$ (where k and ϵ_0 are related by $k = 1/4\pi\epsilon_0$), independent of the distance from the plane. For positive x, the electric field is

$$E_x = \frac{\sigma}{2\epsilon_0}$$

Let $V(x)$ be the potential at some distance x from the yz plane and $V(0)$ be that on the yz plane at $x = 0$. Integrating from $x = 0$ to some general point x, we obtain from Equation 22-4

$$V(x) - V(0) = -\int_0^x E_x \, dx = -\int_0^x \frac{\sigma}{2\epsilon_0} \, dx = -\frac{\sigma}{2\epsilon_0} x$$

or

$$V(x) = V(0) - \frac{\sigma}{2\epsilon_0} x \qquad\qquad 22\text{-}15$$

The potential decreases linearly with distance from the plane. Since it does not approach any limiting value as x approaches infinity, we cannot choose the potential to be zero at $x = \infty$. We can, however, choose V to be zero at $x = 0$; then $V(0) = 0$ and $V(x) = -(\sigma/2\epsilon_0)x$. Alternatively we could choose the potential to have some other value at $x = 0$, say 100 V, in which case the constant $V(0)$ would be 100 V. Since only differences in potential are important, it does not matter what value is chosen for the potential at $x = 0$.

The second method of finding the potential due to a given charge distribution is to compute the potential directly from Equation 22-12. In general, for a system of point charges, the potential at a point P is given by

$$V = \sum_i \frac{kq_i}{r_{i0}} \qquad\qquad 22\text{-}16$$

where r_{i0} is the distance between the field point P and the ith charge q_i. If we have a continuous charge distribution of finite size, the potential at some point P can be found by treating a small element of the charge dq as a point charge and integrating:

$$V = \int \frac{k \, dq}{r} \qquad\qquad 22\text{-}17$$

where r is the distance from the charge element dq to point P. This result is based on the assumption that the potential is zero at infinite distance from the charge distribution. Equation 22-17 can be used only if the charge distribution is of finite extent, so that the potential approaches a limiting value at infinity which can be chosen to be zero. Sufficiently far from any finite charge distribution, the charge distribution looks like a point charge. Then the electric field decreases as $1/r^2$, and the potential decreases as $1/r$, which approaches zero as r approaches infinity. If the charge distribution is not finite, however, the electric field does not decrease as $1/r^2$ and in general the potential does not approach any limiting value at infinity. We saw an example of this for the field of an infinite plane charge with electric field independent of the distance from the plane. Another example, which we consider below, is the infinite line charge. For such charge distributions of infinite extent we cannot use Equation 22-17 to find the potential V. Instead we find the electric field $\mathbf{E}$ first and then find V from Equation 22-4, choosing V to be zero at any convenient point.

Example 22-4 Find the potential on the axis of a ring of radius a and charge Q.

An element of charge dq is shown in Figure 22-7. The distance from this charge element to the field point on the axis of the ring is $s = \sqrt{x^2 + a^2}$, which is the same for all elements of charge on the ring. The potential due to the ring is thus

$$V = \int \frac{k\,dq}{\sqrt{x^2 + a^2}} = \frac{k}{\sqrt{x^2 + a^2}} \int dq = \frac{kQ}{\sqrt{x^2 + a^2}} \qquad \text{22-18}$$

The electric field can be calculated from Equation 22-14:

$$E_x = -\frac{dV}{dx} = -kQ(-\tfrac{1}{2})(x^2 + a^2)^{-3/2}2x = \frac{kQx}{(x^2 + a^2)^{3/2}}$$

which agrees with the direct calculation from Coulomb's law in Chapter 20.

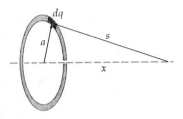

Figure 22-7
Geometry for the calculation of electric potential at a point on the axis of a uniformly charged ring.

Example 22-5 Find the potential on the axis of a disk of uniform surface charge density σ.

Let the disk axis be the x axis and consider positive x only. We can treat the disk as a set of ring charges and use our result from Example 22-4. Consider a ring of radius r and width dr (Figure 22-8). The area of this ring is $2\pi r\,dr$, and its charge is $dq = \sigma\,dA = \sigma 2\pi r\,dr$. The potential on the axis of the disk is found by summing from $r = 0$ to $r = R$:

$$V = \int_{r=0}^{r=R} \frac{k\,dq}{\sqrt{x^2 + r^2}} = \int_0^R \frac{k\sigma 2\pi r\,dr}{\sqrt{x^2 + r^2}} = k\sigma\pi \int_0^R (x^2 + r^2)^{-1/2}2r\,dr$$

This integral is of the form $\int u^n\,du$, with $u = x^2 + r^2$ and $n = -\tfrac{1}{2}$:

$$V = k\sigma\pi \left.\frac{(x^2 + r^2)^{+1/2}}{\tfrac{1}{2}}\right]_0^R = 2\pi k\sigma[(x^2 + R^2)^{1/2} - x] \qquad \text{22-19}$$

Again, the electric field can be obtained from Equation 22-14:

$$E_x = -\frac{dV}{dx} = 2\pi k\sigma \left(1 - \frac{x}{\sqrt{x^2 + R^2}}\right) \qquad \text{22-20}$$

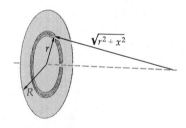

Figure 22-8
Geometry for the calculation of the electric potential at a point on the axis of a uniformly charged disk. The disk is divided into rings of area $2\pi r\,dr$.

which agrees with the result obtained directly from Coulomb's law in Chapter 20. The potential function (Equation 22-19) becomes infinite as R becomes infinite. As discussed, we cannot use Equation 22-17 with the potential at infinity defined to be zero when the charge distribution is of infinite extent. The potential function for an infinite-plane charge distribution is found from the electric field, as before.

Example 22-6 Find the potential due to a spherical shell of charge Q.

We consider a spherical shell of radius R having a uniform surface charge density σ. We are interested in the potential at all points inside and outside the shell. Since this shell is of finite extent, we can use Equation 22-17. The calculation is quite difficult and is given at the end of this section. For this problem it is much easier to find the potential from the electric field, which we already know.*

Outside the spherical shell, the electric field is radial and the same as if all the charge were at the origin:

$$E_r = \frac{kQ}{r^2} \qquad r > R$$

where $Q = 4\pi R^2 \sigma$ is the total charge on the shell. The work needed to bring a test charge from infinity to a point a distance r from the center of the shell for $r > R$ is the same as that for a point charge at the origin; the potential is therefore given by

$$V(r) = \frac{kQ}{r} \qquad r > R$$

where we have chosen the potential to be zero at infinite r. Inside the spherical shell the electric field is zero. It therefore takes no work to move a test charge from point to point inside the shell, and the potential inside the shell is constant. As r approaches R from outside the shell, the potential approaches kQ/R. Hence the constant value of V inside must be kQ/R to make $V(r)$ continuous. Thus

$$V(r) = \begin{cases} \dfrac{kQ}{R} & r \le R \\ \dfrac{kQ}{r} & r \ge R \end{cases}$$

22-21   *Potential of a spherical shell*

This potential function is sketched in Figure 22-9.

Figure 22-9
Electric potential of a uniformly charged spherical shell of radius R as a function of distance r from the center of the shell. Inside the shell the potential has the constant value kQ/R. Outside the shell the potential is the same as that due to a point charge at the center of the sphere.

A common mistake is to think that the potential must be zero inside a spherical shell because the electric field is zero. Actually zero electric field merely implies that the potential does not change. Consider a spherical shell with a small hole so that we can move a test charge in and out of the shell. If we move the test charge from an infinite distance to the shell, the work per charge we must do against the electric force is kQ/R. Inside the shell there is no electric field, and so it takes no work to move the test charge inside the shell. The total amount of work per charge it takes to bring the test charge from infinity to any point inside the shell is just the work it takes to bring it up to the shell radius R: kQ/R. The potential is therefore kQ/R everywhere inside the shell.

---

* The direct calculation of the electric field for a spherical shell is even more difficult (see Example 20-13); however, we were able to infer the properties of this electric field from the consideration of symmetry and the lines of force. If neither the electric field nor the electric potential were known, it would be easiest to find the potential directly, as shown at the end of this section, and then find E from $E_r = -dV/dr$.

Example 22-7 Find the potential due to a spherical ball of uniform charge density.

Let the radius of the ball be R and the total charge be $Q = \frac{4}{3}\pi R^3 \rho$, where ρ is the charge per unit volume. We found the electric field for this charge distribution in Example 20-14. Outside the ball the electric field is kQ/r^2, the same as that due to a point charge. Thus the potential outside is kQ/r. Inside the charged ball the electric field is

$$E_r = \frac{kQ}{R^3}\, r$$

Since the electric field inside the ball is not zero, the potential will not be constant; and since the electric field points radially outward inside the ball, the potential must increase as we move inward; i.e., work must be done on a test charge to move it from the outside edge of the ball inward. The potential therefore has its greatest value at the origin. We can find the potential inside from Equation 22-4. Integrating from $r = 0$ to a general point r inside the ball gives

$$V(r) - V(0) = -\int_0^r E_r\, dr = -\int_0^r \frac{kQ}{R^3}\, r\, dr = -\frac{1}{2}\frac{kQ}{R^3}\, r^2 \qquad 22\text{-}22$$

The potential at the origin, $V(0)$, cannot be chosen to be zero because we have already chosen the potential to be zero at $r = \infty$. We find $V(0)$ by requiring the potential to be continuous at $r = R$. At this point at the edge of the ball the potential is kQ/R. Setting $r = R$ in Equation 22-22, we have

$$\frac{kQ}{R} - V(0) = -\frac{1}{2}\frac{kQ}{R^3}\, R^2 = -\frac{1}{2}\frac{kQ}{R}$$

or

$$V(0) = \frac{3}{2}\frac{kQ}{R}$$

The potential inside the ball is then

$$V = \frac{3kQ}{2R} - \frac{kQr^2}{2R^3} = \frac{kQ}{2R}\left(3 - \frac{r^2}{R^2}\right) \qquad 22\text{-}23$$

The complete potential function is sketched in Figure 22-10.

Example 22-8 Find the potential due to a uniform infinite line charge.

Let the charge per unit length be λ. Since this charge distribution extends to infinity, we cannot use Equation 22-17 to find the potential. The potential does not approach a limiting value at infinity and therefore cannot be chosen to be zero there. We use Equation 22-4. Let E_r be the component of the electric field vector that points away from the line charge. E_r is most easily found using Gauss' law, as in Chapter 21. The result is $E_r = 2k\lambda/r$. If $V(a)$ is the potential at some distance a, the potential at distance r is obtained from Equation 22-4:

$$V(r) - V(a) = -\int_a^r E_r\, dr = -\int_a^r \frac{2k\lambda}{r}\, dr = -2k\lambda\, \ln\frac{r}{a}$$

For a positive line charge, the electric field lines point away from the line and the potential decreases with increasing distance from the line charge. At large values of r the potential decreases without limit. The potential therefore cannot be chosen to be zero at $r = \infty$. Instead, it is

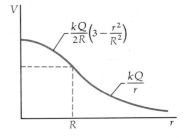

Figure 22-10
Plot of the electric potential V of a ball of charge of uniform density versus distance r from the center of the ball. V has its maximum value at the center of the ball. Outside the ball, V is the same as that due to a point charge at the center of the sphere.

chosen to be zero at any convenient point $r = a$. Then $V(a) = 0$, and the potential is given by

$$V(r) = -2k\lambda \ln \frac{r}{a}$$

22-24

Questions

6. In Example 22-4, does it matter whether the charge Q is uniformly distributed around the ring? Would either V or E_x be different if it were not?

7. If the electric potential is constant throughout a region of space, what can you say about the electric field in that region?

Optional

Direct Calculation of $V(r)$ for a Spherical Shell

This calculation is similar to the calculation of the electric field given in Example 20-13, but slightly easier. The geometry is shown in Figure 22-11. We first consider a point P outside the shell. By symmetry, the potential can depend only on the distance r from the center of the shell. We choose for our charge element the strip shown, which has circumference $2\pi R \sin \theta$ and width $R\, d\theta$. The area of this strip is $dA = 2\pi R^2 \times \sin \theta\, d\theta$, and the charge is $dq = \sigma\, dA = \sigma 2\pi R^2 \sin \theta\, d\theta$. The potential due to this element of charge is

$$dV = \frac{k\, dq}{s} = \frac{k\sigma 2\pi R^2 \sin \theta\, d\theta}{s}$$

22-25

where s is the distance from this ring to the point P, as shown in the figure. The total potential produced by the complete shell is obtained by summing over all the circular strips on the shell, i.e., by integrating from $\theta = 0$ to $\theta = 180°$. The quantities θ and s are related by the law of cosines:

$$s^2 = r^2 + R^2 - 2rR \cos \theta$$

22-26

Differentiating gives

$$2s\, ds = +2rR \sin \theta\, d\theta$$

22-27

Instead of substituting s from Equation 22-26 into Equation 22-25 and integrating over θ, it is easier to substitute $\sin \theta\, d\theta$ from Equation 22-27 and integrate from $s = r - R$ at $\theta = 0$ to $s = r + R$ at $\theta = 180°$. We then have

$$dV = \frac{k\sigma 2\pi R^2 (s\, ds/rR)}{s} = \frac{k\sigma 2\pi R\, ds}{r}$$

Integrating from $s = r - R$ to $s = r + R$ gives

$$V = \int_{r-R}^{r+R} \frac{k\sigma 2\pi R}{r}\, ds = \frac{k\sigma 2\pi R}{r}[(r + R) - (r - R)]$$

$$= \frac{k\sigma 2\pi R}{r} 2R = \frac{k(4\pi R^2 \sigma)}{r} = \frac{kQ}{r}$$

The potential at a point outside the spherical shell is the same as that due to a point charge Q at the origin. The calculation of the potential at a point inside the shell is the same except that now s varies from

Figure 22-11
Geometry for calculating the potential of a spherical shell of charge. The area of the element indicated is $dA = 2\pi R^2 \sin \theta\, d\theta$, and its distance from the field point P is related to r, R, and θ by the law of cosines: $s^2 = r^2 + R^2 - 2rR \cos \theta$.

$s = R - r$ at $\theta = 0$ to $s = R + r$ at $\theta = 180°$. The potential at a point inside the shell is thus given by

$$V = \int_{R-r}^{R+r} \frac{k\sigma 2\pi R}{r}\, ds = \frac{k\sigma 2\pi R}{r}[(R + r) - (R - r)]$$

$$= \frac{k\sigma 2\pi R}{r}(2r) = \frac{k\sigma 4\pi R^2}{R} = \frac{kQ}{R}$$

Inside the sphere the potential is constant and independent of the distance r and has the value kQ/R.

22-5   Charge Sharing

In general, two conductors which are separated in space will not be at the same potential. The potential difference between the conductors depends on the geometrical shapes of the conductors, their separation, and the net charge on each conductor. When two conductors are brought into contact, the charge on the conductors distributes itself so that in electrostatic equilibrium the electric field is zero inside both conductors. In this situation the two conductors in contact may be considered a single conductor. In equilibrium each conductor has the same potential. The transfer of charge from one conductor to another is called *charge sharing*. Coulomb used the method of charge sharing to produce various charges of known ratios to some original charge, in his experiment to find the force law between two small (point) charges.

Charge sharing

Consider a spherical conductor carrying a charge $+Q$. The lines of force outside the conductor point radially outward, and the potential of the conductor relative to infinity is kQ/R. If we bring up a second identical but uncharged conductor, the potential and field lines will change: negative electrons on the uncharged conductor will be attracted to the positive charge Q, leaving the near side of the uncharged conductor with a negative charge and the far side with a positive charge (Figure 22-12). This charge separation on the neutral conductor will affect the originally uniform charge distribution on the positive conductor.

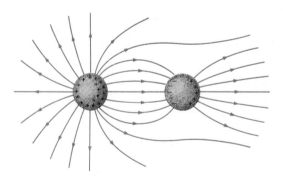

Figure 22-12
Electric field lines for a charged spherical conductor near an uncharged spherical conductor.

Although the detailed calculation of the charge distributions and potential in this case is quite complicated, we can see that some of the field lines leaving the positive conductor will end on the negative charge on the near side of the neutral conductor, and an equal number of lines will leave the far side of that conductor. Since the potential decreases as we move along a field line, the positively charged conductor is at a greater chosen to be zero at any convenient point $r = a$. Then $V(a) = 0$, and the potential is given by

$$V(r) = -2k\lambda \ln \frac{r}{a} \qquad\qquad 22\text{-}24$$

Questions

6. In Example 22-4, does it matter whether the charge Q is uniformly distributed around the ring? Would either V or E_x be different if it were not?

7. If the electric potential is constant throughout a region of space, what can you say about the electric field in that region?

Optional

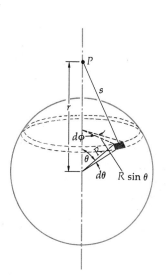

Figure 22-11
Geometry for calculating the potential of a spherical shell of charge. The area of the element indicated is $dA = 2\pi R^2 \sin \theta\, d\theta$, and its distance from the field point P is related to r, R, and θ by the law of cosines: $s^2 = r^2 + R^2 - 2rR \cos \theta$.

Direct Calculation of $V(r)$ for a Spherical Shell

This calculation is similar to the calculation of the electric field given in Example 20-13, but slightly easier. The geometry is shown in Figure 22-11. We first consider a point P outside the shell. By symmetry, the potential can depend only on the distance r from the center of the shell. We choose for our charge element the strip shown, which has circumference $2\pi R \sin \theta$ and width $R\, d\theta$. The area of this strip is $dA = 2\pi R^2 \times \sin \theta\, d\theta$, and the charge is $dq = \sigma\, dA = \sigma 2\pi R^2 \sin \theta\, d\theta$. The potential due to this element of charge is

$$dV = \frac{k\, dq}{s} = \frac{k\sigma 2\pi R^2 \sin \theta\, d\theta}{s} \qquad\qquad 22\text{-}25$$

where s is the distance from this ring to the point P, as shown in the figure. The total potential produced by the complete shell is obtained by summing over all the circular strips on the shell, i.e., by integrating from $\theta = 0$ to $\theta = 180°$. The quantities θ and s are related by the law of cosines:

$$s^2 = r^2 + R^2 - 2rR \cos \theta \qquad\qquad 22\text{-}26$$

Differentiating gives

$$2s\, ds = +2rR \sin \theta\, d\theta \qquad\qquad 22\text{-}27$$

Instead of substituting s from Equation 22-26 into Equation 22-25 and integrating over θ, it is easier to substitute $\sin \theta\, d\theta$ from Equation 22-27 and integrate from $s = r - R$ at $\theta = 0$ to $s = r + R$ at $\theta = 180°$. We then have

$$dV = \frac{k\sigma 2\pi R^2(s\, ds/rR)}{s} = \frac{k\sigma 2\pi R\, ds}{r}$$

Integrating from $s = r - R$ to $s = r + R$ gives

$$V = \int_{r-R}^{r+R} \frac{k\sigma 2\pi R}{r}\, ds = \frac{k\sigma 2\pi R}{r} [(r + R) - (r - R)]$$

$$= \frac{k\sigma 2\pi R}{r} 2R = \frac{k(4\pi R^2 \sigma)}{r} = \frac{kQ}{r}$$

The potential at a point outside the spherical shell is the same as that due to a point charge Q at the origin. The calculation of the potential at a point inside the shell is the same except that now s varies from

$s = R - r$ at $\theta = 0$ to $s = R + r$ at $\theta = 180°$. The potential at a point inside the shell is thus given by

$$V = \int_{R-r}^{R+r} \frac{k\sigma 2\pi R}{r} \, ds = \frac{k\sigma 2\pi R}{r} [(R + r) - (R - r)]$$

$$= \frac{k\sigma 2\pi R}{r} (2r) = \frac{k\sigma 4\pi R^2}{R} = \frac{kQ}{R}$$

Inside the sphere the potential is constant and independent of the distance r and has the value kQ/R.

22-5   Charge Sharing

In general, two conductors which are separated in space will not be at the same potential. The potential difference between the conductors depends on the geometrical shapes of the conductors, their separation, and the net charge on each conductor. When two conductors are brought into contact, the charge on the conductors distributes itself so that in electrostatic equilibrium the electric field is zero inside both conductors. In this situation the two conductors in contact may be considered a single conductor. In equilibrium each conductor has the same potential. The transfer of charge from one conductor to another is called *charge sharing*. Coulomb used the method of charge sharing to produce various charges of known ratios to some original charge, in his experiment to find the force law between two small (point) charges.

Charge sharing

Consider a spherical conductor carrying a charge $+Q$. The lines of force outside the conductor point radially outward, and the potential of the conductor relative to infinity is kQ/R. If we bring up a second identical but uncharged conductor, the potential and field lines will change: negative electrons on the uncharged conductor will be attracted to the positive charge Q, leaving the near side of the uncharged conductor with a negative charge and the far side with a positive charge (Figure 22-12). This charge separation on the neutral conductor will affect the originally uniform charge distribution on the positive conductor.

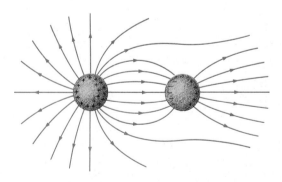

Figure 22-12
Electric field lines for a charged spherical conductor near an uncharged spherical conductor.

Although the detailed calculation of the charge distributions and potential in this case is quite complicated, we can see that some of the field lines leaving the positive conductor will end on the negative charge on the near side of the neutral conductor, and an equal number of lines will leave the far side of that conductor. Since the potential decreases as we move along a field line, the positively charged conductor is at a greater potential than the neutral conductor. If we put the two conductors in contact, positive charge will flow to the neutral conductor until both conductors are at the same potential. (Actually, negative electrons flow from the neutral conductor to the positive conductor. It is slightly more convenient to think of this as a flow of positive charge in the opposite direction.) By symmetry, since the conductors are identical, they will share the original charge equally. If the conductors are now separated, each will carry charge $\frac{1}{2}Q$ and both will be at the same potential.

In Figure 22-13 a small conductor carrying a positive charge q is inside the cavity of a second larger conductor. In equilibrium, the electric field is zero inside the conducting material. The lines of force that leave the positive charge q must end on the inner surface of the large conductor. A negative charge $-q$ must therefore be induced on the inner surface of this conductor. This must occur no matter what the charge is on the outside surface of this conductor. Regardless of the charge on the larger conductor, the small conductor in the cavity is at a greater potential because the lines of force go from this conductor to the larger conductor. If the conductors are now connected, say with a fine conducting wire, *all* the charge originally on the smaller conductor will flow to the larger one. When the connection is broken, there is no charge on the small conductor in the cavity and there are no field lines anywhere within the outer surface of the larger conductor. The positive charge transferred from the smaller conductor to the larger one resides completely on the outside surface. If we bring up a second small positively charged conductor into the cavity (through the small opening shown in the figure) and touch it to the inner surface, we again transfer all the charge to the outer conductor. This procedure can be repeated indefinitely. This method is used to produce large potentials in the Van de Graaff generator, where the charge is brought to the inner surface of a larger spherical conductor by a continuous belt (Figure 22-14). The greater the net charge on the outer conductor, the greater its potential. For example, if the conductor is a spherical shell, its potential is kQ/R. The maximum potential obtainable in this way is limited only by the fact that air molecules become ionized in very high electric fields and the air becomes a conductor. This phenomenon, which is called *dielectric breakdown*, occurs in air at electric field strengths of about $E_{\max} \approx$ 3 MN/C = 3 MV/m. The resulting discharge through the conducting

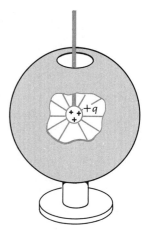

Figure 22-13
Small conductor carrying a positive charge inside a cavity in a larger conductor.

Dielectric breakdown

Figure 22-14
Small demonstration Van de Graaff generator. The top is removed in the photograph on the right to show the charging belt. (*Courtesy of Larry Langrill.*)

air is called *corona discharge*. Since the potential of a spherical conductor is kQ/R and the electric field just outside such a conductor is kQ/R^2, the maximum potential is related to the field by

$$V_{max} = RE_{max}$$

A sphere of radius 1 m can therefore be raised to a potential of about 3 MV in air before breakdown.

When a charge is placed on a conductor of nonspherical shape like that in Figure 22-15, the conductor will be an equipotential surface but the charge density and the electric field $E = \sigma/\epsilon_0$ just outside the conductor will vary from point to point. Near a point where the radius of curvature is small (A in Figure 22-15) the charge density and electric field will be large, whereas near a point where the radius of curvature is large (B in Figure 22-15) the charge density and electric field will be small. For an arbitrarily shaped conductor, the potential at which dielectric breakdown occurs depends on the smallest radius of curvature of any part of the conductor. If the conductor has sharp points of very small radius of curvature, dielectric breakdown will occur at relatively low potentials.

Figure 22-15
Nonspherical conductor. The charge density σ is greatest at points of small radius of curvature (near A) and least at points of large radius of curvature (near B). The electric field $E = \sigma/\epsilon_0$ at the conductor surface is greatest where the radius of curvature is smallest.

Corona discharge

William Vandivert

Corona discharge from overloaded power lines at General Electric test facilities in Pittsfield, Massachusetts. Because of the intense electric field near the wire, the air molecules are ionized and the air becomes conducting. The light is given off when the ions and electrons recombine.

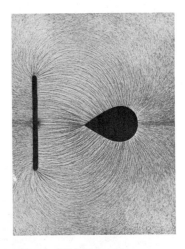

Electric field lines near a nonspherical conductor and plate carrying equal and opposite charges. The lines are shown by small bits of thread suspended in oil. The electric field is strongest near points of small radius of curvature, such as at the ends of the plate and at the pointed left side of the conductor. (*Courtesy of Harold M. Waage, Princeton University.*)

Review

A. Define, explain, or otherwise identify:

Potential difference, 633
Electric potential, 633
Volt, 633
Electronvolt, 635
Electrostatic potential energy, 637

Gradient, 637
Equipotential surface, 638
Charge sharing, 644
Dielectric breakdown, 645
Corona discharge, 646

B. True or false:

1. If the electric field is zero in some region of space, the electric potential must also be zero in that region.

2. If the electric potential is zero in some region of space, the electric field must also be zero in that region.

3. If the electric field is zero at a point, the potential must also be zero at that point.

4. Lines of electric field point toward regions of lower potential.

5. The value of the electric potential can be chosen to be zero at any convenient point in space.

Exercises

Section 22-1, Potential Difference

1. A uniform electric field of 2 kN/C is in the x direction. A point charge $Q = 3 \, \mu C$ initially at rest at the origin is released. (a) What is the kinetic energy of the charge when it is at $x = 4$ m? (b) What is the change in potential energy of the charge from $x = 0$ to $x = 4$ m? (c) What is the potential difference $V(4 \text{ m}) - V(0)$? Find the potential $V(x)$ if $V(x)$ is chosen to be (d) zero at $x = 0$, (e) 4 kV at $x = 0$, (f) zero at $x = 1$ m.

2. An infinite plane sheet of charge density $\sigma = +1 \, \mu C/m^2$ is in the yz plane. (a) What is the magnitude of the electric field in newtons per coulomb? In volts per metre? What is the direction of $\mathbf{E}$ for positive x? (b) What is the potential difference $V_b - V_a$ when point b is at $x = 20$ cm and point a is at $x = 50$ cm? (c) How much work is required by an outside agent to move a test charge $q_0 = +1$ nC from point a to point b without acceleration?

3. A uniform electric field is in the negative x direction. Points a and b are on the x axis, a at $x = 2$ m and b at $x = 6$ m. (a) Is the potential difference $V_b - V_a$ positive or negative? (b) If the magnitude of $V_b - V_a$ is 10^5 V, what is the magnitude E of the electric field?

4. Two parallel conducting plates carry equal and opposite charge densities so that the electric field between them is approximately uniform. The difference in potential between the plates is 500 V, and they are separated by 10 cm. An electron is released from rest at the negative plate. (a) What is the magnitude of the electric field between the plates? Is the positive or negative plate at the higher potential? (b) Find the work done by the electric field as the electron moves from the negative plate to the positive plate. Express your answers in both electron-volts and joules. (c) What is the change in potential energy of the electron when it moves from the negative plate to the positive plate? What is its kinetic energy when it reaches the positive plate?

5. An electric field is given by $\mathbf{E} = ax\mathbf{i}$, where $\mathbf{E}$ is in newtons per coulomb, x is in metres, and a is a positive constant. (a) What are the SI units of a? (b) How much work is done by this field on a positive point charge q_0 when the charge moves from the origin to some point x? (c) Find the potential function $V(x)$ such that $V = 0$ at $x = 0$.

6. A positive charge of magnitude 2 μC is at the origin. (a) What is the electric potential V at a point 4 m from the origin relative to $V = 0$ at infinity? (b) What is the potential energy when a $+3$-μC charge is placed at $r = 4$ m? (c) How much work must be done by an outside agent to bring the 3-μC charge from infinity to $r = 4$ m, assuming that the 2-μC charge is held fixed at the origin? (d) How much work must be done by an outside agent to bring the 2-μC charge from infinity to the origin if the 3-μC charge is first placed at $r = 4$ m and held fixed?

Section 22-2, Potential of a System of Point Charges and Electrostatic Potential Energy

7. Four 2-μC point charges are at the corners of a square of side 4 m. Find the potential at the center of the square (relative to zero potential at infinity) if (a) all the charges are positive, (b) three of the charges are positive and one is negative, (c) two are positive and two are negative.

8. Three point charges are on the x axis, q_1 at the origin, q_2 at $x = 3$ m, and q_3 at $x = 6$ m. Find the potential at the point $x = 0$, $y = 3$ m if (a) $q_1 = q_2 = q_3 = 2 \, \mu C$, (b) $q_1 = q_2 = 2 \, \mu C$ and $q_3 = -2 \, \mu C$, (c) $q_1 = q_3 = 2 \, \mu C$ and $q_3 = -2 \, \mu C$.

9. Find the electrostatic potential energy for each of the charge distributions of Exercise 7.

10. Find the electrostatic potential energy for each of the charge distributions of Exercise 8.

11. Points A, B, and C are at the corners of an equilateral triangle of side 3 m. Equal positive charges of 2 μC are at A and B. (a) What is the potential at point C? (b) How much work is required to bring a positive charge of 5 μC from infinity to point C if the other charges are held fixed? (c) Answer parts (a) and (b) if the charge at B is replaced by a charge of -2 μC.

Section 22-3, Electric Field and Potential: Equipotential Surfaces

12. A point charge $q = 3.00$ μC is at the origin. (a) Find the potential V on the x axis at $x = 3.00$ m and at $x = 3.01$ m. (b) Does the potential increase or decrease as x increases? Compute $-\Delta V/\Delta x$, where ΔV is the change in potential from $x = 3.00$ m to $x = 3.01$ m and $\Delta x = 0.01$ m. (c) Find the electric field at $x = 3.00$ m and compare its magnitude with $-\Delta V/\Delta x$ found in part (b). (d) Find the potential (to three significant figures) at the point $x = 3.00$ m and $y = 0.01$ m and compare your result with the potential on the x axis at $x = 3.00$ m. Discuss the significance of this result.

13. A charge of $+3.00$ μC is at the origin, and a charge of -3.00 μC is on the x axis at $x = 6.00$ m. (a) Find the potential on the x axis at $x = 3.00$ m. (b) Find the electric field on the x axis at $x = 3.00$ m. (c) Find the potential on the x axis at $x = 3.01$ m and compute $-\Delta V/\Delta x$, where ΔV is the change in potential from $x = 3.00$ m to $x = 3.01$ m and $\Delta x = 0.01$ m; compare your result with your answer to part (b).

14. Sketch $V(x)$ versus x for points on the x axis for the charge distribution in Exercise 13.

15. In the following, V is in volts and x is in metres. Find E_x when (a) $V(x) = 2000 + 3000x$; (b) $V(x) = 4000 + 3000x$; (c) $V(x) = 2000 - 3000x$; (d) $V(x) = -2000$ independent of x.

16. The electric potential in some region of space is given by $V(x) = C_1 + C_2 x^2$, where V is in volts, x is in metres, and C_1 and C_2 are positive constants. (a) What is the significance of the constant C_1? (b) Find the electric field $\mathbf{E}$ in this region. In what direction is $\mathbf{E}$?

17. An infinite sheet of charge has surface charge density 1 μC/m^2. How far apart are the equipotential planes whose potentials differ by 100 V?

18. A point charge $q = +\frac{1}{9} \times 10^{-8}$ C is at the origin. Taking the potential to be zero at $r = \infty$, locate the equipotential surfaces at 20-V intervals from 20 to 100 V and sketch to scale. Are these surfaces equally spaced?

19. Two equal positive charges are separated by a small distance. Sketch the lines of force and the equipotential surfaces for this system.

Section 22-4, Calculation of Electric Potential

20. Two positive charges $+q$ are on the x axis at $x = +a$ and $x = -a$. (a) Find the potential $V(x)$ as a function of x for points on the x axis. (b) Sketch $V(x)$ versus x. (c) What is the significance of the minimum in your curve?

21. Sketch $V(x)$ versus x for the uniformly charged ring in the yz plane of Example 22-4. At what point is $V(x)$ a maximum? What is E_x at this point?

22. Consider a ball of charge of uniform charge density with radius $R = 10^{-15}$ m and total charge $+e$. (This is a model of a proton.) If the center of the charge is at the origin, find the electric field and the potential at (a) $r = R$ and (b) $r = 0$.

23. A charge of $q = +10^{-8}$ C is uniformly distributed on a spherical shell of radius 10 cm. (*a*) What is the magnitude of the electric field just outside and just inside the shell? (*b*) What is the magnitude of the electric potential just outside and just inside the shell? (*c*) What is the electric potential at the center of the shell? What is the electric field at that point?

24. An electric field is given by $E_x = 2.0x^3$ kN/C. Find the potential difference between the points on the x axis at $x = 1$ m and $x = 2$ m.

Section 22-5, Charge Sharing

25. Suppose that a Van de Graaff generator has a potential difference of 1 MV between the belt and the outer shell, and that charge is supplied at the rate of 200 μC/s. What minimum power is needed to drive the moving belt?

26. Sketch lines of force and equipotential surfaces both near to and far from the conductor shown in Figure 22-15, assuming that the conductor carries some charge q.

27. To what potential can a small Van de Graaff sphere of radius 10.0 cm be raised before dielectric breakdown?

Problems

1. In the Bohr model of the hydrogen atom (Example 22-1) the electron moves in a circular orbit of radius r around the proton. (*a*) Find an expression for the kinetic energy of the electron as a function of r by setting the force on the electron (given by Coulomb's law) equal to ma, where a is the centripetal acceleration. Show that at any distance r the kinetic energy is half the magnitude of the potential energy. (*b*) Evaluate $\frac{1}{2}mv^2$ and the total energy $E = \frac{1}{2}mv^2 + U$ for $r = 0.529 \times 10^{-10}$ m. (*c*) How much energy (in electronvolts) must be supplied to the hydrogen atom to ionize it, i.e., to remove the electron to infinity with zero energy?

2. In a Van de Graaff accelerator, protons are released from rest at a potential of 5 MV and travel through a vacuum to a region at zero potential. (*a*) Find the speed of the 5-MeV protons. (*b*) If the potential change occurs over a distance of 2.0 m, find the accelerating electric field.

3. Consider two infinite parallel sheets of charge, one in the yz plane and the other at distance $x = a$. (*a*) Find the potential everywhere in space with $V = 0$ at $x = 0$ if the sheets carry equal positive charge density $+\sigma$. (*b*) Do the same if the charge densities are equal and opposite, with the sheet in the yz plane positive.

4. When uranium ^{235}U captures a neutron, it splits into two nuclei (and emits several neutrons, which can cause other uranium nuclei to split). Assume that the fission products are equally charged nuclei with charge $+46e$ and that these nuclei are at rest just after fission and separated by twice their radius $2R \approx 1.3 \times 10^{-14}$ m. (*a*) Using $U = kq_1q_2/2R$, calculate the electrostatic potential energy of the fission fragments. This is approximately the energy released per fission. (*b*) About how many fissions per second are needed to produce 1 MW of power in a reactor?

5. Radioactive ^{210}Po emits alpha particles with energy 5.30 MeV. Assume that just after the alpha particle is formed and escapes from the nucleus, the alpha particle with charge $+2e$ is a distance R from the center of the daughter nucleus ^{206}Pb with charge $+82e$. Calculate R by setting the electrostatic potential energy of the two particles at this separation equal to 5.30 MeV.

6. Four charges are at the corners of a square centered at the origin as follows: q at $x = -a$, $y = +a$ $(-a, +a)$; $2q$ at (a, a); $-3q$ at $(a, -a)$; and $6q$ at $(-a, -a)$. Find (*a*) the electric field at the origin and (*b*) the potential at the origin. (*c*) A

fifth charge $+q$ is placed at the origin and released from rest. Find its speed when it is a great distance from the origin.

7. Find the electrostatic energy of the four point charges in Problem 6.

8. Show that Equation 22-19 for the potential on the axis of a disk charge gives the expected limit $V \rightarrow kq/x$ for x much greater than R, where q is the total charge in the disk.

9. Two large parallel metal sheets carry equal and opposite charge densities of magnitude σ. The sheets have area A and are separated by a distance d. (a) Find the potential difference between the sheets. (b) A third sheet of the same area and thickness a inserted between the original two sheets carries no net charge. Find the potential difference between the original two sheets and sketch the lines of $\mathbf{E}$ in the region between the original two sheets.

10. A point charge $+3e$ is at the origin and a second charge $-2e$ is on the x axis at $x = a$. Sketch the potential function $V(x)$ versus x for all x. At what point or points is $V(x)$ zero? Are these points equilibrium points? How much work is needed to bring a third charge $+e$ to the point $x = \frac{1}{2}a$ on the x axis?

11. An electric dipole consists of a negative charge $-q$ at the origin and a positive charge $+q$ on the z axis at $z = a$ (Figure 22-16). (a) Find the potential on the z axis for $z > a$ and show that when z is much greater than a, $V \approx kp/z^2$, where $p = qa$ is the dipole moment. (b) Show that this result can be obtained by considering a single charge $+q$ at the origin and finding $V(z) - V(z + \Delta z) \approx -(dV/dz)\,\Delta z$, where $\Delta z = a$.

12. For the dipole of Problem 11, show that the potential at a point off-axis a great distance r from the origin is given approximately by

$$V = \frac{kaq \cos \theta}{r^2} = \frac{kp \cos \theta}{r^2} = \frac{kpz}{r^3}$$

Hint: If r_1 is the distance from the positive charge to the field point, as in Figure 22-16, use the law of cosines to show that $r_1 \approx r[1 - (2a \cos \theta)/r]^{1/2}$ for $r \gg a$. Then use the binomial expansion to approximate $kq/r_1 = kqr_1^{-1}$.

13. Two concentric spherical-shell conductors carry equal and opposite charges. The inner shell has radius a and charge $+q$; the outer shell has radius b and charge $-q$. Find the potential difference between the shells, $V_a - V_b$.

14. Two very long coaxial cylindrical-shell conductors carry equal and opposite charges. The inner cylinder has radius a and charge $+q$; the outer shell has radius b and charge $-q$. The length of each cylinder is L. Find the potential difference between the shells.

15. A hollow spherical conductor has inner radius a and outer radius b. A positive point charge $+q$ is at the center of the sphere, and the conductor is uncharged. Find the potential $V(r)$ everywhere, assuming that $V = 0$ at $r = \infty$, and sketch $V(r)$ versus r.

16. A nonconducting sphere of radius R has charge density $\rho = \rho_0 r/R$, where ρ_0 is a constant. (a) Show that the total charge is $Q = \pi R^3 \rho_0$. (b) Show that the total charge inside a sphere of radius $r < R$ is $q = Qr^4/R^4$. (c) Use Gauss' law to find the electric field E_r everywhere. (d) Use $dV = -E_r\,dr$ to find the potential $V(r)$ everywhere, assuming that $V = 0$ at $r = \infty$.

17. Starting with $a + e$ charge, a line of n charges is assembled, each separated by a distance a from its two neighbors. The charges are equal in magnitude and they alternate in sign. (a) Find an expression for the potential at the $(n + 1)$th site when the row contains n charges. (b) If $a = 0.1$ nm, find the binding energy (in eV) of a single charge in the middle of an infinite row. *Note:* $1 - \frac{1}{2} + \frac{1}{3} - \frac{1}{4}$ . . . $= \ln 2$.

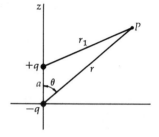

Figure 22-16
Electric dipole on the z axis for Problem 11. The law of cosines can be used to relate the distance r_1 to r and θ.

CHAPTER 23   Capacitance, Electrostatic
Energy, and Dielectrics

Objectives   After studying this chapter, you should:

1. Be able to derive expressions for the capacitance of a parallel-plate capacitor, cylindrical capacitor, and spherical capacitor.

2. Be able to calculate the effective capacitance of systems of parallel and series capacitors.

3. Be able to derive the expression $U = \frac{1}{2}QV$ for the energy stored in a charged capacitor.

4. Be able to discuss the concept of electrostatic field energy.

5. Be able to discuss the effect of a dielectric on the capacitance, charge, potential difference, and electric field in a parallel-plate capacitor.

6. Know what is meant by bound charge and be able to discuss how bound charge arises and what effect it has.

A capacitor* is a useful device for storing charge and energy. It consists of two conductors insulated from each other. A typical capacitor, called a parallel-plate capacitor, consists of two large conducting plates of area A separated by a small distance d. When the plates are connected to a charging device, e.g., a battery, as in Figure 23-1, charge is transferred from one conductor to the other until the potential difference between the conductors due to their equal and opposite charges equals the potential difference between the battery terminals. The amount of charge separated (which equals the magnitude of the charge on either conductor) depends on the geometry of the capacitor—e.g., on the area and separation of the plates in a parallel-plate capacitor—and is directly proportional to the potential difference V. The proportionality constant is called the *capacitance:*

$$Q = CV \hspace{4cm} \text{23-1} \quad \textit{Capacitance defined}$$

---

* A capacitor was formerly called a condenser, a term little used today.

(We write V rather than ΔV for the potential difference to simplify the appearance of the formulas involving capacitance.) The capacitance depends on the size, shape, and geometrical arrangement of the conductors. For example, as we shall see below, the capacitance of a parallel-plate capacitor is proportional to the area of the plates and inversely proportional to their separation distance. The SI unit of capacitance is the coulomb per volt, called a *farad* (F):

$$1 \text{ farad} = 1 \text{ coulomb/volt} \qquad \text{23-2}$$

Since the farad is a rather large unit, submultiples such as the microfarad (1 μF = 10^{-6} F) or the picofarad (1 pF = 10^{-12} F) are often used.

Figure 23-1
Capacitor consisting of two closely spaced parallel-plate conductors. When the conductors are connected to the terminals of a battery, the battery transfers charge from one conductor to the other until the potential difference between the conductors equals that of the battery terminals. The amount of charge transferred is proportional to the potential difference.

23-1   Calculation of Capacitance

The calculation of the capacitance of a capacitor is not difficult in principle. Given any two conductors, we find the potential difference V when there is a charge $+Q$ on one conductor and $-Q$ on the other, and then calculate the capacitance C from

$$C = Q/V \qquad \text{23-3}$$

For simple geometries, e.g., the parallel-plate capacitor or the cylindrical capacitor, we can find the potential difference by first finding the electric field using either Gauss' or Coulomb's law, whichever is more convenient. The potential difference is then found by integrating the electric field along any path connecting the conductors, according to Equation 22-4. We shall illustrate this calculation for three simple arrangements of practical importance. Finding the potential difference between two conductors in more complicated arrangements, such as two nearby (nonconcentric) spherical conductors, is a problem in intermediate or advanced electrostatics and will not be considered here.

Example 23-1 *The Parallel-Plate Capacitor* A parallel-plate capacitor consists of two parallel conducting plates very close together. Let each plate have area A, with charge $+Q$ on one plate and $-Q$ on the other. Each plate then has a surface charge density $\sigma = Q/A$, and the field will be nearly uniform between the plates, as indicated by the equally spaced field lines in Figure 23-2. The electric field in the region of space between the plates and far from the edges will essentially be that due to two infinite plane sheets of charge. The field due to each sheet has the magnitude $\sigma/2\epsilon_0$. Outside the plates these fields cancel, but in the space between the plates they add, and the electric field between the plates is

$$E = \frac{\sigma}{\epsilon_0} = \frac{Q}{\epsilon_0 A}$$

23-4

Figure 23-2
Parallel-plate capacitor. If the plates are closely spaced, the electric field is approximately uniform between the plates.

This result could also be obtained from Gauss' law applied to a pillbox gaussian surface with one face between the plates and the other inside one of the conductors. This result is only approximate because effects near the edges of the plates have been neglected. In practice, plates are often very close to each other. Since this field is constant in the region between the plates, the potential difference between the plates is Ed, where d is the separation distance between the plates. Thus

$$V = Ed = \frac{Qd}{\epsilon_0 A}$$

The capacitance is

$$C = \frac{Q}{V} = \frac{Q}{Qd/\epsilon_0 A} = \frac{\epsilon_0 A}{d}$$

23-5   *Parallel-plate capacitor*

The capacitance is proportional to the area of the plates and inversely proportional to the separation distance. We see from the dimensions of Equation 23-5 that the SI unit of ϵ_0 can be written as a farad per metre:

$$\epsilon_0 = \frac{1}{4\pi k} = 8.85 \times 10^{-12} \text{ F/m}$$

23-6

A numerical calculation illustrates how large the farad unit of capacitance is. Let each plate be a square of side 10 cm and let the plate separation be 1 mm. The capacitance is then

$$C = \frac{\epsilon_0 A}{d} = \frac{(8.85 \times 10^{-12} \text{ F/m}) (0.1 \text{ m}) (0.1 \text{ m})}{0.001 \text{ m}}$$

$$= 8.85 \times 10^{-11} \text{ F} \approx 90 \text{ pF}$$

Example 23-2 The 90-pF capacitor of Example 23-1 is connected to a 12-V battery and charged to 12 V. How many electrons are transferred from one plate to the other?

From the definition of capacitance (Equation 23-1) the charge transferred is

$$Q = CV = (90 \times 10^{-12} \text{ F}) (12 \text{ V}) = 1.1 \times 10^{-9} \text{ C}$$

This is the magnitude of the charge on either plate. The number of electrons in a charge of 1.1×10^{-9} C is

$$N = \frac{Q}{e} = \frac{1.1 \times 10^{-9} \text{ C}}{1.6 \times 10^{-19} \text{ C}} = 6.9 \times 10^9$$

Electric field lines between plates of a parallel-plate capacitor, shown by small bits of thread suspended in oil. When the plates are very close together, the fringing of the field near the edges can be neglected. (*Courtesy of Harold M. Waage, Princeton University.*)

Example 23-3 *The Cylindrical Capacitor* A cylindrical capacitor consists of a small cylinder or wire of radius a and a larger concentric cylindrical shell of radius b. Let $+Q$ be the charge on the inner conductor and $-Q$ be on the outer conductor. The electric field between the conductors is found most easily from Gauss' law. Let us consider a cylindrical gaussian surface of radius r and length L_1, as shown in Figure 23-3. According to Gauss' law, the flux through this surface equals q/ϵ_0, where q is the net charge inside the surface. If we assume that L is much larger than a or b so that we can neglect edge effects, the electric field will be perpendicular to the axis of the conductors and depend only on the perpendicular distance from the axis of the conductors. Then there will be flux through only the cylindrical part of the gaussian surface; i.e., there will be no flux through the faces at each end of the surface. Since the electric field is constant in magnitude on the gaussian surface and perpendicular to this surface, the flux equals the magnitude E_r times the area, which is $2\pi r L_1$. The net charge inside the gaussian surface is $q = \lambda L_1 = QL_1/L$, where $\lambda = Q/L$ is the charge per unit length of the inner conductor. Thus Gauss' law gives for the electric field

A variable parallel-plate capacitor. (*Courtesy of Larry Langrill, Oakland University.*)

$$\phi_{\text{net}} = E_r 2\pi r L_1 = \frac{q}{\epsilon_0} = \frac{\lambda L_1}{\epsilon_0}$$

$$E_r = \frac{\lambda}{2\pi\epsilon_0 r} = \frac{Q}{2\pi\epsilon_0 L}\frac{1}{r} \qquad\qquad 23\text{-}7$$

Figure 23-3
Cylindrical capacitor consisting of an inner wire of radius a and a concentric outer shell of radius b. The electric field is calculated from Gauss' law using a cylindrical gaussian surface of radius r and length L_1 between the conductors.

This result is the same as that for an infinite line charge. The potential difference between the conductors is found from Equation 22-4. Let V_a be the potential of the inner conductor and V_b be that of the outer conductor. Then

$$V_b - V_a = -\int_a^b E_r\, dr = -\frac{Q}{2\pi\epsilon_0 L}\int_a^b \frac{dr}{r} = -\frac{Q}{2\pi\epsilon_0 L}\ln\frac{b}{a}$$

The potential is of course greater on the inner conductor, which carries the positive charge, since the electric field lines point from this conductor to the outer conductor. The magnitude of this potential difference is

$$V = V_a - V_b = \frac{Q \ln (b/a)}{2\pi\epsilon_0 L} \qquad\qquad 23\text{-}8$$

and the capacitance is (Equation 23-3)

$$C = \frac{Q}{V} = \frac{2\pi\epsilon_0 L}{\ln(b/a)} \qquad\qquad 23\text{-}9$$

As expected, the capacitance is proportional to the length of the cylinders. The greater the length, the greater the amount of charge that can be put on the conductors for a given potential difference, since the electric field, and therefore the potential difference, depends only on the charge per unit length.

Example 23-4 *The Spherical Capacitor* A spherical capacitor consists of a small inner conducting sphere of radius R_1 and a larger concentric spherical shell of radius R_2 (Figure 23-4). The inner sphere can be supported on an insulator. A small opening is made in the outer shell so that charge can be placed on the inner sphere. If this opening is small enough, it will have a negligible effect on the spherical symmetry of the capacitor. To find the capacitance we place a charge $+Q$ on the inner sphere and $-Q$ on the outer sphere and find the potential difference. The electric field between the conductors is the same as that due to a point charge Q at the origin. As we have seen, the charge on the outer shell does not contribute to the electric field inside it. (We do not need to know the electric field in the region $r < R_1$ or, outside the outer shell, $r > R_2$, but it is easy to show from Gauss' law that $\mathbf{E} = 0$ in both these regions.) The electric field between the conductors is thus

$$E_r = \frac{kQ}{r^2}$$

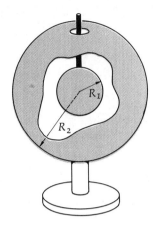

Figure 23-4
Spherical capacitor consisting of an inner sphere of radius R_1 and a concentric outer spherical shell of radius R_2. The electric field between the conductors is radial.

If V_1 is the potential on the inner sphere and V_2 on the outer sphere, the potential difference is

$$V_2 - V_1 = -\int_{R_1}^{R_2} E_r \, dr = -\int_{R_1}^{R_2} \frac{kQ}{r^2} \, dr$$

$$= \frac{kQ}{R_2} - \frac{kQ}{R_1} = -\frac{kQ(R_2 - R_1)}{R_1 R_2}$$

The potential of the inner sphere V_1 is greater than that of the outer sphere, as evidenced by the fact that the electric field lines point radially outward from the inner to the outer sphere. The magnitude of the potential difference is

$$V = V_1 - V_2 = \frac{kQ(R_2 - R_1)}{R_1 R_2} \qquad\qquad 23\text{-}10$$

The capacitance is

$$C = \frac{Q}{V} = \frac{R_1 R_2}{k(R_2 - R_1)} = 4\pi\epsilon_0 R_1 \frac{R_2}{R_2 - R_1} \qquad\qquad 23\text{-}11$$

As the radius of the outer sphere approaches infinity, the capacitance approaches $4\pi\epsilon_0 R_1$. We can think of this result with the outer spherical shell at infinity as the capacitance of a single isolated sphere. For a sphere of radius R, its capacitance, defined as the ratio of its charge to its potential (relative to zero potential at infinity), is

$$C = 4\pi\epsilon_0 R \qquad\qquad 23\text{-}12$$

The capacitance of an isolated spherical conductor is proportional to its radius. Note that in all three examples the capacitance is proportional to the constant ϵ_0 and to some characteristic length of the system.

23-2 Parallel and Series Combinations of Capacitors

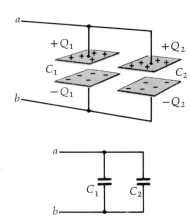

Two or more capacitors are often used in combination. In electric circuits a capacitor is indicated by the symbol $+$. Figure 23-5 shows two capacitors connected in parallel. The upper plates of the two capacitors are connected together by a conducting wire and are therefore at the same potential. The lower plates are also connected together and are at a common potential. It is clear that the effect of adding a second capacitor connected in this way is to increase the capacitance; i.e., the area is essentially increased, allowing more charge to be stored for the same potential difference $V = V_a - V_b$. If the capacitances are C_1 and C_2, the charges Q_1 and Q_2 stored on the plates are given by

$$Q_1 = C_1 V \quad \text{and} \quad Q_2 = C_2 V$$

where V is the potential difference across either capacitor, and the total charge stored is

$$Q = Q_1 + Q_2 = C_1 V + C_2 V = (C_1 + C_2)V \qquad 23\text{-}13$$

The effective capacitance of two capacitors in parallel is defined to be the ratio of the total charge stored to the potential. Thus

$$C_{\text{eff}} = \frac{Q}{V} = C_1 + C_2 \qquad 23\text{-}14$$

Figure 23-5
Two capacitors in parallel. The potential difference across the capacitors is the same, $V_a - V_b$, for each capacitor.

Capacitors in parallel

The effective capacitance is that of a single capacitor which could replace the parallel combination and store the same amount of charge for a given potential difference V. This reasoning can be extended to three or more capacitors connected in parallel, as in Figure 23-6. The effective capacitance equals the sum of the individual capacitances.

In Figure 23-7 two capacitors C_1 and C_2 are connected in series with a potential difference $V = V_a - V_b$ between the upper plate of the first capacitor and the lower plate of the second. Such a situation can be realized in practice by connecting points a and b to the terminals of a battery. If a charge $+Q$ is placed on the upper plate of the first capacitor, there will be an equal negative charge $-Q$ induced on its lower plate. This charge comes from electrons drawn from the upper plate of the second capacitor. Thus there will be an equal charge $+Q$ on the upper plate of the second capacitor and $-Q$ on its lower plate. The potential difference across the upper capacitor is $V_a - V_c = Q/C_1$. Similarly, the potential difference across the second capacitor is $V_c - V_b = Q/C_2$. The

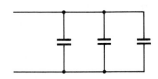

Figure 23-6
Three capacitors in parallel. The effect of adding a parallel capacitor is to increase the effective capacitance.

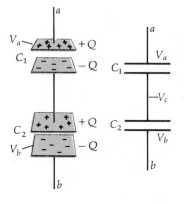

Figure 23-7
Two capacitors in series. The potential difference $V_a - V_b$ is the sum of the potential differences across the capacitors. The charge on each capacitor is the same.

potential difference across the two capacitors in series is the sum of these potential differences:

$$V_a - V_b = (V_a - V_c) + (V_c - V_b) = \frac{Q}{C_1} + \frac{Q}{C_2}$$

Calling this potential difference V, we have

$$V = Q \left(\frac{1}{C_1} + \frac{1}{C_2} \right) = \frac{Q}{C_{\text{eff}}} \qquad \text{23-15}$$

where $C_{\text{eff}} = Q/V$ is the ratio of the charge to the total potential difference across the two capacitors connected in series. From Equation 23-15,

$$\frac{1}{C_{\text{eff}}} = \frac{1}{C_1} + \frac{1}{C_2} \qquad \text{23-16} \qquad \textit{Capacitors in series}$$

or

$$C_{\text{eff}} = \frac{C_1 C_2}{C_1 + C_2} \qquad \text{23-17}$$

The effective capacitance of two capacitors in series is less than that of either of the individual capacitors. For example, in Equation 23-17, C_{eff} can be considered the product of C_1 and $C_2/(C_1 + C_2)$, which is less than 1. Equation 23-16 can be generalized to three or more capacitors in series:

$$\frac{1}{C_{\text{eff}}} = \frac{1}{C_1} + \frac{1}{C_2} + \frac{1}{C_3} + \cdots \qquad \text{23-18}$$

23-3 Electrostatic Energy in a Capacitor

While a capacitor is being charged, positive charge is transferred from the negatively charged conductor to the positively charged conductor. Since the positive conductor is at a greater potential than the negative conductor, the potential energy of the charge being transferred is increased. For example, if a small amount of charge Δq is transferred through a potential difference V, the potential energy of the charge is increased by the amount $\Delta q \, V$ (by definition, potential difference is the potential-energy difference per unit charge). Work must therefore be done to charge a capacitor. Some of this work (or all of it, depending on the charging process) is stored as electrostatic potential energy. (If the capacitor is charged by connecting it to a battery, for example, the battery does twice as much work as is stored as electrostatic potential energy. Half the work done by the battery is wasted as heat in the connecting wires and the battery itself. We shall discuss this in detail in Chapter 25.)

Consider the charging of a parallel-plate capacitor. Since only the potential *difference* between the plates is important, we are free to choose the potential to be zero at any point. It is convenient to choose the potential of the negative plate to be zero. At the beginning of the charging process, neither plate is charged. There is no electric field, and both plates are at the same potential. After the charging process a charge Q_0 has been transferred from one plate to the other, and the potential difference is $V_0 = Q_0/C$, where C is the capacitance. We then have a negative charge $-Q_0$ on one plate, which we have chosen to be at zero potential, and $+Q_0$ on the other plate at potential V_0. One might expect that the work needed to accomplish this would be just the charge Q_0 times the potential energy per unit charge V_0, but only the last bit of charge must be raised by the full potential difference V_0. The potential difference between the plates increases from zero originally to its final value V_0. The average value of the potential difference during the charging process is just $\frac{1}{2}V_0$, and the work needed is $\frac{1}{2}Q_0V_0$. We can see this as follows.

Let q be the charge that has been transferred at some time during the process. The potential difference is then $V = q/C$. If a small amount of charge dq is now transferred from the plate with charge $-q$ at zero potential to the plate with charge q at potential V, its potential energy is increased by

$$dU = V \, dq = \frac{q}{C} \, dq$$

The total increase in potential energy in charging from $q = 0$ to $q = Q_0$ is the energy stored in the capacitor (Figure 23-8):

$$U = \int dU = \int_0^{Q_0} \frac{q}{C} \, dq = \frac{1}{C} \int_0^{Q_0} q \, dq = \frac{1}{2} \frac{Q_0^2}{C}$$

Using $C = Q_0/V_0$, we can write this in a variety of other ways:

$$U = \frac{1}{2} \frac{Q_0^2}{C} = \tfrac{1}{2}Q_0V_0 = \tfrac{1}{2}CV_0^2 \qquad \text{23-19}$$

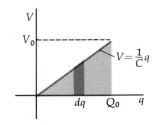

Figure 23-8
The work needed to charge a capacitor is the integral of $V \, dq$ from the original charge $q = 0$ to the final charge $q = Q_0$. This work is the area $\frac{1}{2}Q_0V_0$ under the curve.

Energy in a charged capacitor

The derivation of Equation 23-19 was carried out for the parallel-plate capacitor, but a review of the steps will show that the geometry of the capacitor plays no role in the argument. The expression $C = \epsilon_0 A/d$ was never used. The argument is applicable to any capacitor, and Equation 23-19 is a general expression for the energy stored in a charged capacitor as electrostatic potential energy.

Questions

1. The potential difference of a capacitor is doubled. By what factor does its stored electric energy change?

2. Half the charge is removed from a capacitor. What fraction of its stored energy was removed along with the charge?

23-4   Electrostatic Field Energy

In the process of charging a capacitor, an electric field is created. For example, in a parallel-plate capacitor, there is no electric field when the plates are uncharged. When they have their final charge Q_0, there is a field $E_0 = \sigma/\epsilon_0 = Q_0/\epsilon_0 A$, where A is the area of the plates. The work done in charging the capacitor is therefore the same as the work needed to create an electric field. Consider again the charging of a parallel-plate capacitor. When one plate has charge $+q$ and the other $-q$, the field between the plates is

$$E = \frac{\sigma}{\epsilon_0} = \frac{q}{\epsilon_0 A} \qquad \text{23-20}$$

The work needed to transfer an additional charge dq is force $dq\,E$ times the distance s^*, since the field (and therefore the force) is constant. This is the work done to push dq against the field from $-q$ to $+q$,

$$dW = dq\,Es \qquad\qquad 23\text{-}21$$

From Equation 23-20

$$dE = \frac{dq}{\epsilon_0 A}$$

Substituting $dq = \epsilon_0 A\,dE$ into Equation 23-21 gives

$$dW = (\epsilon_0 As)E\,dE \qquad\qquad 23\text{-}22$$

Therefore the work done to increase the field from $E = 0$ to $E = E_0 = Q_0/\epsilon_0 A$ is

$$W = \epsilon_0 As \int_0^{E_0} E\,dE = \tfrac{1}{2}\epsilon_0 E_0^2 As$$

This work appears as electrostatic potential energy U:

$$U = \tfrac{1}{2}\epsilon_0 E_0^2 As \qquad\qquad 23\text{-}23$$

This result could have been obtained directly from Equation 23-19 using $Q_0 = \epsilon_0 AE_0$ and $C = \epsilon_0 A/s$. It is convenient to think of the energy as being stored in the electric field. Note that, neglecting the edge effects, the field is constant in the region between the plates and zero outside. The volume $\mathcal{V}$ between the plates is As. The energy per unit volume in the field, called the *energy density* η, is thus

$$\eta = \frac{U}{\mathcal{V}} = \tfrac{1}{2}\epsilon_0 E^2 \qquad\qquad 23\text{-}24$$

Electric field energy density

Although we have been considering the simple case of a parallel-plate capacitor, the result (Equation 23-24) for the energy per unit volume in our electrostatic field is generally valid. We shall illustrate the generality of Equation 23-24 by using it to calculate the energy in an electrostatic field that is not constant in space.

Example 23-5 *Energy of a Single Isolated Spherical Conductor of Radius R Carrying a Charge Q* We noted in Example 23-4 that a single spherical conductor can be thought of as a spherical capacitor with the outer shell at infinity, where the potential is chosen to be zero. The potential of this conductor is kQ/R, and its capacitance is $C = 4\pi\epsilon_0 R = R/k$, where k is the Coulomb constant. According to Equation 23-19, the electrostatic energy is

$$U = \tfrac{1}{2}QV = \tfrac{1}{2}Q\,\frac{kQ}{R} = \frac{1}{2}\frac{kQ^2}{R} \qquad\qquad 23\text{-}25$$

A direct derivation of this result by considering the work done in bringing charge from infinity and placing it on the conductor is given as an exercise. Here we give a derivation using the result that the energy per unit volume in an electric field is $\tfrac{1}{2}\epsilon_0 E^2$, as in Equation 23-24. At a distance r from the center of the sphere the electric field is

$$E_r = \frac{kQ}{r^2}$$

---

* We use s here for the plate separation rather than d to avoid confusion with the d used for the differentials dq and dE.

The energy per unit volume is then

$$\eta = \tfrac{1}{2}\epsilon_0 E^2 = \tfrac{1}{2}\epsilon_0 \frac{k^2 Q^2}{r^4} \qquad\qquad \text{23-26}$$

Consider a spherical shell volume element of radius r, thickness dr, and volume $d\mathcal{V} = 4\pi r^2\, dr$ (Figure 23-9). The energy in this volume element is

$$dU = \eta\, d\mathcal{V} = \tfrac{1}{2}\epsilon_0 k^2 \frac{Q^2}{r^4} 4\pi r^2\, dr$$

$$= 2\pi\epsilon_0 k^2 \frac{Q^2}{r^2}\, dr = \frac{1}{2}\frac{kQ^2}{r^2}\, dr \qquad\qquad \text{23-27}$$

using $k = 1/4\pi\epsilon_0$. We calculate the total energy by integrating from $r = R$ to $r = \infty$ since $E = 0$ for $r < R$. Thus

$$U = \int_R^\infty \frac{1}{2}\frac{kQ^2}{r^2}\, dr = \frac{1}{2}\frac{kQ^2}{R}$$

in agreement with Equation 23-25.

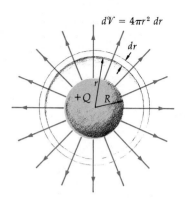

Figure 23-9
Calculation of the electrostatic energy of a spherical conductor with charge Q. The volume of space between r and $r + dr$ is $d\mathcal{V} = 4\pi r^2\, dr$. The electrostatic field energy in this volume element is $\eta\, d\mathcal{V}$, where $\eta = \tfrac{1}{2}\epsilon_0 E^2$ is the energy density.

23-5   Dielectrics

A nonconducting material, e.g., glass or wood, is called a *dielectric*. Faraday discovered that when the space between the two conductors of a capacitor is occupied by a dielectric, the capacitance is increased. If the space (between the plates of a parallel-plate capacitor, for example) is completely filled by the dielectric, the capacitance increases by a factor K which is characteristic of the dielectric and called the *dielectric constant*.

Dielectric constant

Suppose a capacitor of capacitance C_0 is connected to a battery, which charges it to a potential difference V_0 by placing a charge $Q_0 = C_0 V_0$ on the plates. If the battery is disconnected and a dielectric is inserted, filling the space between the plates, the potential difference decreases to a new value,

$$V = \frac{V_0}{K} \qquad\qquad \text{23-28}$$

Since the original charge Q_0 is still on the plates, the new capacitance is

$$C = \frac{Q_0}{V} = \frac{KQ_0}{V_0} = KC_0 \qquad\qquad \text{23-29}$$

If, on the other hand, the dielectric is inserted while the battery is still connected, the battery must supply more charge to maintain the original potential difference. The total charge on the plates is then $Q = KQ_0$. In either case, the capacitance is increased by the factor K.

Since the potential difference between the plates of a parallel-plate capacitor equals the electric field between the plates times the separation d, the effect of the dielectric (with the battery disconnected) is to decrease the electric field by the factor K. If E_0 is the original field without the dielectric, the new field E is

$$E = \frac{E_0}{K} \qquad\qquad \text{23-30}$$

Dielectric reduces electric field

We can understand this result in terms of molecular polarization of the dielectric. If the molecules of the dielectric are polar molecules, i.e., have permanent dipole moments, these moments are originally randomly oriented. In the presence of the field between the capacitor plates, these dipole moments experience a torque, which tends to align them in the direction of the field (Figure 23-10). The amount of alignment depends on the strength of the field and the temperature. At high temperatures, the random thermal motion of the molecules tends to counteract the alignment. In any case, the alignment of the molecular dipoles produces an additional electric field due to the dipoles which is in the direction opposite the original field. The original field is thus weakened. Even if the molecules of the dielectric are nonpolar, they will have induced dipole moments in the presence of the electric field between the plates. The induced dipole moments are in the direction of the original field. Again, the additional electric field due to these induced moments weakens the original field.

A dielectric which has electric dipole moments predominantly in the direction of the external field is said to be *polarized* by the field, whether the polarization is due to alignment of permanent dipole moments of polar molecules or to the creation of induced dipole moments in nonpolar molecules. The net effect of the polarization of a homogeneous dielectric is the creation of a surface charge on the dielectric faces near the plates. Figure 23-11 shows a rectangular slab of homogeneous dielectric which is in a uniform electric field to the right. The molecular dipole moments are indicated schematically. As can be seen from the figure, the net effect of the polarization is to produce a positive surface charge density on the right face.* The charge densities on the dielectric faces are due to the displacement of positive and negative molecular charges near the faces. This displacement is due to the external electric field of the capacitor. The charge on the dielectric, called *bound charge,* is not free to move about like the ordinary free charge on the conducting capacitor plates. Although it disappears when the external electric field disappears, it produces an electric field just like any other charge. We shall now relate the bound charge density σ_b to the dielectric constant K and to the surface charge density σ_f on the capacitor plates, which we call the free charge density because it is free to move about the conductor rather than being bound to the molecules, as in the case of the bound charge density.

(b)

Figure 23-10
(a) Randomly oriented electric dipoles in the absence of an external electric field. (b) In the presence of an external field the dipoles are partially aligned parallel to the field.

(a)

(b)

Figure 23-11
The net effect of alignment of electric dipole moments distributed uniformly in a volume (a) is a positive bound surface charge density on one side of the dielectric and a negative bound surface charge density on the other side, as shown in (b). The result is to weaken the electric field in the dielectric.

The electric field inside the dielectric slab due to the bound charge densities $+\sigma_b$ on the right and $-\sigma_b$ on the left is just the field due to two

---

* If the dielectric is not homogeneous or the field is not uniform, there may be a volume charge density created inside the dielectric because the amount of positive charge pushed out the right side of some volume element may not equal the negative charge pushed out the left side of the element. We shall neglect such complications.

infinite-plane charge densities (assuming that the slab is very thin; i.e., the plates of the capacitor are close together). The field E' thus has the magnitude

$$E' = \frac{\sigma_b}{\epsilon_0} \qquad\qquad 23\text{-}31$$

This field is to the left and subtracts from the electric field due to the ordinary free charge density on the capacitor plates. The original field E_0 has the magnitude

$$E_0 = \frac{\sigma_f}{\epsilon_0} \qquad\qquad 23\text{-}32$$

The magnitude of the resultant field E is the difference of these magnitudes. It also equals E_0/K:

$$E = E_0 - E' = \frac{E_0}{K} \qquad\qquad 23\text{-}33$$

or

$$E' = E_0 \left(1 - \frac{1}{K}\right) = \frac{K-1}{K} E_0 \qquad\qquad 23\text{-}34$$

Writing σ_b/ϵ_0 for E' and σ_f/ϵ_0 for E_0, we have

$$\sigma_b = \frac{K-1}{K} \sigma_f \qquad\qquad 23\text{-}35$$

Bound and free charge densities

The bound charge density σ_b is always less than the free charge density σ_f on the capacitor plates and is zero if $K = 1$, which is the case of no dielectric. Figure 23-12 shows the lines of force for a parallel-plate capacitor with a dielectric of constant $K = 2$ which does not completely fill the space between the plates. According to Equation 23-35, the bound charge density on the dielectric is half the free charge density on the plates. Half the lines that leave the positive charge on the lower plate end on the negative bound charge on the dielectric. Similarly, half the lines which end on the negative charge on the upper plate begin on the positive bound charge on the dielectric. The density of field lines in this diagram shows that the electric field inside the dielectric is just half that outside. The effect of the dielectric on the capacitance depends on how much space is filled.

Figure 23-12
Electric field lines for a dielectric of constant $K = 2$ between the plates of a parallel-plate capacitor. The electric field is reduced to half its original strength, as indicated by the fact that half the original field lines end on one side of the dielectric and begin again on the other side.

Example 23-6 A slab of material of dielectric constant K has the same area as the plates of a parallel-plate capacitor but has a thickness $\frac{3}{4}d$, where d is the separation of the plates. How is the capacitance changed when the slab is inserted between the plates?

Let $E_0 = V_0/d$ be the electric field between the plates when there is no dielectric and the potential difference is V_0. If the dielectric is inserted, the electric field in the dielectric will be $E = E_0/K$ and the potential difference will be

$$V = E_0(\tfrac{1}{4}d) + \frac{E_0}{K}\left(\frac{3}{4}d\right) = E_0 d\left(\frac{1}{4} + \frac{3}{4K}\right) = V_0\frac{K+3}{4K}$$

The potential difference decreases by the factor $(K+3)/4K$ while the free charge Q_0 on the plates remains unchanged. The capacitance thus increases:

$$C = \frac{Q_0}{V} = \frac{4K}{K+3}\frac{Q_0}{V_0} = \frac{4K}{K+3}C_0$$

In addition to increasing the capacitance, a dielectric has two other functions in a capacitor: (1) it provides a mechanical means of separating the two conductors, which must be very close together in order to obtain a large capacitance; (2) the dielectric strength is increased because the dielectric strength of a dielectric is usually greater than that of air. We have already mentioned that the dielectric strength of air is 3 MV/m = 3 kV/mm. Fields greater than this magnitude cannot be maintained in air because of dielectric breakdown; i.e., the air becomes ionized and conducts. Many materials have dielectric strengths greater than that of air, allowing greater potential differences between the conductors of a capacitor.

Three functions of a dielectric

An example of these three dielectric functions is a parallel-plate capacitor made from two sheets of metal foil of large area (to increase the capacitance) separated by a sheet of paper. The paper increases the capacitance because of its polarization; that is, K is greater than 1. It also provides a mechanical separation so that the sheets can be very close together without being in electrical contact. (A small separation is important because the capacitance varies inversely with separation.) Finally, the dielectric strength of paper is greater than that of air, so that greater potential differences can be attained without breakdown. Table 23-1 lists the dielectric constant and dielectric strength of some dielectrics.

Table 23-1
Dielectric constant and strength of various materials

Material	Dielectric constant K	Dielectric strength, kV/mm
Air	1.00059	3
Bakelite	4.9	24
Glass (Pyrex)	5.6	14
Mica	5.4	10–100
Neoprene	6.9	12
Paper	3.7	16
Paraffin	2.1–2.5	10
Plexiglas	3.4	40
Polystyrene	2.55	24
Porcelain	7	5.7
Transformer oil	2.24	12
Water (20°C)	80	

Electrostatics and Xerography

Richard Zallen
Xerox Research Laboratories, Webster, N.Y.

There are many important and beneficial technological applications which could be included in a discussion of uses of electrostatic phenomena. For example, a powerful air-pollution preventer is the electrostatic precipitator, which years ago made life livable near cement mills and ore-processing plants, and which is currently credited with extracting better than 99 percent of the ash and dust from the gases about to issue from chimneys of coal-burning power plants. The basic idea of this very effective antipollution technique is shown in Figure 1. The outer wall of a vertical metal duct is grounded, while a wire running down the center of the duct is kept at a very large negative voltage. In this concentric geometry (which corresponds to the cylindrical capacitor in Example 23-3) a very nonuniform electric field is set up, with lines of force directed radially inward toward the negative wire electrode. Close to the wire the field attains enormous values, large enough to produce an electrical breakdown of air, and the normal placid mixture of neutral gas molecules is replaced by a turmoil of free electrons and positive ions. The electrons from this corona discharge are driven outward from the wire by the electric field. Most of them quickly become attached to oxygen molecules to produce negative O_2^- ions, which are also accelerated outward. As this stream of ions passes across the hot waste gas rising in the duct, small particles carried by the gas become charged by capturing ions and are pulled by the field to the outer wall. If the noxious particles are solid, they are periodically shaken down off the duct into a hopper; if they are liquid, the residue simply runs down the wall and is collected below.

Besides electrostatic precipitation, other technological examples include electrocoating with spray paints and the electrostatic separation of granular mixtures used for the removal of rock particles from minerals, garlic seeds from wheat, even rodent excreta from rice. However, the application which is

Figure 1
Schematic diagram of the use of a corona discharge in an electrostatic precipitator.

Clean
gas out

Dirty
gas in

Discharge
electrode

the main focus of this essay is xerography, the most widely used form of electrostatic imaging, or electrophotography. This is the most familiar use of electrostatics in terms of the number of people who have occasion to use plain-paper copying machines in offices, libraries, and schools, and it also provides a fine example of a process utilizing a sequence of distinct electrostatic events.

The xerographic process was invented in 1937 by Chester Carlson. The term xerography, literally "dry writing," was actually adopted a bit later to emphasize the distinction from wet chemical processes. Carlson's innovative concept did not find early acceptance, and a practical realization of his idea became available only after a small company (in a famous entrepreneurial success story) risked its future in its intensive efforts to develop the process.

Four of the main steps involved in xerography are illustrated in Figure 2. In the interest of clarity the process has been oversimplified, and several subtleties (as well as gaps in our understanding) have been suppressed. Electrostatic imaging takes place on a large thin plate of a photoconducting material supported by a grounded metal backing. A photoconductor is a solid which is a good insulator *in the dark* but which becomes capable of conducting electric current when exposed to light. The unilluminated, insulating state is indicated by shading in Figure 2. In the dark, a uniform electrostatic charge is laid down on the surface of the photoconductor. This charging step (Figure 2a) is accomplished by means of a positive corona discharge surrounding a fine wire held at about +5000 V. This corona (a miniature version of, and opposite in sign to, the intense precipitator corona of Figure 1) is passed over the photoconductor surface, spraying positive ions onto it and charging it to a potential of the order of +1000 V. Since charge is free to flow within the grounded metal backing, an equal and opposite induced charge (see Chapter 21) develops at the metal-photoconductor interface. In the dark the photoconductor contains no mobile charge, and the large potential difference persists across this dielectric layer, which is only 0.005 cm thick.

The photoconductor plate is next exposed to light in the form of an image reflected from the document being copied. What happens now is indicated in Figure 2b. Where light strikes the photoconductor, light quanta (photons) are absorbed, and pairs of mobile charges are created. Each photogenerated pair consists of a negative charge (an electron) and a positive charge (a hole; crudely, a missing electron). Photogeneration of this free charge depends not only on the photoconductor used and on the wavelength and intensity of the incident light but also on the electric field present. This large field (1000 V/0.005 cm $= 2 \times 10^5$ V/cm $= 2 \times 10^7$ V/m) helps to pull apart the mutually attracting electron-hole pairs so that they are free to move separately. The electrons then move under the influence of the field to the surface, where they neutralize positive charges, while the holes move to the photo-

Figure 2

Steps in the xerographic process: (a) charging, (b) exposure, (c) development, and (d) transfer.

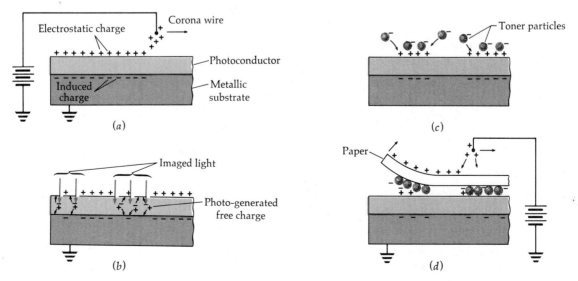

conductor-substrate interface and neutralize negative charges there. Where intense light strikes the photoconductor, the charging step is totally undone; where weak light strikes it, the charge is partially reduced; and where no light strikes it, the original electrostatic charge remains on the surface. The critical task of converting an optical image into an electrostatic image, which is now recorded on the plate, has been completed. This latent image consists of an electrostatic potential distribution which replicates the light and dark pattern of the original document.

To develop the electrostatic image, fine negatively charged pigmented particles are brought into contact with the plate. These *toner particles* are attracted to positively charged surface regions, as shown in Figure 2c, and a visible image appears. The toner is then transferred (Figure 2d) to a sheet of paper which has been positively charged in order to attract them. Brief heating of the paper fuses the toner to it and produces a permanent photocopy ready for use.

Finally, to prepare the photoconductor plate for a repetition of the process, any toner particles remaining on its surface are mechanically cleaned off, and the residual electrostatic image is erased, i.e., discharged, by flooding with light. The photoconductor is now ready for a new cycle, starting with the charging step. In high-speed duplicators the photoconductor layer is often in the form of a moving continuous drum or belt, around the perimeter of which are located stations for performing the various functions of Figure 2. The speed of xerographic printing technology is presently on the order of a few copies per second.*

* For further information on electrostatics in xerography, consult J. H. Dessauer and H. E. Clark (eds.), *Xerography and Related Processes,* Focal Press, New York, 1965, and R. M. Schaffert, *Electrophotography,* rev. ed., Focal Press, New York, 1973. Other applications of electrostatics are discussed in A. D. Moore, *Scientific American,* March 1972.

Review

A. Define, explain, or otherwise identify:

Capacitor, 651	Energy density, 659
Capacitance of a conductor, 655	Dielectric, 660
Capacitors in parallel, 656	Dielectric constant, 660
Effective capacitance, 656	Polarization of a dielectric, 661
Capacitors in series, 657	Bound charge, 662

B. True or false:

1. The capacitance of a conductor is defined to be the total amount of charge it can hold.

2. The capacitance of a parallel-plate capacitor depends on the voltage difference between the plates.

3. The capacitance of a parallel-plate capacitor is proportional to the charge on the plates.

4. The effective capacitance of two capacitors in parallel equals the sum of the individual capacitances.

5. The effective capacitance of two capacitors in series is less than that of either capacitor.

6. Since V is energy per charge, the total energy in a capacitor is QV.

7. The electrostatic energy per unit volume at some point is proportional to the square of the electric field at that point.

Exercises

Section 23-1, Calculation of Capacitance

1. If a parallel-plate capacitor has 0.1-mm separation, what must its area be to have a capacitance of 1 F? If the plates are square, what is the length of their sides?

2. A parallel-plate capacitor has capacitance of 2.0 μF and plate separation of 1.0 mm. (a) How much potential difference can be placed across the capacitor before dielectric breakdown of air occurs (E_{max} = 3 MV/m)? (b) What is the magnitude of the greatest charge the capacitor can store before breakdown?

3. A coaxial cable between two cities has an inner radius of 1.0 cm and an outer radius of 1.1 cm. Its length is 8×10^5 m (about 500 mi). Treat this cable as a cylindrical capacitor and calculate its capacitance.

4. A Geiger tube consists of a wire of radius 0.2 mm and length 12 cm with a coaxial cylindrical shell conductor of the same length and radius 1.5 cm. (a) Find the capacitance assuming that the gas in the tube has a dielectric constant of 1. (b) Find the charge per unit length on the wire when the capacitor is charged to 1.2 kV.

5. What is the capacitance of an isolated spherical conductor of radius (a) 1.8 cm, (b) 1.8 m, (c) 1.8 km?

6. Considering the earth to be a spherical conductor of radius 6400 km, calculate its capacitance.

7. (a) Find the radius of an isolated spherical conductor which has a capacitance of 1 F. (b) What is the ratio of the radius of this sphere to the radius of the earth?

8. Calculate the capacitance of a parallel-plate capacitor of plate area 1.0 m and separation 1.0 cm.

Section 23-2, Parallel and Series Combinations of Capacitors

9. A 10.0-μF capacitor is connected in series with a 20.0-μF capacitor across a 6.0-V battery. (a) What is the equivalent capacitance of this combination? (b) Find the charge on each capacitor. (c) Find the potential difference across each capacitor.

10. A 10.0- and a 20.0-μF capacitor are connected in parallel across a 6.0-V battery. (a) What is the equivalent capacitance of this combination? (b) What is the potential difference across each capacitor? (c) Find the charge on each capacitor.

11. Three capacitors have capacitance 2.0, 4.0, and 8.0 μF. Find the equivalent capacitance (a) if the capacitors are in parallel and (b) if they are in series.

12. A 2.0-μF capacitor is charged to a potential difference of 12.0 V and disconnected from the battery. (a) How much charge is on the plates? (b) When a second capacitor (initially uncharged) is connected in parallel across this capacitor, the potential difference drops to 4.0 V. What is the capacitance of the second capacitor?

13. (a) How many 1.0-μF capacitors would have to be connected in parallel to store 1 mC of charge with a potential difference of 10 V across each? (b) What would be the potential difference across the combination? (c) If these capacitors are connected in series and the potential difference across each is 10 V, find the charge on each and the potential difference across the combination.

14. A 1.0-μF capacitor is connected in parallel with a 2.0-μF capacitor, and the combination is connected in series with a 6.0-μF capacitor. What is the equivalent capacitance of this combination?

15. A 3.0- and a 6.0-μF capacitor are connected in series, and the combination is connected in parallel with an 8.0-μF capacitor. What is the equivalent capacitance of this combination?

Section 23-3, Electrostatic Energy in a Capacitor

16. (a) A 3-μF capacitor is charged to 100 V. How much energy is stored in the capacitor? (b) How much additional energy is required to charge the capacitor from 100 to 200 V?

17. How much energy is stored in an isolated spherical conductor of radius 10 cm charged to 3 kV?

18. (a) A 10-μF capacitor is charged to $Q = 4\ \mu$C. How much energy is stored? (b) If half the charge is removed, how much energy remains?

19. (a) Find the energy stored in a 20-pF capacitor when it is charged to 5 μC. (b) How much additional energy is required to increase the charge from 5 to 10 μC?

20. A conducting sphere of radius R carries charge q. (a) Show that the work needed to bring additional charge dq from infinity to the sphere is $(kq/R)\,dq$. (b) Use this result to show that the work needed to increase the charge from 0 to Q is $\frac{1}{2}(kQ^2/R)$.

21. A parallel-plate capacitor of area A and separation d is charged to a potential difference V and then disconnected from the charging source. The plates are then pulled apart until the separation is $2d$. Find expressions in terms of A, d, and V for (a) the new capacitance, (b) the new potential difference, and (c) the new stored energy. (d) How much work was required to change the plate separation from d to $2d$?

Section 23-4, Electrostatic Field Energy

22. Find the energy per unit volume in an electric field equal to the breakdown field in air (3 MV/m).

23. A parallel-plate capacitor with plate area 2 m² and separation 1.0 mm is charged to 100 V. (a) What is the electric field between the plates? (b) What is the energy per unit volume in the space between the plates? (c) Find the total energy by multiplying your answer to part (b) by the total volume between the plates. (d) Find the capacitance C, calculate the total energy from $U = \frac{1}{2}CV^2$, and compare with your answer to part (c).

Section 23-5, Dielectrics

24. A parallel-plate capacitor is made by placing polyethylene ($K = 2.3$) between sheets of aluminum foil. The area of each sheet is 400 cm², and the spacing is 0.3 mm. Find the capacitance.

25. A parallel-plate capacitor has plates of area 600 cm² and separation 4 mm. It is charged to 100 V and disconnected from the battery. (a) Find the electric field E, the charge density σ, and the energy U. A dielectric of constant $K = 4$ is inserted, completely filling the space between the plates. (b) Find the new electric field E and the potential difference V. (c) Find the new energy. (d) Find the bound charge density.

26. What is the dielectric constant of a dielectric on which the induced bound charge density is (a) 80 percent of the free charge density, (b) 20 percent of the free charge density, and (c) 98 percent of the free charge density?

Problems

1. For the arrangement shown in Figure 23-13, find (a) the total effective capacitance between the terminals, (b) the charge stored on each capacitor, and (c) the total stored energy.

2. A spherical capacitor consists of two spherical shells of radii R_1 and R_2. Show that when these radii are nearly equal, the capacitance is given approximately

Figure 23-13
Problem 1.

by the expression for the capacitance of a parallel-plate capacitor, $C = \epsilon_0 A/d$, where A is the area of the sphere and $d = R_2 - R_1$.

3. For the arrangement in Figure 23-14, find (a) the total effective capacitance between the terminals, (b) the charge stored on each capacitor, and (c) the total stored energy when the potential difference across the terminals is 200 V.

Figure 23-14
Problem 3.

4. The effective capacitance of two capacitors in series can be written $C = C_1 C_2/(C_1 + C_2)$. Show that the correct expression for the effective capacitance of three capacitors in series is $C = C_1 C_2 C_3/(C_1 C_2 + C_2 C_3 + C_1 C_3)$.

5. A 20-pF capacitor is charged to 3.0 kV and is then removed and connected in parallel with an uncharged 50-pF capacitor. (a) What is the new charge on each capacitor? (b) Find the initial energy stored in the 20-pF capacitor and the final energy stored in the two capacitors. Is energy gained or lost in connecting the two capacitors?

6. Find all the different possible effective capacitances that can be obtained using a 1.0-, a 2.0-, and a 4.0-μF capacitor in any combination which includes all three or any two capacitors.

7. Three identical capacitors are connected so that their maximum equivalent capacitance is 15 μF. Find the three other combinations possible (using all three capacitors) and their equivalent capacitances.

8. Two capacitors of capacitance $C_1 = 4$ μF and $C_2 = 12$ μF are connected in series across a 12-V battery. They are then carefully disconnected without being discharged and connected in parallel with positive side to positive side and negative side to negative side. (a) Find the potential difference across each capacitor after they are connected in parallel. (b) Find the energy lost as they are connected in parallel.

9. Work Problem 8 for the two capacitors first connected in parallel across the 12-V battery and then connected in parallel, with the positive side of one capacitor connected to the negative side of the other.

10. A 100- and a 400-pF capacitor are both charged to 2.0 kV. They are then disconnected from the voltage source and connected together in parallel, positive side to positive side and negative side to negative side. (a) Find the resulting potential difference across each capacitor. (b) Find the energy lost as the connections are made.

11. Work Problem 10 for the capacitors connected in parallel with the positive side of one connected to the negative side of the other after they have been charged to 2.0 kV.

12. A parallel-plate capacitor has capacitance C_0 and plate separation d. Two dielectric slabs, of constants K_1 and K_2, each of thickness $\frac{1}{2}d$, and of the same area as the plates, are inserted between the plates as indicated in Figure 23-15. Show that the capacitance is then given by

$$C = \frac{2K_1 K_2}{K_1 + K_2} C_0$$

Figure 23-15
Problem 12.

13. A parallel-plate capacitor of area A and separation d is charged to potential difference V and removed from the charging source. A dielectric slab of constant $K = 2$, thickness d, and area $\frac{1}{2}A$ is inserted, as shown in Figure 23-16. Let σ_1 be the free charge density at the conductor-dielectric surface and σ_2 be the charge density at the conductor-vacuum surface. (a) Why must the electric field have the same value inside the dielectric as in the free space between the plates? (b) Show that $\sigma_1 = 2\sigma_2$. (c) Show that the new capacitance is $3\epsilon_0 A/2d$ and the new potential difference is $\frac{2}{3}V$.

Figure 23-16
Problem 13.

14. A parallel-plate capacitor has plate area A and separation d. A metal slab of thickness t and area A is inserted between the plates. Show that the capacitance is given by $C = \epsilon_0 A/(d - t)$ regardless of where the metal plate is placed. Show that this arrangement can be considered to be a capacitor of separation a in series with one of separation b, where $a + b + t = d$.

15. In Figure 23-17, $C_1 = 2\ \mu F$, $C_2 = 6\ \mu F$, and $C_3 = 3.5\ \mu F$. (a) Find the equivalent capacitance of this combination. (b) If the breakdown voltages of the individual capacitors are $V_1 = 100$ V, $V_2 = 50$ V, and $V_3 = 400$ V, what maximum voltage can be placed across points a and b?

16. A parallel-plate capacitor is filled with two dielectrics of equal size, as shown in Figure 23-18. Show that the capacitance is increased by the factor $(K_1 + K_2)/2$.

17. Early capacitors, called Leyden jars, were actually glass jars coated inside and outside with metal foil. Suppose that the jar is a cylinder 40 cm high with 2.0-mm-thick walls of inner diameter 8 cm. Ignore any field fringing. (a) Find the capacitance of this jar. (b) What maximum charge can it take without collapse? The dielectric constant of the glass is 5.0, and its dielectric strength is 15 MV/m.

18. A parallel-plate capacitor of plate area A and separation x is given a charge Q and removed from the charging source. (a) Find the electrostatic energy stored as a function of x. (b) Find the increase in energy dU due to an increase in plate separation dx from $dU = (dU/dx)\ dx$. (c) If F is the force exerted by one plate on the other, the work done to move one plate a distance dx is $F\ dx = dU$. Show that $F = Q^2/2\epsilon_0 A$. (d) Show that the force found in part (c) equals $\frac{1}{2}EQ$, where Q is the charge on one plate and E is the electric field between the plates. Discuss the reason for the factor $\frac{1}{2}$ in this result.

19. Design a circuit of capacitors which has a capacitance of 2 μF and breakdown voltage of 400 V using as many 2-μF capacitors as needed, each with a breakdown voltage of 100 V.

20. A parallel-plate capacitor of plate area 1.0 m^2 and plate separation distance 0.5 cm has a glass plate of the same area and thickness between its plates. The glass has dielectric constant 5.0. The capacitor is charged to a potential difference of 12.0 V and removed from its charging source. How much work is required to pull the glass plate out of the capacitor?

21. A spherical capacitor has an inner sphere of radius R_1 with charge $+Q$ and an outer concentric spherical shell of radius R_2 with charge $-Q$. (a) Find the electric field and the energy density at any point in space. (b) How much energy is in the volume of the spherical shell of radius r, thickness dr, and volume $4\pi r^2\ dr$ between the conductors? (c) Integrate your expression in part (b) to find the total energy stored in the capacitor and compare your result with that obtained from $U = \frac{1}{2}QV$.

22. A cylindrical capacitor consists of a long wire of radius R_1 and length L with positive charge Q and a concentric outer cylindrical shell of radius R_2, length L, and charge $-Q$. (a) Find the electric field and the energy density at any point in space. (b) How much energy is in the cylindrical shell of radius r, thickness dr, and volume $2\pi rL\ dr$ between the conductors? (c) Integrate your expression in part (b) to find the total energy stored in the capacitor and compare your result with that obtained from $U = \frac{1}{2}QV$.

23. A ball of charge of radius R has a uniform charge density ρ and total charge $Q = \frac{4}{3}\pi R^3\rho$. (a) Find the electrostatic energy density at distance r from the center of the charge for $r < R$ and for $r > R$. (b) Find the energy in a spherical shell of volume $4\pi r^2\ dr$ for both $r < R$ and $r > R$. (c) Compute the total electrostatic energy by integrating your expressions in part (b) and show that your result can be written $U = \frac{3}{5}kQ^2/R$. Explain why this result is greater than that for a spherical conductor of radius R carrying a total charge Q.

Figure 23-17
Problem 15.

Figure 23-18
Problem 16.

CHAPTER 24　　Electric Current

Objectives　After studying this chapter, you should:

1.　Be able to define and discuss the concepts of electric current, current density, drift velocity, resistance, and emf.

2.　Be able to state Ohm's law and distinguish between it and the definition of resistance.

3.　Be able to give the definition of resistivity and describe its temperature dependence.

4.　Be able to discuss the simple model of a real battery in terms of an ideal emf and an internal resistance and find the terminal voltage of a battery when it delivers a current I.

5.　Be able to give the general relationship between potential difference, current, and power.

Figure 24-1
The current through area A at point P is the rate of flow of charge through that area. In time Δt all the charge in the shaded volume $Av_d \Delta t$ flows past point P, where v_d is the speed of the carriers. If there are n carriers per unit volume each with charge q, the total charge is $\Delta Q = nqv_dA \Delta t$ and the current is $I = \Delta Q/\Delta t = nqv_dA$.

An electric current is the rate of flow of electric charge past a point. Although the flow of charge often takes place inside a conductor, this is not necessary; e.g., a beam of charged ions in vacuum from an accelerator constitutes a current. We define electric current as follows. Consider a small area element A, as in Figure 24-1, which might be the cross-sectional area of a conducting wire. The current through the area is defined as the amount of charge flowing through the area per unit time. If ΔQ is the charge that flows through the area in time Δt, the current is

$$I = \frac{\Delta Q}{\Delta t} \qquad\qquad 24\text{-}1$$

Electric current defined

The SI unit of current is the ampere (A):

$$1 \text{ A} = 1 \text{ C/s} \qquad\qquad 24\text{-}2$$

The direction of current is taken to be the direction of flow of positive charge. If the current is due to particles such as electrons which have negative charge, the direction of current is in the direction opposite the flow of the negatively charged particles. This definition of the direction of the current is purely arbitrary. In a conducting wire it is electrons which are free to move and produce the flow of charge. By our definition, the current is in the direction opposite to the motion of the electrons.

There are many examples of electric current outside of conducting wires. For example, a beam of protons from an accelerator produces a current in the direction of the motion of the positively charged protons. In electrolysis, the current is produced by the motion of both electrons and positive ions. Since these particles move in opposite directions, both produce current in the same direction. In nearly all applications, the motion of negative charges to the left is indistinguishable from the motion of positive charges to the right. [An exception is the Hall effect (Chapter 26).] We can always think of current as motion of positive charges in the direction of the current and remember (if we need to) that in conducting wires, for example, the electrons are moving in the direction opposite to the current.

24-1   Current and Motion of Charges

An electric current can be related to the motion of the charged particles responsible for it. Let us first consider the case of current in a conducting wire of cross-sectional area A. Let n be the number of free charge-carrying particles per unit volume. For a conducting wire, these charge carriers are the free electrons. We shall assume that each particle carries a charge q and moves with velocity v_d. In a time Δt all the particles in the volume $Av_d\,\Delta t$, shaded in Figure 24-1, pass through the area element at point P. The number of particles in this volume is $nAv_d\,\Delta t$, and the total charge is

$$\Delta Q = qnAv_d\,\Delta t$$

The current at point P is thus

$$I = \frac{\Delta Q}{\Delta t} = nqv_dA \qquad\qquad 24\text{-}3$$

The current per unit area is the *current density J*:

$$J = \frac{I}{A} = nqv_d \qquad\qquad 24\text{-}4$$

We can generalize the idea of current density to apply to any type of current whether confined to a wire or not. We define the current-density vector **J** by

$$\mathbf{J} = nq\mathbf{v}_d \qquad\qquad 24\text{-}5 \qquad \textit{Current density}$$

The current-density vector is in the direction of $\mathbf{v}_d$ if the charge q is positive, and in the opposite direction if q is negative. The quantity $\mathbf{v}_d$ is the average velocity of the charge carriers. For electrons in conductors this velocity is called the *drift velocity*. The current density is a fundamental quantity related by Equation 24-5 to the number density of charge carriers n, the charge q, and the average velocity of the carriers $\mathbf{v}_d$. If the current is due to more than one type of charged particle (as in electrolysis), the current density is found by summing $n_iq_i(\mathbf{v}_d)_i$ over the particles:

$$\mathbf{J} = \sum_i n_iq_i(\mathbf{v}_d)_i \qquad\qquad 24\text{-}6$$

If **J** is constant over the area A, the current through A is

$$I = \mathbf{J} \cdot \hat{\mathbf{n}} A = J_n A \qquad \text{24-7}$$

where $\hat{\mathbf{n}}$ is the unit vector perpendicular to the plane of the area A and J_n is the component of the current density parallel to $\hat{\mathbf{n}}$. If the current density is not constant, the current through a surface is found by integration:

$$I = \int \mathbf{J} \cdot \hat{\mathbf{n}} \, dA \qquad \text{24-8}$$

We can get an idea of the order of magnitude of the drift velocity for electrons in a conducting wire by putting typical magnitudes into Equation 24-3.

Example 24-1 What is the drift velocity of electrons in a typical copper wire (14 gauge) of radius 0.0814 cm carrying a current of 1 A?

If we assume one free electron per copper atom, the density of free electrons is the same as the density of atoms. Then

$$n = \frac{(6.02 \times 10^{23} \text{ atoms/mol}) (8.92 \text{ g/cm}^3)}{63.5 \text{ g/mol}}$$
$$= 8.46 \times 10^{22} \text{ atoms/cm}^3$$

and

$$v_d = \frac{I}{Ane} = \frac{1 \text{ C/s}}{\pi (0.0814 \text{ cm})^2 (8.46 \times 10^{22} \text{ cm}^{-3}) (1.60 \times 10^{-19} \text{ C})}$$
$$\approx 3.55 \times 10^{-3} \text{ cm/s}$$

We see that typical drift velocities are very small.

The actual instantaneous velocity of an electron in a metal is much larger than the drift velocity, which is the average velocity. The behavior of electrons in a metal is similar to that of gas molecules in air. In still air, the gas molecules move with large instantaneous velocities between collisions, but the average velocity is zero. When there is a breeze, the air molecules have a small drift velocity in the direction of the breeze superimposed on the much larger instantaneous velocity. We discuss this point further in Section 24-4 when we consider the classical model of electric conduction in metals.

Question

1. Two wires of the same diameter but different materials are joined together to carry the same electric current. In one wire the density of charge carriers is twice that in the other. How do the drift velocities in the two wires compare?

24-2 Ohm's Law and Resistance

In our study of conductors in electrostatics we argued that the electric field inside a conductor must be zero in electrostatic equilibrium. If this were not so, the free charges inside the conductor would move about. We are now considering nonelectrostatic equilibrium situations, in which the free charge *does* move in a conductor. When a conductor carries a current, there is an electric field inside the conductor. The current is in the direction of the electric field. Figure 24-2 shows a segment of wire of length L and cross-sectional area A carrying a current I. Since there is an electric field in the wire, the potential at point a is greater than that at point b. If the segment is short enough for us to neglect any variation of the electric field over the distance L, the potential difference between points a and b is

$$\Delta V = V_a - V_b = EL \qquad\qquad 24\text{-}9$$

where E is the electric field. For most materials such as wire conductors, the potential difference between two points is found experimentally to be proportional to the current I in the conductor. This experimental result is known as *Ohm's law*. The constant of proportionality is called the resistance R:

$$\Delta V = RI \qquad\qquad 24\text{-}10$$

Ohm's law

V_a V_b
 a L b

A E I

Figure 24-2
Segment of wire carrying current I. The potential difference is related to the electric field by $V_a - V_b = EL$.

The SI unit of resistance, the volt per ampere, is called an ohm (Ω). The resistance of a conducting wire depends on the length of the wire, its cross-sectional area, the type of material, and the temperature, but for materials obeying Ohm's law, it does not depend on the current I. Such materials, which include most metals, are called *ohmic* materials. For many other, nonohmic materials the potential difference is not proportional to the current. For such materials, the resistance between two points is defined by

$$R = \frac{\Delta V}{I} \qquad\qquad 24\text{-}11$$

Resistance defined

where ΔV is the potential difference between the points. Although the definition of resistance for nonohmic materials (Equation 24-11) is the same as that for ohmic materials, the resistance R so defined is independent of the current I for ohmic materials, whereas for nonohmic materials, the resistance does depend on the current. Figure 24-3 shows the potential difference ΔV versus current I for ohmic and nonohmic materials. In both cases, the resistance R is $\Delta V/I$, but for nonohmic materials R depends on the current I and for ohmic materials it does not. Ohm's law therefore is not a fundamental law of nature like Newton's laws or the laws of thermodynamics but an empirical description of a property shared by many materials.

For ohmic materials, the resistance of a conducting wire is found to be proportional to the length of the wire and inversely proportional to its cross-sectional area. The proportionality constant is called the *resistivity* ρ of the conductor:*

Figure 24-3
V versus I for ohmic and nonohmic materials. The resistance $R = V/I$ is independent of I for ohmic materials, as indicated by the constant slope of the line.

$$R = \rho \frac{L}{A} \qquad\qquad 24\text{-}12$$

Resistivity defined

* Care must be taken to distinguish between the resistivity ρ and the volume charge density, also designated by ρ. It should be clear from the context which interpretation of ρ is intended.

The unit of resistivity is the ohm-metre ($\Omega \cdot$m). The reciprocal of the resistivity is the conductivity σ.*

$$\sigma = \frac{1}{\rho} \qquad\qquad 24\text{-}13$$

Resistivity and conductivity

In terms of the conductivity, the resistance of an ohmic conducting wire is

$$R = \frac{L}{\sigma A} \qquad\qquad 24\text{-}14$$

Note that Equations 24-10 and 24-14 are the same as the equations for thermal conduction and resistance except that ΔT for thermal conduction is replaced by ΔV and the thermal conductivity k is replaced by the electric conductivity σ.

The resistivity (and conductivity) of any given metal depends on the temperature. Figure 24-4 shows the temperature dependence of the resistivity for copper. Except at very low temperatures, the resistivity varies nearly linearly with temperature. The resistivity is usually given in tables in terms of its value ρ_{20} at 20°C and the temperature coefficient of resistivity α, which is the slope of the ρ-versus-T curve. The resistivity ρ at some other Celsius temperature t is then given by

$$\rho = \rho_{20}[1 + \alpha(t - 20°C)] \qquad\qquad 24\text{-}15$$

(Since the Celsius and absolute temperatures differ only in the choice of zero, the resistivity has the same slope whether plotted against t or T.)

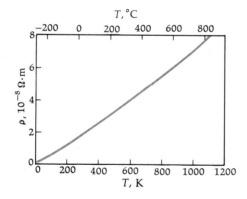

Figure 24-4
Plot of the resistivity ρ versus absolute temperature T for copper. (*By permission from D. Halliday and R. Resnick, Fundamentals of Physics, p. 512, John Wiley & Sons, Inc., New York, 1974.*)

The resistivity ρ and temperature coefficient α are listed in Table 24-1 for various materials. This table shows that there is a wide range of values for the resistivity and an enormous difference between conductors and insulators.

Wires used to carry electric current are manufactured in standard sizes. The diameter of the circular cross section is indicated by a gauge number; higher numbers correspond to smaller diameters. For example, the diameter of a 10-gauge copper wire is 2.588 mm and that of 14-gauge wire is 1.628 mm. Handbooks give the combination ρ/A in ohms per centimetre or ohms per foot.

* Again, the same symbol σ is used in two different contexts, for conductivity and for surface charge density.

Table 24-1
Resistivities and temperature coefficients

Material	Resistivity ρ at 20°C, $\Omega \cdot m$	Temperature coefficient α at 20°C, K^{-1}
Silver	1.6×10^{-8}	3.8×10^{-3}
Copper	1.7×10^{-8}	3.9×10^{-3}
Aluminum	2.8×10^{-8}	3.9×10^{-3}
Tungsten	5.5×10^{-8}	4.5×10^{-3}
Iron	10×10^{-8}	5.0×10^{-3}
Lead	22×10^{-8}	4.3×10^{-3}
Mercury	96×10^{-8}	0.9×10^{-3}
Nichrome	100×10^{-8}	0.4×10^{-3}
Carbon	3500×10^{-8}	-0.5×10^{-3}
Germanium	0.45	-4.8×10^{-2}
Silicon	640	-7.5×10^{-2}
Wood	$10^{8}-10^{14}$	
Glass	$10^{10}-10^{14}$	
Hard rubber	$10^{13}-10^{16}$	
Amber	5×10^{14}	
Sulfur	1×10^{15}	

The Bettmann Archive

Georg Simon Ohm (1781–1854)

Example 24-2 Calculate ρ/A in ohms per metre for 14-gauge copper wire.
From Table 24-1 we have for the resistivity of copper

$$\rho = 1.7 \times 10^{-6} \ \Omega \cdot cm$$

The cross-sectional area of 14-gauge wire is

$$A = \frac{\pi d^2}{4} = \frac{\pi}{4} (0.163 \ cm)^2 = 2.09 \times 10^{-2} \ cm^2$$

Thus

$$\frac{\rho}{A} = \frac{1.7 \times 10^{-6} \ \Omega \cdot cm}{2.09 \times 10^{-2} \ cm^2}$$
$$= (8.13 \times 10^{-5} \ \Omega/cm) = 8.13 \times 10^{-3} \ \Omega/m$$

This example shows that the copper connecting wires used in the laboratory have a very small resistance.

Example 24-3 What is the electric field in a 14-gauge copper wire that carries a current of 1 A?
According to Example 24-2, the resistance of a 1-m length of 14-gauge copper wire is $8.13 \times 10^{-3} \ \Omega$. The voltage drop across 1 m of this wire is

$$\Delta V = IR = 8.13 \times 10^{-3} \ V$$

and the electric field is

$$E = \frac{\Delta V}{L} = 8.13 \times 10^{-3} \ V/m$$

Note that the electric field in a conducting wire is very small.

Ohm's law can also be written in terms of the electric field E and the current density J. Substituting $\Delta V = EL$ (Equation 24-9) and $I = AJ$ into Equation 24-10, we have

$$EL = R(AJ) = \rho \frac{L}{A}(AJ)$$

or

$$\mathbf{J} = \frac{1}{\rho}\mathbf{E} = \sigma\mathbf{E} \qquad\qquad 24\text{-}16$$

We have written this as a vector equation because the current density and the electric field vectors are in the same direction.

There are many metals for which the resistivity is zero below a certain temperature T_c, called the *critical temperature*. This phenomenon, called *superconductivity*, was discovered in 1911 by the Dutch physicist H. Kamerlingh Onnes. Figure 24-5 shows his plot of the resistance of mercury versus temperature. The critical temperature for mercury is 4.2 K. Critical temperatures for other superconductors range from less than 0.1 K for hafnium and iridium to 9.2 K for niobium. Many alloys are also superconducting. The highest critical temperature yet known (23 K) is for the alloy Nb_3Ge. The search continues for higher-temperature superconductors.

The conductivity of a superconductor cannot be defined since there can be a finite current density even when the electric field in the superconductor is zero. Steady currents have been observed to persist in superconducting rings for years with no electric field and no apparent loss. The phenomenon of superconductivity can be understood only with the help of quantum mechanics, which is beyond the scope of this book. The first successful theory of superconductivity was published by Bardeen, Cooper, and Schrieffer in 1957 and is known as the BCS theory.

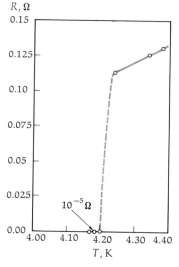

Figure 24-5
Plot by Kamerlingh Onnes of resistance of mercury versus temperature, showing sudden decrease at the critical temperature $T = 4.2$ K. (*By permission from C. Kittel, Introduction to Solid State Physics, 3d ed., John Wiley & Sons, Inc., New York, 1966.*)

Questions

2. Wire a and wire b have the same electric resistance and are made of the same material. Wire a has twice the diameter of wire b. How do the lengths of the wires compare?

3. If wires a and b in Question 2 carry the same currents, how do their current densities compare? The voltage drops across them? The electric fields inside the wires?

4. In our study of electrostatics we concluded that there is no electric field within the material of a conductor. Why do we now find it possible to discuss electric fields inside conducting material?

24-3 Energy in Electric Circuits

When there is electric current in a conductor, electric energy is continually converted into thermal energy in the conductor. For example, in the simple classical model of conduction, covered in detail in Section 24-4, the electric field in the conductor accelerates the free electrons for a short time, giving them an increased kinetic energy which is quickly transferred into thermal energy of the conductor by collisions between the electrons and the lattice ions of the conductor. Thus, though the electrons continually gain energy from the electric field, this energy is

Demonstration of persistent currents. Oppositely directed superconducting current loops are set up in the ring and ball so that the magnetic force between the currents is repulsive. The ball floats above the ring, its weight balanced by the magnetic force of repulsion. (*Courtesy of Cryogenic Technology, Inc.*)

immediately transferred to thermal energy of the conductor and the electrons maintain a steady drift velocity on the average.

In general when (positive) charge flows in a conductor, it flows from high potential to low potential in the direction of the electric field (the negatively charged electrons of course move in the opposite direction). The charge thus loses potential energy. This potential-energy loss does not appear as kinetic energy of the charge carriers except momentarily, before it is transferred to the lattice ions by collisions. Consider the charge ΔQ which passes point P_1 in Figure 24-6 during a time Δt. If the potential at that point is V_1, the charge has potential energy $\Delta Q\, V_1$. During that time interval, the same amount of charge passes point P_2, where the potential is V_2. It has potential energy $\Delta Q\, V_2$, which is less than the original potential energy. The energy lost by the charge passing through this segment of the conductor is $-\Delta W$, given by

$$-\Delta W = \Delta Q(V_1 - V_2) = \Delta Q\, V$$

where $V = V_1 - V_2$ is the potential drop from point 1 to point 2. (Here we write V rather than ΔV to simplify the notation.) The rate of energy loss by the charge is

$$-\frac{\Delta W}{\Delta t} = \frac{\Delta Q}{\Delta t}\, V = IV$$

where I is the current. The power loss in the conductor is thus

$$P = IV \qquad\qquad\qquad\qquad\qquad \text{24-17} \qquad \textit{Power}$$

If I is in amperes and V in volts, the power loss is in watts.

This expression for electric-power loss can easily be remembered by recalling the definitions of V and I. The voltage drop is the potential-energy decrease per unit charge, and the current is the charge flowing per unit time. The product IV is thus the energy loss per unit time, or the power loss. As we have seen, this power goes into heating the conductor.

Using the definition of resistance $R = V/I$, we can write Equation 24-17 in several other useful forms by eliminating either V or I:

$$P = (IR)I = I^2R \qquad \text{or} \qquad P = V\frac{V}{R} = \frac{V^2}{R} \qquad\qquad \text{24-18}$$

The energy put into a conductor is called *Joule heat*.

To have a steady current in a conductor we need to have a supply of electric energy. A device which supplies electric energy is called a *seat of electromotive force* or simply a *seat of emf*. It converts chemical, mechanical, or other forms of energy into electric energy. It is often a battery, which converts chemical energy into electric energy, or a generator, which converts mechanical energy into electric energy. A seat of emf does work on the charge passing through it, raising the potential energy of the charge. The increase in the potential energy per unit charge is called the emf ε of the seat. When a charge ΔQ flows through a seat of emf, its potential energy is increased by the amount $\Delta Q\, \varepsilon$. The unit of emf is the volt, the same as the unit of potential difference. The symbol for a seat of emf in a circuit diagram is ⊣⊦. The longer line indicates the higher-potential side. Ideally, a seat of emf maintains a constant potential difference ε between its two sides independent of the rate of flow of charge, i.e., independent of the current in a circuit.

Figure 24-7 shows a simple circuit consisting of a resistance R connected to a seat of emf. The resistance is indicated by the symbol ⌁.

Figure 24-6
At point P_1 the charge ΔQ has potential energy $\Delta Q\, V_1$, and at point P_2 its potential energy is $\Delta Q\, V_2$. The energy loss in the conductor is $\Delta Q\, V$, where $V = V_1 - V_2$, and the power loss is IV.

emf

Figure 24-7
Simple circuit with an ideal emf, a resistance R, and a connecting wire that is assumed to be resistanceless.

Russ Kinne/Photo Researchers, Inc.

An unusual source of emf is the electric eel (*Electrophorus electricus*), which lives in the Amazon and Orinoco rivers of South America. When attacking, the eel can develop up to 600 V between its head (positive) and its tail (negative).

The straight lines in the circuit diagram indicate connecting wires of negligible resistance. Since the choice of zero potential is always arbitrary, we are interested only in the potential differences between various points in the circuit. The seat of emf maintains a constant potential difference ε between points a and b, with point a at the higher potential. There is no potential difference between points a and c or between points d and b because the connecting wire is assumed to have negligible resistance. The potential difference between points c and d is therefore also ε, and the current in the resistor is given by $I = \varepsilon/R$. The direction of the current in this circuit is clockwise, as indicated in the figure. Note that *in the seat of emf the charge flows from low potential to high potential.** When charge ΔQ flows through the seat of emf, its potential energy is increased by the amount $\Delta Q\,\varepsilon$. It then flows through the conductor, where it loses this potential energy to heat. The rate at which energy is supplied by the emf is

$$P = \frac{\Delta W}{\Delta t} = \frac{\Delta Q\,\varepsilon}{\Delta t} = \varepsilon\,I \qquad\qquad 24\text{-}19$$

In this simple circuit, the power input by the seat of emf equals that dissipated in the resistor. An emf can be thought of as a sort of charge pump. It pumps the charge from low potential energy to high potential energy, analogously to a water pump pumping water from low to high gravitational potential energy.

Figure 24-8 shows a mechanical analog to the simple electric circuit discussed above. Marbles of mass m roll along an inclined board with many nails in it. The marbles are accelerated by the gravitational field between collisions with the nails. They transfer the kinetic energy obtained between collisions to the nails during the collisions. Because of the many collisions, the marbles move with a small drift velocity toward the ground. When they reach the bottom, a boy picks them up and starts them again. The boy is the analog of the emf. He does work mgh on each marble. The work per mass is gh, which is analogous to the work per charge done by the emf. The energy source in this case is the internal chemical energy of the boy.

A real battery is more than merely a seat of emf. The potential difference across the battery terminals, called the *terminal voltage*, is not simply the emf of the battery. Consider the simple circuit of a battery

(a)

(b)

Figure 24-8
Mechanical analog of resistance and emf. (a) As the marbles roll down the incline, their potential energy is converted into kinetic energy, which is quickly converted into heat because of collisions with the nails in the board. (b) A boy lifts the marbles up from low potential energy to high potential energy, converting his internal chemical energy into potential energy of the marbles.

Terminal voltage

* In some cases studied in the next chapter, such as charging one battery by another, the charge can flow in the opposite direction through a seat of emf.

Brown Brothers

The first practical method of maintaining a steady electric current was the battery, invented by Alessandro Volta in 1800. The earliest battery was a series of electric cells containing a copper disk and a zinc disk separated by a moistened pasteboard. Volta is shown here demonstrating his electric cell to Napoleon.

and resistor. Figure 24-9 shows a typical dependence of the terminal voltage of the battery on the current in the battery. The terminal voltage decreases slightly as the current increases.* As we have seen, an ideal seat of emf maintains a constant potential difference between its sides, independent of the current. Because the decrease in terminal voltage is approximately linear, we can represent a real battery by a seat of emf plus a small resistance r, called the *internal resistance* of the battery. Figure 24-10 shows a simple circuit containing a battery, a resistor, and connecting wires. As before, we ignore any resistance in the connecting wires. The circuit diagram for this circuit is shown in Figure 24-11. As charge passes from point b to point a, its potential energy is first increased as it passes through the seat of emf and then decreased slightly as it passes through the internal resistance of the battery. If the current in the circuit is I, the potential at point a is related to that at point b by

$$V_a = V_b + \varepsilon - Ir$$

Figure 24-9
Terminal voltage V versus current I through a real battery. The dashed line shows the expected result for an ideal emf ε.

* This is assuming that the current through the battery is in the same direction as the emf of the battery, which is the case for a circuit with only one seat of emf. When a battery is being charged (by another seat of emf), the current is opposite to the direction of the emf and the terminal voltage increases slightly with increasing current.

Figure 24-10
A simple circuit consisting of a battery, a resistor, and connecting wires. (*Courtesy of Larry Langrill, Oakland University.*)

Figure 24-11
Circuit diagram representing the circuit with battery and resistor in Figure 24-10. The battery can be represented by an ideal emf ε and a small resistance r.

The terminal voltage is thus

$$V_a - V_b = \varepsilon - Ir \qquad\qquad 24\text{-}20$$

The terminal voltage of the battery decreases linearly with current, as indicated in Figure 24-9. The potential drop across the resistor R is the same as the terminal voltage. The current I is thus given by

$$I = \frac{V_a - V_b}{R} = \frac{\varepsilon - Ir}{R}$$

or

$$I = \frac{\varepsilon}{R + r} \qquad\qquad 24\text{-}21$$

Questions

5. What are several common kinds of emf's? What sort of energy is converted into electric energy by the emf in each?

6. In a simple electric circuit like that shown in Figure 24-11 the charge outside the emf flows from positive voltage toward negative voltage, but in the same circuit the current inside the emf flows from minus to plus. Explain how this is possible.

7. Figure 24-8 illustrates a mechanical analog to a simple electric circuit. Devise another in which the current is a flow of water instead of marbles.

8. An emf is a device which converts chemical, mechanical, or another form of energy into electric energy. Could a device which uses electric energy as its source of energy be considered an emf? Can you think of any examples? What about an electronic power supply which serves the same function as a battery?

24-4 Classical Model of Electric Conduction

A classical model of electric conduction was first proposed by P. Drude in 1900 and developed by Hendrik A. Lorentz about 1909. This model successfully predicts Ohm's law and relates the conductivity and resistivity to the motion of free electrons in conductors. This classical theory is useful in understanding conduction even though it has been replaced by a modern theory based on quantum mechanics. We present the classical theory of electric conduction in this section and then discuss how some of the parameters are reinterpreted in the modern theory in Section 24-5.

In the classical model of electric conduction, a metal or other conductor is pictured as a regular three-dimensional array of atoms or ions with a large number of electrons free to move about the whole metal. In copper, for example, there is approximately one free electron per copper atom. The number of free electrons per atom can be measured using the Hall effect (Chapter 26). A copper atom with one electron missing is a copper ion with a positive charge $+e$. The regular arrangement of copper ions in metallic copper is called a *lattice*. In the absence of an electric field, the free electrons move about the metal much like gas molecules in a container. The free electrons make collisions with the ions of the lattice and are in thermal equilibrium with it. The rms speed of the electrons can be calculated from the equipartition theorem. The result is the same as that for an ideal-gas molecule with the electron mass replacing the molecular mass in Equation 16-23:

$$v_{\text{rms}} = \sqrt{\frac{3kT}{m}}$$ 24-22

For example, at temperature $T = 300$ K the rms speed is

$$v_{\text{rms}} = \sqrt{\frac{3(1.38 \times 10^{-23} \text{ J/K}) (300 \text{ K})}{9.11 \times 10^{-31} \text{ kg}}} = 1.17 \times 10^5 \text{ m/s}$$

This speed is much larger than the drift velocity calculated in Example 24-1.

At first glance it is surprising that any material obeys Ohm's law. In the presence of an electric field, a free electron experiences a force $q\mathbf{E}$ which would cause an acceleration $q\mathbf{E}/m$ if it were the only force acting on the electron. But Ohm's law implies a steady drift velocity (not acceleration) of the electron which is proportional to the electric field. The current density $\mathbf{J} = nqv_d$ is proportional to the drift velocity, and according to Ohm's law, $\mathbf{J}$ is proportional to the electric field $\mathbf{E}$. Of course there are other forces acting on the electrons because they make collisions with the lattice ions. In the classical model of conduction, it is assumed that after an electron makes a collision with a lattice ion, its velocity is completely unrelated to that before the collision. The justification of this assumption is that the drift velocity is very small compared with the random thermal velocity of the electron. We thus picture the electron as being accelerated by the electric field for a short time between collisions with the lattice ions, thereby acquiring a small drift velocity. After each collision, which gives it a random velocity, the electron is again accelerated by the field. The motion of such an electron is shown in Figure 24-12, in which the drift velocity is greatly exaggerated. In Figure 24-13b a small drift velocity $\mathbf{v}_d$ has been added to the electron velocities, which in Figure 24-13a are random. We can relate the drift velocity to the electric field by ignoring the random thermal velocities of the electrons and assuming that the electron starts from rest after a collision. After a time τ the electron will have a velocity of magnitude equal to the acceleration $q\mathbf{E}/m$ times the time:

$$v = \frac{qE}{m} \tau$$ 24-23

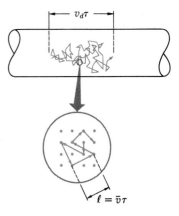

Figure 24-12
Electron path in a wire. Superimposed on the random thermal motion is a slow drift in the direction of the electric force $q\mathbf{E}$. The mean free path ℓ, mean time between collisions τ, and mean speed $\bar{v}$ are related by $\ell = \bar{v}\tau$.

(a) Average electron velocity = 0   (b) Average electron velocity = —

Figure 24-13
(a) Stationary positive ions, and electrons with a random distribution of velocities. (b) A small drift velocity to the right has been added to each electron velocity, as indicated for the electron in the lower left corner. (*Adapted from E. M. Purcell, Electricity and Magnetism, p. 123, Berkeley Physics Course, McGraw-Hill Book Company, New York, 1965. Courtesy of Education Development Center, Inc., Newton, Mass.*)

If we use this expression for the drift velocity and insert it in Equation 24-4 for the current density, we have

$$J = nqv_d = nq \frac{qE}{m} \tau = \frac{nq^2\tau}{m} E \qquad \qquad \text{24-24}$$

Comparing this result with the definition of electric conductivity and its reciprocal, the resistivity, gives

$$\sigma = \frac{nq^2\tau}{m} \quad \text{and} \quad \rho = \frac{m}{nq^2\tau} \qquad \qquad \text{24-25}$$

The average distance the electron travels between collisions is called the *mean free path* ℓ. It is the product of the mean speed $\bar{v}$ and the mean time between collisions τ: *Mean free path*

$$\ell = \bar{v}\tau \qquad \qquad \text{24-26}$$

In terms of the mean free path and the mean speed, the conductivity and resistivity are

$$\sigma = \frac{nq^2\ell}{m\bar{v}} \quad \rho = \frac{m\bar{v}}{nq^2\ell} \qquad \qquad \text{24-27}$$

According to Ohm's law, the conductivity and resistivity are independent of the electric field E. The quantities in Equation 24-27 that might depend on the electric field are the mean speed $\bar{v}$ and the mean free path ℓ. As we have seen, the drift velocity is very much smaller than the rms speed of the electrons in thermal equilibrium with the lattice ions. Thus the electric field has essentially no effect on the mean speed of the electrons. The mean free path of the electron depends on the chance that in traveling some distance the electron will collide with a lattice ion. According to the classical model, this chance depends on the size of the lattice ion and on the density of ions, neither of which depends on the electric field E. Thus this model predicts Ohm's law with the resistivity and conductivity given by Equations 24-27.

Example 24-4 Use Equation 24-23 and the results of Examples 24-1 and 24-3 to estimate the mean time between collisions for electrons in 14-gauge copper wire when the current is 1 A.

From these examples we have $v_d = 3.55 \times 10^{-3}$ cm/s = 3.55×10^{-5} m/s and $E \approx 8.13 \times 10^{-3}$ V/m. Using $q = 1.60 \times 10^{-19}$ C and $m = 9.11 \times 10^{-31}$ kg for the charge and mass of the electron, we have

$$\tau = \frac{mv_d}{qE} = \frac{(9.11 \times 10^{-31}\text{ kg}) (3.55 \times 10^{-5}\text{ m/s})}{(1.60 \times 10^{-19}\text{ C}) (8.13 \times 10^{-3}\text{ V/m})}$$

$$= 2.49 \times 10^{-14}\text{ s}$$

The time between collisions is very short.

Example 24-5 Assuming that the mean speed is about 10^5 m/s, the same order of magnitude as the rms speed calculated in this section, find the order of magnitude of the mean free path of an electron in a copper wire.

Using $\tau \approx 2.5 \times 10^{-14}$ s from Example 24-4, we have

$$\ell = \bar{v}\tau = (10^5\text{ m/s}) (2.5 \times 10^{-14}\text{ s})$$

$$= 2.5 \times 10^{-9}\text{ m}$$

Although successful in predicting Ohm's law, the classical theory of conduction has several defects. The numerical magnitudes of the conductivity and resistivity calculated from Equation 24-27 using classical methods for finding the mean free path and the mean speed differ from measured values by a factor of 10 or so, depending on the temperature at which these values are measured. The temperature dependence of resistivity is given completely by the mean speed $\bar{v}$ in Equation 24-27, which is proportional to $\sqrt{T}$. Thus this calculation does not give a linear dependence on temperature. Finally, the classical model says nothing about why some materials are conductors, others insulators, and still others semiconductors, a subject we discuss briefly in Section 24-6.

24-5 Corrections to the Classical Theory of Conduction

The results of the modern theory of electric conduction based on quantum mechanics can be discussed briefly and qualitatively. The important features of this theory for our discussion are that rms speed v_{rms} is not given by Equation 24-22, because the equipartition theorem does not hold for electrons in a metal, and that the mean free path must be calculated using the wave nature of propagation of electrons in the metal. The calculation of the mean speed $\bar{v}$ is quite complicated. The results are that $\bar{v}$ is essentially independent of temperature and equals about 1.6 Mm/s for copper, roughly 16 times that calculated at $T = 300$ K from the equipartition theorem. The wave analog of electron collisions with the lattice ions is the scattering of electron waves by the ions. A detailed calculation of electron-wave scattering by a perfectly periodic lattice of identical ions gives the result that there is no scattering; the mean free path is infinite. Thus for such a perfect crystal, the resistance is zero. Electron waves are scattered only if the lattice is not perfectly periodic. There are two main causes for lattice deviations from perfect periodicity. One is impurities. For example, if some zinc is introduced into pure copper, the previously perfect periodicity is destroyed. At very low temperatures the resistance of a metal is primarily due to impurities. The other cause of deviations is the displacement of the lattice ions due to vibrations. This effect is dominant at ordinary temperatures. Thus the mean free path is not determined by the size of the lattice ions. At very low temperatures the ions effectively look like points to the electron so far as scattering is concerned. At normal temperatures, the effective area of an ion is proportional to the square of the amplitude of vibration. This in turn is proportional to the energy of vibration, which is proportional to T (except at very low temperatures). The modern theory accurately accounts for the temperature dependence of resistance.

As an illustration of these ideas it is instructive to compare the resistivity of copper and brass (an alloy of copper and zinc) at temperatures of 300 and 4 K. The resistivity of copper drops by a factor of about 50 to 100 (depending on the purity) when the temperature drops from 300 to 4 K, indicating that most of the resistivity at 300 K is due to thermal motion. However, the resistivity of brass drops by only a factor of 4 with temperatures from 300 to 4 K. A large part of the resistivity of brass is due to the temperature-independent irregularity of the copper-zinc lattice.

24-6 Conductors, Insulators, and Semiconductors

The classical model we have discussed for electric conduction with the modifications due to quantum mechanics gives a good account of Ohm's law and the temperature dependence of resistivity and conductivity, but it does not deal with the important question of why some materials are good conductors and others are insulators. As we have seen, there is an enormous variation in the conductivity and resistivity from the best conductors to the best insulators. The difference between a conductor and an insulator can be characterized empirically by the density of charge carriers n. For a good conductor the number of charge carriers is about one per atom; for insulators the number of charge carriers is nearly zero.

Energy quantization

To discuss why this is so, even qualitatively, we need two important ideas from quantum mechanics. One is that the energy of an electron in an atom cannot take on just any value but is quantized. This idea was first put forward by Bohr, who proposed that the possible energy values or levels for the hydrogen atom are given by the simple formula

$$E_n = -\frac{13.6}{n^2}\ \text{eV}$$

where n is any integer, $n = 1, 2, 3,\ldots$. The lowest possible energy for the hydrogen atom corresponds to $n = 1$; the other possible energies are given by this expression with other values of n. No simple formulas exist for the possible energies of electrons in other more complicated atoms, but the important point for this discussion is that for any atom there is a discrete set of possible energies.

Pauli exclusion principle

The other quantum-mechanical result needed to understand why some materials are conductors and others are not is the *Pauli exclusion principle*. Only two electrons (with opposite spins) can occupy a given quantum state. Consider for a moment two identical atoms very far apart. Each atom has the same set of allowed energy levels. Let us focus our attention on one such level (such as $n = 2$ for hydrogen). There is a level of the same energy for each atom. If the atoms are brought close together, the energy of this level changes because of the influence of the other atom. If the atoms are close enough together, we can no longer associate the energy levels with one of the atoms but must look at the two-atom system. The two previously identical energy levels split into two levels of slightly different energy for the two-atom system. Similarly, if we have N identical atoms, a particular energy level of the isolated atom splits into N different nearly equal energy levels when the atoms are sufficiently close together. The number of copper atoms in a piece of copper is very large, on the order of Avogadro's number. A single atomic energy level identified with an isolated copper atom splits into a *band* of a very large number of energy levels which are closely spaced in energy. Because the number of levels in the band is so large, they are spaced almost continuously within the band. There is a separate band of levels for each particular energy level of the isolated atom. In the isolated atom the different levels are often far apart. (For example, in hydrogen the energy for $n = 1$ is -13.6 eV, and for $n = 2$ it is $-13.6/4 = -3.4$ eV.) The energy bands corresponding to these individual levels for a large number of atoms in a solid may be widely separated in energy, may be close together, or may even overlap in energy, depending on the kind of atom and the type of bonding in the crystal.

Figure 24-14 shows four possible kinds of band structure for a solid. The band structure for copper is shown in Figure 24-14a. The lower bands are filled with the inner electrons of the atoms; i.e., according to the Pauli exclusion principle, no more electrons can occupy levels in these bands. The uppermost band containing electrons is only about half full. In the normal state, at low temperatures, the lower half of the energy band is filled, and the upper half is empty. At higher temperatures, a few of the electrons are in the higher energy states in that band because of thermal excitation, but there are still many unfilled energy states above the filled ones.

When an electric field is established in the conductor, the electrons in the upper band are accelerated, which means that their energy is increased. This is consistent with the Pauli exclusion principle because there are many empty energy states just above those occupied by electrons in this band. These electrons are thus the conduction electrons, and the band is called the *conduction band.*

Conduction band

Figure 24-14b shows the band structure for a typical insulator. At $T = 0$ K the highest energy band that contains electrons is completely full. The next energy band, containing empty energy states, is separated from the last filled band by an energy gap which is quite large compared with thermal energies (of the order of 0.02 eV). This energy gap is sometimes referred to as a *forbidden energy band.* Very few electrons can be thermally excited to the nearly empty conduction band, even at fairly high temperatures. When an electric field is established in the solid, electrons cannot be accelerated because there are no empty energy states at nearby energies. In the classical model, we describe this by saying that there are no free electrons. The small conductivity that is observed is due to the very few electrons that are thermally excited into the upper nearly empty conduction band.

Figure 24-14
Four possible band structures for a solid. In (a) an allowed band is only partially full, whereas in (c) two allowed bands overlap. In both cases nearby energy states are available for excitation of electrons, and so these materials are conductors. (b) The band structure of an insulator, where there is a forbidden band with a large energy gap between a filled allowed band and the next allowed band. (d) Since the energy gap between the filled band and the next allowed band is small, some electrons are excited to the next allowed band at normal temperatures, making this material a semiconductor.

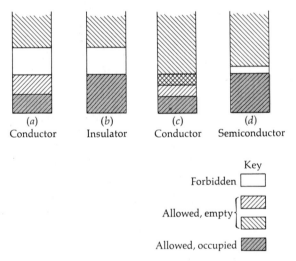

These results—that the conduction band is essentially empty and that the gap between it and the lower energy band that is nearly filled is large—can be derived by the methods of quantum mechanics. In some materials the energy gap between the top filled band and the empty conduction band is very small (Figure 24-14d). At ordinary tem-

Semiconductor

peratures there are an appreciable number of electrons in the conduction band due to thermal excitation. Such a material is called a *semiconductor*. In the presence of an electric field, the electrons in the conduction band can be accelerated. Also, for each electron in the conduction band there is a vacancy, or hole, in the nearly filled band (this band is called the *valence band*). In the presence of a field, electrons in this band can be excited to a vacant energy level. This contributes to the electric current and is most easily described as the motion of a hole in the direction of the field and opposite the motion of the electrons. The hole thus acts like a positive charge. (An analogy is a line of cars with a space the size of one car: as the cars move to fill the space, the space moves backward, in the direction opposite the motion of the cars.) An interesting feature of a semiconductor is that as the temperature increases, the conductivity increases (and the resistivity decreases), contrary to the usual behavior of conductors. The reason is that as the temperature is increased, the number of free electrons n is increased because there are more electrons in the conduction band. (The number of holes of course also increases.) This increase in n outweighs the natural decrease in conductivity due to a smaller mean free path at higher temperatures.

A semiconductor can also be made from an insulator by introducing certain impurities (this procedure is called *doping*) which slightly modify the band structure. These slight modifications have a great effect on the conductivity. In one type of doped semiconductor the impurity atoms provide electrons in energy states which lie in the original forbidden band just below the empty conduction band (Figure 24-15a).

Figure 24-15
(*a*) Energy bands of an *n*-type semiconductor. Impurity atoms provide filled energy levels just below the empty conduction band and donate electrons to the conduction band. (*b*) Energy bands of a *p*-type semiconductor. Impurity atoms provide empty energy levels just above the filled valence band and accept electrons from it.

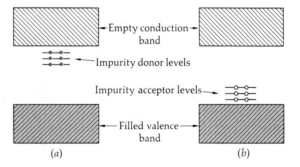

These impurity energy levels are called *donor levels*. The impurity atoms donate electrons to the conduction band. This is called an *n-type semiconductor* because the charge carriers are negative electrons. In a second kind of doped semiconductor the impurity atoms have empty energy levels just above the filled band in the original solid, as in Figure 24-15b. These levels accept electrons from the filled valence band when these electrons are thermally excited to a higher energy state. Thus a hole is created in the valence band, which is free to propagate in the direction of an electric field. This is called a *p-type semiconductor* because the charge carriers are positive holes.

Transistors

Reuben E. Alley, Jr.
U.S. Naval Academy

In 1949, John Bardeen, Walter Brattain, and William Shockley, all of Bell Telephone Laboratories, initiated a revolution in electronics with the invention of the transistor. In 1956 they received a Nobel prize for their work.

Transistors, semiconductor diodes, and the integrated circuit, which combines the functions of many transistors and their associated resistors and capacitors into one small package, have almost completely replaced vacuum tubes in all but a few special applications, e.g., television picture tubes and high-power radio transmitters. These solid-state devices can be made very small, very light, and from readily available materials. They are cheap to manufacture, require little power, and operate at room temperatures. A spectacular example of reduction in size and cost is the pocket calculator, available in some models for under $10. The most expensive parts of the modern pocket calculator are said to be the control-button switches. Their size, which is limited by the size of the human finger, prevents further miniaturization by a factor of perhaps 25.

At least one company offers a programmable calculator whose memory capacity and computational abilities equal or perhaps surpass those of the best vacuum-tube computer of the early 1950s. Its cost (a few hundred dollars) is less than one-hundredth and its volume is less than one-thirty-thousandth that of the vacuum-tube counterpart. It is portable, operates on tiny batteries, and is highly reliable. In contrast, the vacuum-tube computer required an enormous power supply and air conditioning to remove the great quantities of waste heat resulting from the high operating temperature of the vacuum tubes. This high temperature seriously limited the life and reliability of the individual components.

Manufacture of diodes, transistors, and integrated circuits requires a very pure sample of one element throughout which a carefully controlled, exceedingly small amount (a few parts per million) of another element has been uniformly distributed. The basic material usually is germanium or silicon. Both these elements are semiconductors; i.e., at room temperature they have relatively few free charges (compared with the practically unlimited supply of free charge in a conductor such as copper). The second element (the *impurity* or *doping* material) must lie near silicon or germanium in the periodic table but have one less or one more valence electron than silicon or germanium. Commonly used are boron (valence 3) and arsenic (valence 5). Adding boron produces a material that acts as if it had a large number of free positive charges. Such material is called a *p*-type semiconductor. Similarly, the introduction of arsenic into the pure material produces a sample in which current is carried by flow of negative charges (an *n*-type semiconductor).

A single crystal of silicon made to contain a *p*-type region and an adjacent *n*-type region with a fairly abrupt discontinuity between them is a diode. Within a specified voltage range (below the breakdown voltage indicated in Figure 1) a diode is a good conductor for one polarity of applied voltage and a very poor conductor for applied voltage of opposite polarity.

Semiconductor diodes are used in *rectifiers,* which convert the alternating voltage supplied by all power companies into unidirectional voltage. With *filters* and *regulating circuits* one can construct a *power supply* that provides a direct voltage of constant amplitude.

Radios, television sets, sound-reproducing systems, and many other devices require a circuit that will accept a time-dependent voltage or current of very small amplitude (the *input*) and provide a greatly magnified replica of the input voltage or current as its *output.* Such a circuit, which must be supplied with energy from a battery or a power supply, is called an *amplifier.*

Figure 1
Diode characteristic. Note the
different scales in the first
and third quadrants.

Figure 2

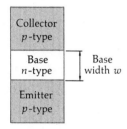

Figure 3
A *pnp* transistor biased for
normal operation in the
common-emitter connection.

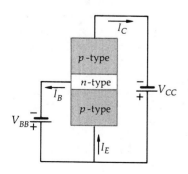

The basic element of most amplifiers is the transistor, which contains three distinct regions of semiconductor material. Figure 2 is an idealized representation of a *pnp* transistor, where the three regions are labeled with their common designations. In a properly designed transistor the level of doping is much greater in the emitter than in the other two regions, and the base width w is very narrow.

The device represented in Figure 2 contains two *pn* junctions. In a practical circuit (Figure 3) voltages are supplied so that the emitter-base (*EB*) junction is forward-biased and the collector-base (*CB*) junction is reverse-biased. The difference in doping levels of the emitter and the base means that practically all the current across the *EB* junction consists of positive charges from the emitter. Because the base width is small and the *CB* junction is reverse-biased, most of the positive charges from the emitter that enter the base diffuse across its narrow width and enter the collector. The few positive charges that are not collected by the *CB* junction leave the base through the external connection. In Figure 3, therefore, I_C is almost but not quite equal to I_E, and I_B is much smaller than either I_C or I_E. It is customary to write

$$I_C = \beta I_B$$

where β, the *current gain*, is an important parameter of the transistor. Transistors can be designed to have values of β as low as 10 or as high as several hundred.

In a useful transistor, β will be constant; i.e., the device will be linear, over a restricted range of base current, provided that proper biases are maintained on the two junctions. Figure 4 shows a simple amplifier. A small time-varying voltage, say from the motion of a needle in the groove of a phonograph record, is in series with a bias voltage V_{BB}. The base current is then the sum of a steady current I_B produced by V_{BB} and a varying current i_b produced by v_s, the voltage to be amplified. Because v_s may at any instant be either positive or negative, V_{BB} must be included and must be large enough to ensure that there is always a forward bias on the *EB* junction. Since the device is linear, the collector current will consist of two parts, a steady current $I_C = \beta I_B$ and a time-varying current $i_c = \beta i_b$. We thus have a *current amplifier* where the time-varying output current i_c is β times the input current i_b. In such an amplifier, the steady currents I_C and I_B, while essential to proper operation, are usually not of interest. (If an *npn* transistor is used, the bias voltages must be reversed and the currents will all be in the other direction. Otherwise, operation is identical with that described here.)

(a)

(b)

(c)

Figure 4
(a) Simple amplifier circuit using *pnp* transistor. (b) The same circuit with the transistor represented by a standard symbol. (c) Amplifier using *npn* transistor. Battery polarities and current directions are opposite those in part (b).

Figure 5
(a) Five-watt audio amplifier using an integrated circuit. The integrated circuit is partly hidden by a piece of metal 3.3 cm long. The five cylinders are capacitors.
(b) Photomicrograph of an integrated circuit with more than 50 components, on a silicon chip approximately 0.25 cm by 0.2 cm.

(a)

(b)

Ohm's law tells us that v_s and i_b are related by

$$i_b = \frac{v_s}{R_b + r_b}$$

where r_b is the internal resistance of the transistor between base and emitter. Likewise, in the collector circuit the current i_c produces a voltage v_L across the resistor R_L,

$$v_L = i_c R_L$$

We already have

$$i_c = \beta i_b = \beta \frac{v_s}{R_b + r_b}$$

and therefore,

$$v_L = \beta \frac{R_L}{R_b + r_b} v_s$$

and the voltage gain or voltage amplification is

$$\frac{v_L}{v_s} = \beta \frac{R_L}{R_b + r_b}$$

In a practical case β may be 100, and the ratio $R_L/(R_b + r_b)$ may be $\frac{1}{2}$. Then the voltage gain is 50. A more detailed derivation shows that v_L and v_s are exactly *out of phase*; i.e., when v_s has its most positive value, v_L has its most negative value. For a simple amplifier this phase shift is not important because all input voltages, regardless of frequency, are affected identically. The simple voltage amplifier is the basis of many circuits such as oscillators and modulators that constitute communication systems.

The complete amplifier in a record or tape player generally consists of several *stages* similar to Figure 4 connected in *cascade* so that the output of one becomes the input of the next. Thus the very small voltage produced by the motion of the needle or by the passage of magnetized tape through the player head controls the large amounts of power required to drive a system of loudspeakers. The energy delivered by the speakers is supplied by the sources of direct voltage connected to each transistor.

The resistance of a semiconductor material depends upon the impurity concentration. Reverse-biased diodes have capacitance that can be controlled by controlling the bias voltage. These two facts are exploited in the manufacture of *integrated circuits,* in which several transistors along with associated resistors and capacitors are interconnected on a single small piece of silicon. Figure 5a shows a 5-W audiofrequency amplifier that uses one integrated circuit containing 14 transistors, 3 diodes, and 10 resistors. Figure 5b shows a different integrated circuit containing even more circuit elements.

In addition to greater reductions in their size and weight, the introduction of integrated circuits has changed the design of electronics devices radically. For example, one need no longer design an amplifier but can choose from those available a complete amplifier with the desired characteristics. It is then procured and installed as a complete unit. Such possibilities have enabled small companies with little expertise in design to build rather complex electronics equipment and compete favorably in price and quality.

Review

A. Define, explain, or otherwise identify:

Current, 671   Superconductivity, 677
Ampere, 671   Joule heat, 678
Current density, 672   emf, 678
Drift velocity, 672   Terminal voltage, 679
Ohm's law, 674   Internal resistance, 680
Resistance, 674   Mean free path, 683
Resistivity, 674   Energy bands, 685
Conductivity, 675   Conduction band, 686

B. True or false:

1. The net motion of electrons is in the direction of the current.

2. Current density is proportional to the drift velocity.

3. $R = \Delta V/I$ is Ohm's law.

4. The current in a conductor is always proportional to the potential drop across the conductor.

5. $R = \Delta V/I$ holds for all materials for which resistance can be defined.

6. The resistance of a metal is due to irregularities in the spacing of the lattice ions.

Exercises

Section 24-1, Current and Motion of Charges

1. A wire carries a steady current of 2.0 A. (*a*) How much charge flows past a point in the wire in 5.0 min? (*b*) If the current is due to flow of electrons, how many electrons flow past a point in this time?

2. A 10-gauge copper wire (diameter 2.59 mm) carries a current of 20 A. Assuming one free electron per copper atom, calculate the electron drift velocity.

3. In a 3.0-cm-diameter fluorescent tube, 2.0×10^{18} electrons and 0.5×10^{18} positive ions (charge $+e$) flow past a point each second. (*a*) What is the current in the tube? (*b*) What is the current density?

4. In a certain electron beam, there are 5.0×10^6 electrons per cubic centimetre. The kinetic energy of the electrons is 10.0 keV, and the beam is cylindrical with diameter 1.00 mm. (*a*) What is the velocity of the electrons? (*b*) Find the current density. (*c*) Find the beam current.

5. A charge $+q$ moves in a circle of radius r with speed v. (*a*) Express the frequency f with which the charge passes a point in terms of r and v. (*b*) Show that the average current is qf and express it in terms of v and r.

6. A ring of radius R has charge per unit length λ which rotates with angular velocity ω about its axis. Find an expression for the current.

7. A 20.0-MeV proton beam of diameter 2.0 mm in a certain accelerator constitutes a current of 1.0 mA. The beam strikes a metal target and is absorbed by it. (*a*) What is the density of protons in the beam? (*b*) How many protons strike the target in 1.0 min? (*c*) If the target is initially uncharged, express the charge of the target as a function of time.

8. A 10-gauge wire (diameter 2.59 mm) is welded end to end to a 14-gauge wire (diameter 1.63 mm). The wires carry a current of 15 A. (*a*) What is the current density in each wire? (*b*) If both wires are copper with one free electron per atom, find the drift velocity in each wire.

Section 24-2, Ohm's Law and Resistance

9. A 10-m wire of resistance 0.2 Ω carries a current of 5 A. (a) What is the potential difference across the wire? (b) What is the magnitude of the electric field in the wire?

10. A potential difference of 100 V produces a current of 3 A in a certain resistor. (a) What is its resistance? (b) What is the current when the potential difference is 25 V?

11. A copper wire and an iron wire have the same length and diameter and carry the same current I. (a) Find the potential drop across each wire and the ratio of these drops. (b) In which wire is the electric field greater?

12. A piece of carbon is 3.0 cm long and has a square cross section 0.5 cm on a side. A potential difference of 8.4 V is maintained across its long dimension. (a) What is the resistance of the block? (b) What is the current in this resistor? (c) What is the current density?

13. A tungsten rod is 50 cm long and has a square cross section 1.0 mm on a side. (a) What is its resistance at 20°C? (b) What is its resistance at 40°C?

14. The third (current-carrying) rail of a subway track is made of steel and has a cross-sectional area of about 55 cm². What is the resistance of 10 km of this track?

15. A wire of length 1 m has a resistance of 0.3 Ω. It is uniformly stretched to a length of 2 m. What is its new resistance?

16. What is the potential difference across one wire of a 30-m extension cord of 16-gauge copper wire (diameter 0.130 cm) carrying a current of 5.0 A?

17. At what temperature will the resistance of a copper wire be 10 percent greater than it is at 20°C?

18. How long is a 14-gauge copper wire which has a resistance of 2 Ω?

19. A cube of copper has side 2.0 cm. If it is drawn out into a 14-gauge wire, what will its resistance be?

Section 24-3, Energy in Electric Circuits

20. What is the power dissipated in a 10.0-Ω resistor if the potential difference across it is 50 V?

21. Find the power dissipated in a resistor of resistance (a) 5 Ω and (b) 10 Ω connected across a constant potential difference of 110 V.

22. A 200-W heater is used to heat water in a cup. Assume that 90 percent of the energy goes into heating the water. (a) How long does it take to heat 0.25 kg of water from 15 to 100°C? (b) How long does it take to boil this water away after it reaches 100°C?

23. A 1-kW heater is designed to operate at 220 V. (a) What is its resistance, and what current does it draw? (b) What is the power of this resistor if it operates at 110 V?

24. A 10.0-Ω resistor is rated capable of dissipating 5.0 W. (a) What maximum current can this resistor tolerate? (b) What voltage across this resistor will produce this current?

25. If energy costs 6 cents per kilowatt-hour, (a) how much does it cost to operate an electric toaster for 4 min if the toaster has resistance 11.0 Ω and is connected across 100 V? (b) How much does it cost to operate a heater of resistance 5.0 Ω across 100 V for 8 h?

26. A battery has an emf of 12.0 V. How much work does it do in 5 s if it delivers 3 A?

27. A battery with 12-V emf has a terminal voltage of 11.4 V when it delivers 20 A to the starter of a car. What is the internal resistance r of the battery?

28. (a) How much power is delivered by the emf of the battery in Exercise 27 when it delivers 20 A? How much of this power is delivered to the starter? (b) By how much does the chemical energy of the battery decrease when it delivers 20 A for 3 min in starting the car? (c) How much heat is developed in the battery when it delivers 20 A for 3 min?

29. A 12-V car battery has an internal resistance of 0.4 Ω. (a) What is the current if the battery is shorted momentarily? (b) What is the terminal voltage when the battery delivers 20 A to start the car?

Section 24-4, Classical Model of Electric Conduction

30. A current of 30 mA exists in a copper wire of cross-sectional area 2.0 mm². (a) What is the current density? (b) What is the drift velocity of the electrons? (c) What is the average time between collisions? (d) What is the mean free path of electrons in this wire (assume $\bar{v} = 10^5$ m/s)?

31. Repeat Exercise 30 for a current of 5.0 mA in a copper wire of cross-sectional area 1.0 mm².

Section 24-5, Corrections to the Classical Theory of Conduction, and Section 24-6, Conductors, Insulators, and Semiconductors

There are no exercises for these sections.

Problems

1. A 16-gauge copper wire (diameter 1.29 mm) can safely carry a maximum current of 6 A (assuming rubber insulation). (a) How great a potential difference can be applied across 40 m of such a wire? (b) Find the current density and the electric field in the wire when it carries 6 A. (c) Find the power dissipated in the wire when it carries 6 A.

2. A Van de Graaff accelerator belt carries a surface charge density of 5 μC/m². The belt is 0.5 m wide and moves at 20 m/s. (a) What current does it carry? (b) If this charge is raised to a potential of 100 kV, what is the minimum power of the motor needed to drive the belt?

3. A linear accelerator produces a beam of electrons in which the current is not constant but consists of a pulsed beam of particles. Suppose that the pulse current is 1.6 A for a 0.1-μs duration. (a) How many electrons are accelerated in each pulse? (b) What is the average beam current if there are 1000 pulses per second? (c) If the electrons acquire an energy of 400 MeV, what is the average (minimum) power input to the accelerator? (d) What is the peak power input? (e) What fraction of the time is the accelerator actually accelerating particles? (This is called the *duty factor* of the accelerator.)

4. A coil of nichrome wire is to be used as the heating element in a water boiler required to generate 8.0 g of steam per second. The wire has a diameter of 1.80 mm and is connected to a 115-V power supply. Find the length of wire required.

5. An 80.0-m copper wire 1.0 mm in diameter is joined end to end with a 49.0-m iron wire of the same diameter. The current in each is 2.0 A. (a) Find the current density in each wire. (b) Find the electric field in each wire. (c) Find the potential difference across each wire. (d) Find the equivalent resistance which would carry 2.0 A at a potential difference equal to the sum of that across the two, and compare it with the sum of the resistances of the two.

6. Two wires have the same length but different cross-sectional areas and are made from different materials. They are laid side by side and joined at each end so that the potential difference across each wire is the same. Let R_1 be the resistance of one wire and R_2 be that of the other. (a) If the potential difference across each is V, find the current in each and find the total current $I = I_1 + I_2$. (b) Find an expression for the resistance of a single wire which would carry the same total current when the potential difference is V.

7. A wire of area A, length l_1, resistivity ρ_1, and temperature coefficient α_1 is connected to a second wire of length l_2, resistivity ρ_2, temperature coefficient α_2, and the same area A, so that the wires carry the same current. (a) Show that if the current is I, the potential drop across the two wires is IR, where $R = R_1 + R_2$ is the sum of the resistances. (b) Show that if $\rho_1 l_1 \alpha_1 + \rho_2 l_2 \alpha_2 = 0$, the total resistance R is independent of temperature for small temperature changes. (c) If one wire is made of carbon and the other copper, find the ratio of their lengths such that R is approximately independent of temperature.

8. A 100-W heater resistor is designed to operate across 100 V. (a) What is its resistance and what current does it draw? (b) Show that if the potential difference across the resistor changes by a small amount ΔV, the power changes by a small amount ΔP, where $\Delta P/P \approx 2\,\Delta V/V$. Hint: Approximate the changes with differentials. (c) Find the approximate power dissipated in the resistor if the potential difference increases to 115 V.

CHAPTER 25   Direct-Current Circuits

Objectives   After studying this chapter you should:

1. Be able to determine the equivalent resistances of series or parallel resistors to simplify various combinations of resistors.

2. Be able to state Kirchhoff's rules and use them to analyze various dc circuits.

3. Be able to find the time constant for an *RC* circuit and sketch both the charge *Q* on the capacitor and the current *I* as functions of time for charging and discharging a capacitor.

4. Be able to draw the circuit diagrams and calculate the proper series or shunt resistors needed to make an ammeter, voltmeter, and ohmmeter from a given galvanometer.

In this chapter we analyze some simple circuits consisting of batteries, resistors, and capacitors in various combinations. These circuits are called direct-current (dc) circuits because the direction of the current in any one part of the circuit does not vary. Circuits in which the current alternates in direction (ac circuits) will be discussed in Chapter 30.

25-1   Series and Parallel Resistors

Two or more resistors connected so that the same charge must flow through each are said to be connected in series. Resistors R_1 and R_2 in Figure 25-1*a* are examples of resistors in series. They must carry the

Figure 25-1
(*a*) Two resistors in series. (*b*) The resistors in part (*a*) can be replaced by a single equivalent resistance $R_{eq} = R_1 + R_2$, which gives the same total potential drop when carrying the same current as in part (*a*).

(*a*)   (*b*)

same current I. The potential drop across the two resistors (from point a to point c in Figure 25-1) is

$$V = IR_1 + IR_2 = I(R_1 + R_2)$$

We can often simplify the analysis of a circuit with resistors in series by replacing such resistors with a single equivalent resistance which gives the same total potential drop V when carrying the same current I. The equivalent resistance for series resistances is the sum of the original resistances:

$$R_{eq} = \frac{V}{I} = R_1 + R_2$$

When there are more than two resistances in series, the equivalent resistance is

$$R_{eq} = R_1 + R_2 + R_3 + \cdots \qquad 25\text{-}1$$

Series resistors

Two resistors connected as in Figure 25-2a so that they have the same potential difference across them are said to be connected in parallel. Let I be the current from point a to point b. At point a the current splits into two parts, I_1 through resistor R_1 and I_2 through R_2. The total current is the sum of the individual currents:

$$I = I_1 + I_2 \qquad 25\text{-}2$$

Let $V = V_a - V_b$ be the potential drop across either resistor. In terms of the currents and resistances,

$$V = I_1 R_1 = I_2 R_2$$

(a)

(b)

Figure 25-2
(a) Two resistors in parallel.
(b) The two resistors in part (a) can be replaced by a single equivalent resistance R_{eq} related to R_1 and R_2 by $1/R_{eq} = 1/R_1 + 1/R_2$.

The ratio of the two currents is therefore the inverse ratio of the resistances; the smaller resistance carries the larger current and vice versa:

$$\frac{I_1}{I_2} = \frac{R_2}{R_1} \qquad 25\text{-}3$$

The equivalent resistance of a combination of parallel resistors is defined to be that resistance R_{eq} for which the same total current I produces the potential drop V:

$$R_{eq} = \frac{V}{I}$$

Equation 25-2 can then be written

$$I = \frac{V}{R_{eq}} = I_1 + I_2 = \frac{V}{R_1} + \frac{V}{R_2}$$

or

$$\frac{1}{R_{eq}} = \frac{1}{R_1} + \frac{1}{R_2} \qquad 25\text{-}4$$

This result can be generalized to any number of resistances in parallel:

$$\frac{1}{R_{eq}} = \frac{1}{R_1} + \frac{1}{R_2} + \frac{1}{R_3} + \cdots \qquad\qquad 25\text{-}5$$

The equivalent resistance of a combination of parallel resistors is less than that of any resistor. Consider, for example, any one resistor such as R_1 carrying current I_1 with potential drop $V = I_1 R_1$. The effect of another resistor or other resistors in parallel with R_1 is to increase the total current for the same potential drop V. The equivalent resistance V/I is thereby decreased.

The equivalent resistance for two parallel resistors can be found from Equation 25-4 by putting the two terms on the right side of the equation over a common denominator and taking the reciprocal of each side. The result is

$$R_{eq} = \frac{R_1 R_2}{R_1 + R_2} \qquad\qquad 25\text{-}6$$

This result explicitly demonstrates that R_{eq} is less than either R_1 or R_2 since the right side of the equation can be considered to be the product of either resistance (such as R_1) and a factor less than 1, such as $R_2/(R_1 + R_2)$. When there are three or more parallel resistors, it is usually easier to solve Equation 25-5 for $1/R_{eq}$ than to remember a general result analogous to Equation 25-6. For example, the equivalent resistance of three parallel resistors is given by the somewhat complicated expression (see Problem 16)

$$R_{eq} = \frac{R_1 R_2 R_3}{R_1 R_2 + R_2 R_3 + R_3 R_1} \qquad\qquad 25\text{-}7$$

The current in each of two parallel resistors can be found from the fact that the potential drop across the combination is $I R_{eq}$ and is also equal to $I_1 R_1$ and to $I_2 R_2$:

$$I_1 R_1 = I_2 R_2 = I R_{eq} = I \frac{R_1 R_2}{R_1 + R_2}$$

Then

$$I_1 = \frac{R_2}{R_1 + R_2} I \quad \text{and} \quad I_2 = \frac{R_1}{R_1 + R_2} I \qquad\qquad 25\text{-}8$$

Example 25-1 For the circuit shown in Figure 25-3 find the equivalent resistance of the parallel combination of resistors, the current carried by each, and the total current through the emf.

We first find the equivalent resistance of the 6- and 12-Ω resistors in parallel:

$$\frac{1}{R_{eq}} = \frac{1}{6\ \Omega} + \frac{1}{12\ \Omega} = \frac{3}{12\ \Omega} = \frac{1}{4\ \Omega}$$

$$R_{eq} = 4\ \Omega$$

Note that this equivalent resistance is less than that of either of the parallel resistors. Figure 25-4 shows the circuit with R_{eq} replacing the parallel combination. The resistances $R_{eq} = 4\ \Omega$ and $r = 2\ \Omega$ are in series. The equivalent resistance of this series combination is $R'_{eq} = R_{eq} + r = 6\ \Omega$. The current in the circuit is therefore

$$I = \frac{\mathcal{E}}{R'_{eq}} = \frac{6\ \text{V}}{6\ \Omega} = 1\ \text{A}$$

This is the total current in the seat of emf.

Figure 25-3
Circuit for Example 25-1.

Figure 25-4
Simplification of the circuit in Figure 25-3 by replacing the parallel resistors by a single equivalent resistance.

The currents in the 6- and 12-Ω resistors could be found from Equation 25-8 or more simply by noting that since the total current from point a to b in Figure 25-4 is 1 A and the resistance R_{eq} is 4 Ω, the voltage drop from point a to b is $IR_{eq} = 4$ V. The current in the 6-Ω resistor needed to produce a 4-V potential drop is $I_6 = V/R_1 = (4 \text{ V})/(6 \text{ Ω}) = \frac{2}{3}$ A. Similarly, the current in the 12-Ω resistor is $I_{12} = (4 \text{ V})/(12 \text{ Ω}) = \frac{1}{3}$ A. The 6-Ω resistor carries twice as much current as the 12-Ω resistor.

Example 25-2 Find the equivalent resistance between points a and b for the combination of resistors shown in Figure 25-5.

This combination of resistors may look complicated, but it is easily treated step by step. We first find the equivalent resistance of the 4- and 12-Ω resistors, which are in parallel. Using Equation 25-4, we obtain

$$\frac{1}{R_{eq}} = \frac{1}{4 \text{ Ω}} + \frac{1}{12 \text{ Ω}} = \frac{4}{12 \text{ Ω}} = \frac{1}{3 \text{ Ω}}$$

or $R_{eq} = 3$ Ω. In Figure 25-6 we have replaced the 4- and 12-Ω resistors by their equivalent, a 3-Ω resistor. Since this 3-Ω resistor is in series with a 5-Ω resistor, the equivalent resistance of the bottom branch of this combination is 8 Ω. We are now left with an 8-Ω resistor in parallel with a 24-Ω resistor. The equivalent resistance of these two parallel resistors is again found from Equation 25-4. The result is $R_{eq} = 6$ Ω for the complete combination.

Question

1. A common mistake is to generalize Equation 25-6 to $R_1 R_2 R_3 / (R_1 + R_2 + R_3)$ for the equivalent resistance of three parallel resistors. What simple argument can you give to show that such an expression could not be correct?

25-2   Kirchhoff's Rules

Although the methods discussed in the previous section for replacing series and parallel combinations of resistors by their equivalent resistances are very useful for simplifying many combinations of resistors, they are not sufficient for the analysis of many simple circuits, particularly those containing more than one battery. For example, the two resistors R_1 and R_2 in the circuit of Figure 25-7 cannot be replaced by an equivalent resistance for parallel resistors because the potential difference is not the same across each *resistor*. Two simple rules, called *Kirchhoff's rules*, are applicable to any dc circuit containing batteries and resistors connected in any way:

1. The algebraic sum of the increases or decreases in potential around any closed circuit loop must be zero.

2. At any junction point in a circuit where the current can divide, the sum of the currents into the junction must equal the sum of the currents out of the junction.

Rule 1, called the loop rule, follows from the simple fact that in the steady state the potential difference between any two points is con-

Figure 25-5
Combination of resistors for Example 25-2.

Figure 25-6
Figure 25-5 simplified by replacing the parallel combination of 4- and 12-Ω resistances with the equivalent 3-Ω resistor.

Figure 25-7
A simple circuit that cannot be analyzed by replacing series or parallel resistors by their equivalents. The potential differences across the resistors R_1 and R_2 are not equal because of the emf $\mathcal{E}_2$. Therefore these resistors cannot be replaced by an equivalent resistance.

Kirchhoff's rules

stant. As we move around a circuit loop, the potential may decrease or increase as we pass through a resistor or battery, but when we have completely traversed a loop and arrive back at our starting point, the net change in the potential must be zero. There are several ways of applying this rule in practice. If we call the potential increases positive and the decreases negative, we can put all the changes on one side of an equation and set the algebraic sum of these changes equal to zero. Alternatively, we can list the potential increases separately from the potential decreases and set the sum of the increases equal to the sum of the decreases.

Rule 2 follows from the conservation of charge. This rule is needed for multiloop circuits containing points where the current can divide, as in Figure 25-8. We have labeled the currents in the three wires I_1, I_2, and I_3. In a time interval Δt, charge $I_1 \Delta t$ flows into the junction from the left. In the same time, charges $I_2 \Delta t$ and $I_3 \Delta t$ flow out of the junction to the right. Since there is no way that charge can originate at this point or collect there, conservation of charge implies rule 2, which for this case gives

$$I_1 = I_2 + I_3 \qquad\qquad 25\text{-}9$$

In this section we shall illustrate the application of Kirchhoff's laws to various dc circuits. We begin with a simple circuit with one battery and one loop and then give examples of more complicated circuits containing more than one battery and more than one loop.

Figure 25-9 shows a simple circuit consisting of a battery of emf $\mathcal{E}$, internal resistance r, and external resistance R across the terminals of the battery. The current I is in the direction shown. The connecting wires indicated by straight lines in this diagram are assumed to have negligible resistance. Any two points along such a line are therefore always at the same potential. If we start from point a and move around the circuit with the current (clockwise in this case), we first encounter a potential drop IR from point b to point c. We then encounter a potential increase $\mathcal{E}$ from point d to point e across the emf of the battery and finally another potential drop Ir from point e to point a across the internal resistance of the battery. Kirchhoff's first rule then gives

$$\mathcal{E} - IR - Ir = 0$$

or

$$I = \frac{\mathcal{E}}{r + R} \qquad\qquad 25\text{-}10$$

Note that the power output $\mathcal{E}I$ of the emf goes partly into Joule heating of the external resistor and partly into Joule heating of the battery because of its internal resistance. We can see this by multiplying each term in Equation 25-10 by I, giving

$$\mathcal{E}I = I^2 r + I^2 R \qquad\qquad 25\text{-}11$$

Example 25-3 An 11-Ω resistance is connected across a battery of emf 6 V and internal resistance 1 Ω.* Find the current, the terminal voltage of the battery, the power delivered by the emf, and the power delivered to the external resistance.

---

* Real batteries usually have internal resistance of the order of a few hundredths of an ohm. To simplify calculations we may exaggerate the value of the internal resistance, as in this example; in other examples we may ignore it completely.

The Bettman Archive

Gustav Robert Kirchhoff (1824–1887).

Figure 25-8
If charge cannot originate or collect at the junction point a, conservation of charge implies that the current I_1 into the point must equal the current $I_2 + I_3$ out of the point.

Figure 25-9
Simple circuit containing a battery of emf $\mathcal{E}$, an internal resistance r, and an external resistance R.

From Equation 25-10 the current is

$$I = \frac{6 \text{ V}}{1 \text{ }\Omega + 11 \text{ }\Omega} = 0.5 \text{ A}$$

The terminal voltage across the battery, $V_a - V_d$, is equal to the voltage drop $V_b - V_c$ across the external resistance, which is $IR = (0.5 \text{ A})(11) = 5.5$ V. This differs from the emf of 6 V of the battery because of the potential drop across the internal resistance, $Ir = (0.5 \text{ A})(1 \text{ }\Omega) = 0.5$ V. The power delivered by the emf is $P = (6 \text{ V})(0.5 \text{ A}) = 3$ W. The power delivered to the external resistance is $I^2R = (0.5 \text{ A})^2(11 \text{ }\Omega) = 2.75$ W. The other 0.25 W of power is dissipated as heat in the internal resistance of the battery.

Example 25-4 For a battery of given emf and internal resistance r, what value of external resistance R should be placed across the terminals to obtain the greatest Joule heating in R?

The resistance R is sometimes called the *load resistance*. The power into R is I^2R, where I is given by Equation 25-10. Thus the power is

$$P = I^2R = \frac{\mathcal{E}^2}{(r + R)^2} R \qquad 25\text{-}12$$

Figure 25-10 shows a sketch of P versus R. The maximum value of P occurs when $R = r$. Thus the maximum power is obtained when the load resistance equals the internal resistance. A similar result also occurs for more complicated ac circuits and is known as *impedance matching*. It is left as an applied calculus problem for the student to derive the result that P is a maximum when $R = r$.

Example 25-5 Find the current in the circuit shown in Figure 25-11.

In this circuit we have two seats of emf and three external resistors. We cannot predict the direction of the current unless we know which emf is greater, but we do not have to know the direction of the current before solving the problem. We can assume any direction and solve the problem with that assumption. If the assumption is incorrect, we shall get a negative number for the current, indicating that its direction is opposite that assumed.

Let us assume I to be clockwise as indicated in the figure. For this example we shall apply Kirchhoff's first rule by finding the potential increases and setting their sum equal to the sum of the potential decreases. The potential drops and increases as we traverse the circuit in the assumed direction of the current, starting at point a, are given in Table 25-1. We encounter a potential drop traversing one of the emfs and an increase traversing the other. Kirchhoff's first rule gives

$$\mathcal{E}_1 = IR_1 + IR_2 + \mathcal{E}_2 + Ir_2 + IR_3 + Ir_1 \qquad 25\text{-}13$$

Solving for the current I, we obtain

$$I = \frac{\mathcal{E}_1 - \mathcal{E}_2}{R_1 + R_2 + R_3 + r_1 + r_2} \qquad 25\text{-}14$$

Note that if $\mathcal{E}_2$ is greater than $\mathcal{E}_1$, we get a negative number for the current I, indicating that we have chosen the wrong direction for I. For $\mathcal{E}_2$ greater than $\mathcal{E}_1$ the current is in the counterclockwise direction. On the other hand, if $\mathcal{E}_1$ is the greater emf, we get a positive number for I, indicating that its assumed direction is correct.

Figure 25-10
Plot of the power into resistor I^2R versus the resistance R. The power is a maximum when the load resistance equals the internal resistance of the battery.

Impedance matching

Figure 25-11
Circuit for Example 25-5.

Table 25-1

$a \rightarrow b$	Drop IR_1
$b \rightarrow c$	Drop IR_2
$c \rightarrow d$	Drop $\mathcal{E}_2$
$d \rightarrow e$	Drop Ir_2
$e \rightarrow f$	Drop IR_3
$f \rightarrow g$	Increase $\mathcal{E}_1$
$g \rightarrow a$	Drop Ir_1

Let us assume for this example that ε_1 is the greater emf and that the current is clockwise, as indicated. A charge ΔQ moving through battery 2 from point c to point d loses energy $\varepsilon_2 \Delta Q$. In this battery, electric energy is converted into chemical energy and stored. Battery 2 is *charging*. We can account for the energy balance in this circuit by multiplying each term in Equation 25-13 by the current I. Then the term on the left, $\varepsilon_1 I$, is the rate at which battery 1 puts energy into the circuit. This energy comes from internal chemical energy of the battery. The first term on the right, $I^2 R_1$, is the rate of production of Joule heat in resistor R_1. There are similar terms for each of the other resistors. The term $\varepsilon_2 I$ is the rate at which electric energy is converted into chemical energy in the second battery.

Example 25-6 The quantities in the circuit of Example 25-5 have the magnitudes $\varepsilon_1 = 12$ V, $\varepsilon_2 = 4$ V, $r_1 = r_2 = 1\ \Omega$, $R_1 = R_2 = 5\ \Omega$, $R_3 = 4\ \Omega$. Find the potentials at the points indicated in Figure 25-12, assuming that the potential at point f is zero, and discuss the energy balance in the circuit.

Since only potential differences are important, any point in a circuit can be chosen to have zero potential. In this example we have chosen point f in Figure 25-12. The potentials of the other points are then found relative to point f.* This procedure often simplifies the analysis of a problem.

We first find the current in the circuit. From Equation 25-14 we have

$$I = \frac{12 - 4}{5 + 5 + 4 + 1 + 1} = \frac{8\ \text{V}}{16\ \Omega} = 0.5\ \text{A}$$

Figure 25-12
Circuit for Example 25-6.

We can now find the potentials at the indicated points relative to zero potential at point f. Since, by definition, the emf maintains a constant potential difference $\varepsilon_1 = 12$ V between point g and point f, the potential at point g is 12 V. The potential at point a is less than at g by the potential drop $Ir_1 = 0.5(1) = 0.5$ V. Thus the potential at point a is $12 - 0.5 = 11.5$ V. Similarly the potential drops across the 5-Ω resistors R_1 and R_2 are each $IR_1 = 0.5(5) = 2.5$ V. The potential at point b is then $11.5 - 2.5 = 9$ V, and that at c is 6.5 V. The potential drop across ε_2 is 4 V. Thus point d is at a potential of 2.5 V. Since the drop across the 1-Ω resistance r_2 is 0.5 V, the potential at e is 2 V. The potential drop across the 4-Ω resistance R_3 is $IR_3 = 2$ V. This gives zero for the potential at f, consistent with our original assumption.

The power delivered by the emf ε_1 is $\varepsilon_1 I = 12(0.5) = 6$ W. The power into the internal resistance of battery 1 is $I^2 r_1 = 0.5^2(1) = 0.25$ W. Thus the power delivered by the battery to the external circuit is $6 - 0.25 = 5.75$ W. This also equals $V_1 I$, where $V_1 = V_a - V_f = 11.5$ V is the terminal voltage of that battery. The total power into the external resistances in the circuit is $I^2 R_1 + I^2 R_2 + I^2 R_3 = I^2(R_1 + R_2 + R_3) = 0.5^2(5 + 5 + 4) = 3.5$ W. The power into the battery being charged is $(V_c - V_e)I = (6.5 - 2)(0.5) = 2.25$ W. Part of this, $I^2 r_2 = 0.25$ W, is dissipated in the internal resistance, and part, $\varepsilon_2 I = 2$ W, is the rate at which energy is stored in that battery.

---

* In more complicated circuits, we may wish to keep some point such as point f at a fixed potential relative to the earth even though some of the elements of the circuit change. This can be done by connecting that point to the earth, i.e., grounding it. Since the earth is such a large conductor, it has nearly infinite capacitance. Thus its potential remains essentially constant and is usually chosen to be zero.

Note that the terminal voltage of the battery that is being charged in Example 25-6 is $V_c - V_e = 4.5$ V, which is greater than the emf of the battery. Because of its internal resistance, a battery is not completely reversible. If the same 4-V battery were to deliver 0.5 A to an external circuit, its terminal voltage would be 3.5 V (again assuming the value of 1 for its internal resistance). If the internal resistance is very small, the terminal voltage of the battery is nearly equal to its emf, whether the battery is delivering current to an external circuit or being charged. Some real batteries, such as the storage batteries used in automobiles, are nearly reversible and can easily be recharged; others (such as dry cells) are not. If you attempt to recharge a dry cell by driving current through it from its positive to its negative terminal, most, if not all, of the energy expended goes into heat rather than into chemical energy of the dry cell.

We now consider an example of a circuit which contains more than one loop. In such circuits we need to apply Kirchhoff's second rule at junction points where the current splits into two or more parts.

Example 25-7 Find the current in each part of the circuit shown in Figure 25-13.

Let I be the current through the 18-V battery in the direction shown. At point b, this current divides into currents I_1 and I_2, as shown. Until we know the solution for the currents, we cannot be sure of their direction. For example, we need to know whether point b or point c is at the higher potential in order to know the direction of current through the 6-Ω resistor. The current directions indicated are merely guesses. Applying the junction rule (rule 2) to the point b, we obtain

$$I = I_1 + I_2$$

There are three possible loops for applying rule 1, loops $abcd$, $befc$, and $abefcd$. We need only two more equations to determine the three unknown currents. Equations for any two of the loops will be sufficient. (The third loop will then give redundant information.) Traversing the loop $abcd$ in the clockwise direction and applying Kirchhoff's first rule gives

$$(12\ \Omega)I + (6\ \Omega)I_1 = 18\ \text{V}$$

or

$$2I + I_1 = 3\ \text{A} \qquad\qquad 25\text{-}15$$

Applying rule 1 to loop $befc$, we obtain another equation,

$$(1\ \Omega)I_2 + (1\ \Omega)I_2 = 12\ \text{V} + (6\ \Omega)I_1$$

Note that in moving from c to b we encounter a voltage increase because the (assumed) current I_1 is in the opposite direction. Combining terms and substituting $I - I_1$ for I_2 in this equation, we obtain

$$2(I - I_1) = 12\ \text{A} + 6I_1$$
$$2I = 12\ \text{A} + 8I_1 \qquad\qquad 25\text{-}16$$

Equations 25-15 and 25-16 can be solved for the unknown currents I and I_1. The results are

$$I = 2\ \text{A} \qquad I_1 = -1\ \text{A}$$

Then

$$I_2 = I - I_1 = 2 - (-1) = 3\ \text{A}$$

Figure 25-13
Circuit for Example 25-7.

Our original guess about the direction of I_1 was incorrect. The current through the 6-Ω resistor is in the direction from point c to point b.

Example 25-8 Find the currents I, I_1, and I_2 for the circuit in Figure 25-14.

This circuit is the same as in Figure 25-7, with $\varepsilon_1 = 12$ V, $\varepsilon_2 = 5$ V, $R_1 = 4\ \Omega$, $R_2 = 2\ \Omega$, and $R_3 = 3\Omega$. We choose directions for the currents as shown in the figure, and we obtain from the junction rule:

$$I = I_1 + I_2$$

Using this to eliminate I and applying the loop rule to the outer loop, we get

$$12 - 2I_1 - 5 - 3(I_1 + I_2) = 0 \qquad \text{25-17}$$

Applying the loop rule to the first loop on the left gives

$$12 - 4I_2 - 3(I_1 + I_2) = 0 \qquad \text{25-18}$$

When we solve Equations 25-17 and 25-18 for the currents I_1 and I_2, we obtain $I_1 = 1.5$ A and $I_2 = 0.5$ A. The total current through the 12-V emf is then $I = 1.5$ A $+ 0.5$ A $= 2.0$ A.

Figure 25-14
Circuit for Example 25-8.

25-3 RC Circuits

In this section we consider circuits containing capacitors, in which the current is not steady but varies with time. Figure 25-15 shows a capacitor with an initial charge $+Q_0$ on the upper plate and $-Q_0$ on the lower plate. It is connected to a resistor R and a switch S, which is open to prevent the charge from flowing through the resistor. The potential difference across the capacitor is initially $V_0 = Q_0/C$, where C is the capacitance. Since there is no current with the switch open, there is no potential drop across the resistor. Thus there is also a potential difference V_0 across the switch.

We close the switch at time $t = 0$. Since there is now a potential difference across the resistor, there must be a current in it. The initial current is $I_0 = V_0/R$. The current is due to the flow of charge from the positive plate to the negative plate through the resistor, and so after a time the charge on the capacitor is reduced. The rate of decrease of the charge equals the current. If Q is the charge on the capacitor at any time, the current at that time is

$$I = -\frac{dQ}{dt} \qquad \text{25-19}$$

We can write an equation for the charge and current as functions of time by applying Kirchhoff's first rule to the circuit after the switch is closed. Traversing the circuit in the direction of the current, we encounter a potential drop IR across the resistor and a potential increase Q/C across the capacitor. Kirchhoff's first rule gives

$$IR = \frac{Q}{C} \qquad \text{25-20}$$

where both Q and I are functions of time and are related by Equation

Figure 25-15
(a) Parallel-plate capacitor in series with a switch and a resistor R. (b) Circuit diagram for (a).

25-19. Substituting $-dQ/dt$ for I in this equation, we have

$$-R\frac{dQ}{dt} = \frac{Q}{C}$$

or

$$\frac{dQ}{dt} = -\frac{1}{RC}Q \qquad\qquad 25\text{-}21$$

The function $Q(t)$ whose rate of change is proportional to itself is the exponential. The solution of Equation 25-21 is

$$Q(t) = Q_0 e^{-t/RC} = Q_0 e^{-t/t_c} \qquad\qquad 25\text{-}22$$

where the time

$$t_c = RC \qquad\qquad 25\text{-}23$$

Time constant

is the time in which the charge decreases to $1/e$ of its original value; it is called the *time constant* of the circuit. After a time t_c the charge is $Q = Q_0 e^{-1} = 0.37 Q_0$. After a time $2t_c$, the charge is $Q = Q_0 e^{-2} = 0.135 Q_0$, and so forth. We can understand the time constant as follows. Originally, the charge on the capacitor is Q_0, and its rate of change as given by Equation 25-21 is $-Q_0/RC$. If this rate of change were constant in time, the charge would decrease to zero in a time $t = t_c$, as indicated by the dashed line in Figure 25-16. However, the rate of change of the charge is not constant but proportional to the charge itself. As the charge decreases, the magnitude of the rate of change decreases. The function $Q(t)$ is shown by the solid curve in Figure 25-16. After a long time both the value of Q and its slope, which is proportional to Q, approach zero as shown.

The current is obtained from Equation 25-19 by differentiating Equation 25-22:

$$I = -\frac{dQ}{dt} = \frac{Q_0}{RC} e^{-t/RC} = I_0 e^{-t/RC} \qquad\qquad 25\text{-}24$$

where $I_0 = Q_0/RC = V_0/R$ is the initial current. The current also decreases exponentially with time and falls to $1/e$ of its initial value after a time $t_c = RC$.

Figure 25-17 shows a circuit used for charging a capacitor, which we assume to be originally uncharged. The switch, originally open, is closed at time $t = 0$. Charge immediately begins to flow into the capacitor. If the charge on the capacitor at some time is Q and the current in the circuit is I, Kirchhoff's first rule gives

$$\varepsilon = V_R + V_C = IR + \frac{Q}{C} \qquad\qquad 25\text{-}25$$

In this circuit, the current equals the rate of increase of charge on the capacitor:

$$I = \frac{dQ}{dt}$$

Substituting this into Equation 25-25 gives

$$\varepsilon = R\frac{dQ}{dt} + \frac{Q}{C} \qquad\qquad 25\text{-}26$$

At time $t = 0$ the charge is zero and the current is $I_0 = \varepsilon/R$. The charge

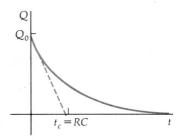

Figure 25-16
Plot of charge Q on capacitor versus time t for the circuit in Figure 25-15. The charge decreases by a factor $1/e$ in the time $t_c = RC$. The time constant t_c is also the time in which the capacitor would be fully discharged if the discharge rate were constant, as indicated by the dashed line.

Figure 25-17
Circuit for charging a capacitor to a potential difference ε.

then increases and the current decreases, as can be seen from Equation 25-25. The charge reaches a maximum value of $C\varepsilon$, again from Equation 25-25, when the current I equals zero. Figures 25-18 and 25-19 show the behavior of the charge versus time and the current versus time. The analytical expressions for $Q(t)$ and $I(t)$ are obtained by solving the differential equation 25-26:

$$Q(t) = C\varepsilon(1 - e^{-t/RC}) \qquad\qquad 25\text{-}27$$

$$I = \frac{dQ}{dt} = \frac{\varepsilon}{R} e^{-t/RC} \qquad\qquad 25\text{-}28$$

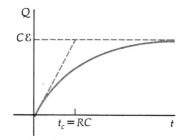

Figure 25-18
Plot of charge on the capacitor versus time for the circuit in Figure 25-17. After time $t = t_c$ the charge is 63 percent of its final value of $C\varepsilon$. If the charging rate were constant and equal to its initial value, the capacitor would be fully charged after time $t = t_c$, as indicated by the dashed line.

Example 25-9 A 6-V battery of negligible internal resistance is used to charge a 2-μF capacitor through a 100-Ω resistor. Find the initial current, the final charge, and the time required to obtain 90 percent of the final charge.

The initial current is $I_0 = \varepsilon/R = (6\text{ V})/(100\ \Omega) = 0.06$ A. The final charge is $Q_f = \varepsilon C = (6\text{ V})(2\ \mu\text{F}) = 12\ \mu$C. The time constant for this circuit is $RC = (100\ \Omega)(2\ \mu\text{F}) = 200\ \mu$s. We expect the charge to reach 90 percent of its final value in a time of the order of several time constants. We can find the exact solution from Equation 25-27, using $Q = 0.9\varepsilon C$:

$$Q = 0.9\varepsilon C = \varepsilon C(1 - e^{-t/RC})$$

$$0.9 = 1 - e^{-t/RC}$$

$$e^{-t/RC} = 1 - 0.9 = 0.1$$

$$\ln e^{-t/RC} = -\frac{t}{RC} = \ln 0.1 = \ln 1 - \ln 10 = -\ln 10 = -2.3$$

since $\ln 0.1 = \ln \frac{1}{10} = \ln 1 - \ln 10$ and $\ln 1 = 0$. Thus

$$t = RC \ln 10 = 2.3RC = 2.3(200\ \mu\text{s}) = 460\ \mu\text{s}$$

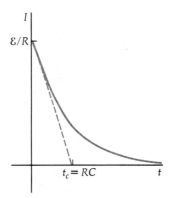

Figure 25-19
Plot of current versus time for the circuit in Figure 25-17. The current decreases by a factor $1/e$ in the time $t = t_c = RC$.

During the charging process a total charge $Q_f = \varepsilon C$ flows through the battery. The battery does work

$$W = Q_f\varepsilon = \varepsilon^2 C$$

The energy stored in the capacitor is just half this amount. By Equation 23-19,

$$U = \tfrac{1}{2}QV = \tfrac{1}{2}Q_f\varepsilon = \tfrac{1}{2}\varepsilon^2 C$$

We now show that the other half of the energy provided by the battery goes into Joule heat in the resistor. The rate of energy put into the resistor is

$$\frac{dW_R}{dt} = I^2 R$$

Using Equation 25-28 for the current, we have

$$\frac{dW_R}{dt} = \left(\frac{\varepsilon}{R} e^{-t/RC}\right)^2 R = \frac{\varepsilon^2}{R} e^{-2t/RC}$$

We find the total heat produced by integrating from $t = 0$ to $t = \infty$:

$$W_R = \int_{t=0}^{t=\infty} dW_R = \int_0^\infty \frac{\varepsilon^2}{R} e^{-2t/RC}\ dt$$

The integration can be done by substituting $x = 2t/RC$. Then

$$dt = \frac{RC}{2}\,dx$$

$$W_R = \frac{\mathcal{E}^2}{R}\frac{RC}{2}\int_0^\infty e^{-x}\,dx = \tfrac{1}{2}\mathcal{E}^2 C$$

since the integral is 1. This answer is independent of the resistance R. When a capacitor is charged by a constant emf, half the energy provided by the battery is stored in the capacitor and half goes into heat, independent of the resistance. (This heat energy includes power into the internal resistance of the battery.)

25-4 Ammeters, Voltmeters, and Ohmmeters

We turn now to the consideration of the measurement of electrical quantities in dc circuits. Devices which measure current, potential difference, and resistance are called ammeters, voltmeters, and ohmmeters, respectively. To measure the current through the resistor in the simple circuit in Figure 25-20 we place an ammeter in series with the resistor, as indicated in the figure. Since the ammeter has some resistance, the current in the circuit is changed when the ammeter is inserted. Ideally, the ammeter should have a very small resistance so that only a small change will be introduced in the current to be measured. The potential difference across the resistor is measured by placing a voltmeter across the resistor in parallel with it, as shown in Figure 25-21. The voltmeter reduces the resistance between points a and b, thus increasing the total current in the circuit and changing the potential drop across the resistor. An ideal voltmeter has a very large resistance, to minimize its effect on the circuit.

The principal component of an ammeter or voltmeter is a galvanometer, a device which detects a small current through it. A common type, the *d'Arsonval galvanometer*, consists of a coil of wire free to turn, an indicator of some kind, and a scale. The galvanometer is designed so that the scale reading is proportional to the current in the galvanometer. The galvanometer operates on the principle that a coil carrying a current in a magnetic field experiences a torque which is proportional to the current. This torque rotates the coil until it is balanced by the restoring torque provided by the mechanical suspension of the coil. We shall study the effects of magnetic field on current-carrying wires in Chapter 26; here we merely needed to know that such a torque exists, that it is proportional to the current in the coil, and that with proper design of the magnet the torque does not depend on the angle of rotation of the coil. Figure 25-22 shows a sketch of a galvanometer coil in a magnetic field.

Since the restoring torque of the suspension is proportional to the angle of rotation of the coil, the equilibrium angle of rotation will be proportional to the current in the coil. The resistance of the galvanometer and the current needed to produce full-scale deflection are the two parameters important for the construction of an ammeter or voltmeter from a galvanometer. Typical values of these parameters for a portable pivoted-coil laboratory galvanometer are $R_g = 20\ \Omega$ and $I_g = 0.5$ mA. The voltage drop across a galvanometer with these parameters

Figure 25-20
To measure the current in the resistor R an ammeter Ⓐ is placed in series with the resistor.

Figure 25-21
To measure the voltage drop across a resistor, a voltmeter Ⓥ is placed in parallel with the resistor.

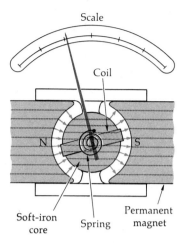

Figure 25-22
D'Arsonval galvanometer.
When the coil carries a current, the magnet exerts a
torque on the coil proportional to the current, causing
the coil to twist. The deflection read on the scale is proportional to the current in the
coil.

is thus $I_g R_g = 10^{-2}$ V for full-scale deflection. To construct an ammeter
from a galvanometer, we place a small resistance, called a *shunt resistor,*
in parallel with the galvanometer. Since the shunt resistance is usually
much smaller than the resistance of the galvanometer, most of the current is carried by the shunt and the effective resistance of the ammeter
is much smaller than the galvanometer resistance.

Resistors are added in series with a galvanometer to construct a voltmeter. Figure 25-23 illustrates the construction of an ammeter and voltmeter from a galvanometer. The choice of the appropriate resistors for
the construction of an ammeter or voltmeter from a galvanometer is best
illustrated by example.

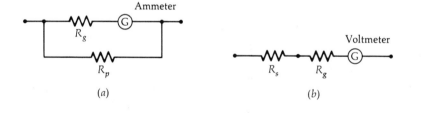

Figure 25-23
(*a*) An ammeter consists of a
galvanometer Ⓖ and a small
parallel resistance R_p, called a
shunt resistance. (*b*) A voltmeter consists of a gal-
vanometer and a large series
resistance R_s.

Example 25-10 Using a galvanometer with a resistance of 20 Ω, for
which 5×10^{-4} A gives full-scale deflection, design an ammeter which
will read full scale when the current is 5 A.

Ammeter

Since the total current through the ammeter must be 5 A when the
current through the galvanometer is just 5×10^{-4} A, most of the current
must go through the shunt resistor. Let R_p be the shunt resistance and
I_p be the current through the shunt. Since the galvanometer and shunt
are in parallel, we have $I_g R_g = I_p R_p$ and $I_p + I_g = 5$ A or

$$I_p = 5 \text{ A} - I_g = 5 - 5 \times 10^{-4} \text{ A} \approx 5 \text{ A}$$

Thus the value of the shunt resistor should be

$$R_p = \frac{I_g}{I_p} R_g = \frac{5 \times 10^{-4}}{5} \times 20 = 2 \times 10^{-3} \ \Omega$$

Since the resistance of the shunt is so much smaller than the resistance
of the galvanometer, the effective resistance of the parallel combination
is approximately equal to the shunt resistance.

Example 25-11 Using the same galvanometer as in Example 25-10, design a voltmeter which will read 10 V at full-scale deflection.

Voltmeter

Let R_s be the value of a resistor in series with the galvanometer. We want to choose R_s so that a current of $I_g = 5 \times 10^{-4}$ A gives a potential drop of 10 V. Thus

$$I_g(R_s + R_g) = 10 \text{ V}$$

$$R_s + R_g = \frac{10 \text{ V}}{5 \times 10^{-4} \text{ A}} = 2 \times 10^4 \ \Omega$$

$$R_s = 2 \times 10^4 - R_g = 2 \times 10^4 - 20 = 19,980 \ \Omega \approx 20 \text{ k}\Omega$$

Figure 25-24 shows an ohmmeter consisting of a battery, a galvanometer, and a resistor, which can be used to measure resistance. The resistance R_s is chosen so that the galvanometer reads full scale when the terminals a and b are shorted (touched together). Thus full scale on the galvanometer is marked zero resistance. When the terminals are connected across an unknown resistance R, the current is less than I_g and the galvanometer reads less than full scale. The current in this case is

Ohmmeter

$$I = \frac{\varepsilon}{R + R_s + R_g} \qquad\qquad 25\text{-}29$$

Since this depends on R, the scale can be calibrated in terms of the resistance measured, from zero at full scale to infinite resistance at zero deflection. Since the calibration of the scale is far from linear and depends on the constancy of the emf of the battery, such an ohmmeter is not a high-precision instrument, but it is quite useful for making quick, rough determinations of resistance.

(a)

(b)

Figure 25-24
(a) Ohmmeter consisting of a battery in series with a galvanometer and a resistor R_s chosen so that the galvanometer reads full-scale deflection when points a and b are shorted. (b) Calibration of the galvanometer scale.

Some caution must be exercised in the use of an ohmmeter since it sends a current through the resistance to be measured. For example, consider an ohmmeter with a 1.5-V battery and a galvanometer similar to that in the previous examples. The series resistance needed is

$$I_g(R_s + R_g) = 1.5 \text{ V}$$

or

$$R_s = \frac{1.5}{5 \times 10^{-4}} - R_g = 3000 - 20 = 2980 \ \Omega$$

Suppose we were to use the ohmmeter to measure the resistance of a more sensitive galvanometer which gives a full-scale reading with a current of 10^{-5} A and has a resistance of about 20 Ω. When the terminals a and b are placed across this more sensitive galvanometer, the current will be just slightly less than 5×10^{-4} A because the total resistance is 3020 Ω, which is just slightly more than 3000 Ω. Such a current, about 50 times that needed to produce full-scale deflection, would ruin a sensitive galvanometer.

Questions

2. Under what conditions might it be advantageous to use a galvanometer which is less sensitive, i.e., requires a greater current I_g for full-scale deflection, than the one discussed in Examples 25-10 and 25-11?

3. When the series resistance R_s is properly chosen for a particular emf ε of an ohmmeter, any value of resistance from zero to infinity can be measured. Why are there different scales on a practical ohmmeter for different ranges of resistance to be measured?

25-5   The Wheatstone Bridge

An accurate method of measuring resistance uses a circuit known as a *Wheatstone bridge* (Figure 25-25). In this figure R_x is an unknown resistance, and R_1, R_2, and R_4 are presumed known. The galvanometer is used as a null detector. The resistances R_1 and R_2 are varied until there is no current in the galvanometer; the bridge is then said to be balanced. Under this condition, the potential at point a is the same as that at point b. Thus the potential drop across R_x must equal that across R_1:

$$I_2 R_x = I_1 R_1$$

Similarly, the drop across R_2 must equal that across R_4:

$$I_2 R_4 = I_1 R_2$$

We can eliminate the currents from these two equations by dividing one equation by the other, and then we can solve for the unknown resistance R_x:

$$\frac{R_x}{R_4} = \frac{R_1}{R_2}$$

or

$$R_x = R_4 \frac{R_1}{R_2} \qquad\qquad 25\text{-}30$$

In practice, a large variable resistance is placed in series with the galvanometer to reduce its sensitivity and protect it when the bridge is unbalanced. When the bridge is nearly balanced, the series resistance is decreased and eventually shorted out for maximum sensitivity of the galvanometer. Often a 1-m wire is used for the resistors R_1 and R_2 (Figure 25-26). Point a is a sliding contact, which is varied until the bridge is balanced. If the wire is uniform, the ratio R_1/R_2 is just the ratio of the lengths L_1/L_2. The measurement does not depend on the emf used to supply the current.

Question

4. What are the advantages and disadvantages of a Wheatstone bridge compared with an ohmmeter?

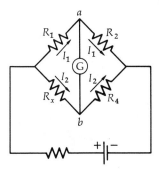

Figure 25-25
Wheatstone-bridge circuit for measuring an unknown resistance R_x in terms of known resistances R_1, R_2, and R_4.

Figure 25-26
Slide-wire Wheatstone bridge. The resistances R_1 and R_2 are proportional to the lengths L_1 and L_2. Their ratio is varied by moving a sliding contact along the wire until the bridge is balanced.

Exponential Growth

Albert A. Bartlett
University of Colorado at Boulder

If a town with a population of 10,000 has a steady annual population growth of 5 percent, how large will the population be in 10, 30, and 100 y? If I invest \$100 in an account that pays 15 percent annual interest (compounded continuously), how large will the account be in 10, 30, and 100 y? These questions are real-life applications of what may be the single most important differential equation:

$$\frac{dN}{dt} = kN \tag{1}$$

This equation describes the situation in which the rate of change of a quantity dN/dt is proportional to the size N of the quantity. In our first example N is the population of the town and $k = 0.05\ y^{-1}$; in the second example N is the number of dollars in the account and $k = 0.15\ y^{-1}$. In both cases t is the time in years. The meaning of k can be seen by rearranging Equation 1 to read

$$k = \frac{1}{N}\frac{dN}{dt} \tag{2}$$

Thus k is the fractional change in N, dN/N, per unit time dt. It follows that $100k$ is P, the percent growth per unit time. These two examples are cases of steady growth in which the quantity always increases by a fixed fraction (for example, 5 percent) in a fixed length of time (each year).

The solution of Equation 1 is the exponential function

$$N = N_0 e^{kt} \tag{3}$$

where e is the base of natural logarithms ($2.718 \cdots$) and N_0 is the size of N at the time $t = 0$. If $k > 0$, Equation 3 describes exponential growth (steady growth), and if $k < 0$, it describes exponential decay. Exponential decay occurs in Newton's law of cooling (Chapter 18), where N is the temperature difference between an object and its surroundings; in the study of RC circuits (Chapter 25), where N is the charge on the capacitor; in LR circuits (Chapter 28), where N is the current in the circuit; and in radioactive decay (Chapter 37), where N is the number of radioactive nuclei which have not yet decayed. Here we focus our attention on exponential or steady growth, where k is a positive constant. Exponential growth occurs whenever the rate of increase in a quantity is proportional to that quantity.

In our two examples in the first paragraph, the town's population will be

$$N = 10{,}000\, e^{0.05t}$$

and the size of the bank account will be

$$N = 100\, e^{0.15t}$$

where t is the time in years.

Using Equation 3 one can show that for a given value of k, a constant time T_c is required for N to increase in size from its present value N_0 to CN_0; where C is a constant:

$$T_c = \frac{1}{k}\ln C \tag{4}$$

The time for N to double in size ($C = 2$) is

$$T_2 = \frac{1}{k}\ln 2 = \frac{0.693}{k} \tag{5}$$

Table 1

Doubling times for various rates of steady growth

k, y^{-1}	P, $\%/\text{y}$	T_2, y
0.01	1	69.3
0.02	2	34.7
0.03	3	23.1
0.05	5	13.9
0.07	7	9.90
0.10	10	6.93
0.15	15	4.62
0.20	20	3.47
0.30	30	2.31

which can be expressed as

$$T_2 = \frac{69.3}{P} \approx \frac{70}{P} \qquad\qquad 6$$

where $P = 100k$ is the percent growth per unit time. Table 1 shows doubling times for different steady rates of growth calculated from Equation 6.

From Table 1 we see that a quantity growing steadily at 5 percent per year will double in size in 13.9 y (steady growth at a rate of 5 percent per month will cause doubling to take place in 13.9 months). A quantity growing steadily at 15 percent per year will double in 4.62 y. We should develop the habit of doing the mental arithmetic of Equation 6 every time we see a percent growth rate in the news or elsewhere. People who do not grasp the meaning of 15 percent annual inflation will understand when you tell them that this means that prices will double every 4.6 years!

Equation 3 can be rewritten in a very useful form using T_2:

$$N = N_0 e^{kt} = N_0 2^{t/T_2} \qquad\qquad 7$$

In this form we can estimate the answers to the questions that opened this essay. Since the doubling time for an annual growth rate of 5 percent is roughly 14 y, it can be seen that 10 y is less than one doubling time, 30 y is approximately two doubling times, and 100 y is approximately 7 doubling times. Thus in 10 y the population will increase from 10,000 to roughly 16,000; in 30 y it will have doubled twice (to 40,000); and in 100 y it will have doubled 7 times (a factor of $2^7 = 128$), and so the population will be 1,280,000. For a 15 percent annual interest rate compounded continuously ($T_2 = 4.62$ y) the figures are shown in Table 2.

Table 2

The value of $100 deposited at 15% annual interest (compounded continuously) after a time t

t, y	t/T_2	$2^{t/T_2}$	Size of account
10	$10/4.62 = 2.16$	4.48	$448
30	$30/4.62 = 6.49$	90	$9000
100	$100/4.62 = 21.6$	3.27×10^6	$327,000,000

We can estimate the value of large powers of 2 by noting that

$$2^{10} = 1024 \approx 10^3 \qquad\qquad 8$$

For example,

$$2^{22} = 2^{10} \times 2^{10} \times 2^2 \approx 10^3 \times 10^3 \times 4 = 4 \times 10^6$$

Some very important properties of steady growth can be illustrated by the following example. Legend has it that the game of chess was invented by a mathematician who worked for a king. As a reward the mathematician asked the king for the quantity of wheat determined by placing one grain of wheat on the first square of the chessboard, on the next square doubling the one grain to make two, on the next square doubling the two grains to make four, and continuing until the doubling has been done for all the squares (see Table 3). We can use Equation 8 to get an approximate expression for this total number of grains:

$$2^{64} \approx 10^3 \times 10^3 \times 10^3 \times 10^3 \times 10^3 \times 10^3 \times 16 = 1.6 \times 10^{19} \text{ grains}$$

The actual value of 2^{64} is 1.84×10^{19}. This is approximately 500 times the 1980 worldwide harvest of wheat, which may be more wheat than people have harvested in the entire history of the earth. How did we get such a large number? It was simple. We started with one grain and let the number grow until it had doubled a mere 63 times. A study of Table 3 reveals another important point.

Table 3
Grains of wheat on a chessboard

No. of square	No. on square	Total no. on board
1	1	1
2	2	3
3	4	7
4	8	15
5	16	31
6	32	63
. . . .
64	2^{63}	$2^{64} - 1$

The growth in any doubling time is greater than the sum of all of the preceding growth. Thus when we put 32 grains on the sixth square, the 32 is larger than the total of 31 already on the board.

In his speech on energy (April 18, 1977) President Carter said, "in each of those decades [the 1950s and 1960s] more oil was consumed [worldwide] than in all of mankind's previous history." This statement sounds incredible, but a look at Table 3 shows that this is a simple consequence of steady growth with a doubling time of 10 years, i.e., an annual growth rate of 7 percent. This was the growth rate of world oil consumption for nearly 100 years before 1970. This feature of exponential growth is extremely important. Suppose, for example, that the annual growth rate of 7 percent continues and that just as the world's supply of oil is to be exhausted, a new source of oil is discovered equal to the total amount of oil consumed worldwide in the history of mankind. If oil consumption continues to grow at the annual rate of 7 percent, this fantastic new supply will last one doubling time, or 10 years. This example makes it clear that, contrary to what prominent people often suggest, exploration and production of fossil fuels cannot meet the needs of continued growth for very long.

It is interesting to calculate the length of time a given finite resource will last if its rate of use grows steadily (k = a constant) until the last of the resource is used. The rate of use of the resource R will then be

$$R = R_0 e^{kt} \qquad\qquad 9$$

where R_0 is the rate at $t = 0$. If we have a given quantity M_0 of the resource at time $t = 0$, we can find the time T_e when this quantity will be exhausted by integrating the rate $R\ dt$ from time $t = 0$ to $t = T_e$ to find the total amount used in T_e, and setting this equal to M_0:

$$\int_0^{T_e} R\ dt = \frac{R_0}{k}(e^{kT_e} - 1) = M_0$$

Solving for T_e, we obtain

$$T_e = \frac{1}{k} \ln\left(\frac{kM_0}{R_0} + 1\right) \qquad\qquad 10$$

We can use Equation 10 to estimate the lifetime T_e of United States coal reserves for different steady rates of growth of coal production. The U.S. Geological Survey* estimates the Reserve Base of United States coal to be $M_0 = 379 \times 10^9$ tonnes (1 tonne = 10^3 kg). The rate of production in 1978 was $R_0 = 0.6 \times 10^9$ tonnes/y. Using these numbers in Equation 10 we find the results shown in Table 4.

If we have zero growth of coal production, United States coal would last $T_e = M_0/R_0 = 632$ y. This is the basis for news stories which say "at present rates of consumption" United States coal will last about 600 y.** These stories go on then to stress the national urgency of achieving rapid growth of United States coal production, but the stories don't tell how growth reduces T_e (see Table 4). For example, President Gerald Ford called for United States coal production to grow at an annual rate of about 10 percent ($T_e = 42$ y); President Carter called for 5 percent ($T_e = 70$ y); and President Reagan is continuing and enlarging President Carter's program of rapid growth of United States coal exports.*** Part of our problem with oil in the United States in 1981 arises from the fact that for decades the United States was a major exporter of oil. Now we seem to be repeating the process with coal.

The exponential arithmetic that arises so naturally in physics and in many other real-life situations is incredibly important—in part because it is so poorly

Table 4
Lifetime of United States coal reserves for different steady rates of growth of coal production

Annual growth rate, %	Expiration time of the reserve base
0	632
1	199
2	131
3	100
5	70
10	42
15	30

* P. Averitt, Coal Resources of the United States, *U.S. Geological Survey Bulletin* 1412, Jan. 1, 1974.

** *Newsweek Magazine*, July 16, 1979, p. 23.

*** National Geographic Magazine, Feb. 1981, p. 19.

understood by the people who make our policies (both local and national), who follow the dictum that "growth is good," and who never ask how long growth can continue. One can see that the energy shortage is caused by growth in consumption; the energy crisis is caused by our failure to understand the arithmetic and consequences of this growth. In the words of Aldous Huxley, "Facts do not cease to exist because they are ignored."

Let us summarize the points we have made about the arithmetic of steady growth (the exponential function):

1. Steady growth happens whenever the rate of increase of a quantity is proportional to the quantity (Equation 1).

2. It is characterized by a fixed doubling time (Equations 5 and 6).

3. In every doubling time the amount of growth is greater than the total of all of the preceding growth.

4. Growth for a modest number of doubling times gives astronomically large numbers.

5. If consumption of a resource is growing steadily, enormous new supplies of the resource will allow only very short extensions of the period of steady growth.

It has been said that the greatest shortcoming of the human race is our inability to understand the exponential function.*

* These topics have been developed in more detail in *The American Journal of Physics,* vol. 46, September 1978, p. 876; *Journal of Geological Education,* vol. 28, January 1980, p. 4; "The Exponential Function," *The Physics Teacher,* October, November 1976; January, March, April 1977; January to March 1978; January 1979; and "The Forgotten Fundamentals of the Energy Crisis," videotape, the Media Center, University of Colorado, Boulder, Colorado.

Review

A. Define, explain, or otherwise identify:

Series resistors, 696   Galvanometer, 706
Parallel resistors, 697   Ammeter, 707
Kirchhoff's rules, 698   Voltmeter, 708
Load resistance, 700   Ohmmeter, 708
Impedance matching, 700   Wheatstone bridge, 709
Time constant, 704

B. True or false:

1. The effective resistance of two series resistors is always greater than that of either resistor.

2. The effective resistance of two parallel resistors is always less than that of either resistor.

3. To measure the current through a resistor, an ammeter is placed in series with the resistor.

4. To measure the potential drop across a resistor a voltmeter is placed in parallel with the resistor.

5. The time constant $t_c = RC$ is the time needed to discharge a capacitor completely.

6. A Wheatstone bridge is used to measure emf.

Figure 25-27
Exercise 1.

Exercises

Section 25-1, Series and Parallel Resistors

1. (a) Find the equivalent resistance between points a and b in Figure 25-27. (b) If the potential drop between a and b is 12 V, find the current in each resistor.

2. Repeat Exercise 1 for the resistors shown in Figure 25-28.

3. Repeat Exercise 1 for the resistors shown in Figure 25-29.

4. Repeat Exercise 1 for the resistors shown in Figure 25-30.

Figure 25-28
Exercise 2.

Figure 25-29
Exercises 3 and 5.

Figure 25-30
Exercise 4.

5. In Figure 25-29 the current in the 6-Ω resistor is 3 A. (a) What is the potential drop between a and b? (b) What is the current in the 2-Ω resistor?

6. (a) Show that the equivalent resistance between points a and b in Figure 25-31 is R. (b) What would be the effect of adding a resistance R between points c and d?

7. Repeat Exercise 1 for the resistors shown in Figure 25-32.

8. Consider the equivalent resistance of two parallel resistors R_1 and R_2 as a function of the ratio $x = R_2/R_1$. (a) Show that $R_{eq} = R_1 x/(1 + x)$. (b) Sketch R_{eq} as a function of x.

9. The battery in Figure 25-33 has negligible resistance. Find (a) the current in each resistor and (b) the power delivered by the battery.

Figure 25-31
Exercise 6.

Figure 25-32
Exercise 7.

Figure 25-33
Exercise 9.

Section 25-2, Kirchhoff's Rules

10. A 6-V (emf) battery with internal resistance 0.3 Ω is connected to a variable resistance R. Find the current and power delivered by the battery when R is (*a*) 5 Ω and (*b*) 10 Ω. (*c*) What is the maximum power that can be dissipated in the resistance R?

11. A variable resistance R is connected across a potential difference V which remains constant independent of R. At one value $R = R_1$, the current is 6.0 A. When R is increased to $R_2 = R_1 + 10.0$ Ω, the current drops to 2.0 A. Find (*a*) R_1 and (*b*) V.

12. A battery has emf ε and internal resistance r. When a 5.0-Ω resistor is connected across the terminals, the current is 0.5 A. When this resistor is replaced by an 11.0-Ω resistor, the current is 0.25 A. Find the emf ε and internal resistance r.

13. In Figure 25-34 the emf is 6 V and $R = 0.5$ Ω. The rate of Joule heating in R is 8 W. (*a*) What is the current in the circuit? (*b*) What is the potential difference across R? (*c*) What is r?

14. For the circuit shown in Figure 25-35 find (*a*) the current, (*b*) the power delivered or absorbed by each emf, and (*c*) the rate of Joule heat production in each resistor. (Assume the batteries have negligible internal resistance.)

15. In the circuit shown in Figure 25-34 let $\varepsilon = 18.0$ V and $r = 3.0$ Ω. (*a*) Graph the current through R as a function of R for the range $R = 0$ to $R = 6.0$ Ω. (*b*) For the same range of R, graph the power supplied to R as a function of R. (*c*) At what value of R is the power supplied to R a maximum? What is the value of this maximum power?

16. In the circuit shown in Figure 25-36 the batteries have negligible internal resistance, and the ammeter has negligible resistance. (*a*) Find the current through the ammeter. (*b*) Find the energy delivered by the 12-V battery in 3 s. (*c*) Find the total heat produced in that time. (*d*) Account for the difference in your answers to parts (*b*) and (*c*).

17. In the circuit shown in Figure 25-37 the batteries have negligible internal resistance. (*a*) Find the current in each resistor, (*b*) the potential difference between points *a* and *b*, and (*c*) the power supplied by each battery.

18. Repeat Exercise 17 for the circuit shown in Figure 25-38.

Figure 25-34
Exercises 13 and 15.

Figure 25-35
Exercise 14.

Figure 25-36
Exercise 16.

Figure 25-37
Exercise 17.

Figure 25-38
Exercise 18.

Section 25-3, *RC* Circuits

19. A 6-μF capacitor is initially charged to 100 V and then connected across a 500-Ω resistor. (*a*) What is the initial charge on the capacitor? (*b*) What is the initial current just after the capacitor is connected to the resistor? (*c*) What is the time constant of this circuit? (*d*) How much charge is on the capacitor after 6 ms?

20. (a) For Exercise 19, find the initial energy stored in the capacitor. (b) Show that the energy stored in the capacitor is given by $U = U_0 e^{-2t/t_c}$, where U_0 is the initial energy and $t_c = RC$ is the time constant. (c) Make a careful sketch of the energy U in the capacitor versus time t.

21. A 0.10-μF capacitor is given a charge Q_0. After 4 s its charge is observed to be $\frac{1}{2}Q_0$. What is the effective resistance across this capacitor?

22. A 1.0-μF capacitor, initially uncharged, is connected in series with a 10-kΩ resistor and a 5.0-V battery of negligible internal resistance. (a) What is the charge on the capacitor after a very long time? (b) How long does it take for the capacitor to reach 99 percent of its maximum charge?

23. A 1-MΩ resistor is connected in series with a 1.0-μF capacitor and a 6.0-V battery of negligible internal resistance. The capacitor is initially uncharged. After a time $t = t_c = RC$, find (a) the charge on the capacitor, (b) the rate at which the charge is increasing, (c) the current, (d) the power supplied by the battery, (e) the power dissipated in the resistor, and (f) the rate at which the energy stored in the capacitor is increasing.

24. Repeat Exercise 23 for the time $t = 2t_c$.

Section 25-4, Ammeters, Voltmeters, and Ohmmeters

25. A galvanometer has a resistance of 100 Ω. It requires 1 mA to give full-scale deflection. (a) What resistance should be placed in parallel with the galvanometer to make an ammeter that reads 2 A at full-scale deflection? (b) What resistance should be placed in series to make a voltmeter that reads 5 V at full-scale deflection?

26. Sensitive galvanometers can detect currents as small as 1 pA. How many electrons per second does this current represent?

27. A sensitive galvanometer has resistance 100 Ω and requires 1.0 μA of current to produce full-scale deflection. (a) Find the shunt resistance needed to construct an ammeter which reads 1.0 mA full scale. (b) What is the resistance of the ammeter? (c) What resistance would be required to construct a voltmeter reading 3.0 V full scale from this galvanometer?

28. A galvanometer of resistance 100 Ω gives full-scale deflection when its current is 1.0 mA. It is used to construct an ammeter whose full-scale reading is 100 A. (a) Find the shunt resistance needed. (b) What is the resistance of the ammeter? (c) If the shunt resistor consists of a piece of 10-gauge copper wire (diameter 2.59 mm), what should its length be?

29. The galvanometer in Exercise 28 is used with a 1.5-V battery of negligible internal resistance to make an ohmmeter. (a) What resistance R_s should be placed in series with the galvanometer? (b) What resistance R will give half-scale deflection? (c) What resistance will give a deflection of one-tenth full scale?

30. For the ohmmeter in Exercise 29, indicate how the galvanometer scale should be calibrated by representing the galvanometer scale on a straight line of some length L where the end of the line ($x = L$) represents full-scale reading at $R = 0$. Divide the line into 10 equal divisions and indicate values of the resistance at each division.

31. A galvanometer of resistance 100 Ω gives full-scale reading when its current is 0.1 mA. It is to be used in a multirange voltmeter as shown in Figure 25-39, where the connections refer to full-scale readings. Determine R_1, R_2, and R_3.

Figure 25-39
Exercise 31.

1 V   10 V   100 V

32. The galvanometer of Exercise 31 is to be used in a multirange ammeter with full-scale readings as indicated in Figure 25-40. Determine R_1, R_2, and R_3.

Section 25-5, The Wheatstone Bridge

33. In a slide-wire Wheatstone bridge using a uniform wire of length 1 m, R_1 is proportional to the length of wire from 0 to the balance point and R_2 is proportional to the length from the balance point to 100 cm (the end of the wire). If the fixed resistor is $R_4 = 200$ Ω, find the unknown resistance if (a) the bridge balances at the 18-cm mark, (b) the bridge balances at the 60-cm mark, and (c) the bridge balances at the 95-cm mark.

34. In the Wheatstone bridge of Exercise 33, with $R_4 = 200$ Ω, the bridge balances at the 98-cm mark. (a) What is the unknown resistance? (b) What effect would an error of 2 mm have on the measured value of the unknown resistance? (c) How should R_4 be changed so that this unknown resistor will give a balance point nearer the 50-cm mark?

Problems

1. A battery with emf ε and internal resistance r is connected to an external resistance R. Show that (a) the power P supplied by the emf is given by $\varepsilon^2/(r + R)$, (b) the power P_R supplied to the external resistor is $\varepsilon^2 R/(r + R)^2$, and (c) the power P_r dissipated in the battery is $\varepsilon^2 r/(r + R)^2$. Sketch on the same graph P, P_R, and P_r versus R.

2. Show that as R varies in Problem 1, P_R is maximum at $R = r$, and that the maximum power is $P_R = \varepsilon^2/4r$.

3. A sick car battery of emf 11.4 V and internal resistance 0.01 Ω is connected to a load of 2.0 Ω. To help the ailing battery, a second battery of emf 12.6 V and internal resistance 0.01 is connected by jumper cables to the terminals of the first battery. Draw a circuit diagram for this situation and find the current in each part of the circuit. Find the power delivered by the second battery and discuss where this power goes, assuming that both emfs and internal resistances are constant.

4. In Problem 3 assume that the first emf increases at a constant rate of 0.2 V/h, but that that of the second battery and the internal resistances are constant. Find the current in each part of the circuit as a function of time. Sketch a graph of the power delivered to the first battery as a function of time.

5. In the circuit shown in Figure 25-41 find (a) the current in each resistor, (b) the power supplied by each emf, and (c) the power dissipated in each resistor.

6. Two identical batteries with emf ε and internal resistance r may be connected across a resistance R either in series or in parallel. Which method of connection supplies the greater power to R when (a) $R < r$, (b) $R > r$?

7. In the circuit shown in Figure 25-42 find the potential difference between points a and b.

8. The space between the plates of a parallel-plate capacitor is filled with a dielectric of constant K and resistivity ρ. (a) Show that the charge on the plates decays with time constant $t_c = \varepsilon_0 K\rho$. (b) If the dielectric is mica with $K = 5.0$ and $\rho = 9 \times 10^{13}$ $\Omega \cdot$m, find the time for the charge to decrease to $1/e^2 \approx 14$ percent of its initial value.

9. In the circuit shown in Figure 25-43 the battery has internal resistance 0.01 Ω. An ammeter of resistance 0.01 Ω is inserted at point a. (a) What is the reading of the ammeter? (b) By what percentage is the current changed because of the ammeter? (c) The ammeter is removed and a voltmeter of resistance 1 kΩ is connected from a to b. What is the reading of the voltmeter? (d) By what percentage is the voltage drop from a to b changed by the presence of the voltmeter?

Figure 25-40
Exercise 32.

Figure 25-41
Problem 5.

Figure 25-42
Problem 7.

Figure 25-43
Problem 9.

10. If you have two batteries, one with ε = 9.0 V and r = 0.8 Ω and the other with ε = 3.0 V and r = 0.4 Ω, how would you connect them to give the largest current through a resistor R for (*a*) R = 0.2 Ω, (*b*) R = 0.6 Ω, (*c*) R = 1.0 Ω, and (*d*) R = 1.5 Ω?

11. In the circuit shown in Figure 25-44 the reading of the ammeter is the same with both switches open as with both closed. Find the resistance R.

Figure 25-44
Problem 11.

12. In Figure 25-45 the galvanometer has resistance 10 Ω and is connected across a 90-Ω resistor. The value of the current which causes full-scale deflection can be chosen by using connections *ab*, *ac*, *ad*, or *ae*. (*a*) How should the 90-Ω resistor be divided so that the current causing full-scale deflection decreases by a factor of 10 for each successive connection *ab*, *ac*, etc.? (*b*) What should be the full-scale deflection current in the galvanometer I_g so that this ammeter has ranges 1.0 mA, 10 mA, 100 mA, and 1 A?

Figure 25-45
Problem 12.

13. Figure 25-46 shows two possible ways to use a voltmeter and ammeter to measure an unknown resistance. Assume that the internal resistance of the battery is negligible and that the resistance of the voltmeter is 1000 times that of the ammeter: R_v = 1000 R_a. The calculated value of R is taken to be R_c = V/I, where V and I are the readings of the voltmeter and ammeter. (*a*) Discuss which circuit is preferred for values of R in the range from 10R_a to 0.9R_v. (*b*) Find R_c for each circuit if R_a = 0.1 Ω, R_v = 100 Ω, and R = 0.5, 3, and 80 Ω.

14. See Problem 13. For the circuits shown in Figure 25-46 show that R_c = V/I is related to the true value R by

$$\frac{1}{R_c} = \frac{1}{R} + \frac{1}{R_v}$$

for circuit *a* and R_c = $R + R_a$ for circuit *b*. If ε = 1.5 V, R_a = 0.01 Ω, and R_v = 10 kΩ, what range of values of R can be measured such that R_c is within 5 percent of R using circuit *a*? Using circuit *b*?

Figure 25-46
Problems 13 and 14.

(*a*)   (*b*)

15. Consider the circuit in Figure 25-47, where r is the internal resistance of the emf and R_a the resistance of the ammeter. (a) Show that the ammeter reading is given by

$$\varepsilon \left(R_2 + R_a + r + \frac{R_2 + R_a}{R_1} r \right)^{-1}$$

(b) Show that if the ammeter and emf are interchanged, the ammeter reading is

$$\varepsilon \left(R_2 + R_a + r + \frac{R_2 + r}{R_1} R_a \right)^{-1}$$

Note that if $R_a = r$, or if both are negligible, the reading is the same. (When one can neglect R_a and r, this symmetry can be very useful in analyzing circuits with only one emf; it is not valid for more than one emf.)

16. Derive Equation 25-7 for the equivalent resistance of three parallel resistors.

17. In the circuit shown in Figure 25-48 the capacitor is originally uncharged with the switch open. At $t = 0$ the switch is closed. (a) What is the current supplied by the emf just after the switch is closed? (b) What is the current a long time after the switch is closed? (c) Derive an expression for the current through the emf for any time after the switch is closed. (d) After a long time t' the switch is opened. How long does it take for the charge on the capacitor to decrease to 10 percent of its value at $t = t'$ if $R_1 = R_2 = 5$ kΩ and $C = 1.0$ μF?

18. Two batteries with emfs ε_1 and ε_2 and with internal resistances r_1 and r_2 are connected in parallel. Prove that the optimal load resistance (for maximum power delivery) R connected in parallel with this combination is $R = r_1 r_2/(r_1 + r_2)$.

Figure 25-47
Problem 15.

Figure 25-48
Problem 17.

CHAPTER 26   The Magnetic Field

Objectives   After studying this chapter, you should:

1. Be able to calculate the magnetic force on a current element and on a moving charge in a given magnetic field.

2. Be able to calculate the magnetic moment of a current loop and the torque exerted on a current loop in a magnetic field.

3. Be able to discuss the Thomson experiment for measuring q/m for electrons.

4. Be able to describe a velocity selector, a mass spectrograph, and a cyclotron.

5. Be able to discuss the Hall effect.

It was known to the early Greeks more than 2000 years ago that certain stones from Magnesia, now called magnetite, attract pieces of iron.* Early experiments in magnetism were concerned chiefly with the behavior of permanent magnets. It is not known when a magnet was first used for navigation, but there is written reference to such use as early as the twelfth century. In 1269 an important discovery was made by Pierre de Maricourt, when he laid a needle on a spherical natural magnet at various positions and marked the directions taken by the needle. The lines encircled the magnet just as the meridian lines encircle the earth, passing through two points at opposite ends of the sphere. Because of the analogy with the meridian lines of the earth, he called these points the *poles* of the magnet. It was noted by many experimenters that every magnet of whatever shape had two poles, a north pole and a south pole, where the magnetic force exerted by the magnet is strongest. Like poles of two magnets repel each other, and unlike poles attract.

* Much of this discussion follows Sir Edmund Whittaker's excellent *A History of the Theories of Aether and Electricity*, Nelson, London, 1953; Torchbook edition, Harper, New York, 1960.

In 1600 William Gilbert discovered why a compass needle orients itself in definite directions: the earth itself is a permanent magnet. Since
the north pole of a compass needle is attracted to the north geographic
pole of the earth, this north geographic pole of the earth is actually a
south magnetic pole. The attraction and repulsion of magnetic poles
was studied quantitatively by John Michell in 1750. Using a torsion balance, Michell showed that the attraction and repulsion of the poles of
two magnets are of equal strength and vary inversely as the square of
the distance between the poles. These results were confirmed shortly
thereafter by Coulomb. The law of force between two magnetic poles is
similar to that between two electric charges, but there is an important
difference: magnetic poles always occur in pairs. It is impossible to iso-
late a single magnetic pole. If a magnet is broken in half, equal and opposite poles appear at the break point, so that there are two magnets,
each with equal and opposite poles. Coulomb explained this result by
assuming that the magnetism is contained in each molecule of the
magnet.

The connection between electricity and magnetism was not known
until the nineteenth century, when Hans Christian Oersted discovered
that an electric current affects the orientation of a compass needle. Subsequent experiments by André Marie Ampère and others showed that
electric currents attract bits of iron and that parallel currents attract each
other. Ampère proposed the theory that electric currents are the source
of all magnetism; in ferromagnets these currents were thought to be
"molecular" current loops which are aligned when the material is magnetized. Ampère's model is the basis of the theory of magnetism today,
though in many cases the "currents" in magnetic materials are related
to the intrinsic electron spin, a quantum property of the electron which

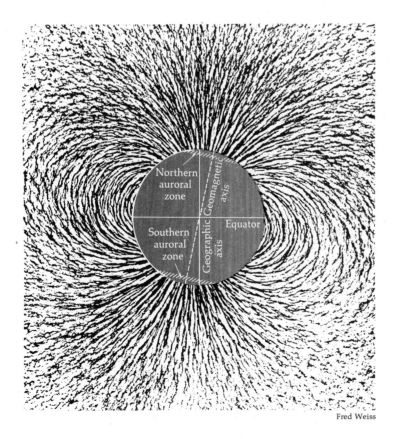

Magnetic field lines of the
earth indicated by iron filings
around a uniformly magne-
tized sphere.

Fred Weiss

could not be envisioned in the nineteenth century. Further connections between magnetism and electricity were demonstrated by experiments of Michael Faraday and Joseph Henry showing that a changing magnetic field produces a nonconservative electric field, and by James Clerk Maxwell's theory showing that a changing electric field produces a magnetic field.

The basic magnetic interaction is one between two charges in motion. As with the electrostatic interaction, it is convenient to consider the magnetic force exerted by one moving charge on another to be transmitted by a third agent, the magnetic field. A moving charge produces a magnetic field, and that field in turn exerts a force on a second moving charge. Since a moving charge constitutes an electric current, the magnetic interaction can also be thought of as an interaction between two currents.

In this chapter we consider only the effects of given magnetic fields on moving charges and currents in wires. We postpone consideration of the origin of the magnetic field until Chapter 27.

26-1 Definition of the Magnetic Field **B**

We can define the magnetic field vector **B** at a point in space similar to the way we defined the electric field **E** (recall that we defined **E** at a point in space to be the force per unit charge on a test charge q_0 placed at that point). It is observed experimentally that when a charge has a velocity **v** in the vicinity of a magnet or a current-carrying wire, there is an additional force on it which depends on the magnitude and direction of the velocity. We can easily separate these two forces by measuring the force on the charge when it is at rest and subtracting this *electric* force from the total force acting when the charge is moving. For simplicity, let us assume that there is no electric field at the point in space we are considering. Experiments with various charges moving with various velocities at such a point in space give the following results for the magnetic force:

1. The force is proportional to the magnitude of the charge q.

2. The force is proportional to the speed v.

3. The magnitude and direction of the force depend on the direction of the velocity **v**.

4. If the velocity of the particle is along a certain line in space, the force is zero.

5. If the velocity is not along this line, there is a force which is perpendicular to this line and also perpendicular to the velocity.

6. If the velocity makes an angle θ with this line, the force is proportional to sin θ.

7. The force on a negative charge is in the direction opposite that on a positive charge with the same velocity.

Magnetic force on a moving charge

We can summarize these experimental results by defining a magnetic field vector **B** to be along the line described in result 4, and by writing

$$\mathbf{F} = q\mathbf{v} \times \mathbf{B} \qquad\qquad\qquad 26\text{-}1$$

For historical reasons, the vector **B** is usually called the *magnetic-induction vector* or the *magnetic flux density*. (A related vector **H**, the magnetic field intensity, will be introduced in Chapter 29 when we study magnetic materials. In free space, the vectors **B** and **H** are proportional to each other. In keeping with custom we shall refer to this **B** field as the *magnetic induction*, although the phrase magnetic field **B** may also be used.)

Figure 26-1 shows the force exerted on various moving charges when the magnetic-induction vector **B** is in the vertical direction. Note that the direction of **B** can be found by experiment using Equation 26-1.

The SI unit of magnetic induction is the tesla (T).* A charge of one coulomb moving with a velocity of one metre per second perpendicular to a magnetic field of one tesla experiences a force of one newton:

$$1\,\text{T} = 1\,\text{N·s/C·m} = 1\,\text{N/A·m}$$

This unit is rather large; a common unit, derived from the cgs system, is the gauss (G), related to the tesla as follows:

$$1\,\text{T} = 10^4\,\text{G} \qquad\qquad\qquad 26\text{-}2$$

Since magnetic fields are often given in gauss, which is not an SI unit, it is important to remember to convert to teslas when making calculations.

Figure 26-1
Direction of the magnetic force on a charged particle with velocity **v** at various orientations in a magnetic field **B**.

Example 26-1 The magnetic induction of the earth has a magnitude of 0.6 G and is directed downward and northward, making an angle of about 70° with the horizontal. (These data are approximately correct for the central United States.) A proton moves horizontally in the northward direction with speed $v = 10^7$ m/s. Calculate the force on the proton.

Figure 26-2 shows the directions of the magnetic induction **B** and the proton's velocity **v**. The angle between them is $\theta = 70°$. The direction of the force is $\mathbf{v} \times \mathbf{B}$, which is west for a proton moving north. The magnitude of the magnetic force is

$$F = qvB \sin\theta = (1.6 \times 10^{-19}\,\text{C})\,(10^7\,\text{m/s})\,(0.6 \times 10^{-4}\,\text{T})\,(0.94)$$
$$= 9.02 \times 10^{-17}\,\text{N}$$

The acceleration of the proton is

$$a = \frac{F}{m} = \frac{9.02 \times 10^{-17}\,\text{N}}{1.67 \times 10^{-27}\,\text{kg}} = 5.40 \times 10^{10}\,\text{m/s}^2$$

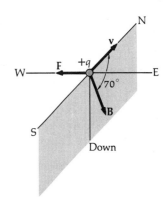

Figure 26-2
Force on a proton moving north in the magnetic field of the earth, which makes an angle of 70° with the horizontal north direction. The force is directed toward the west.

When a wire carries a current in a magnetic field, there is a force on the wire which is the sum of the magnetic forces on the charged particles whose motion produces the current. Figure 26-3 shows a short segment of wire of cross-sectional area A and length ℓ carrying a current I. If the wire is in a magnetic field **B**, the magnetic force on each charge is $q\mathbf{v}_d \times \mathbf{B}$, where $\mathbf{v}_d$ is the drift velocity of the charge carriers. The number of charges in the wire segment is the number n per unit volume times the volume $A\ell$. Thus the total force **F** on the wire segment is

$$\mathbf{F} = (q\mathbf{v}_d \times \mathbf{B})nA\ell$$

---

* This unit is also called a weber per square metre, because the unit for magnetic flux is called the weber (Wb).

Figure 26-3
Current element of length ℓ. Each charge experiences a magnetic force $q\mathbf{v}_d \times \mathbf{B}$.

By Equation 24-3, the current in the wire is

$$I = nqv_d A$$

Hence the force can be written

$$\mathbf{F} = I\boldsymbol{\ell} \times \mathbf{B} \qquad\qquad 26\text{-}3$$

where $\boldsymbol{\ell}$ is a vector whose magnitude is the length of the wire and whose direction is parallel to $q\mathbf{v}_d$, the direction of the current I. Equation 26-3 assumes that the wire is straight and that the magnetic induction does not vary over the length of the wire. It is easily generalized for an arbitrarily shaped wire in any magnetic field. We merely choose a very small wire segment $d\boldsymbol{\ell}$ and write the force on this segment $d\mathbf{F}$:

$$d\mathbf{F} = I\,d\boldsymbol{\ell} \times \mathbf{B} \qquad\qquad 26\text{-}4$$

Force on a current element

where $\mathbf{B}$ is the magnetic-induction vector at the segment. The quantity $I\,d\boldsymbol{\ell}$ is called a *current element*. We find the total force on the wire by summing (or integrating) over all the current elements using the appropriate field $\mathbf{B}$ at each element.

Equation 26-4 may be taken as an alternate definition of $\mathbf{B}$ equivalent to Equation 26-1. The magnitude and direction of $\mathbf{B}$ can be found using this definition by measuring the force on a current element $I\,d\boldsymbol{\ell}$ for various orientations of the element.

Questions

1. Charge q moves with velocity $\mathbf{v}$ through a magnetic field $\mathbf{B}$. At this instant it experiences a magnetic force $\mathbf{F}$. How would the force differ if the charge had the opposite sign? If the velocity had the opposite direction? If the magnetic field had the opposite direction?

2. For what angle between $\mathbf{B}$ and $\mathbf{v}$ is the magnetic force on q greatest? Least?

3. A moving electric charge may experience both electric and magnetic forces. How could you distinguish whether a force causing a charge to deviate from a straight path is an electric or a magnetic force?

4. How can a charge move through a region of magnetic induction without ever experiencing any magnetic force?

5. Show that the force on a current element is the same in direction and magnitude regardless of whether positive charges, negative charges, or a mixture of positive and negative charges create the current.

6. A current-carrying wire passes through a magnetic field, but the wire does not experience any magnetic force. How is this possible?

26-2 Magnets in Magnetic Fields

When a small permanent magnet such as a compass needle is placed in a magnetic field, it tends to orient itself so that the north pole points in the direction of $\mathbf{B}$. This effect also occurs with previously unmagnetized iron filings, which become magnetized in the presence of a $\mathbf{B}$ field. Figure 26-4 shows a small magnet which makes an arbitrary angle with a magnetic field $\mathbf{B}$. There is a force $\mathbf{F}_1$ on the north pole in the direction of $\mathbf{B}$ and an equal but opposite force $\mathbf{F}_2$ on the south pole. The *pole strength* *of the magnet* q_m is defined* to be the ratio of the magnitude of the force **F** on the pole and the magnitude of the magnetic induction **B**:

$$q_m = \frac{F}{B} \qquad\qquad 26\text{-}5$$

Since a tesla is equal to a newton per ampere-metre, the SI unit of magnetic pole strength is the ampere-metre (A·m). If we adopt the sign convention that the north pole is positive and the south pole negative, the force on a pole can be written as a vector equation:

$$\mathbf{F} = q_m\mathbf{B} \qquad\qquad 26\text{-}6$$

From Figure 26-4 we see that there is a torque acting on a magnet in a magnetic field. If $\boldsymbol{\ell}$ is a vector pointing from the south pole to the north pole with the magnitude of the distance between the poles, the torque is

$$\boldsymbol{\tau} = \boldsymbol{\ell} \times \mathbf{F} = \boldsymbol{\ell} \times q_m\mathbf{B} = q_m\boldsymbol{\ell} \times \mathbf{B} \qquad\qquad 26\text{-}7$$

The magnetic moment **m** of a magnet is defined to be

$$\mathbf{m} = q_m\boldsymbol{\ell} \qquad\qquad 26\text{-}8$$

The SI unit of magnetic moment is the ampere-metre² (A·m²). In terms of the magnetic moment, the torque on a magnet is

$$\boldsymbol{\tau} = \mathbf{m} \times \mathbf{B} \qquad\qquad 26\text{-}9$$

These equations are completely analogous to those for the torque on an electric dipole in an electric field (Section 20-7). However, there is an important difference between the magnetic pole strength q_m and the electric charge q. An isolated magnetic pole has never been experimentally detected. Magnetic poles always come in pairs; i.e., the fundamental unit of magnetism is the magnetic dipole. Experimentally, it is the magnetic dipole moment **m** of a magnet that can be easily measured by placing the magnet in a magnetic field of known strength and measuring the torque. The magnetic pole strength q_m is then found from Equation 26-8 by dividing the magnitude of the magnetic moment by the length of the magnet.

It is customary to indicate the direction of the magnetic induction **B** by drawing lines which are parallel to **B** at each point in space and to indicate the magnitude of **B** by the density of the lines, just as for the electric field **E**. Such lines are called *lines of magnetic induction*. For a given magnetic induction **B** the lines can be found using a compass needle or iron filings since these small magnets align themselves in the direction of **B**. Figure 26-5 shows the lines of magnetic induction near a bar magnet and near a long straight current-carrying wire. External to the magnet, the lines of **B** leave the north pole and enter the south pole.

---

* The notation for magnetic pole strength q_m is used so that the magnetic equations resemble the corresponding equations for electric charges in electric fields. The subscript m reminds us that q_m denotes a magnetic pole and not an electric charge.

Fred Weiss

Education Development Center, Inc., Newton, Mass.

Figure 26-5
Magnetic field lines indicated by the alignment of iron filings for (*left*) a magnet and (*right*) a long straight current-carrying wire.

This is consistent with the fact that the force on a north pole is in the direction of **B** and two north poles repel each other. We shall investigate how to calculate **B** for these cases in later chapters.

Questions

7. Although the force on a north pole of a magnet is in the direction of the lines of magnetic induction, the force on an electric charge q moving through a region of magnetic induction **B** is not. Is it possible to draw a set of lines which would indicate the direction of the magnetic force on a moving charge at each point in space?

8. The lines of magnetic induction near the long straight current-carrying wire shown in Figure 26-5 (*right*) are circles concentric with the wire. If a single isolated magnetic pole did exist, what would happen to it if it were placed near such a wire?

26-3 Torque on a Current Loop in a Uniform Magnetic Field

When a wire carrying a current I is placed in a uniform magnetic field, there are forces on each part of the wire. If the wire is in the shape of a closed loop, there is no net force on the loop because the forces on the different parts of the wire sum to zero (assuming that the magnetic induction **B** is the same at all parts of the wire). In general, however, the magnetic forces produce a torque on the wire which tends to rotate the loop so that the area of the loop is perpendicular to the magnetic induction **B**.

Figure 26-6 shows a rectangular loop carrying a current I. The loop is in a region of uniform magnetic field parallel to the plane of the loop. The forces on each segment of the loop are indicated in the figure. There are no forces on the top or bottom of the loop because for those segments the current is parallel or antiparallel to the magnetic induction **B**, so that $I \, d\ell \times \mathbf{B}$ is zero. The forces on the sides of the loop have the magnitude

$$F_1 = F_2 = IaB$$

where a is the length of the side. These forces are equal and opposite.

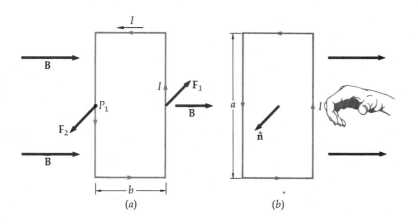

Figure 26-6
(*a*) Forces exerted on a rectangular current loop in a uniform magnetic field **B** parallel to the plane of the loop. (*b*) Unit vector $\hat{\mathbf{n}}$ perpendicular to the plane of the loop. The sense of $\hat{\mathbf{n}}$ is defined by the right-hand rule. With the fingers curling in the direction of the current, the thumb points in the direction of $\hat{\mathbf{n}}$. The forces in (*a*) tend to twist the loop so that $\hat{\mathbf{n}}$ rotates into **B**.

Hence they form a couple. The net force is zero, and the torque about any point is independent of the location of the point. Point P_1 is a convenient point about which to compute the torque. The magnitude of the torque is

$$F_1 b = IabB = IAB$$

where *b* is the width of the loop and $A = ab$ is the area of the loop. The torque is therefore the product of the current, the area of the loop, and the magnetic induction *B*. This torque tends to twist the loop so that its plane is perpendicular to **B**. The orientation of the loop can be described conveniently by a unit vector $\hat{\mathbf{n}}$ perpendicular to the plane of the loop. The sense of $\hat{\mathbf{n}}$ is chosen to be that given by the right-hand rule applied to the circulating current, as illustrated in Figure 26-6. The torque thus tends to rotate $\hat{\mathbf{n}}$ into the direction of **B**.

Figure 26-7 shows the forces exerted by a uniform magnetic field on a rectangular loop whose normal unit vector $\hat{\mathbf{n}}$ makes an angle θ with the magnetic induction **B**. Again, the net force on the loop is zero. The torque about any point (point *P*, for example) is the product of the force $F_2 = IaB$ and the lever arm $b \sin \theta$. The torque thus has the magnitude

$$\tau = IaBb \sin \theta = IAB \sin \theta$$

where again $A = ab$ is the area of the loop. This torque can be written conveniently in terms of the cross product of $\hat{\mathbf{n}}$ and **B**:

$$\boldsymbol{\tau} = IA\hat{\mathbf{n}} \times \mathbf{B} \qquad\qquad 26\text{-}10$$

Equation 26-10 is of the same form as Equation 26-9 for the torque on a bar magnet in a uniform magnetic field, with the quantity $IA\hat{\mathbf{n}}$ for the loop playing the role of the magnetic moment $q_m\boldsymbol{\ell}$ of the bar magnet. We thus define the magnetic moment of a current loop to be $IA\hat{\mathbf{n}}$. Often a loop contains several turns of wire, each turn having the same area and carrying the same current. The total magnetic moment of such a loop is the product of the number of turns and the magnetic moment of each turn. The total magnetic moment of a loop of *N* turns is then

$$\mathbf{m} = NIA\hat{\mathbf{n}} \qquad\qquad 26\text{-}11$$

With this definition, the torque on the current loop is given by Equation 26-9:

$$\boldsymbol{\tau} = \mathbf{m} \times \mathbf{B}$$

Equation 26-10, which we have derived for a rectangular loop, holds in general for a loop of any shape. The torque on any loop is the cross

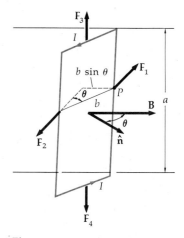

Figure 26-7
Rectangular current loop whose unit normal $\hat{\mathbf{n}}$ makes an angle θ with a uniform magnetic field **B**. The torque on the loop is $IA\hat{\mathbf{n}} \times \mathbf{B} = \mathbf{m} \times \mathbf{B}$, where $\mathbf{m} = IA\hat{\mathbf{n}}$ is the magnetic moment of the loop.

Magnetic moment of current loop

product of the magnetic moment of the loop and the magnetic induction **B**, where the magnetic moment is defined to be a vector whose magnitude is the current times the area of the loop and whose direction is perpendicular to the area of the loop (Figure 26-8). A current loop thus behaves like a small bar magnet. We shall see in the next chapter that the magnetic field produced by a current loop at points far from the loop is identical to that produced by a bar magnet at points far from the magnet if the magnetic moments are the same. Since such a field is of the same form as the electric field produced by an electric dipole, the analogy of a current loop with a bar magnet is useful for understanding the behavior of the loop in an external magnetic field and for explaining the magnetic field produced by the loop. We say more about the significance of this analogy in Chapter 27.

An important application of the torque exerted on a coil in a magnetic field is the d'Arsonval galvanometer discussed in Section 25-4 (see Figure 25-22). In this case, the magnetic field produced by a permanent magnet is radial at the coil, so that the angle θ between **B** and the normal to the plane of the coil is 90° independent of the orientation of the coil. The torque produced by the magnetic field is proportional to the current in the coil and is balanced by a restoring torque due to the spring suspension. If this restoring torque is proportional to the angular deflection of the coil, the angular deflection will be proportional to the current I. In any case, the angular deflection of the coil can be calibrated to indicate the current in the coil.

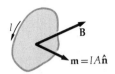

Figure 26-8
A current loop of arbitrary shape experiences a torque **m** × **B** in a magnetic field.

Question

9. A current loop has its magnetic moment antiparallel to a uniform **B** field. What is the torque on the loop? Is this equilibrium stable or unstable?

26-4 Motion of a Point Charge in a Magnetic Field

An important characteristic of the magnetic force on a moving charged particle described by Equation 26-1 is that the force is always perpendicular to the velocity of the particle. The magnetic force therefore does no work on the particle, and the kinetic energy of the particle is unaffected by this force. The magnetic force changes the direction of the velocity but not its magnitude.

In the special case where the velocity of a particle is perpendicular to a uniform magnetic field, as shown in Figure 26-9, the particle moves in a circular orbit. The magnetic force provides the centripetal force necessary for circular motion. We can relate the radius of the circle to the magnetic field and the velocity of the particle by setting the resultant force equal to the mass m times the centripetal acceleration v^2/r. The resultant force in this case is just qvB since **v** and **B** are perpendicular. Thus Newton's second law gives

$$qvB = \frac{mv^2}{r}$$

or

$$r = \frac{mv}{qB} \qquad\qquad 26\text{-}12$$

Figure 26-9
Charged particle moving in a plane perpendicular to a uniform magnetic field which is into the plane of the paper (indicated by the crosses). The magnetic force is perpendicular to the velocity of the particle, causing it to move in a circle of radius r, which can be found from $qvB = mv^2/r$.

Circular path of electrons moving in a magnetic field produced by two large coils. The electrons ionize the gas in the tube, causing it to give off a bluish glow that indicates the path of the beam. (*Courtesy of Larry Langrill, Oakland University.*)

The angular frequency of the circular motion is

$$\omega = \frac{v}{r} = \frac{qB}{m} \qquad \qquad 26\text{-}13$$

and the period is

$$T = \frac{2\pi}{\omega} = \frac{2\pi m}{qB} \qquad \qquad 26\text{-}14$$

Note that the frequency given by Equation 26-13 does not depend on the radius of the orbit or the velocity of the particle. This frequency is called the *cyclotron frequency*. Two of the many interesting applications of the circular motion of charged particles in a uniform magnetic field, the mass spectrograph and the cyclotron, will be discussed in Examples 26-3 and 26-4.

Cyclotron frequency

If a charged particle enters a region of uniform magnetic induction with a velocity which is not perpendicular to **B**, the path of the particle is a helix. The component of velocity parallel to **B** is not affected by the magnetic field. Consider, for example, a uniform magnetic field in the z direction, and let v_z be the component of velocity of the particle parallel to the field. In a reference frame moving in the z direction with speed v_z, the particle has a velocity perpendicular to the field and moves in a circle in the xy plane. In the original frame the path of the particle is a helix which winds around the lines of **B**, as shown in Figure 26-10.

The magnetic force on a charged particle moving in a uniform magnetic field can be balanced by an electrostatic force if the magnitudes and directions of the magnetic and electric fields are properly chosen. Since the electric force is in the direction of the electric field (for positive particles) and the magnetic force is perpendicular to the magnetic field, the electric and magnetic fields must be perpendicular to each other if

Figure 26-10
When a charged particle has a component of velocity parallel to a magnetic field **B**, it moves in a helical path.

Cloud-chamber photograph of the helical path of an electron moving in a magnetic field. The path of the electron is made visible by the condensation of water droplets in the cloud chamber. (*Courtesy of Carl E. Nielsen, Ohio State University.*)

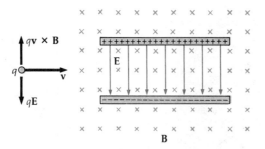

Figure 26-11
Crossed electric and magnetic
fields. When a positive particle moves to the right, it
experiences a downward electric force $q\mathbf{E}$ and an upward
magnetic force $q\mathbf{v} \times \mathbf{B}$, which
balance if $vB = E$.

the forces are to balance. Figure 26-11 shows a region of space between
the plates of a capacitor where there is an electric field and a perpendicular magnetic field (which can be produced by a magnet not shown).
Such an arrangement of perpendicular fields is called *crossed fields*. Consider a particle of charge q entering this space from the left. If q is posi-
tive, the electric force $q\mathbf{E}$ is down and the magnetic force $q\mathbf{v} \times \mathbf{B}$ is up. If
the charge is negative, each of the forces is reversed. Since $\mathbf{v}$ is perpendicular to $\mathbf{B}$, the magnitude of the magnetic force is just qvB. These two
forces will balance if $qE = qvB$ or

$$v = \frac{E}{B} \qquad\qquad 26\text{-}15$$

For given magnitudes of the electric and magnetic fields, the forces will
balance only for particles with the speed given by Equation 26-15. Any
particle, no matter what its mass or charge, will traverse the space undeflected if its speed is given by Equation 26-15. A particle of greater
speed will be deflected in the direction of the magnetic force; one of less
speed will be deflected in the direction of the electric force. Such an
arrangement of fields is called a *velocity selector*; its use is illustrated in   *Velocity selector*
Examples 26-2 and 26-3.

The motion of charged particles in nonuniform magnetic fields is
quite complicated. We shall discuss qualitatively some cases in which
the magnetic field $\mathbf{B}$ is nearly uniform so that the path of the particle is
approximately a helix. Consider a magnetic field indicated by the lines
of induction in Figure 26-12, which represents a three-dimensional
field axially symmetric about the horizontal axis. The field is strongest
at the ends and weakest at the center, as indicated by the line spacing.
Such a field is sometimes called a *magnetic bottle* because charged par-   *Magnetic bottle*
ticles can be trapped in it, as we shall show.

Let a positive particle have an initial velocity into the paper at point
P_1. If the magnetic induction $\mathbf{B}$ were horizontal at this point, the particle would move in a circle. However, $\mathbf{B}$ has a small vertical component, and the force $q\mathbf{v} \times \mathbf{B}$ has a small component to the right. The particle thus accelerates to the right as it moves in a nearly circular path
about the axis of symmetry of the field. According to Equation 26-12,

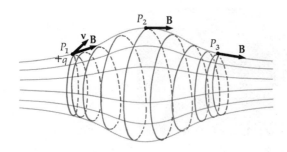

Figure 26-12
Magnetic bottle. The positive
particle moves back and forth
between P_1 and P_3 along the
indicated path.

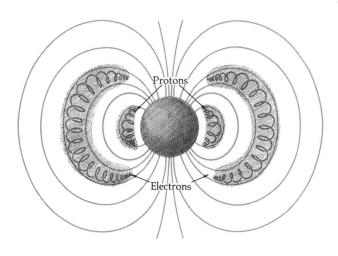

Figure 26-13
Van Allen belts. Protons (inner belts) and electrons (outer belts) are trapped in the earth's magnetic field and spiral along the field lines between the north and south poles, as in a magnetic bottle.

the radius of the circular motion is proportional to the velocity component perpendicular to **B** and inversely proportional to **B**. (Since the speed of the particle cannot be changed by the magnetic force, which is perpendicular to the velocity, an increase in the axial component of the velocity is accompanied by a slight decrease in the component perpendicular to the axis.) As the particle moves to the right, it enters a region of weaker magnetic field. The radius of its nearly circular path thus increases. At point P_2 the magnetic field is horizontal. The particle has no axial acceleration at this point but continues to move to the right because of its axial velocity. To the right of point P_2 the magnetic force has a component to the left which decreases the axial velocity until it is zero at point P_3, where the particle begins to move to the left. Since the field is stronger at point P_3, the radius of the circular path is smaller than at point P_2. The particle thus oscillates back and forth between points P_1 and P_3. A motion somewhat similar to this is the oscillation of charged ions back and forth between the earth's magnetic poles in the Van Allen belts (Figure 26-13).

Example 26-2 *Thomson's Measurement of q/m for Electrons* An example of the use of a velocity selector and the measurement of the deflection of charged particles in an electric field is the famous experiment of J. J. Thomson in 1897, in which he showed that rays in cathode-ray

J. J. Thomson in his laboratory. (*Courtesy of the Cavendish Laboratory, University of Cambridge.*)

tubes can be deflected by electric and magnetic fields and therefore consist of charged particles. By observing the deflection of these rays with various combinations of electric and magnetic fields, Thomson was able to prove that all the particles had the same charge-to-mass ratio q/m and to determine the ratio. He showed that particles with this charge-to-mass ratio can be obtained using any material for the cathode, which means that these particles, now called electrons, are a fundamental constituent of all matter.

Figure 26-14 shows the cathode-ray tube he used. Electrons are emitted from a cathode C, which is at a negative potential relative to the

Figure 26-14
Thomson's tube for measuring q/m. Electrons from the cathode C pass through the slits at A and B and strike a phosphorescent screen S. The beam can be deflected by an electric field between the plates D and F or by a magnetic field (not shown). [*From J. J. Thomson,* Philosophical Magazine, *Ser. 5, vol. 44 (1897).*]

slits A and B. An electric field in the direction from A to C accelerates the electrons. They pass through the slits A and B into a field-free region and then encounter between the plates D and F an electric field that is perpendicular to the velocity of the electrons. The acceleration produced by this electric field gives the electrons a vertical component of velocity when they leave the region between the plates. They strike the phosphorescent screen S at the far right side of the tube at some displacement Δy from the point at which they strike when there is no field between the plates D and F. When the electrons strike the screen, it glows, indicating the location of the beam. The deflection Δy consists of two parts, Δy_1, which occurs while the electrons are between the plates, and Δy_2, which occurs after the electrons leave the region between the plates (Figure 26-15).

Figure 26-15
Deflection of the electron beam in the Thomson experiment.

Let x_1 be the horizontal distance across the deflection plates D and F. If the velocity of the electron is v_0 when it enters the plates, the time spent between the plates will be

$$t_1 = \frac{x_1}{v_0}$$

Assuming that the electron was moving horizontally when it entered the region between the deflection plates, the vertical velocity when it leaves this region will be

$$v_y = at_1 = \frac{qE}{m}\, t_1 = \frac{qE}{m}\frac{x_1}{v_0}$$

where E is the electric field between the plates. The deflection in this region will be

$$\Delta y_1 = \tfrac{1}{2} a t_1^2 = \frac{1}{2} \frac{qE}{m} \left(\frac{x_1}{v_0}\right)^2$$

The electron then travels an additional horizontal distance x_2 in the field-free region between the deflection plates and the screen. Since there is no force on the electron (except the negligible gravitational force), the velocity of the electron is constant in this region. It therefore takes a time

$$t_2 = \frac{x_2}{v_0}$$

to reach the screen and suffers an additional vertical deflection given by

$$\Delta y_2 = v_y t_2 = \frac{qE}{m} \frac{x_1}{v_0} \frac{x_2}{v_0}$$

The total deflection at the screen is

$$\Delta y = \Delta y_1 + \Delta y_2 = \frac{1}{2} \frac{qE}{m} \left(\frac{x_1}{v_0}\right)^2 + \frac{qE}{m} \frac{x_1 x_2}{v_0^2} \qquad \text{26-16}$$

If the initial velocity v_0 is known, a measurement of the deflection for a given electric field E yields a value for q/m. The initial velocity v_0 is determined by introducing a magnetic field **B** between the plates perpendicular to both the electric field and the initial velocity and adjusting its magnitude so that the beam is undeflected. The velocity of the electrons is then given by Equation 26-15. The magnetic field is then turned off, and the deflection is measured.

Example 26-3 *The Mass Spectrograph* The mass spectrograph, first made by Francis William Aston in 1919 and improved by Kenneth Bainbridge and others, was designed to measure the masses of isotopes. It measures the mass-to-charge ratio of ions by determining the velocity of the ions and then measuring the radius of their circular orbit in a uniform magnetic field. According to Equation 26-12, the mass-to-charge ratio is given by

$$\frac{m}{q} = \frac{Br}{v} \qquad \text{26-17}$$

where B is the magnetic induction, r the radius of the circular orbit, and v the speed of the particle. In Figure 26-16, a simple schematic drawing of a mass spectrograph, ions from the source are accelerated by an electric field and enter a uniform magnetic field produced by an electromagnet. If the ions start from rest and move through a potential drop V, their kinetic energy when they enter the magnet equals the loss in potential energy qV:

$$\tfrac{1}{2} m v^2 = qV \qquad \text{26-18}$$

The ions move in a semicircle of radius r given by Equation 26-17 and strike a photographic film at point P_2, a distance $2r$ from the point where they entered the magnet. The speed v can be eliminated from Equations 26-17 and 26-18 to find m/q in terms of the known quantities V, B, and r. The result is

$$\frac{m}{q} = \frac{B^2 r^2}{2V} \qquad \text{26-19}$$

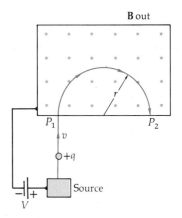

Figure 26-16
Schematic drawing of a mass spectrograph. The outward magnetic field is indicated by the dots.

Strictly speaking, the mass of an ion can be determined only if the charge q is known. However this is not important because the charge q is either one or two electron charges and the spectrograph is used to compare the masses of several isotopes which have nearly equal mass. In Aston's original spectrograph this could be done to a precision of about 1 part in 10,000. The precision is improved by the introduction of a velocity selector between the ion source and the magnet, making it possible to determine the velocity of the ions accurately and to limit the range of velocities of the incoming ions.

Example 26-4   *The Cyclotron*   The cyclotron was invented by E. O. Lawrence and M. S. Livingston in 1934 to accelerate particles such as protons or deuterons to high kinetic energy. The high-energy particles are then used to bombard nuclei, causing nuclear reactions, which are studied to obtain information about the nucleus. High-energy protons or deuterons are also used to produce radioactive materials and for medical purposes. The operation of the cyclotron is based on the fact that the period of motion of the charged particle in a uniform magnetic field is independent of the velocity of the particle.

Figure 26-17
Schematic drawing of a cyclotron. The upper (south) pole face of the magnet has been omitted.

Figure 26-17 is a schematic drawing of a cyclotron. The particles move in two semicircular metal containers called *dees* (because of their shape). The containers lie in a vacuum chamber which is in a uniform magnetic field provided by an electromagnet. (The region in which the particles move must be evacuated so that the particle will not lose energy and be scattered in collisions with air molecules.) The dees are maintained at a potential difference V, which alternates in time with the period T, chosen to be equal to the cyclotron period given by Equation 26-14. The charged particle is initially injected with a small velocity from an ion source S near the center of the magnetic field. It moves in a semicircle in one of the dees and arrives at the gap between the dees after a time $\frac{1}{2}T$, where T is the cyclotron period and also the period of the alternating potential across the dees. Let us call the dee in which the particle traverses this first semicircle dee 1. The alternating potential is phased so that when the particle arrives at the gap between the dees, dee 1 is at a higher potential than dee 2 so that the particle is accelerated across the gap by the electric field across the gap and gains kinetic energy qV. The region inside each dee is shielded from electric fields by the metal dee. Thus the particle moves in a semicircle of larger radius in dee 2 and again arrives at the gap after a time $\frac{1}{2}T$. By this time the poten-

M. S. Livingston and E. O. Lawrence standing in front of their 27-in cyclotron in 1934. Lawrence won the Nobel prize (1939) for the invention of the cyclotron. (*Courtesy of Lawrence Radiation Laboratory, University of California, Berkeley.*)

tial between the dees has been reversed, and dee 2 is at the higher potential. Once more the particle is accelerated across the gap and gains kinetic energy qV. In each half revolution, the particle gains kinetic energy qV and moves into a semicircular orbit of larger and larger radius until it leaves the magnetic field. In a typical cyclotron, a particle may make 50 to 100 revolutions.

Questions

10. By observing the path of a particle how can you distinguish whether the particle is deflected by a magnetic field or an electric field?

11. A beam of positively charged particles passes undeflected from left to right through a velocity selector in which the electric field is up. The beam is then reversed so that it travels from right to left. Will the beam be deflected in the velocity selector? If so, in which direction?

A modern 83-in cyclotron at the University of Michigan. (A deflecting magnet is shown in the foreground.) (*Courtesy of University of Michigan.*)

26-5 The Hall Effect

In Section 26-1 we calculated the force exerted by a magnetic field on a current-carrying wire. This force is actually exerted directly on the charge carriers in the wire, the electrons. The force is transferred to the wire by the forces which bind the electrons to the wire at the surface. Since the charge carriers themselves experience the magnetic force when a current-carrying wire is in a magnetic field, the carriers are accelerated toward one side of the wire. This phenomenon, called the *Hall effect*, allows us to determine the sign of the charge on the carrier and the number density n of charge carriers in a conductor. It also provides a convenient method for measuring magnetic fields.

Figure 26-18*a* shows a conducting strip carrying a current I to the right in a magnetic field which we have chosen to be into the paper. Let us assume for the moment that the current consists of positively charged particles moving to the right. The magnetic force on these particles will be in the direction $q\mathbf{v}_d \times \mathbf{B}$, which is up in the figure. The positive particles will thus move up to the top of the strip, leaving the bottom of the strip with an excess negative charge. This charge separation causes an electrostatic field in the strip which opposes the magnetic force on the charge carriers. When the electrostatic and magnetic forces cancel, the charge carriers will no longer move upward. In this equilibrium situation, the upper part of the strip will be positively charged and be at a greater potential than the negatively charged lower part. On the other hand, if the current consists of negatively charged particles, they must be moving to the left for a current to the right. The magnetic force $q\mathbf{v}_d \times \mathbf{B}$ will again be up since we have changed the sign of both q and $\mathbf{v}_d$. Again the carriers will be forced to the upper part of the strip, but the upper part of the strip now carries a negative charge (because the charge carriers are negative) and the lower part a positive charge. A measurement of the sign of the potential difference between the upper and lower part of the strip tells us the sign of the charge carriers. Experiments for the current and magnetic field in the directions indicated in the figure show that the upper part of the strip carries a negative charge and is at a lower potential than the lower part. This observation led to the discovery that the charge carriers in metallic conductors are negative.

If we connect the upper and lower portions of the strip with a wire of resistance R, the negative electrons will flow from the upper part of the strip through the wire to the lower part. As soon as some electrons leave the upper part of the strip, more carriers in the strip will be driven up by the magnetic force because the electrostatic force is momentarily weakened by the reduced charge separation. Thus the magnetic force maintains the potential difference across the strip. The strip is thus a seat of emf. The potential difference between the top and bottom of the strip is called the *Hall emf*. Its magnitude is not hard to calculate. The magnetic force on the charge carriers in the strip has the magnitude qv_dB. This magnetic force is balanced by the electrostatic force qE, where E is the electric field due to the charge separation. Thus $E = v_dB$. If the width of the strip is w, the potential difference is Ew. The Hall emf is therefore

$$\mathcal{E}_H = Ew = v_dBw \qquad\qquad 26\text{-}20$$

Hall emf

We can see from Equation 26-20 that for ordinary-sized strips and magnetic fields, the Hall emf is very small since the drift velocity for ordinary currents is very small.

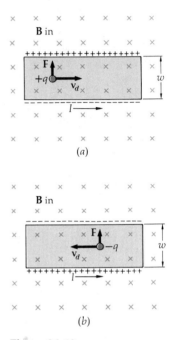

Figure 26-18
The Hall effect. Whether the current to the right is due to positive particles moving to the right as in (*a*) or to negative particles moving to the left as in (*b*), the magnetic force $q\mathbf{v}_d \times \mathbf{B}$ will be upward on the charge carriers. The sign of the carriers can be determined by observing whether the top of the strip is (*a*) positive or (*b*) negative.

From measurements of the size of the Hall emf for a strip of a given size carrying a known current in a known magnetic field we can determine the number of charge carriers per unit volume in the strip. By Equation 24-5 the current density is

$$J = nqv_d \qquad\qquad 26\text{-}21$$

The current density is determined by measuring the current I and the cross-sectional area of the strip A. The quantity q is just one electron charge. The drift velocity is determined from Equation 26-20 by measuring the Hall emf. Thus the number density of charge carriers n is given by

$$n = \frac{J}{qv_d} = \frac{I}{Aqv_d} \qquad\qquad 26\text{-}22$$

(In Example 24-1 we estimated the drift velocity v_d by assuming a value for n, namely one electron per atom for copper. This assumption for n is justified by Hall emf measurements for copper.)

Although the Hall emf is ordinarily very small, it provides a convenient method for measuring magnetic fields. Combining Equations 26-20 and 26-21, we can write for the Hall emf

$$\mathcal{E}_H = \frac{J}{nq} wB = \frac{Iw}{nqA} B \qquad\qquad 26\text{-}23$$

A given strip can be calibrated by measuring the Hall emf for a given current in a known magnetic field. The strength of the magnetic induction B of an unknown field can then be measured by placing the strip in the unknown field, sending a current through the strip, and measuring $\mathcal{E}_H$.

Review

A. Define, explain, or otherwise identify:

Magnetic induction, 723
Magnetic flux density, 723
Tesla, 723
Gauss, 723
Current element, 724
Pole strength, 725
Magnetic moment of magnet, 725
Lines of magnetic induction, 725
Magnetic moment of current loop, 727

Cyclotron frequency, 729
Crossed fields, 730
Velocity selector, 730
Magnetic bottle, 730
Mass spectrograph, 733
Cyclotron, 734
Hall effect, 736
Hall emf, 736

B. True or false:

1. The magnetic force is always perpendicular to the velocity of the particle.

2. The torque on a magnet tends to align the magnetic moment parallel to **B**.

3. A current loop in a uniform magnetic field behaves like a small magnet.

4. The period of a particle moving in a circle in a magnetic field is proportional to the radius of the circle.

5. The drift velocity of electrons in a wire can be determined from the Hall effect.

Exercises

Section 26-1, Definition of the Magnetic Field B

1. Find the magnetic force on a proton moving with velocity 4 Mm/s in the positive x direction in a magnetic field of 2 T in the positive z direction.

2. A charge $q = -2\,nC$ moves with velocity $\mathbf{v} = -3.0 \times 10^6\mathbf{i}$ m/s. Find the force on the charge if the magnetic field is (a) $\mathbf{B} = 0.6\mathbf{j}$ T, (b) $\mathbf{B} = 0.6\mathbf{i} + 0.6\mathbf{j}$ T, (c) $\mathbf{B} = 0.8\mathbf{i}$ T, (d) $\mathbf{B} = 0.6\mathbf{i} + 0.6\mathbf{k}$ T.

3. A uniform magnetic field of magnitude 1.5 T is in the positive z direction. Find the force on a proton if its velocity is (a) $\mathbf{v} = 3.0\mathbf{i}$ Mm/s, (b) $\mathbf{v} = 2.0\mathbf{j}$ Mm/s, (c) $\mathbf{v} = 8\mathbf{k}$ Mm/s, (d) $\mathbf{v} = 3.0\mathbf{i} + 4.0\mathbf{j}$ Mm/s.

4. An electron moves with velocity 5 Mm/s in the xy plane at an angle of 30° to the x axis and 60° to the y axis. A magnetic field of 1.5 T is in the positive y direction. Find the force on the electron.

5. A straight wire segment 2 m long makes an angle of 30° with a uniform magnetic field of 0.5 T. Find the force on the wire if it carries a current of 2 A.

6. A straight wire 10 cm long carrying a current of 3.0 A is in a uniform magnetic field of 1.5 T. The wire makes an angle of 37° with the direction of $\mathbf{B}$. What is the magnitude of the force on the wire?

7. A long wire parallel to the x axis carries a current of 10.0 A in the direction of increasing x. There is a uniform magnetic field of magnitude 2.0 T in the positive y direction. Find the force per unit length on the wire.

8. The wire segment in Figure 26-19 carries a current of 2.0 A from a to b. There is a magnetic field $\mathbf{B} = 1.0\mathbf{k}$ T. Find the total force on the wire and show that it is the same as if the wire were a straight segment from a to b.

Figure 26-19
Exercise 8.

Section 26-2, Magnets in Magnetic Fields

9. A small magnet is placed in a magnetic field of 0.1 T. The maximum torque experienced by the magnet is 0.20 N·m. (a) What is the magnetic moment of the magnet? (b) If the length of the magnet is 4 cm, what is the pole strength q_m?

10. Show that the SI unit of magnetic moment, ampere-metres², is the same as joules per tesla.

11. The cgs unit for magnetic moment is the dyn·cm/G (1 dyn = 10^{-5} N). (a) Find the conversion factor between 1 dyn·cm/G and 1 A·m² = 1 J/T (see Exercise 10). (b) A useful unit in atomic physics for magnetic moment is electron-volts per gauss. Find the conversion factor between electronvolts per gauss and ampere-metres² (= joules per tesla).

12. A small magnet of length 5 cm is placed at an angle of 45° to the direction of a uniform magnetic field of magnitude 0.04 T. The observed torque has the magnitude 0.10 N·m. (a) Find the magnetic moment of the magnet. (b) Find the pole strength q_m.

Section 26-3, Torque on a Current Loop in a Uniform Magnetic Field

13. A small circular coil of 20 turns of wire lies in a uniform magnetic field of 0.5 T so that the normal to the plane of the coil makes an angle of 60° with the direction of $\mathbf{B}$. The radius of the coil is 4 cm, and it carries a current of 3 A. (a) What is the magnitude of the magnetic moment of the coil? (b) What torque is exerted on the coil?

Figure 26-20
Exercises 15 and 16.

14. The SI unit for the magnetic moment of a current loop is ampere-metres². Use this to show that 1 T = 1 N/A·m.

15. A rectangular 50-turn coil has sides 6.0 and 8.0 cm long and carries a current of 2.0 A. It is oriented as shown in Figure 26-20 and pivoted about the z axis. (a) If the wire in the xy plane makes an angle of 37° with the y axis as shown, what angle does the unit normal $\hat{\mathbf{n}}$ make with the x axis? Write an expression for $\hat{\mathbf{n}}$ in terms of the unit vectors $\mathbf{i}$ and $\mathbf{j}$. (b) What is the magnetic moment of the coil? (c) Find the torque on the coil when there is a uniform magnetic field $\mathbf{B} = 1.5\mathbf{j}$ T.

16. The coil of Exercise 15 is pivoted about the z axis and held at various posi-

tions in a uniform magnetic field $\mathbf{B} = 2.0\mathbf{j}$ T. Make a sketch of the coil position and find the torque exerted when the unit normal is (a) $\hat{\mathbf{n}} = \mathbf{i}$, (b) $\hat{\mathbf{n}} = \mathbf{j}$, (c) $\hat{\mathbf{n}} = -\mathbf{j}$, (d) $\hat{\mathbf{n}} = (\mathbf{i} + \mathbf{j})/\sqrt{2}$.

17. What is the maximum torque on a 500-turn circular coil of radius 0.5 cm which carries a current of 1.0 mA and resides in a uniform magnetic field of 0.1 T?

18. A particle of charge q and mass M moves in a circle of radius r with angular velocity ω. (a) Show that the average current is $I = q\omega/2\pi$ and the magnetic moment has the magnitude $m = \frac{1}{2}q\omega r^2$. (b) Show that the angular momentum of this particle has the magnitude $L = Mr^2\omega$ and that magnetic-moment and angular-momentum vectors are related by $\mathbf{m} = (q/2M)\mathbf{L}$.

Section 26-4, Motion of a Point Charge in a Magnetic Field

19. A proton moves in a circular orbit of radius 80 cm perpendicular to a uniform magnetic field of magnitude 0.5 T. (a) What is the period for this motion? (b) Find the speed of the proton. (c) Find the kinetic energy of the proton.

20. A particle of charge q and mass m has momentum $p = mv$ and kinetic energy $E_k = \frac{1}{2}mv^2 = p^2/2m$. If it moves in a circular orbit of radius r in a magnetic field B, show that (a) $p = Bqr$ and (b) $E_k = B^2q^2r^2/2m$.

21. An electron of kinetic energy 25 keV moves in a circular orbit in a magnetic field of 0.2 T. (a) Find the radius of the orbit. (b) Find the angular frequency and the period of the motion.

22. Protons, deuterons, and alpha particles of the same kinetic energy enter a uniform magnetic field $\mathbf{B}$ which is perpendicular to their velocities. Let r_p, r_d, and r_α be the radii of their circular orbits. Find the ratios r_d/r_p and r_α/r_p. Assume that $m_\alpha = 2m_d = 4m_p$.

23. An alpha particle travels in a circular path of radius 0.5 m in a magnetic field of 1.0 T. Find the (a) period, (b) speed, and (c) kinetic energy (in electron-volts) of the alpha particle. Take $m = 6.65 \times 10^{-27}$ kg for the mass of the alpha particle.

24. Show that the cyclotron frequency is the same for deuterons as for alpha particles and that each is half that of a proton in the same magnetic field.

25. A beam of protons moves along the x axis in the positive x direction with speed 10 km/s through a region of crossed fields. (a) If there is a magnetic field of magnitude 1.0 T in the positive y direction, find the magnitude and direction of the electric field. (b) Would electrons of the same velocity be deflected by these fields? If so, in what direction?

26. A velocity selector has a magnetic field of magnitude 0.1 T perpendicular to an electric field of magnitude 0.2 MV/m. (a) What must the speed of a particle be to pass through undeflected? What energy must (b) protons and (c) electrons have to pass through undeflected?

27. The plates of a Thomson q/m apparatus are 5.0 cm long and are separated by 1.0 cm. The end of the plates is 25.0 cm from the tube screen. The kinetic energy of the electrons is 2.0 keV. (a) If a potential of 20.0 V is applied across the deflection plates, by how much will the beam deflect? (b) Find the magnitude of a crossed $\mathbf{B}$ field which will allow the beam to pass through undeflected.

28. A singly ionized ^{24}Mg ion (mass 24.0 u) is accelerated through a 2-kV potential and bent in a magnetic field of 50 mT in a mass spectrometer. (a) Find the radius of curvature of the orbit for the ion. (b) What is the difference in radius for ^{26}Mg and ^{24}Mg ions?

29. A cyclotron for accelerating protons has a magnetic field of 1.5 T and a radius of 0.5 m. (a) What is the cyclotron frequency? (b) Find the maximum energy of the protons when they emerge. (c) How do your answers change if deuterons are used instead of protons?

30. A certain cyclotron has a magnetic field of 2.0 T and is designed to accelerate protons to 20 MeV. (*a*) What is the cyclotron frequency? (*b*) What must the
minimum radius of the magnet be to achieve the 20-MeV emergence energy? (*c*)
If the alternating potential applied to the dees has a maximum value of 50 kV,
how many orbital trips must the protons make before emerging with 20 MeV
energy?

Section 26-5, The Hall Effect

31. A metal strip 2 cm wide and 0.1 cm thick carries a current of 20 A in a uniform magnetic field of 2.0 T, as shown in Figure 26-21. The Hall emf is measured to be 4.27 μV. (*a*) Calculate the drift velocity of the electrons in the strip.
(*b*) Find the number density of charge carriers in the strip.

32. (*a*) In Figure 26-21 which point (*a* or *b*) is at the higher potential? (*b*) If the
metal strip is replaced by a *p*-type semiconductor in which the charge carriers
are positive, which point will be at the higher potential?

33. The number density of free electrons in copper is 8.5×10^{22} electrons per
cubic centimetre. If the metal strip in Figure 26-21 is copper and the current is
10 A, find (*a*) the drift velocity v_d and (*b*) the Hall emf. (Assume that the magnetic field is 2.0 T, as in Exercise 31.)

34. A copper strip ($n = 8.5 \times 10^{22}$ electrons per cubic centimetre) 2 cm wide
and 0.1 cm thick (like that shown in Figure 26-21) is used to measure the magnitudes of unknown magnetic fields which are perpendicular to the strip. Find the
magnitude of B when $I = 20$ A and the Hall emf is (*a*) 2.00 μV, (*b*) 5.25 μV, and
(*c*) 8.00 μV.

Figure 26-21
Exercises 31 to 34.

Problems

1. A 10.0-cm length of wire has mass 5.0 g and is connected to a source of emf
by light flexible leads. A magnetic field $B = 0.5$ T is horizontal and perpendicular to the wire. Find the current necessary to float the wire; i.e., the current
such that the magnetic force balances the weight of the wire.

2. A metal crossbar of mass *M* rides on a pair of long horizontal conducting rails
separated by a distance *L* and connected to a device that supplies constant current *I* to the circuit, as shown in Figure 26-22. A uniform magnetic field **B** is established as shown. (*a*) If there is no friction and the bar starts from rest at
$t = 0$, show that at time *t* the bar has velocity $v = (BIL/M)t$. (*b*) In which direction will the bar move? (*c*) If the coefficient of static friction is μ_s, find the minimum field *B* necessary to start the bar moving.

3. In Figure 26-22 assume the rails are frictionless but tilted upward so that they
make an angle θ with the horizontal. (*a*) What vertical magnetic field *B* is
needed so that the bar will not slide down the rails? (*b*) What is the acceleration
of the bar if *B* is twice the value found in (*a*)?

4. Show that the radius of the orbit in a cyclotron is proportional to the square
root of the number of orbits completed.

5. A mass spectrograph is preceded by a velocity selector consisting of parallel
plates separated by 2.0 mm and having a potential difference of 160 V. The
magnetic field between the plates is 0.42 T. The magnetic field in the mass spectrograph is 1.2 T. Find (*a*) the speed of the ions entering the mass spectrograph
and (*b*) the separation distance of the peaks on the photographic plate for singly
ionized ^{238}U and ^{235}U.

6. A wire bent into some arbitrary shape carries a current *I* in a uniform magnetic field **B**. Show that the total force on the part of the wire from some point *a*
to some point *b* is **F** = *I***L** × **B**, where **L** is the vector from *a* to *b*. *Hint:* Integrate
$d\mathbf{F} = I\,d\boldsymbol{\ell} \times \mathbf{B}$ for constant **B**.

Figure 26-22
Problems 2 and 3.

7. If you have a wire of fixed length L and make a coil of N turns, the larger the number of turns, the smaller the area enclosed by the wire. Show that for a wire of given length carrying current I the maximum magnetic moment is achieved with a coil of just one turn and the magnitude of this magnetic moment is $IL^2/4\pi$. (You need consider only circular coils. Why?)

8. A nonconducting rod of length ℓ has a uniform charge per unit length λ and is rotated with angular velocity ω about an axis through one end and perpendicular to the rod. (a) Consider a small segment of length dx and charge $dq = \lambda\,dx$ at a distance x from the pivot. Show that the magnetic moment of this segment is $\frac{1}{2}\lambda\omega x^2\,dx$. (b) Integrate your result to show that the total magnetic moment of the rod is $m = \frac{1}{6}\lambda\omega\ell^3$. (c) Show that the magnetic moment $\mathbf{m}$ and angular momentum $\mathbf{L}$ are related by $\mathbf{m} = (Q2M)\mathbf{L}$, where M is the mass of the rod and Q is the total charge.

9. A nonconducting disk of radius R has uniform surface charge density σ and rotates with angular velocity ω about its axis. (a) Consider a ring of radius r and thickness dr. Show that the total current in this ring is $dI = (\omega/2\pi)\,dq = \omega\sigma r\,dr$. (b) Show that the magnetic moment of the ring is $dm = \pi\omega\sigma r^3\,dr$. (c) Integrate your result for part (b) to show that the total magnetic moment of the disk is $m = \frac{1}{4}\pi\omega\sigma R^4$. (d) Show that the magnetic moment $\mathbf{m}$ and angular momentum $\mathbf{L}$ are related by $\mathbf{m} = (Q/2M)\mathbf{L}$, where M is the total mass of the disk and Q is its total charge.

10. Particles of charge q and mass m are accelerated from rest through a potential difference V and enter a region of uniform magnetic field $\mathbf{B}$ perpendicular to the velocity. If r is the radius of curvature of their circular orbit, show that $q/m = 2V/r^2B^2$.

11. A beam of particles enters a region of uniform magnetic field $\mathbf{B}$ with velocity $\mathbf{v}$ which makes a small angle θ with $\mathbf{B}$. Show that after a particle moves a distance $2\pi(m/qB)v\cos\theta$ measured along the direction of $\mathbf{B}$, the velocity of the particle is in the same direction as when it entered the field.

12. A small magnet of moment $\mathbf{m}$ makes an angle θ with a uniform magnetic field $\mathbf{B}$. (a) How much work must be done by an external torque to twist the magnet by a small amount $d\theta$? (b) Show that the work required to rotate the magnet until it is perpendicular to the field is $W = mB\cos\theta$. (c) Use your result for part (b) to show that if the potential energy of the magnet is chosen as zero when the magnet is perpendicular to the field, the potential energy at angle θ is $U(\theta) = -\mathbf{m}\cdot\mathbf{B}$. (d) Would any part of this problem be different if the magnet were replaced by a coil carrying current such that its magnetic moment is $\mathbf{m}$?

CHAPTER 27 Sources of the Magnetic Field

We now turn to a consideration of the origins of the magnetic field **B**. The earliest known sources of magnetism were magnets, but after Hans Christian Oersted described his finding in September 1820 that a compass needle is deflected by an electric current, many scientists investigated the properties of the magnetism associated with electric currents. One month later, Jean Baptiste Biot and Félix Savart announced the results of their measurements of the force on a magnetic pole near a long current-carrying wire and analyzed these results in terms of the magnetic field produced by each element of the current. André Marie Ampère extended these experiments and showed that current elements themselves experience a force in the presence of a magnetic field; in particular, he showed that two currents exert forces on each other.

Ampère also obtained the Biot-Savart result for the magnetic field due to a current element and developed a model explaining the behavior of permanent magnets in terms of microscopic current loops within the magnetic material. In this model it is proposed that all matter contains microscopic current loops. In nonmagnetic materials the loops are randomly oriented, producing no net effect, but in magnetic materials, current loops are aligned and produce a magnetic field with the same characteristics as that produced by aligned current loops of ordinary conduction current in wires. Ampère's model is essentially correct in that we can treat all magnetic fields as arising from an electric current of some kind. However, the microscopic currents (called *amperian currents*) are more complicated than Ampère could have foreseen, being related to the motions of atomic electrons, which cannot be fully described classically. In particular, ferromagnetism is intimately connected with electron spin, an intrinsic property of the electron.

27-1 The Biot-Savart Law

From their investigations of the force produced by a long wire carrying a current on a magnetic pole, Biot and Savart proposed an expression relating the magnetic induction vector **B** at a point in space to an element of the current that produces it. Let $I\,d\boldsymbol{\ell}$ be an element of current (Figure 27-1). The magnetic induction $d\mathbf{B}$ produced by this element at a field point P a distance r away is

$$d\mathbf{B} = k_m \frac{I\,d\boldsymbol{\ell} \times \hat{\mathbf{r}}}{r^2} \qquad\qquad 27\text{-}1 \qquad \textit{Biot-Savart law}$$

where $\hat{\mathbf{r}}$ is the unit vector pointing from the current element to the field point and k_m is a constant which depends on the units. Equation 27-1, the *Biot-Savart law*, was also deduced by Ampère. The magnetic induction due to the total current in a circuit can be found by using the Biot-Savart law for the field due to each current element and summing over all the current elements in the circuit.

The SI unit of current, the ampere, is defined so that the magnetic constant k_m is exactly 10^{-7}. This definition of the ampere will be discussed in more detail in the next section. The units of k_m will be shown below to be newtons per ampere[2]. As with the Coulomb constant, it is customary to write the magnetic constant k_m in terms of another constant μ_0, called the *permeability of free space*. These constants are related by

$$k_m = \frac{\mu_0}{4\pi} = 10^{-7}\ \mathrm{N/A^2}$$

$$\mu_0 = 4\pi k_m = 4\pi \times 10^{-7}\ \mathrm{N/A^2} \qquad\qquad 27\text{-}2$$

In terms of the constant μ_0 the Biot-Savart law is written

$$d\mathbf{B} = \frac{\mu_0}{4\pi} \frac{I\,d\boldsymbol{\ell} \times \hat{\mathbf{r}}}{r^2} \qquad\qquad 27\text{-}3$$

The Biot-Savart law is analogous to Coulomb's law. The source of the magnetic field is the current element $I\,d\boldsymbol{\ell}$, just as the charge q is the source of the electrostatic field. The magnetic field decreases as the square of the distance from the current element, like the decrease in the electric field with distance from a point charge, but the directional aspects of these fields are quite different. Whereas the electrostatic field points in the radial direction $\mathbf{r}$ from the point charge to the field point (assuming a positive charge), the magnetic field is perpendicular both to the radial direction and to the direction of the current element $I\,d\boldsymbol{\ell}$. At a point along the line of the current element, e.g., point P_2 in Figure 27-1, the magnetic field is zero.

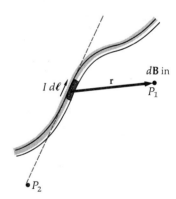

Figure 27-1
The magnetic field $d\mathbf{B}$ at point P_1 due to the current element $I\,d\boldsymbol{\ell}$ is $d\mathbf{B} = k_m(I\,d\boldsymbol{\ell} \times \hat{\mathbf{r}})/r^2$. At point P_2 the field due to this element is zero because $d\boldsymbol{\ell} \times \hat{\mathbf{r}}$ is zero.

By combining the Biot-Savart law with the expression for the force on a current element in a magnetic field (Equation 26-4) we can write an equation for the force exerted by one current element on another. The force on current element $I_2 \, d\boldsymbol{\ell}_2$ exerted by element $I_1 \, d\boldsymbol{\ell}_1$ is

$$d\mathbf{F}_{12} = I_2 \, d\boldsymbol{\ell}_2 \times \left(k_m \frac{I_1 \, d\boldsymbol{\ell}_1 \times \hat{\mathbf{r}}}{r^2} \right) \qquad 27\text{-}4$$

From this equation we can see that if the force is in newtons, the distances $d\boldsymbol{\ell}_1$, $d\boldsymbol{\ell}_2$, and r are in metres, and the currents are in amperes, the units of k_m are newtons per ampere^2.

Optional

Figure 27-2
The forces exerted by current elements on each other are not equal and opposite. Here **B** at element 2 due to element 1 is into the paper, giving a force on element 2 to the left, but **B** due to element 2 at element 1 is zero.

This relation is remarkable in that the force exerted by element 1 on element 2 is not equal and opposite to that exerted by element 2 on element 1. That is, these forces do not obey Newton's action-reaction law, as can be demonstrated by considering the special case illustrated in Figure 27-2. Here, the magnetic field at element 2 due to element 1 is into the paper, and the force on element 2 is to the left. However, the magnetic field at element 1 due to element 2 is zero because $d\boldsymbol{\ell}_2 \times \hat{\mathbf{r}}$ is zero. Thus no force is exerted on element 1. If the force $\mathbf{F}_{12}$ is the only force acting on the two-current-element system, the system will accelerate in the direction of $\mathbf{F}_{12}$ and linear momentum of the system will apparently not be conserved. We recall that it was the experimental observation of conservation of momentum in collisions that originally led Newton to the law of action and reaction. In most situations, current elements are but a part of a complete circuit. If the current elements shown are parts of complete circuits, as in Figure 27-3, there will be forces on the elements from other parts of the circuits. A detailed analysis of the total force exerted on one circuit by the other shows that the total forces *do* obey Newton's third law; i.e., the force exerted on circuit 1 by circuit 2 is equal and opposite to the total force exerted on circuit 2 by circuit 1. This analysis can be found in intermediate books on electricity and magnetism.

It is possible to produce the equivalent of isolated current elements by accelerating charges for a short time and then stopping them. Then the problem of the apparent violation of Newton's action-reaction law and of conservation of momentum is real. It is this type of situation for which our discussion of action at a distance in Chapter 4 is pertinent. We need to include in our system the electric and magnetic fields as well as the two current elements. The acceleration of electric charges necessary to produce isolated current elements also produces electromagnetic radiation, which carries momentum. A detailed analysis of a situation like that pictured in Figure 27-2 shows that indeed the current element system is accelerated to the left while the electromagnetic radiation carries momentum to the right. When we include the field and its momentum in the system, the total momentum of the system is again conserved.

Figure 27-3
Complete circuits containing the current elements shown in Figure 27-2. The total force which circuit 1 exerts on circuit 2 is equal and opposite to the total force which circuit 2 exerts on circuit 1.

We now illustrate how the Biot-Savart law is used to calculate the magnetic induction for two important current configurations, a long straight wire and a current loop.

Magnetic Induction of a Long Straight Current

Figure 27-4 shows the geometry needed to calculate the magnetic induction **B** at point P due to the current in the long wire shown. We choose the wire to be the x axis and the y axis to be perpendicular to the wire through point P. A typical current element $I\,d\ell$ at a distance x is shown.

Figure 27-4
(a) Geometry for calculation from the Biot-Savart law of the magnetic field at point P due to a straight current segment. Each element gives a contribution to the magnetic field at point P directed out of the paper. (b) The result is given by Equation 27-6 in terms of the angles θ_1 and θ_2.

The vector **r** points from the element to the field point P. The direction of the magnetic field at P due to this element is out of the paper, as determined by the direction of $I\,d\ell \times \mathbf{r}$. We note that all such current elements of the wire give contributions in this same direction, and so we need only compute the magnitude of the field. The field due to the current element shown has the magnitude

$$dB = k_m \frac{I\,dx}{r^2} \sin \phi = k_m \frac{I\,dx \cos \theta}{r^2} \qquad 27\text{-}5$$

where ϕ and θ are the angles shown. In order to sum over all the current elements, we need to relate the variables θ, r, and x. It turns out to be easiest to eliminate x and r in favor of θ. We have

$$x = y \tan \theta$$

Then $dx = y \sec^2 \theta\, d\theta = y(r^2/y^2)\, d\theta = (r^2/y)\, d\theta$ using $\sec \theta = r/y$. Substituting this expression for dx into Equation 27-5, we have

$$dB = \frac{k_m I}{r^2} \frac{r^2\,d\theta}{y} \cos \theta = \frac{k_m I}{y} \cos \theta\, d\theta$$

Let us first calculate the contribution from the current elements to the right of the point $x = 0$. We sum over these elements by integrating from $\theta = 0$ to $\theta = \theta_1$, where θ_1 is the angle between the line perpendicular to the wire and the line from P to the right end of the wire, as shown. We have for this contribution

$$B_1 = \int_0^{\theta_1} \frac{k_m I}{y} \cos \theta\, d\theta = \frac{k_m I}{y} \sin \theta_1$$

Similarly, the contribution from elements to the left of the point $x = 0$ is

$$B_2 = \frac{k_m I}{y} \sin \theta_2$$

We thus have for the total magnetic field due to the wire

$$B = \frac{k_m I}{y}(\sin\theta_1 + \sin\theta_2) \qquad 27\text{-}6$$

Magnetic field of a straight wire

This result gives the magnetic field due to any wire segment in terms of the angles subtended at the field point by the ends of the wire. If the wire is very long, these angles are nearly 90°. The result for an infinitely long wire is obtained from Equation 27-6 by using $\theta_1 = \theta_2 = 90°$:

$$B = \frac{2k_m I}{y} = \frac{\mu_0 I}{2\pi y} \qquad 27\text{-}7$$

B due to an infinite straight wire

The direction of **B** is such that the lines of **B** encircle the wire as shown in Figure 27-5. This direction can be remembered by a right-hand rule, as shown in that figure. This result for the magnetic field **B** due to a current in a long straight wire was found experimentally by Biot and Savart in 1820. From the analysis of this experimental result they were able to discover the expression given in Equation 27-1 for the magnetic induction due to an element of the current.

Figure 27-5
The magnetic field due to a very long straight wire is tangent to a circle around the wire. The direction of **B** is related to the direction of the current I by the right-hand rule illustrated.

Example 27-1 Find the magnetic induction field at the center of a square current loop of side 1 m carrying a current of 1 A.

From Figure 27-6 we see that each side of the loop contributes a field in the direction out of the paper. By the symmetry of the situation we need only calculate the field due to one side and multiply by 4. The distance between one side and the field point y is $\frac{1}{2}\ell = \frac{1}{2}$m, and the angles θ_1 and θ_2 are 45°. Thus the field due to one segment is

$$B_1 = \frac{k_m I}{\frac{1}{2}\ell}(\sin 45° + \sin 45°) = \frac{2k_m I}{\ell}2\frac{1}{\sqrt{2}} = 2\sqrt{2}\,\frac{k_m I}{\ell}$$

and the total field is

$$B = 4B_1 = 8\sqrt{2}\,\frac{k_m I}{\ell} = 8\sqrt{2}\,\frac{10^{-7}\times 1}{1} = 11.3\times 10^{-7}\text{ T}$$

Figure 27-6
Square current loop for Example 27-1.

Magnetic Induction on the Axis of a Circular Current Loop

The geometry needed for this calculation is shown in Figure 27-7. We first consider the current element at the top of the loop. Here $I\,d\ell$ is out of the paper and perpendicular to the vector **r**, which lies in the plane of the paper. The magnetic field due to this element is in the direction shown. The field $d\mathbf{B}$ is perpendicular to **r**, and since it is perpendicular to $I\,d\ell$, it must be in the plane of the paper as shown. The magnitude of $d\mathbf{B}$ is

$$|d\mathbf{B}| = k_m\frac{|I\,d\ell\times\hat{\mathbf{r}}|}{r^2} = k_m\frac{I\,d\ell}{x^2+R^2}$$

since $d\ell$ and $\hat{\mathbf{r}}$ are perpendicular and $r^2 = x^2 + R^2$.

When we sum around the current elements in the loop, components of $d\mathbf{B}$ perpendicular to the axis of the loop sum to zero, leaving only the components parallel to the axis. We thus compute only the x component. From the figure,

$$dB_x = dB\sin\theta = dB\frac{R}{\sqrt{x^2+R^2}} = \frac{k_m I\,d\ell}{x^2+R^2}\frac{R}{\sqrt{x^2+R^2}}$$

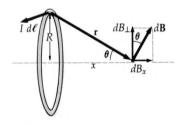

Figure 27-7
Geometry for the calculation of **B** on the axis of a current loop.

Since neither x nor R varies as we sum over the elements in the loop, this sum merely gives $\oint d\ell = 2\pi R$:

$$B_x = \frac{k_m I(2\pi R)R}{(x^2 + R^2)^{3/2}} = \frac{\mu_0}{2} \frac{IR^2}{(x^2 + R^2)^{3/2}} \qquad 27\text{-}8$$

B on axis of a circular loop

An important special case of this result is the magnetic field at the center of the loop, which we obtain by setting $x = 0$:

$$B_x = \frac{\mu_0 I}{2R} \qquad 27\text{-}9$$

At great distances from the loop, x is much greater than R and we can neglect R^2 compared with x^2 in the denominator of Equation 27-8. Then

$$B_x \to \frac{2k_m I(\pi R^2)}{x^3} = \frac{2k_m m}{x^3} = \frac{\mu_0}{2\pi} \frac{m}{x^3} \qquad 27\text{-}10$$

Dipole field of a current loop

where $m = I(\pi R^2)$ is the magnetic moment of the loop. Note the similarity of this expression to that for the electrostatic field on the axis of an electric dipole of moment p:

$$E_x \to \frac{2kp}{x^3}$$

A current loop behaves like a magnetic dipole both in experiencing a torque $\mathbf{m} \times \mathbf{B}$ when placed in an external magnetic field and in producing a dipole field at great distances from the loop. Figure 27-8 shows the lines of $\mathbf{B}$ for a current loop. Our result that a current loop produces a dipole field is not confined to the axis of the loop.

Questions

1. For a given distance r from a current element, in what direction does the contribution of the current element to the total magnetic field have its greatest value?

2. What is the effect of replacing the single loop in Figure 27-7 with a coil of N turns of the same radius, each carrying the current I?

3. A common method for cutting down on the magnetic fields created by leads carrying current to electrical devices is to twist together the wires carrying the incoming and outgoing currents. Explain.

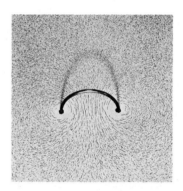

Figure 27-8
Lines of magnetic induction $\mathbf{B}$ due to a circular current loop indicated by iron filings. (*Avco Research Laboratory.*)

27-2 Definition of the Ampere and the Coulomb

One week after Ampère heard of Oersted's discovery of the effect of a current on a compass needle, he showed that two parallel currents attract each other if they are in the same direction and repel each other if they are in opposite directions. Figure 27-9 shows two long straight wires separated by a distance r and carrying currents I_1 and I_2 in the same direction. We can compute the force exerted by one wire on the other from our result for the magnetic field of a long current-carrying wire and Equation 26-3 for the force exerted by a magnetic field on a current element. Let us consider a segment of the second wire of length ℓ_2. The magnetic field $\mathbf{B}_1$ at this segment is perpendicular to the segment

Figure 27-9
Two long straight wires carrying parallel currents. The magnetic field at the current element $I_2\ell_2$ due to current I_1 is into the paper, giving an attractive force $\mathbf{F}_2 = I_2\ell_2 \times \mathbf{B}_1$.

and has the magnitude $2k_mI_1/r$ if the segment is sufficiently close to wire 1 and wire 1 is long enough for the angles in Equation 27-6 to be approximately 90°. The force on this segment has the magnitude

$$F_2 = |I_2\boldsymbol{\ell}_2 \times \mathbf{B}_1| = \frac{2k_mI_1I_2\ell_2}{r}$$

From the direction of the cross product $\boldsymbol{\ell}_2 \times \mathbf{B}_1$ we see that this force on the segment of wire 2 is toward wire 1. The force per unit length is

$$\frac{F_2}{\ell_2} = \frac{2k_mI_1I_2}{r} \qquad\qquad 27\text{-}11$$

In Chapter 20 we deferred the definition of the coulomb as a unit of charge, mentioning that it is defined in terms of the ampere. The ampere is defined as follows:

Ampere defined

This definition allows the unit of current (and therefore of electric charge) to be determined by a mechanical experiment. In practice of course, the currents are chosen to be much closer together than 1 m so that the wires need not be so long and the force is large enough to measure accurately. Figure 27-10 shows a current balance which can be used to calibrate an ammeter from the fundamental definition of the ampere. The upper conductor is free to rotate about the knife edges and is balanced so that the wires are a small distance apart. Since the conductors are wired in series to carry the same current but in opposite directions, the wires repel rather than attract. The force of repulsion can be measured by placing weights on the upper conductor until it balances again at the original separation. This definition of the ampere makes the magnetic constant $k_m = \mu_0/4\pi$ exactly equal to 10^{-7}. The ratio of the Coulomb constant k and the magnetic constant k_m is

$$\frac{k}{k_m} = \frac{1/4\pi\epsilon_0}{\mu_0/4\pi} = \frac{1}{\epsilon_0\mu_0} = \frac{9 \times 10^9 \text{ N·m}^2/\text{C}^2}{10^{-7} \text{ N·s}^2/\text{C}^2} = 9 \times 10^{16} \text{ m}^2/\text{s}^2$$

This ratio equals the square of the speed of light, a fact noted by James Clerk Maxwell in 1860. He showed that the laws of electricity and magne-

Figure 27-10
Current balance used in elementary physics laboratory to calibrate an ammeter. (*Courtesy of Larry Langrill, Oakland University.*)

tism imply that an accelerated charge radiates energy in the form of waves that travel with speed

$$c = \sqrt{\frac{k}{k_m}} = \frac{1}{\sqrt{\epsilon_0 \mu_0}} = 3 \times 10^8 \text{ m/s} \qquad 27\text{-}12$$

Since this speed is the same as that of light, Maxwell speculated that light is an electromagnetic wave produced by the acceleration of atomic charges.

27-3 Ampère's Law

We have noted that the lines of **B** for a long straight current-carrying wire encircle the wire. These lines are quite different from any lines of electric field we have studied. The electrostatic field is conservative. The work done on a test charge by the electrostatic field when the charge is moved in a complete circle is always zero. This work per charge is the sum of $\mathbf{E} \cdot d\ell$ around the path. The line integral of the electrostatic field around any closed path is zero because the electrostatic field is conservative:

$$\oint_C \mathbf{E} \cdot d\ell = 0 \qquad C \text{ any closed curve}$$

Clearly the sum of $\mathbf{B} \cdot d\ell$ around a closed path is not necessarily zero. If we take this sum along a circular path enclosing a long wire carrying a current, the magnetic-induction vector **B** is everywhere tangent to the path (Figure 27-11). Then $\mathbf{B} \cdot d\ell$ is everywhere positive if we travel in the direction of the field lines. Since **B** is in fact parallel to $d\ell$ and has the constant magnitude given by equation 27-7, we can easily compute this sum:

$$\oint \mathbf{B} \cdot d\ell = \frac{2k_m I}{r} \oint d\ell = \frac{2k_m I}{r} 2\pi r = 4\pi k_m I \qquad 27\text{-}13a$$

This integral is independent of the radius chosen for the circular path.

Brown Brothers

André Marie Ampere (1775–1836).

Figure 27-11
Circular path around a long current-carrying wire. The integral of $\mathbf{B} \cdot d\ell$ around this curve is $4\pi k_m I = \mu_0 I$.

This result, known as *Ampère's law*, is more general than indicated by our calculation. It holds for the line integral of $\mathbf{B} \cdot d\ell$ around any closed path which encloses a steady current.

The line integral around any closed path equals $4\pi k_m$ times the total current which crosses any area bounded by the path. In terms of the constant $\mu_0 = 4\pi k_m$, Ampère's law is written

$$\oint_C \mathbf{B} \cdot d\ell = \mu_0 I \qquad 27\text{-}13b \qquad \textit{Ampère's law}$$

(It is because of the 4π factor in Equation 27-13*a* that the magnetic constant k_m is written $\mu_0/4\pi$.)

Ampère's law holds for any curve C as long as the currents are steady. Like Gauss' law, it is *useful* for the calculation of the magnetic field only in situations with considerable symmetry. If the symmetry is great enough, the line integral can be written as the product of B and some distance, thus relating B to the current enclosed by the curve. Also, like Gauss' law, if the currents are steady, Ampère's law holds even if there is no symmetry; but in those cases it is no help in finding an expression for the magnetic field.

Figure 27-12
Long straight wire of radius a carrying current I uniformly distributed with current density $J = I/\pi a^2$. The magnetic field at distance r inside the wire can be calculated from Ampère's law applied to the circle of radius r.

Example 27-2 A long straight wire of radius a carries a current I which is uniformly distributed over the wire with current density $J = I/\pi a^2$ (Figure 27-12). Find the magnetic field both inside and outside the wire.

We can calculate **B** from Ampère's law because of the high degree of symmetry. Consider a circle of radius r concentric with the axis of the wire. We expect **B** to be tangential to the circle, as with a very thin wire. By symmetry, the magnitude of **B** is constant everywhere on the circle. Thus

$$\oint \mathbf{B} \cdot d\boldsymbol{\ell} = B \oint d\ell = 2\pi r B$$

The current through C depends on whether r is less or greater than the radius of the wire. For r greater than a, the total current I crosses the area bounded by C. Then

$$\oint \mathbf{B} \cdot d\boldsymbol{\ell} = B2\pi r = \mu_0 I$$

$$B = \frac{\mu_0 I}{2\pi r} = 2\frac{\mu_0}{4\pi}\frac{I}{r} \qquad r > a$$

This result is the same as obtained in Equation 27-7. To find **B** inside the wire we choose r less than a. Then the current through C is

$$\pi r^2 J = \frac{r^2}{a^2} I$$

and

$$\oint \mathbf{B} \cdot d\boldsymbol{\ell} = 2\pi r B = \mu_0 \frac{r^2}{a^2} I$$

$$B = \frac{\mu_0}{2\pi}\frac{I}{a^2} r \qquad r < a \qquad\qquad 27\text{-}14$$

Figure 27-13 is a sketch of B versus r.

Figure 27-13
Plot of B versus r calculated in Example 27-2. Inside the wire, B is proportional to r, whereas outside, it is inversely proportional to r.

B inside wire carrying uniform current

For an example in which Ampère's law is not useful in calculating the magnetic induction due to a steady current, consider a current loop as shown in Figure 27-14. We calculated the magnetic induction on the axis of such a loop from the Biot-Savart law (Equation 27-8). According to Ampère's law, the line integral of $\mathbf{B} \cdot d\boldsymbol{\ell}$ around a curve such as curve C in Figure 27-14 equals μ_0 times the current I in the loop. Although Ampère's law is true for this curve, the magnetic induction **B** is not constant along any curve encircling the current, nor is it everywhere tangent to any such curve. Thus there is not enough symmetry in this situation to allow us to calculate B from Ampère's law.

Limitations of Ampère's Law

Suppose we attempt to use Ampère's law to find the magnetic field at a point on the perpendicular bisector of a current segment of length ℓ. Let the point be a distance r from the segment.

Figure 27-14
Ampère's law holds for curve C encircling the current in the circular loop, but it is not useful in finding **B** because **B** is neither constant along the curve nor tangent to it.

This situation is shown in Figure 27-15. A direct application of Ampère's law again gives

$$B = \frac{\mu_0}{2\pi} \frac{I}{r}$$

This result is the same as for an infinitely long wire since we have the same symmetry arguments. The result does not agree with that obtained from the Biot-Savart law. That law gives a smaller result which depends on the length of the current segment and agrees with experiment. There are two possibilities for the discrepancy: (1) Ampère's law is correct but was incorrectly applied; (2) Ampère's law is not correct. In order to understand which explanation applies we must ask how we can obtain a current segment as shown in Figure 27-15. One possibility is shown in Figure 27-16, where the current segment is just one segment of a continuous loop carrying a steady current. For this situation, Ampère's law is correct, but our application was not. If there are other current segments in the problem, as in Figure 27-16, we do not have the symmetry we assumed to compute $\oint \mathbf{B} \cdot d\boldsymbol{\ell}$. For example, $\mathbf{B}$ is not tangential to the curve shown, and its magnitude is not constant along the curve. Then even though Ampère's law is correct, it is not useful for the calculation of the magnetic field for this situation. Figure 27-17 shows another possibility for obtaining a current segment, i.e., a source of charge at point P_1 and a sink at point P_2. For example, we might have a small spherical conductor with charge $+Q$ at the left and one with charge $-Q$ at the right. When they are connected, a current $I = -dQ/dt$ exists in the segment for a short time until the spheres are uncharged. For this case we do have the symmetry needed to assume that $\mathbf{B}$ is tangential to the curve and of constant magnitude along the curve, but the current is not a steady current. *When the current is not steady, Ampère's law does not hold.* We shall see in Section 27-7 how Maxwell generalized Ampère's law to hold for all currents, steady or not. When the generalized form is used, the magnetic field calculated for this situation agrees with that found from the Biot-Savart law.

Figure 27-15
Application of Ampère's law to find the magnetic field on the bisector of a finite current segment gives an incorrect result.

Figure 27-16
If the current segment in Figure 27-15 is a part of a complete circuit carrying a steady current, Ampère's law for curve C is correct, but there is not enough symmetry to use it to find the magnetic field at point P.

Figure 27-17
If the current segment in Figure 27-15 is due to a momentary flow of charge from a small conductor on the left to one at the right, there is enough symmetry to compute $\oint \mathbf{B} \cdot d\boldsymbol{\ell}$ along curve C, but Ampère's law does not hold because the current is not steady.

27-4 The Magnetic Field of a Solenoid

A wire wound into a helix, as in Figure 27-18, called a *solenoid,* is used to produce a strong uniform magnetic field in a small region of space. It plays a role in magnetism analogous to that of the parallel-plate capacitor in providing a strong uniform electrostatic field between its plates. The magnetic field of a solenoid is essentially that of a set of N identical coils placed side by side. Figure 27-19 shows the lines of magnetic induction for two such coils. In the space between the coils near the axis, the fields of the individual coils add; between the coils but at distances from the axis greater than the coil radius the individual fields tend to cancel. Figure 27-20 shows the field lines for a long tightly wound solenoid. Inside the solenoid, the field lines are approximately parallel to

Figure 27-18
A solenoid produces a strong uniform magnetic field in the space inside it.

the axis and closely and uniformly spaced, indicating a strong uniform magnetic field. Outside the solenoid, the lines are much less dense. They diverge from one end and converge at the other end, like the lines of electric field from disks of positive and negative charge (Figure 27-21). Note that *inside* the solenoid the magnetic field lines are very different from electric field lines from the disk charges.

Because the magnetic field is very strong and nearly uniform inside the solenoid, as indicated by the field lines in Figure 27-20, and the field is very weak outside the solenoid, we can use Ampère's law to find a useful approximate result for the magnitude of **B**. Let the solenoid have radius r, length ℓ, and N turns of wire carrying a current I, and assume that ℓ is much greater than r. We shall assume that **B** is uniform and parallel to the axis inside the solenoid and zero outside. We apply Ampère's law to the curve C (Figure 27-22), which is a rectangle with sides of length a and b. The only contribution to the integral $\oint \mathbf{B} \cdot d\boldsymbol{\ell}$ around this curve is along side 1, since along the opposite side (outside the solenoid) the magnetic field is assumed to be zero, and along the two short sides **B** is perpendicular to $d\boldsymbol{\ell}$. Since **B** is assumed to be uniform and axial, the total line integral is merely Bb. The total current through this curve is the current I times the number of turns of wire enclosed.

Figure 27-19
The magnetic fields due to two adjacent loops of a solenoid add in the region inside the solenoid and subtract in the region outside.

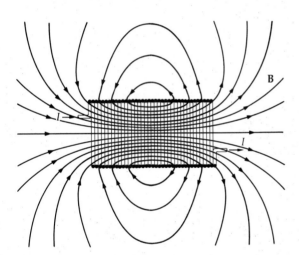

Figure 27-20
Magnetic field lines of a tightly wound solenoid. The field is strong and nearly uniform inside the solenoid. Outside it the field lines are like those of the electric field due to two disk charges (Figure 27-21).

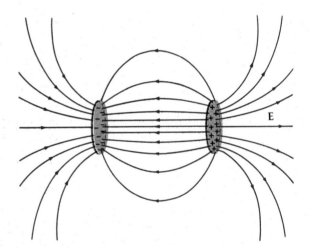

Figure 27-21
Electric field lines due to two oppositely charged disks. The lines outside the cylinder bounded by the disks are like the magnetic field lines of a solenoid (Figure 27-20). Between the disks the **E** lines are directed opposite to the corresponding magnetic field lines of a solenoid.

Since there are N turns in a total length ℓ, the number of turns in the length b is Nb/ℓ and Ampère's law gives

$$\oint \mathbf{B} \cdot d\boldsymbol{\ell} = Bb = \mu_0 \frac{NbI}{\ell}$$

or

$$B = \mu_0 \frac{N}{\ell} I = \mu_0 nI \qquad\qquad 27\text{-}15 \qquad B \text{ inside solenoid}$$

Figure 27-22
The magnetic field inside a solenoid can be calculated by applying Ampere's law to the curve C shown, assuming that **B** is uniform inside and zero outside.

where $n = N/\ell$ is the number of turns per unit length. Figure 27-23 shows how the magnetic field on the axis of a solenoid varies with distance from the ends. The approximation that the field is constant independent of position along the axis is quite good except very near the ends. At a point right at the end, the magnitude of **B** is about half that at the center for a long solenoid.

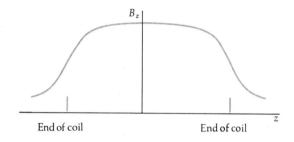

Figure 27-23
Plot of B_z versus z on the axis of a solenoid. Note that B_z is nearly constant inside the solenoid except near the ends.

We shall calculate the magnetic field outside the solenoid only for the special case of a point far from the end compared with the radius of the coils. Then each coil gives a dipole field, since we are far from it compared with its radius. We can calculate the resultant field for N coils by summing the fields of the individual coils. We can anticipate the result. Since each loop produces a field like that of a small bar magnet, i.e., a dipole field, we expect a set of adjacent loops to give a field like that of a set of adjacent small bar magnets, i.e., like a bar magnet of length ℓ. This result is derived at the end of this section.

Figure 27-24 shows the magnetic field lines of a bar magnet of the same shape as the solenoid in Figure 27-18. The lines of **B** are identical for these two cases.

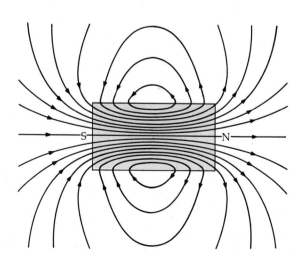

Figure 27-24
Magnetic field lines of a bar magnet. The lines are the same everywhere as those of a solenoid of similar shape.

Question

4. The magnetic field inside a long solenoid has magnitude B. How would the field change if the current in the solenoid were doubled? If the diameter were doubled without changing the length or number of turns? If the number of turns were doubled by making the solenoid twice as long? If the number of turns were doubled without changing the length of the solenoid?

Optional

Fred Weiss

Magnetic field lines of a solenoid, indicated by iron filings.

Figure 27-25
Geometry for calculating the magnetic field on the axis of a tightly wound solenoid at a point far from both ends. The element of length dx gives a magnetic dipole field of a current loop carrying current $di = nI\,dx$, where I is the current in the solenoid.

Let us consider a point on the axis a distance x_0 from the center and distances $x_0 - \frac{1}{2}\ell$ from the close end and $x_0 + \frac{1}{2}\ell$ from the far end, with $x_0 - \frac{1}{2}\ell$ much greater than the solenoid radius r (Figure 27-25). We first consider an element of the solenoid of length dx at a distance x from the center. If $n = N/\ell$ is the number of turns per unit length, there are $n\,dx$ turns of wire in this element, each carrying a current I. The element is thus equivalent to a loop carrying a current $di = nI\,dx$, giving a dipole field

$$dB = 2k_m \frac{di\,A}{x'^3} = 2k_m nIA \frac{dx}{x'^3}$$

where $x' = x_0 - x$ is the distance from the element to the field point and $A = \pi r^2$ is the area of the coil. We find the total field at x_0 due to the solenoid by summing from $x = -\frac{1}{2}\ell$ to $x = +\frac{1}{2}\ell$. Performing the integration between these limits, we obtain

$$B = k_m nIA \left(\frac{1}{r_1^2} - \frac{1}{r_2^2} \right) \qquad\qquad 27\text{-}16$$

where $r_1 = x_0 - \frac{1}{2}\ell$ is the distance to the near end and $r_2 = x_0 + \frac{1}{2}\ell$ is the distance to the far end. We have obtained the expected result. We can put this into a form similar to Coulomb's law by defining a magnetic-pole strength q_m:

$$q_m = nIa$$

Then

$$B = k_m \frac{q_m}{r_1^2} - k_m \frac{q_m}{r_2^2} \qquad\qquad 27\text{-}17$$

Note that the dipole moment of the solenoid is

$$q_m \ell = nIA\ell = NIA \qquad\qquad 27\text{-}18$$

which is the sum of the dipole moments of the individual loops, as expected.

27-5 The Magnetic Field of a Bar Magnet

The first quantitative investigations of magnetic fields were made with permanent magnets. Using a torsion balance, John Michell found that the force of attraction or repulsion between two poles varies as the product of the pole strengths and inversely with the square of the distance between the poles. Since the force acting on a single pole in a magnetic induction field **B** is $q_m\mathbf{B}$, this result implies that the magnetic field due to a pole is proportional to q_m/r^2 and in the radial direction. According to Ampère's model, all magnetic fields are due to currents of some kind. In permanent magnets or other magnetized material these currents are due to the intrinsic motion of atomic electrons. Although these motions are very complicated, for this model we need only assume that the motions are equivalent to closed-circuit loops (Figure 27-26). The magnetic moments of the current loops shown in this figure are in the direction parallel to the axis. If the material is homogeneous, the net current at any point inside the material is zero because of cancellation of neighboring current loops. However, since there is no cancellation on the surface of the material, the result of these current loops is equivalent to a current on the surface of the material, called an *amperian current*. The surface current is similar to the real conduction current in a tightly wound solenoid. The magnetic field due to the surface current is the same as that due to an equivalent "surface" current in a solenoid. Let M be the amperian current per unit length on the surface of a cylindrical bar magnet. The corresponding quantity for a solenoid is nI, where n is the number of turns per unit length and I is the current in each turn. If we replace the current per unit length nI in Equation 27-18 by M, we have for the dipole moment of the magnet

$$q_m\ell = M(A\ell) \qquad 27\text{-}19$$

where A is the cross-sectional area of the magnet and ℓ the length. Since $A\ell$ is the total volume of the magnet, M equals the magnetic moment per unit volume (Figure 27-27). The vector **M**, whose magnitude is the magnetic moment per unit volume and whose direction is that of the dipole moment of the magnet, is called the *magnetization vector* of the magnet. The direction of **M** is from the south (negative) pole to the north (positive) pole. (Any magnetized material, whether it is a permanent magnet or some material magnetized by an external magnetic field, can be characterized by the magnetic moment per unit volume, **M**, which for non-homogeneous material may vary from point to point in the material.) The pole strength of a cylindrical bar magnet of cross-sectional area A is related to its magnetization M through Equation 27-19:

$$q_m = MA \qquad 27\text{-}20$$

The magnetic induction B due to a bar magnet is then given by Equation 27-17 with q_m equal to the magnetization times the area. In our derivation of that equation for a solenoid we assumed that the field point was far away compared with the radius of the solenoid. Thus this equation is only an approximation for the field due to a magnet also. This approximation is equivalent to the assumption that the pole strength of the magnet is concentrated at a single point at each end of the magnet. From the observation that the lines of magnetic induction outside a cylindrical bar magnet are identical to the lines of **E** for two equally but oppositely charged disks, we see that we can improve this approximation in the region just outside the magnet by replacing these

Figure 27-26
Model of atomic current loops when all the atomic dipoles are parallel to the axis of the cylinder. The net current at any point inside the material is zero due to cancellation of neighboring atoms. The result is a surface current similar to that of a solenoid.

Figure 27-27
Disk for relating the surface current to the dipole moment. If M is the current per unit length on the surface, the total current is $M\,dx$ and the dipole moment is $M\,dx\,A = M\,d\mathcal{V}$, where $d\mathcal{V}$ is the volume.

Magnetization vector **M**

point poles with a surface pole density σ_m on the ends, where σ_m is given by

$$\sigma_m = \frac{q_m}{A} = M \qquad\qquad 27\text{-}21$$

In the region inside a solenoid the magnetic induction is approximately $\mu_0 nI$, according to Equation 27-15. This approximation is good as long as the field point is not near the ends, as indicated in Figure 27-23. Replacing the current per unit length of the solenoid, nI, with the corresponding amperian current per unit length of a bar magnet, M, we have for the field inside a magnet far from the ends

$$\mathbf{B} = \mu_0\mathbf{M} \qquad\qquad 27\text{-}22$$

We emphasize that the magnetic induction $\mathbf{B}$ for a bar magnet of magnetization $\mathbf{M}$ is identical everywhere to that of a solenoid of the same shape and carrying a current per unit length nI equal to M.

27-6 Magnetic Flux

The flux of the magnetic field through a surface is defined similarly to the flux of the electric field. Let dA be an element of area on the surface and $\hat{\mathbf{n}}$ be the unit vector perpendicular to the element. The magnetic flux ϕ_m is then defined by

$$\phi_m = \int \mathbf{B} \cdot \hat{\mathbf{n}}\, dA \qquad\qquad 27\text{-}23$$

If the surface is a plane with area A and $\mathbf{B}$ is constant in magnitude and direction over the surface, making an angle θ with the unit normal vector, the flux is

$$\phi_m = BA\cos\theta \qquad\qquad 27\text{-}24$$

The unit of flux is called the weber (Wb):

$$1\ \text{Wb} = 1\ \text{T·m}^2 \qquad\qquad 27\text{-}25$$

We are often interested in the magnetic flux through a coil containing several turns of wire. Figure 27-28 shows a coil with two turns in a uniform magnetic field $\mathbf{B}$ perpendicular to the plane of the coil. The total area enclosed by the coil is twice the area enclosed by each turn. The magnetic flux through the coil is therefore the product of the field B, the area of each coil A, and the number of turns N, which in this case is two.

Figure 27-28
The area enclosed by a coil of two turns is twice the area bound by each turn. In general, the area bound by a coil of N turns is N times that of each turn.

Example 27-3 Find the magnetic flux through a long, tightly wound solenoid the length ℓ, cross-sectional area A, and number of turns N, carrying current I.

We shall use the approximation that the magnetic field inside the solenoid is constant, equal to its value at the center:

$$B = \mu_0 nI = \mu_0 \frac{N}{\ell} I$$

The total area is NA since there are N turns, each of area A. Since the magnetic field is perpendicular to the area of each turn, the flux is

$$\phi_m = BA = \mu_0 \frac{N}{\ell} INA = \mu_0 \frac{N^2}{\ell} AI = \mu_0 n^2 (\ell A) I$$

where $n = N/\ell$ is the number of turns per unit length, and the quantity

ℓA is the volume of the space inside the solenoid containing the magnetic field B.

In Chapter 21 we considered the flux of the electric field through a closed surface and found that the flux, which is proportional to the net number of lines leaving the surface, is proportional to the net electric charge within the surface. The lines of magnetic field differ from those of electric field in that the magnetic lines are continuous and have no beginning or end. Thus for any closed surface, the same number of lines must enter the surface as leave the surface. This means that the net magnetic flux through any *closed* surface is always zero:

$$\oint \mathbf{B} \cdot \hat{\mathbf{n}} \, dA = 0 \qquad\qquad 27\text{-}26$$

The magnetic field differs fundamentally from the electric field in that there are no magnetic charges or poles where the field lines originate or terminate. In Section 27-4 we defined a magnetic-pole strength q_m in order to understand the magnetic field *outside* a solenoid or bar magnet better. Figure 27-29 shows a closed surface surrounding one end of a solenoid or bar magnet (the lines for both are identical). This figure should be compared with Figure 27-30 showing a similar closed surface surrounding the positive charge of an electric dipole. In the region corresponding to that outside the solenoid or bar magnet, the lines of **B** are identical to the lines of **E** for the dipole, and in each case they leave the closed surface indicated. However, inside the solenoid or magnet, the lines of **B** are different from the lines of **E** for the charge distribution. The magnetic field lines enter the closed surface inside the solenoid or magnet and there is no divergence of lines from the region of the pole. The lines which leave the surface also enter it.

Equation 27-26 is a statement of the experimental result that isolated magnetic poles do not exist. There are no regions of space in which lines of magnetic induction begin or end.

Isolated magnetic poles have not been found experimentally

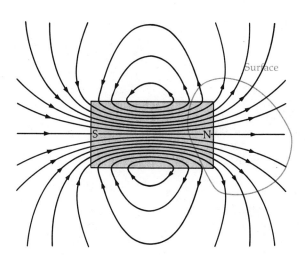

Figure 27-29
The net magnetic flux through a surface enclosing the north pole of a magnet is zero because each line that leaves the surface outside the magnet enters the surface inside the magnet.

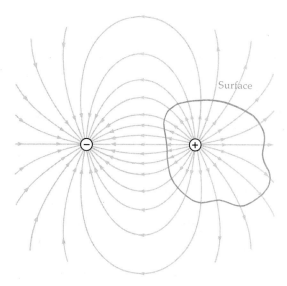

Figure 27-30
Electric flux through a surface enclosing a positive charge, for comparison with Figure 27-29. Here the net flux is not zero.

Questions

5. How can a small disk of area A be placed in a magnetic field so that there is no magnetic flux through this area?

6. How should a small loop of wire be oriented in a magnetic field so that the flux of field lines through it is a maximum? A minimum?

7. How does a positive (north) magnetic pole q_m as defined for a solenoid or magnet in Sections 27-4 and 27-5 differ from a true isolated positive pole? Sketch the lines of **B** near the poles for these cases.

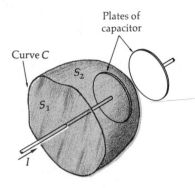

27-7 Maxwell's Displacement Current

When the current is not steady, Ampère's law does not hold. We can see why by considering the charging of a capacitor (Figure 27-31). Consider the curve C. According to Ampère's law, the line integral of the magnetic field around this curve equals μ_0 times the total current through any surface bounded by the curve. Such a surface need not be a plane. Two surfaces bounded by the curve C are indicated in the figure. The current through surface 1 is I. There is no current through surface 2 since the charge stops on the capacitor plate. There is thus an ambiguity in the phrase "current through any surface bounded by the curve." However, for steady currents, charge does not build up at any point, and we get the same current no matter which surface we choose.

Figure 27-31
Two surfaces S_1 and S_2 bounded by the same curve C. The current I passes through surface S_1 but not S_2. Ampère's law, which relates the line integral of B around the curve C to the total current passing through any surface bounded by C, is not valid when the current is not continuous, as here, where it stops at the capacitor plate.

Maxwell recognized this flaw in Ampère's law and showed that the law can be generalized to include all situations if the current I in the equation is replaced by the sum of the true current I and another term I_d, called *Maxwell's displacement current*. The displacement current is defined by

$$I_d = \epsilon_0 \frac{d\phi_e}{dt} \qquad\qquad 27\text{-}27$$

Displacement current

where ϕ_e is the flux of the electric field. The generalized form of Ampère's law is then

$$\oint \mathbf{B} \cdot d\boldsymbol{\ell} = \mu_0(I + I_d) = \mu_0 I + \mu_0\epsilon_0 \frac{d\phi_e}{dt} \qquad\qquad 27\text{-}28$$

We can understand the generalization of Ampère's law by considering Figure 27-31 again. Let us call the sum $I + I_d$ the generalized current. According to our arguments above, the same generalized current must cross any area bounded by the curve C. Thus there can be no net generalized current into or out of the closed volume. If there is a net true current I into the volume, there must be an equal net displacement current I_d out of the volume. In the volume pictured there is a net true current I into the volume which increases the charge within the volume:

$$I = \frac{dQ}{dt}$$

The flux of the electric field out of the volume is related to the charge by Gauss' law:

$$\phi_{e\text{ net out}} = \oint \mathbf{E} \cdot \hat{\mathbf{n}}\, dA = \frac{1}{\epsilon_0} Q$$

The rate of increase of charge is thus proportional to the rate of increase of the net flux out of the volume:

$$\epsilon_0 \frac{d\phi_{e \text{ net out}}}{dt} = \frac{dQ}{dt}$$

and the net current into the volume equals the net displacement current out of the volume.

A significant feature of Maxwell's generalization is that a magnetic field is produced by a changing electric field as well as by true electric currents. Maxwell was undoubtedly led to this generalization by the reciprocal result that an electric field is produced by a changing magnetic flux as well as by electric charges. This latter result, known as *Faraday's law,* preceded Maxwell's generalization. We shall study Faraday's law in the next chapter. These two properties of electric and magnetic fields—that a changing flux of one field produces the other —are connected with electromagnetic radiation, of which light is an example.

Review

A. Define, explain, or otherwise identify:

Biot-Savart law, 743   Amperian current, 755
Current balance, 748   Magnetization vector, 755
Ampère's law, 749   Magnetic flux, 756
Solenoid, 751   Displacement current, 758
Pole strength of a solenoid, 754

B. True or false:

1. The magnetic field due to a current element decreases as the square of the distance from the element.

2. The magnetic field due to a current element is parallel to the element.

3. The magnetic field due to a very long wire carrying a current decreases as the square of the distance from the wire.

4. A solenoid and a uniformly magnetized magnet of the same shape and same magnetic moment produce the same magnetic field everywhere in space.

5. Lines of **B** never diverge from a point in space.

Exercises

Section 27-1, The Biot-Savart Law

1. A small current element $I\,d\ell$, with $d\ell = 2\mathbf{k}$ mm and $I = 2$ A, is centered at the origin. Find the magnetic-induction field $d\mathbf{B}$ at the following points: (a) on the x axis at $x = 3$ m, (b) on the x axis at $x = -6$ m, (c) on the z axis at $z = 3$ m, (d) on the y axis at $y = 3$ m.

2. For the current element of Exercise 1 find the magnitude and indicate the direction of $d\mathbf{B}$ at the point $x = 0$, $y = 3$ m, $z = 4$ m.

3. For the current element of Exercise 1, find the magnitude of $d\mathbf{B}$ and indicate its direction on a diagram for the points (a) $x = 2$ m, $y = 4$ m, $z = 0$; (b) $x = 2$ m, $y = 0$, $z = 4$ m.

4. A long straight wire carries a current of 75 A. Find the magnitude of **B** at distances (a) 10 cm, (b) 50 cm, and (c) 2 m from the center of the wire.

Exercises 5 to 10 refer to Figure 27-32, which shows two long straight wires in the xy plane parallel to the x axis. One wire is at y = −6 cm and the other is at y = +6 cm. The current in each wire is 20 A.

5. If the currents in Figure 27-32 are in the negative x direction, find **B** at the points on the y axis at (a) y = −3 cm, (b) y = 0, (c) y = +3 cm, (d) y = +9 cm.

6. Sketch B_z versus y for points on the y axis when both currents are in the negative x direction.

7. Find **B** at points on the y axis as in Exercise 5 but with the current in the wire at y = −6 cm in the negative x direction and the current in the wire at y = +6 cm in the positive x direction.

8. Sketch B_z versus y for points on the y axis when the currents are in the direction opposite that in Exercise 7.

9. Find **B** at the point on the z axis at z = 8 cm if (a) the currents are parallel, as in Exercise 5; (b) the currents are antiparallel, as in Exercise 7.

10. Find the magnitude of the force per unit length exerted by one wire on the other.

11. The current in the wire of Figure 27-33 is 8.0 A. Find **B** at point P due to each wire segment and sum to find the resultant **B**.

12. Consider a circular loop of radius R carrying current I. Show directly from the Biot-Savart law that the magnetic field at the center of the loop due to each segment of length $d\ell$ has the magnitude

$$dB = \frac{\mu_0}{4\pi}\frac{I\,d\ell}{R^2}$$

and that the resultant magnetic field at the center is $B = \mu_0 I/2R$.

13. Find the magnetic field at point P in Figure 27-34 if the current is 15 A.

14. A single-turn circular loop of radius 10.0 cm is to produce a field at its center which will just cancel the earth's field at the equator, which is 0.7 G directed north. Find the current in the wire and make a sketch showing the orientation of the loop and current.

15. A wire of length L carries a current I. Find the magnetic field B at the center when (a) the wire is bent into a square of side L/4 and (b) the wire is bent into a circle of circumference L. (c) Which gives the greater value of B?

16. In Figure 27-35, find the magnetic field at point P, which is at the common center of the two semicircular arcs.

Section 27-2, Definition of the Ampere and the Coulomb

17. Two long straight parallel wires 10.0 cm apart carry currents of equal magnitude I. They repel each other with a force per unit length of 4.0 nN/m. (a) Are the currents parallel or antiparallel? (b) Find I.

18. A wire of length 12 cm is suspended by flexible leads above a long straight wire. Equal and opposite currents are established in the wires such that the 12-cm wire floats 1.5 mm above the long wire with no tension in its suspension leads. If the mass of the 12-cm wire is 10.0 g, what is the current?

19. In a student experiment with a current balance, the upper wire of length 30 cm is pivoted so that with no current it balances at 2 mm above a fixed parallel wire also 30 cm long. When the wires carry equal and opposite currents I, the upper wire again balances at its original position when a 2.4-g mass is placed on it. What is the current I?

20. Three long parallel straight wires pass through the corners of an equilateral triangle of side 10.0 cm, as shown in Figure 27-36, where a dot means that the current is out of the paper and a cross means that it is into the paper. If each cur-

Figure 27-32
Two long straight parallel wires for Exercises 5 to 10.

Figure 27-33
Exercise 11.

Figure 27-34
Exercise 13.

Figure 27-35
Exercise 16.

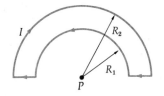

Figure 27-36
Exercises 20 and 21.

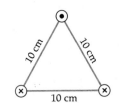

rent is 15.0 A, find (a) the force per unit length on the upper wire and (b) the magnetic field **B** at the upper wire due to the lower two wires. *Hint:* It is easier to find the force per unit length directly from Equation 27-11 and use your result to find **B** than to find **B** first and use it to find the force.

21. Work Exercise 20 with the current in the lower right corner of Figure 27-36 reversed.

Section 27-3, Ampère's Law

22. A long straight thin-walled cylindrical shell of radius R carries a current I. Find **B** inside and outside the cylinder.

23. In Figure 27-37, one current is 10 A into the paper; the other current is 10 A out of the paper; and each curve is a circular path. (a) Find $\oint \mathbf{B} \cdot d\boldsymbol{\ell}$ for each path indicated. (b) Which path, if any, can be used to find **B** at some point due to these currents?

24. A very long coaxial cable has an inner wire and a concentric outer cylindrical conducting shell of radius R. At one end the wire is connected to the shell. At the other end the wire and shell are connected to the opposite terminals of a battery so that there is a current I down the wire and back up the shell. Assume that the cable is straight and find B (a) at points far from the ends and between the wire and shell, and (b) outside the cable.

25. A wire of radius 0.5 cm carries a current of 100 A uniformly distributed over the cross section. Find B (a) at 0.1 cm from the center of the wire, (b) at the surface of the wire, and (c) at a point outside the wire 0.2 cm from the surface of the wire. (d) Make a graph of B versus distance from the center of the wire.

26. Show that a uniform magnetic field with no fringing field as shown in Figure 27-38 is impossible because it violates Ampère's law. Do this by applying Ampère's law to the rectangular curve shown by the dashed lines.

Section 27-4, The Magnetic Field of a Solenoid

27. A solenoid has a length of 25 cm, radius 1 cm, and 400 turns, and carries 3 A of current. Find (a) B on the axis at the center, (b) the magnetic moment of the solenoid, (c) the pole strength q_m, and (d) an approximate value for B on the axis a distance 2 m from one end.

28. Work Exercise 27 for a solenoid of length 30 cm, radius 2 cm, and 800 turns, and carrying 2 A.

29. Two tightly wound solenoids each 10 cm long have 200 turns and a cross-sectional area of 0.5 cm², and carry current of 4.0 A. Each has its axis along the x axis, and the currents are such that each has its north pole on the right and its south pole on the left. Their centers are 60 cm apart. (a) Find the pole strength of each solenoid. (b) Find the net force of attraction of the solenoids from Coulomb's law of attraction or repulsion between magnetic poles, $F = k_m q_{1m} q_{2m} / r^2$.

Section 27-5, The Magnetic Field of a Bar Magnet

30. What is the magnitude of the magnetization vector **M** that gives a magnetic induction field of 5000 G at the center of a long cylindrical bar magnet?

31. A bar magnet has the same shape as the solenoid in Exercise 27. Find the magnetization **M** such that the magnetic field produced by the bar magnet is the same as that produced by the solenoid.

32. A long thin rod of iron is uniformly magnetized with magnetization $M = 2 \times 10^5$ A/m along its axis. Its length is 15 cm, and its cross-sectional area is 0.3 cm². (a) What is the magnitude of **B** at the center? (b) Find the magnetic-pole strength q_m at each end. (c) Find the magnetic moment of the magnet.

Figure 27-37
Exercise 23.

Figure 27-38
Exercise 26.

33. An iron atom is like a small bar magnet of magnetic moment 1.8×10^{-23} A·m². The number of atoms per unit volume in iron is 8.55×10^{28} atoms per cubic metre. (a) If all the iron atoms in an iron bar are aligned with their magnetic moment along the axis, what is the magnitude of the magnetization M? (b) What is B at the center of the magnet, assuming it to be a long rod?

Section 27-6, Magnetic Flux

34. A uniform magnetic field of magnitude 2000 G is parallel to the x axis. A square coil of side 5 cm has a single turn and makes an angle θ with the z axis, as shown in Figure 27-39. Find the magnetic flux through the coil when (a) $\theta = 0$, (b) $\theta = 30°$, (c) $\theta = 60°$, (d) $\theta = 90°$.

Figure 27-39
Exercise 34.

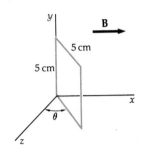

35. A circular coil has 25 turns and a radius of 5 cm. It is at the equator, where the earth's magnetic field is 0.7 G north. Find the magnetic flux through the coil when (a) its plane is horizontal, (b) its plane is vertical and its axis points north, (c) its plane is vertical and its axis points east, and (d) its plane is vertical and its axis makes an angle of 30° with the north.

36. Find the magnetic flux through the solenoid in Exercise 27. (Assume the magnetic field to be constant inside the solenoid.)

37. Find the magnetic flux through the solenoid in Exercise 28.

38. A circular coil of 15 turns of radius 4 cm is in a uniform magnetic field of 4000 G in the positive x direction. Find the flux through the coil when the unit normal vector to the plane of the coil is (a) $\hat{\mathbf{n}} = \mathbf{i}$, (b) $\hat{\mathbf{n}} = \mathbf{j}$, (c) $\hat{\mathbf{n}} = (\mathbf{i} + \mathbf{j})/\sqrt{2}$, (d) $\hat{\mathbf{n}} = \mathbf{k}$, and (e) $\hat{\mathbf{n}} = 0.6\mathbf{i} + 0.8\mathbf{j}$.

Section 27-7, Maxwell's Displacement Current

39. A parallel-plate capacitor has circular plates of radius 1.0 cm separated by 1.0 mm in air. Charge is flowing onto the upper plate and off the lower plate at a rate of 5 A. (a) Find the time rate of change of the electric field between the plates. (b) Compute the displacement current between the plates and show that it equals 5 A. (c) Find the displacement current density J_d between the plates and show that $J_d = \epsilon_0 \, dE/dt$.

40. For Exercise 39, show that at distance r from the axis of the plates the magnetic field between the plates is given by $B = \frac{1}{2}\mu_0 J_d r$ if r is less than the radius of the plates and J_d is the displacement current density between the plates.

41. Show that for a parallel-plate capacitor the displacement current is given by $I_d = C \, dV/dt$, where C is the capacitance and V the voltage across the capacitor.

Problems

1. Three very long parallel wires are at the corners of a square, as shown in Figure 27-40. Find the magnetic field B at the unoccupied corner of the square when (a) all the currents are into the paper, (b) I_1 and I_3 are in and I_2 is out, (c) I_1 and I_2 are in and I_3 is out.

Figure 27-40
Problem 1.

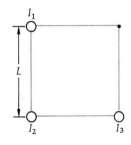

2. Four long straight parallel wires each carry current I. In a plane perpendicular to the wires, the wires are at the corners of a square of side a. Find the force per unit length on one of the wires if (a) all the currents are in the same direction and (b) the currents in the wires at adjacent corners are oppositely directed.

3. A power cable carrying 50.0 A is located 2.0 m below the earth's surface, but its direction and precise position are unknown. Show how you could locate the cable using a compass. Assume that you are at the equator, where the earth's magnetic field is 0.7 G north.

4. A square loop of side L lies in the yz plane with its center at the origin. It carries a current I. Find the magnetic field B at any point on the x axis and show from your expression that for x much larger than L

$$B_x = \frac{\mu_0}{2\pi} \frac{m}{x^3}$$

where $m = IL^2$ is the magnetic moment of the loop.

5. A very long straight wire carries a current of 20.0 A. An electron is 1.0 cm from the center of the wire and is moving with speed 5.0×10^6 m/s. Find the force on the electron when it moves (a) directly away from the wire, (b) parallel to the wire in the direction of the current, and (c) perpendicular to the wire and tangent to a circle around the wire.

6. A relatively inexpensive ammeter, called a *tangent galvanometer*, can be made using the earth's field. A plane circular coil of N turns and radius R is oriented such that the field B_c it produces in the center of the coil is either east or west. A compass is placed at the center of the coil. When there is no current in the coil, the compass points north. When there is a current I, the compass points in the direction of the resultant magnetic field B at an angle θ to the north. Show that the current I is related to θ and the horizontal component of the earth's field B_e by

$$I = \frac{2RB_e}{\mu_0 N} \tan \theta$$

7. A particle of positive charge q moves in a helical path around a long straight wire carrying current I. Its velocity has components $v_\parallel$ parallel to the current and v_t tangent to a circle around the current. Find the relationship between $v_\parallel$ and v_t such that the magnetic force provides the necessary centripetal force.

8. A very long bar magnet of pole strength q_m lies along the axis of a circular loop with its north pole at the center of the loop, as in Figure 27-41. The loop has radius r and carries current I. (a) From the definition $\mathbf{F} = q_m\mathbf{B}$, find the magnitude and direction of the force exerted by the loop on the magnet. Assume that the south pole of the magnet is far enough away to be neglected. (b) Let $\mathbf{B}$ be the magnetic field at the loop due to the north pole of the magnet. Assuming that $\mathbf{B}$ points directly away from the pole, find the magnitude and direction of the total force exerted by the pole on the loop (in terms of $\mathbf{B}$, r, and I). (c) Using Newton's third law, set the magnitudes of the forces obtained in (a) and (b) equal to obtain the relation

$$B = \frac{\mu_0}{4\pi} \frac{q_m}{r^2}$$

for the magnetic field due to a magnetic pole.

Figure 27-41
Problem 8.

9. A very long straight conductor has a circular cross section of radius R and carries a current I. Inside the conductor there is a cylindrical hole of radius a whose axis is parallel to the axis of the conductor and a distance b from it. Let the z axis be the axis of the conductor, and let the axis of the hole be at $x = b$ (Figure 27-42). Find the magnetic field B at the point (a) on the x axis at $x = 2R$, (b) on the y axis at $y = 2R$. Hint: Consider a uniform current distribution throughout the cylinder of radius R plus a current in the opposite direction in the hole.

Figure 27-42
Problems 9 and 10.

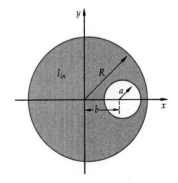

10. For the cylinder with the hole in Problem 9, show that the magnetic field inside the hole is uniform and find its magnitude and direction.

11. A disk of radius R carries a fixed charge density σ and rotates with angular velocity ω. (a) Consider a circular strip of radius r and width dr. Show that the current produced by this strip is $dI = (\omega/2\pi)\, dq = \sigma\omega r\, dr$. ($b$) Use your result for part (a) to show that the magnetic field at the center of the disk is $B = \frac{1}{2}\mu_0\sigma\omega R$. ($c$) Use your result for part (a) to find the magnetic field at a point on the axis of the disk a distance x from the center.

12. A circular loop of radius R carrying a current I is centered at the origin with its axis along the x axis. Its current is such that it produces a magnetic field in the positive x direction. (a) Sketch B_x versus x for points on the x axis. Include both positive and negative x. Compare this sketch with that for E_x due to a charged ring of the same size. (b) A second identical loop carrying the same current in the same sense is in a plane parallel to the xy plane with its center at $x = d$. Sketch the magnetic field on the x axis due to each coil separately and the resultant field due to the two coils. Show from your sketch that dB_x/dx is zero midway between the coils.

13. Two coils which are separated by a distance equal to their radius and carrying equal currents such that their axial fields add are called *Helmholtz coils*. A feature of Helmholtz coils is that the resultant magnetic field between the coils is very uniform. Let $R = 10$ cm, $I = 20$ A, and $N = 300$ turns for each coil. Place one coil with center at the origin and the other at $x = 10$ cm (as in Problem 12 with $d = R = 10$ cm). (a) Calculate the resultant field B_x at points $x = 5$ cm, $x = 7$ cm, $x = 9$ cm, and $x = 11$ cm and from this, sketch B_x versus x. (Note that B_x is symmetric about the midpoint of the coils.) (b) Show that $dB_x/dx = 0$ and $d^2B_x/dx^2 = 0$ at $x = \frac{1}{2}R$. (c) Show that $d^3B_x/dx^3 = 0$ at $x = \frac{1}{2}R$.

14. The magnetic field of a small current-carrying coil can be found at points off axis and a great distance from the coil by considering the coil to be a small magnet of pole strength q_m and length ℓ with $q_m\ell = IA$, where A is the area of the coil and I is its current. Consider a small coil at the origin with its magnetic moment in the positive z direction. Show that the field at a point on the x axis a great distance away is given by

$$\mathbf{B} = -\frac{\mu_0}{4\pi}\frac{IA}{x^3}\,\mathbf{k}$$

Hint: Calculate $\mathbf{B}$ from Coulomb's law, assuming a positive pole on the z axis at $z = +\frac{1}{2}\ell$ and a negative pole at $z = -\frac{1}{2}\ell$. [See part (c) of Problem 8.]

15. A long straight wire carries a current of 20 A, as shown in Figure 27-43. A rectangular coil with two sides parallel to the straight wire has sides 5 and 10 cm with its near side a distance 2 cm from the wire. (a) Compute the magnetic flux through the rectangular coil. *Hint:* Calculate the flux through a strip of area $dA = (10$ cm$)\,dx$ and integrate from $x = 2$ cm to $x = 7$ cm. (b) If the rectangle carries a counterclockwise current of 5 A, find the net force on it.

Figure 27-43
Problem 15.

16. A solenoid has n turns per unit length and radius R and carries current I. Its axis is at the x axis with one end at $x = -x_1$ and the other at $x = +x_2$. Consider a part of the solenoid of length dx at distance x from the origin to be a circular loop carrying current $nI\,dx$. (a) Use Equation 27-8 for the magnetic field at the origin due to this loop and integrate from $x = -x_1$ to $x = +x_2$ to find the total magnetic field at the origin due to the solenoid. *Hint:* You will need to use the indefinite integral

$$\int \frac{dx}{r^3} = \frac{x}{R^2 r} \qquad \text{where } r = \sqrt{x^2 + R^2}$$

(b) Show that the result can be written as $B = \frac{1}{2}\mu_0 nI\,(\cos\theta_1 + \cos\theta_2)$, where $\cos\theta_1 = x_1/r_1$ and $\cos\theta_2 = x_2/r_2$. (c) Use this result to show that $B \approx \mu_0 nI$ if R is much less than x_1 and x_2.

17. A large 50-turn circular loop of radius 10.0 cm carries a current 4.0 A. At the center of the loop is a small 20-turn coil of radius 0.5 cm carrying current 1.0 A. The planes of the two coils are perpendicular. Find the torque exerted by the large coil on the small coil. (Neglect any variation in $\mathbf{B}$ due to the large coil over the region occupied by the small coil.)

18. A solenoid has n turns per unit length, radius R_1, and carries current I. (a) A large circular loop of radius $R_2 > R_1$ and N turns encircles the solenoid at a point far away from the ends of the solenoid. Find the magnetic flux through the loop. (b) A small circular loop of radius $R_3 < R_1$ is completely inside the solenoid far from its ends with its axis parallel to that of the solenoid. Find the magnetic flux through the loop.

These Helmholtz coils at the Kettering Magnetics Laboratory at Oakland University are used to cancel the earth's magnetic field and to provide a uniform magnetic field in a small region.

19. A long cylindrical conductor of radius R carries current I of uniform density $J = I/\pi R^2$. Find the magnetic flux per unit length through the area indicated in Figure 27-44.

Figure 27-44
Problem 19.

20. A long solenoid has n turns per unit length and radius R and carries current I. Its axis is the x axis with one end at $x = -\frac{1}{2}L$ and the other at $x = +\frac{1}{2}L$, where L is the total length of the solenoid. Show that the magnetic field B at any point on the x axis is given by

$$B = \tfrac{1}{2}nI \, (\cos \, \theta_1 + \cos \, \theta_2)$$

where

$$\cos \, \theta_1 = \frac{\tfrac{1}{2}L + x}{[R^2 + (\tfrac{1}{2}L + x)^2]^{1/2}} \quad \text{and} \quad \cos \, \theta_2 = \frac{\tfrac{1}{2}L - x}{[R^2 + (\tfrac{1}{2}L - x)^2]^{1/2}}$$

Note that for points outside the solenoid, $\cos \, \theta_2$ is negative (for x positive). (See Problem 16.)

21. Use the results of Problem 20 to plot the magnetic field B on the axis of a solenoid as a function of x. Let $L = 10$ cm, $R = 0.5$ cm, $I = 4$ A, and $n = 1500$ turns/m. Plot B for positive x ranging from $x = 0$ to $x = 20$ cm.

22. In this problem you are to show that the generalized form of Ampère's law (Equation 27-28) and the Biot-Savart law give the same result in a situation in which they both can be used. Figure 27-45 shows two charges $+Q$ and $-Q$ on the x axis at $x = -a$ and $x = +a$ with a current $I = -dQ/dt$ along the line between them. Point P is on the y axis at $y = R$. (a) Use Equation 27-6, obtained from the Biot-Savart law, to show that the magnitude of B at point P is

$$B = \frac{\mu_0}{2\pi R} \frac{Ia}{\sqrt{a^2 + R^2}}$$

Figure 27-45
Problem 22.

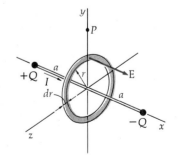

(b) Show that the electric field at any point y on the y axis is in the x direction with magnitude

$$E_x = \frac{1}{2\pi\epsilon_0} \frac{Qa}{(y^2 + a^2)^{3/2}}$$

(c) Consider a circular strip of radius r and width dr in the yz plane with center at the origin. Show that the flux of the electric field through this strip is $E_x \, dA = (Q/\epsilon_0)a(r^2 + a^2)^{-3/2}r \, dr$. (d) Use your result in part (c) to find the total flux ϕ_e through a circular area of radius R. Show that $\epsilon_0\phi_e = Q(1 - a/\sqrt{a^2 + R^2})$. (e) Find the displacement current I_d and show that $I + I_d = Ia/\sqrt{a^2 + R^2}$. Then show that Equation 27-28 gives the same result for B as that found in part (a).

CHAPTER 28   Faraday's Law

Objectives   After studying this chapter you should:

1. Be able to state Faraday's law and use it to find the emf induced by a changing magnetic flux.

2. Be able to state Lenz's law and use it to find the direction of the induced current in various applications of Faraday's law.

3. Be able to discuss the various forces involved and the energy balance in an example of motional emf.

4. Be able to describe the betatron as an example of the application of Faraday's law.

5. Be able to discuss eddy currents.

6. Be able to state the definition of self- and mutual inductance and derive an expression for the self-inductance of a tightly wound solenoid.

7. Be able to apply Kirchhoff's laws to obtain the differential equations for LR, LC, and LRC circuits and to discuss the general behavior of the solution for each circuit.

8. Be able to state the expression for the energy stored in a magnetic field and for the magnetic field energy density.

Figure 28-1
Demonstration of induced emf. When the magnet is moved away from the coil, an emf is induced in the coil, as shown by the galvanometer deflection. No deflection is observed when the magnet is stationary. (*Courtesy of Larry Langrill, Oakland University.*)

The fact that electric currents can be induced by changing magnetic fields was discovered in the early 1830s by Michael Faraday and simultaneously by the American physicist Joseph Henry. In a simple demonstration of induced currents, the ends of a coil of wire are attached to a galvanometer and a strong magnet is moved toward or away from the coil (Figure 28-1). The momentary deflection of the galvanometer *during* the motion indicates an induced electric current in the coil-galvanometer circuit. In another demonstration, a circuit containing a large electromagnet made of many turns of wire wrapped around an iron core is broken by a knife switch. As the circuit is broken, a spark jumps the gap across the switch. The changing magnetic field caused by breaking the circuit induces a large emf across the switch. It results in dielectric breakdown of the air, indicated by the spark. The results of these and many other experiments can be expressed by a single relation known as Faraday's law. Let ϕ_m be the flux of the magnetic field through a circuit (Figure 28-2). If this flux is changed in any way, there is an emf induced in the circuit given by

$$\varepsilon = -\frac{d\phi_m}{dt} \qquad\qquad 28\text{-}1$$

The emf is usually detected by observing a current in the circuit. In our previous discussions, the emf in a circuit has been localized in a specific region of the circuit, namely, between the terminals of the battery. The emf produced by a changing magnetic flux is not localized but must be considered to be distributed throughout the circuit. Since emf is work done per unit charge, there must be a force exerted on the charge associated with the emf. Energy is put into the circuit by the emf, and this force is therefore not conservative. The force per unit charge is a nonconservative electric field $\mathbf{E}$; it differs from electrostatic fields given by Coulomb's law, which are conservative. The integral of $\mathbf{E} \cdot d\ell$ around the complete circuit is the work per charge done by this nonconservative electric field, and is equal to the emf in the circuit. (This is true whether the emf is localized or not. In a battery the nonconservative electric field arises because of chemical forces inside the battery.) In terms of this nonconservative electric field, Faraday's law can be written

$$\varepsilon = \oint \mathbf{E} \cdot d\ell = -\frac{d\phi_m}{dt} \qquad\qquad 28\text{-}2 \qquad \textit{Faraday's law}$$

The magnetic flux through a circuit can be changed in many different ways: the current producing the flux may be increased or decreased; permanent magnets may be moved toward the circuit or away from it; the circuit itself may be moved toward or away from the source of the flux; or the area of the circuit may be increased or decreased in a fixed magnetic field. In every case, an emf is induced in the circuit equal in magnitude to the rate of change of the magnetic flux. We discuss the significance of the negative sign in Faraday's law in Section 28-2.

Figure 28-2
Flux of **B** through a circuit. If the flux changes, an emf is induced in the circuit.

28-1   Motional EMF

Figure 28-3 shows a conducting rod sliding along conducting rails which are connected to a resistor. A uniform magnetic field is directed into the paper. Since the magnetic flux through the circuit is changing (the area of the circuit increases as the rod moves), an emf is induced in the circuit. Let ℓ be the separation of the rails and x be the distance from the left end to the rod at some time. The magnetic flux at this time is then

$$\phi_m = BA = B\ell x$$

The rate of change of the flux is

$$\frac{d\phi_m}{dt} = B\ell\frac{dx}{dt} = B\ell v$$

where $v = dx/dt$ is the speed of the rod. The emf induced in this circuit is therefore

$$\varepsilon = (-)\frac{d\phi_m}{dt} = (-)B\ell v \qquad\qquad 28\text{-}3$$

Figure 28-3
Rod sliding on rails in magnetic field.

An engraving of Michael Faraday (1791–1867) in his laboratory in the basement of the Royal Institution in London, where he carried out his research on electricity and magnetism. (*Courtesy of the Niels Bohr Library, American Institute of Physics.*)

(We defer the discussion of the minus sign in Faraday's law until the next section.)

The direction of the emf in this case is such as to produce a current in the counterclockwise sense. This example of Faraday's law is one in which we can understand the origin of the emf by considering the known forces acting on the electrons in the circuit. Since the current in the rod is upward from b to a, the electrons move downward from a to b. The velocity of a typical electron has a downward vertical component, the drift velocity, and a horizontal component v equal to the speed of the rod. The electron's velocity $\mathbf{v}_e$ thus makes an angle θ with the horizontal, as shown in Figure 28-4. Since the horizontal component of the velocity is v, the speed of the rod, we have

$$v_e \cos \theta = v \qquad\qquad 28\text{-}4$$

The magnetic force $\mathbf{f}_m = -e\mathbf{v}_e \times \mathbf{B}$ is in the plane of the figure perpendicular to $\mathbf{v}_e$, as shown, and has the magnitude

$$f_m = ev_eB \qquad\qquad 28\text{-}5$$

If $\mathbf{f}_m$ were the only force acting on the electron, it could not stay in the rod as the rod moves to the right. The rod exerts a horizontal force $\mathbf{f}_r$ on' the electron to balance the horizontal component of $\mathbf{f}_m$, which is $f_m \sin \theta$:

$$f_r = f_m \sin \theta \qquad\qquad 28\text{-}6$$

Since $\mathbf{f}_m$ is perpendicular to the motion of the electron, it does no work. The work done on the electron is done by the force $\mathbf{f}_r$. As the electron moves down the rod, the rod moves to the right; the electron therefore moves along a diagonal path of length S (Figure 28-4). The lengths S and ℓ are related by

$$\ell = S \sin \theta \qquad\qquad 28\text{-}7$$

Figure 28-4
Forces on an electron in the moving rod of Figure 28-3. The electron velocity v_e has a horizontal component v, the speed of the rod, and a vertical component, its drift velocity along the rod. The magnetic force $\mathbf{f}_m$ is perpendicular to $\mathbf{v}_e$ and does no work. The rod exerts a horizontal force $\mathbf{f}_r$ on the electron with magnitude $f_m \sin \theta$. This force has a component in the direction of motion of the electron and therefore does work on it. The work per charge is shown to be equal to $BL v$.

The work done on the electron as it moves down the complete length of the rod is

$$W = f_r \cos \theta \, S = (f_m \sin \theta) \cos \theta \, S = f_m \cos \theta \, \ell$$

where we have used Equation 28-6 for f_r and $S \sin \theta = \ell$ from Equation 28-7. Substituting $ev_e B$ for f_m, we obtain for the work done on the electron:

$$W = ev_e B \cos \theta \, \ell = eB(v_e \cos \theta)\ell = eBv\ell$$

The work per unit charge is the emf $Bv\ell$, in agreement with our result from Faraday's law.

The force $\mathbf{f}_r$ is exerted by the rod on the electron. The electron exerts an equal but opposite force $-\mathbf{f}_r$ on the rod. This force is to the left in the figure. To keep the rod moving to the right with constant speed, an external agent must exert a force to the right. The work done by this external agent provides the energy that goes into the circuit. (In this case, the energy goes into Joule heat I^2R in the resistor in the circuit.)

Faraday and his wife in the 1850s.

28-2 Lenz's Law

The direction of the induced emf and current can be found from a general statement known as *Lenz's law*.

The emf and induced current are in such a direction as to tend to oppose the change which produced them.

Lenz's law

In our statement of Lenz's law we did not specify just what kind of change it is that causes the induced emf. This was left vague to allow a variety of interpretations, all of which are valid and give the same result for the direction of the induced emf and current. A few illustrations will clarify this point.

In the example of the rod moving on the rails, above, the magnetic field is into the paper. Let us call this direction positive. The movement of the rod then tends to increase the flux in this direction. According to Lenz's law, the induced current is in the direction that will oppose this change. As we have seen, the induced emf and current are counter-clockwise. The flux produced *by the induced current* is out of the paper, opposing the increase in flux through the circuit which results from the motion of the rod. If the current induced were in the clockwise direction, the induced flux would add to the increase produced by the moving rod, contrary to Lenz's law. Lenz's law is thus a statement that the induced current tends to maintain the status quo. If we try to increase the flux through a circuit, a current will be induced which tends to decrease the flux. If we move the rod to the left in our example, decreasing the flux, Lenz's law tells us that the induced current will be clockwise to help maintain the original flux through the circuit.

We can also interpret Lenz's law in terms of forces and the conservation of energy. In our example of the rod moving to the right, there is a magnetic force on the rod due to the induced current. Since the current is upward in the rod and the magnetic field is inward, the force is to the left,

$$\mathbf{F}_m = I\boldsymbol{\ell} \times \mathbf{B}$$

This force tends to oppose the motion of the rod (see Figure 28-5). If the rod is to move to the right with constant velocity, there must be an external force to the right of magnitude $I\ell B$. Lenz's law then tells us that there will be a current induced in such a direction that the magnetic force opposes this external force.

When the rod is moving with constant velocity, the power input by the external force on the rod goes into heating the resistor. The power input is Fv, where F is the external force. Thus

$$P = Fv = I\ell Bv = I^2R \qquad 28\text{-}8$$

or

$$IR = B\ell v \qquad 28\text{-}9$$

By Kirchhoff's first rule the potential drop IR must equal the emf in the circuit. Thus we find by means of energy balance that the emf in the circuit is

$$\varepsilon = B\ell v \qquad 28\text{-}10$$

If we assume the opposite of Lenz's law, energy cannot be conserved. Suppose we start with the rod at rest on the rails and give it a slight push to the right. Assuming the opposite of Lenz's law, we would have a current in the clockwise direction. Then there would be a magnetic force on the rod to the right which would accelerate the rod, increasing its velocity. The emf and induced current would thus be increased. That would further increase the force on the rod to the right. Hence, the opposite of Lenz's law gives a situation in which the rod continues to accelerate to the right and the current continues to increase without limit. With no power input to this circuit, the power I^2R into the resistor would increase without limit.

The negative sign in Faraday's law (Equation 28-1 or 28-2) is the mathematical expression of Lenz's law: the sense of the induced emf is opposite to the sense of the rate of change of magnetic flux. In practice, it is usually easiest to compute the magnitude of the emf by computing the magnitude of $d\phi_m/dt$ (momentarily ignoring the sign) and then finding the direction of the emf and the induced current using whatever interpretation of Lenz's law is most convenient. We shall illustrate this in the next section.

28-3 Applications of Faraday's Law

In most applications of Faraday's law it is not possible to understand the origin of the induced emf from our previous knowledge of magnetic forces. That is, we cannot derive Equation 28-1 by considering magnetic or electric forces on charges, as we did for the special case of the rod moving on rails. Instead, we must take Faraday's law as expressed in Equation 28-1 to be an independent law of nature expressing a wide range of experimental results. In this section we consider more examples to illustrate the use of Faraday's law and Lenz's law to find the direction of the induced current.

Figure 28-6 shows a bar magnet moving toward a loop which has a resistance R. Since the magnetic field from the bar magnet is to the right, out of the north pole of the magnet, the movement of the magnet toward the loop tends to increase the flux through the loop to the right. The induced current in the loop is in the direction shown, to create a flux which opposes the change.

Figure 28-5
The magnetic force $\mathbf{F}_m$ on the rod opposes the motion of the rod. Here the rod is moving to the right, and $\mathbf{F}_m$ is to the left.

Figure 28-6
When the bar magnet moves toward the loop, the induced emf in the loop produces a current in the direction shown.

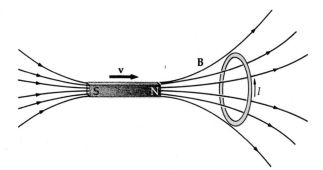

Figure 28-7
The magnetic moment of the loop (indicated by the outlined magnet) due to the induced current is such as to oppose the motion of the bar magnet. Here the bar magnet is moving toward the loop, and so the induced magnetic moment repels the bar magnet.

Figure 28-7 shows the induced magnetic moment of the current loop in Figure 28-6. The loop acts like a small magnet with north pole to the left and south pole to the right. Since opposite poles attract and like poles repel, the induced magnetic moment of the loop exerts a force on the bar magnet to the left to oppose its motion toward the loop. Again, we could use Lenz's law in terms of forces rather than flux. If the bar magnet is moved toward the loop, the induced current must produce a magnetic moment to oppose this change.

In Figure 28-8 the bar magnet is at rest, and the loop is moved away from it. The induced current and magnetic moment are shown in the figure. In this case the magnetic moment of the loop attracts the bar magnet, as required by Lenz's law.

In Figure 28-9 when the current in circuit 1 is changed, there is a change in the flux through circuit 2. Suppose the switch in circuit 1 is open, with no current in the circuit. When the switch is closed (Figure 28-9b), the current in circuit 1 does not reach its steady value ε_1/R_1 instantaneously but takes a short time to change from zero to this value. During this short time, while the current is increasing, the flux in circuit 2 is changing and there is an induced current in that circuit in the direction shown. When the current in the first circuit reaches its steady value, the flux is no longer changing and there is no induced current in circuit 2. An induced current in circuit 2 in the opposite direction appears momentarily when the switch in circuit 1 is opened (Figure 28-9c) and the current is decreasing to zero.

Figure 28-8
When the loop is moved away from a stationary bar magnet, the induced magnetic moment attracts the bar magnet, again opposing the relative motion.

Figure 28-9
(a) Two adjacent circuits. (b) Just after the switch is closed, I_1 is increasing and B is increasing in the direction shown. The changing flux in circuit 2 induces current I_2. The flux due to I_2 opposes the increase in flux due to I_1. (c) As the switch is opened, I_1 decreases and B decreases. The induced current I_2 tends to maintain the flux in the circuit, opposing the decrease.

(a)

(b)

(c)

For our next example we consider a single isolated circuit. When there is a current in the circuit, there is magnetic flux through the circuit due to its own current. When the current is changing, the flux is changing and there is an induced emf in the circuit. This *self-induced* emf opposes the change in the current. It is because of this self-induced emf that the current in a circuit cannot jump instantaneously from zero to some finite value or from some value to zero. Henry first noticed this effect when he was experimenting with a circuit consisting of many turns of a wire (Figure 28-10), an arrangement that gives a large flux through the circuit for even a small current. Henry noticed a spark across the switch when he tried to break the circuit. This spark is due to the large induced emf which occurs when the current varies rapidly, as with opening the switch. In this case the induced emf tries to maintain the original current. The large induced emf produces a large voltage drop across the switch as it is opened. The electric field between the poles of the switch is large enough to tear the electrons from the air molecules, causing dielectric breakdown. When the molecules in the air dielectric are ionized, the air conducts electric current in the form of a spark.

For our final example, we consider a *flip coil*, a device for measuring magnetic fields. The plane of the coil in Figure 28-11 is perpendicular to a uniform magnetic induction **B**, which is to be measured. The coil is connected to a *ballistic galvanometer*, a device designed to measure the total charge passing through it in a short time. The flux through the flip coil is

$$\phi_m = NBA$$

where N is the number of turns and A is the area of the coil. If the coil is suddenly rotated about a diameter through 90°, the flux decreases to zero. While the flux is decreasing, there is an emf in the coil and a current in the coil-galvanometer circuit. The current is

$$I = \frac{\mathcal{E}}{R} = \frac{1}{R}\frac{d\phi_m}{dt}$$

where R is the total resistance of the coil and galvanometer. The total charge that passes through the galvanometer is

$$Q = \int I\,dt = \frac{1}{R}\int d\phi_m = \frac{\phi_m}{R} = \frac{NAB}{R}$$

If the total charge is measured, the magnetic induction B can be found from

$$B = \frac{RQ}{NA} \qquad\qquad 28\text{-}11$$

28-4 Eddy Currents

In the examples we have discussed, the currents produced by a changing flux were set up in definite circuits. Often a changing flux sets up circulating currents, called *eddy currents*, in a piece of bulk metal like the core of a transformer. Consider a conducting slab between the pole faces of an electromagnet (Figure 28-12). If the magnetic induction B between the pole faces is changing with time (as it will if the current in the magnet windings is alternating current), the flux through any closed loop in the slab will be changing. For example, the flux through the

Figure 28-10
Circuit with many turns of wire, giving a large flux for a given current in the circuit. The emf induced in the circuit when the current changes opposes the change, trying to maintain the original current.

Figure 28-11
Flip-coil circuit for measuring the magnetic field **B**. When the coil is flipped over, the total charge flowing through the galvanometer is proportional to B.

Figure 28-12
Eddy currents. If the magnetic field is changing, an emf is induced in any closed path in the metal, such as the curve C shown. The induced emf causes a current in the circuit.

curve C indicated in the figure is just the magnetic induction B times the area enclosed by the curve. If B varies, the flux will vary and there will be an induced emf around curve C. Since path C is in a conductor, there will be a current given by the emf divided by the resistance of the path. In this figure we have indicated just one of the many closed paths which will contain currents if the magnetic field between the pole faces varies. Circulating, or eddy, currents are usually unwanted because the heat produced is not only a power loss but must be dissipated. The power loss can be reduced by increasing the resistance of the possible paths for the eddy currents, as in Figure 28-13. Here the conducting slab is laminated, i.e., made up of small strips glued together. Because of the resistance between the strips, the eddy currents are essentially confined to the strips. The large eddy-current loops are broken up, and the power loss is greatly reduced.

The existence of eddy currents can be demonstrated by pulling a copper or aluminum sheet between the poles of a strong permanent magnet (Figure 28-14). Part of the area enclosed by curve C in this figure is in the magnetic field, and part is outside the field. As the sheet is pulled to the right, the flux through this curve decreases (assuming that the flux into the paper is positive). According to Faraday's law and Lenz's law, a clockwise current will be induced around this curve. Since this current is directed upward in the region between the pole faces, the magnetic field exerts a force on the current to the left, opposing motion of the sheet. You can feel this force on the sheet if you try to pull a sheet suddenly through a strong magnetic field. If the sheet has cuts in it, as in Figure 28-15, the eddy currents are lessened and the force is greatly reduced.

Figure 28-13
The eddy currents in a metal slab can be reduced by constructing the slab from small strips. The resistance of the path indicated by C is now large because of the glue between the strips.

Figure 28-14
Demonstration of eddy currents. When the metal slab is pulled to the right, there is a magnetic force to the left on the induced current, opposing the motion.

Figure 28-15
If the metal slab has cuts as shown, the eddy currents are reduced because of the lack of good conducting paths.

Questions

1. A bar magnet is dropped inside a long vertical pipe. The pipe is evacuated so that there is no air resistance, but the falling magnet still reaches a terminal velocity. Explain why.

2. A sheet of metal fixed to the end of a pivoted rod will swing like a pendulum about the pivot, but if the sheet is made to swing through the gap between two poles of a magnet, the oscillation will rapidly damp out. Why?

28-5   The Betatron

An interesting use of Faraday's law is the betatron (Figure 28-16), invented in 1941 by Donald Kerst at the University of Illinois to accelerate electrons (also called beta particles) to high energies. In the betatron, electrons move in a circular orbit at a constant radius in an evacuated doughnut-shaped tube. They are accelerated by a tangential, nonconservative electric field, which is produced by a changing magnetic flux generated by an electromagnet through the orbit. The centripetal force needed to keep the electrons moving in a circle is provided by the same magnet. This can be done if the magnetic induction B_0 at the orbit radius is just half the average value of the magnetic induction over the complete area of the orbit. To accomplish this the pole pieces of the magnet are shaped as shown in Figure 28-17, so that the magnetic induction decreases with distance from the center of the circle.

Figure 28-16
D. W. Kerst with the first betatron in 1941. This betatron
was capable of accelerating
electrons to an energy of 2.3
MeV. The electron beam
strikes a target inside the
doughnut, producing x-rays
used for research in nuclear
physics. (*Courtesy of the Niels
Bohr Library, American Institute of Physics.*)

We define the average magnetic induction B_{av} to be the flux divided by the area:

$$B_{av} = \frac{1}{\pi R^2} \int \mathbf{B} \cdot \hat{\mathbf{n}} \, dA = \frac{\phi_m}{\pi R^2} \qquad 28\text{-}12$$

The emf around the orbit is

$$\varepsilon = \frac{d\phi_m}{dt} = \pi R^2 \frac{dB_{av}}{dt} \qquad 28\text{-}13$$

This emf is equivalent to a tangential, nonconservative electric field given by Equation 28-2:

$$\varepsilon = \oint \mathbf{E} \cdot d\ell = E2\pi R = \pi R^2 \frac{dB_{av}}{dt}$$

or

$$E = \tfrac{1}{2} R \frac{dB_{av}}{dt} \qquad 28\text{-}14$$

The tangential force qE increases the magnitude of the momentum of the electron according to Newton's second law:

$$\frac{dp}{dt} = qE = \tfrac{1}{2} qR \frac{dB_{av}}{dt} \qquad 28\text{-}15$$

The centripetal force which keeps the electrons moving in a circle is provided by the radial force qvB_0, where B_0 is the value of the magnetic induction at the orbit radius. The condition for circular motion is Equation 26-12, which we rewrite as*

$$p = B_0 qR \qquad 28\text{-}16$$

These equations are consistent with each other if the magnetic induction at the orbit is exactly half the average value over the area

$$B_0 = \tfrac{1}{2} B_{av} \qquad 28\text{-}17$$

Ceramic doughnut

Figure 28-17
The pole faces in a betatron magnet are shaped as shown so that the magnetic field at the orbit radius of the electrons has a magnitude half that of the average magnetic field through the area enclosed by the orbit.

Betatron condition

---

* Because electrons in a betatron move with speeds nearly equal to the speed of light, newtonian mechanics does not apply and the theory of special relativity must be used. We shall see when we study this theory in Chapter 35 that for high-speed particles, Newton's second law in the form $\mathbf{F} = d\mathbf{p}/dt$ is valid relativistically if momentum is redefined in such a way that $\mathbf{p} = m\mathbf{v}$ holds in the low-speed nonrelativistic limit. If the relativistic expression for momentum is used, Equations 28-15 and 28-16 are valid both relativistically and classically.

The magnetic field is produced by coils which carry alternating current. The magnetic induction B then varies as in Figure 28-18. The electrons are injected into orbit just as the magnetic field crosses zero and begins to increase. (The electrons are injected from an electron gun just outside the stable orbit and are quickly brought into the stable orbit by a magnetic field from a current pulse through some auxiliary coils.) Only the first quarter of the cycle can be used. As the flux through the orbit changes, the electrons gain energy and their momentum increases, according to Equation 28-15. They stay in their orbit because the field at the orbit B_0 changes at the same rate as the momentum. Typically, the electrons make several hundred thousand revolutions during the quarter cycle in which they are accelerated. When the magnetic field is near its maximum value, the electrons are ejected from their orbit by a current pulse through a set of auxiliary coils which momentarily weakens the induction at the orbit. Usually the electron beam is moved into a larger orbit and hits the injector gun, producing x-rays. The x-rays are used for nuclear research or for medical purposes.

Figure 28-18
B versus t in the accelerating part of the betatron cycle. The electrons are injected just after B crosses zero and ejected just before maximum B.

Questions

3. Why can't the second quarter of the cycle in Figure 28-18 be used to accelerate the electrons?

4. Why can't the fourth quarter of the cycle be used? Is there any other part of the cycle that can be used?

5. The iron core of the betatron magnet-pole face is made not from one piece of iron but from many pie-shaped wedges. Why?

28-6 Inductance

The flux through a circuit can be related to the current in that circuit and the currents in other nearby circuits. (We shall assume that there are no permanent magnets around.) Consider the two circuits in Figure 28-19. The magnetic field at some point P consists of a part due to I_1 and a part due to I_2. These fields are proportional to the currents producing them and could, in principle, be calculated from the Biot-Savart Law. We can therefore write the flux through circuit 2 as the sum of two parts; one part is proportional to the current I_1 and the other to the current I_2:

Figure 28-19
Two adjacent circuits. The magnetic field at point P is partly due to current I_1 and partly due to I_2. The flux through either circuit is the sum of two terms, one proportional to I_1 and the other to I_2.

Self-inductance and mutual inductance

$$\phi_{m2} = L_2 I_2 + M_{12} I_1 \qquad\qquad 28\text{-}18$$

where L_2 and M_{12} are constants. The constant L_2, called the *self-inductance* of circuit 2, depends on the geometrical arrangement of that circuit. The constant M_{12}, called the *mutual inductance* of the two circuits, depends on the geometrical arrangement of both circuits. In particular, we can see that if the circuits are far apart, the flux through circuit 2 due to the current I_1 will be small and the mutual inductance will be small. An equation similar to Equation 28-18 can be written for the flux through circuit 1:

$$\phi_{m1} = L_1 I_1 + M_{21} I_2 \qquad\qquad 28\text{-}19$$

The self-inductance L_1 depends only on the geometry of circuit 1, whereas the mutual inductance M_{21} depends on the arrangement of

776 CHAPTER 28 Faraday's Law

both circuits. Though it is certainly not obvious, it can be shown in general that these two mutual inductances are equal:

$$M_{12} = M_{21} \qquad\qquad 28\text{-}20$$

When the circuits are fixed and only the currents change, their induced emfs are, from Faraday's law,

$$\mathcal{E}_1 = -\frac{d\phi_{m1}}{dt} = -L_1\frac{dI_1}{dt} - M\frac{dI_2}{dt}$$

$$28\text{-}21$$

$$\mathcal{E}_2 = -L_2\frac{dI_2}{dt} - M\frac{dI_1}{dt}$$

where we have dropped the subscripts on the mutual inductance.

The SI unit of inductance is the henry (H). From Equations 28-19 and 28-21 we see that the henry is related to other SI units by

$$1\ \text{H} = 1\ \text{T·m}^2/\text{A} = 1\ \text{V·s/A}$$

We shall see that inductance can always be written as the product of a μ_0 and some characteristic length. The SI unit of the constant μ_0 can therefore be conveniently expressed as henrys per metre:

$$\mu_0 = 4\pi \times 10^{-7}\ \text{H/m}$$

In principle, the calculation of the self- and mutual inductances is straightforward for any given circuits and arrangement. For example, to calculate L_1 we need only assume some current I_1 and calculate the magnetic field from the Biot-Savart law at every point on some surface bounded by the circuit. This magnetic field will of course be proportional to the current I_1. We then calculate the flux of this field by integrating over the area. Since B is proportional to I_1, the flux will also be proportional to I_1. The proportionality constant is the self-inductance L_1. Similarly, we calculate the mutual inductance by assuming some current I_1 in circuit 1 and calculating the magnetic field everywhere on a surface bounded by circuit 2, and from it the flux through that circuit. Because of the difficulty in calculating the magnetic field in general from the Biot-Savart law and in integrating this field over any surface, it is not surprising that the calculations of self-inductance and mutual inductance are extremely difficult except for a few very simple geometrical arrangements. Consider, for example, the calculation of the self-inductance for a single circular loop of radius R. The magnetic field at the center of the loop is $B = \frac{1}{2}\mu_0 I/R$. But in order to find the flux through the loop, we need to know the value of B everywhere on the circle enclosed by the loop, not just at the center. Such a calculation is too difficult to do here. The self-inductance of a single loop depends not only on the radius of the loop but also on the radius of the conducting wire.

We now discuss some cases in which inductance can be calculated, at least approximately.

Self-Inductance of a Solenoid

This calculation is not difficult because, to a good approximation, the magnetic field inside the solenoid is uniform and given by

$$B = \mu_0\frac{N}{\ell}I = \mu_0 nI$$

where $n = N/\ell$ is the number of turns per unit length, N the total number of turns, ℓ the total length, and I the current in the solenoid.

The area enclosed by this circuit is the area of one turn times the number of turns, and so the flux is

$$\phi_m = BNA = \mu_0 nINA = \mu_0 n \frac{N}{\ell} \ell AI = \mu_0 n^2(\ell A)I \qquad 28\text{-}22$$

where A is the area of a single loop. As expected, the flux is proportional to the current I. The proportionality constant is the self-inductance:

$$L = \frac{\phi_m}{I} = \mu_0 n^2 A\ell = \mu_0 N^2 \frac{A}{\ell} \qquad 28\text{-}23$$

Self-inductance of a solenoid

The self-inductance is proportional to the square of the number of turns per unit length and to the volume $A\ell$. It can also be thought of as the product of μ_0, N^2, and the characteristic length A/ℓ.

Example 28-1 Find the self-inductance of a solenoid of length 10 cm, area 5 cm², and 100 turns.

We can calculate the self-inductance from Equation 28-23 in henrys if we put all quantities in SI units:

$$n = \frac{N}{\ell} = \frac{100 \text{ turns}}{0.1 \text{ m}} = 10^3 \text{ turns/m}$$

$$A\ell = (5 \times 10^{-4} \text{ m}^2)(0.1 \text{ m}) = 5 \times 10^{-5} \text{ m}^3$$

$$\mu_0 = 4\pi \times 10^{-7} \text{ H/m}$$

$$L = (4\pi \times 10^{-7})(10^3)^2(5 \times 10^{-5}) = 2\pi \times 10^{-5} \text{ H}$$

Example 28-2 At what rate must the current in the solenoid of Example 28-1 change to induce an emf of 20 V?

From Equation 28-21 with $M = 0$ we have

$$\varepsilon = -L \frac{dI}{dt} = 20 \text{ V}$$

Then

$$\frac{dI}{dt} = -\frac{\varepsilon}{L} = \frac{20 \text{ V}}{2\pi \times 10^{-5} \text{ H}} = -3.18 \times 10^5 \text{ A/s}$$

Mutual Inductance between a Long Wire and a Rectangular Loop

Two circuits for which the mutual inductance can be calculated are shown in Figure 28-20, where the wire on the left is assumed to be so long that the remaining part of its circuit is far from the rectangular loop and contributes negligible flux. To find the flux through the rectangular loop we must integrate. Since the magnetic field due to a long wire decreases as $1/r$ with the distance from the wire, we choose the strip-shaped area element shown in the figure. The area of this strip is $c\, dx$, and the flux through the strip is

$$d\phi_m = \frac{\mu_0 I}{2\pi x} c\, dx \qquad 28\text{-}24$$

where I is the current in the long wire. The total flux is found by integrating from $x = a$ to $x = b$:

$$\phi_m = \frac{\mu_0}{2\pi} cI \int_a^b \frac{dx}{x} = \frac{\mu_0 Ic}{2\pi} \ln \frac{b}{a} \qquad 28\text{-}25$$

Figure 28-20
Long wire and circuit for calculation of mutual inductance.

The flux through the rectangular loop due to the current I in the long wire is proportional to I, as expected. The proportionality constant is the mutual inductance:

$$M = \frac{\mu_0 c}{2\pi} \ln \frac{b}{a}$$

Questions

6. How would the inductance of a solenoid be changed if the same length of wire were wound onto a cylinder of the same diameter but twice as long? If twice as much wire were wound onto the same cylinder?

7. Two solenoidal inductances are wound on cylinders of the same diameter. Is it possible for them to have the same inductance even though one has twice the total length of wire of the other?

28-7 *LR* Circuits

As we have seen, self-inductance in a circuit prevents the current from rising or falling instantaneously. Circuits containing coils or solenoids of many turns have a large self-inductance. Such a coil or solenoid is called an *inductor*. The symbol for an inductor is ∿. We can often neglect the self-inductance of the rest of the circuit compared with that of an inductor.

A circuit containing batteries, resistors, and inductors is called an *LR* circuit. Since all circuits contain resistance and self-inductance, the analysis can be applied to some extent to all circuits. All circuits also have some capacitance between parts of the circuit at different potentials. We shall include the effects of capacitance in Section 28-9, when we study *LC* and *LCR* circuits. We neglect capacitance here in order to simplify the analysis and to bring out the special features of inductance in a simple situation.

Figure 28-21 shows a typical circuit. We neglect the internal resistance of the battery and treat it as an ideal seat of emf. The current in the circuit is originally zero. At $t = 0$ we close the switch. As discussed before, while the current is increasing, there is an induced emf in the inductance in the direction opposing the current increase. In the figure the current will be clockwise, as shown. We can relate the current, its rate of change dI/dt, and the emf of the battery by applying Kirchhoff's first rule to the circuit shown. Because of the negative sign in Equation 28-21, which reduces to $\varepsilon = -L\, dI/dt$ for this circuit, and the possibility that the dI/dt itself may be positive or negative, there can be confusion in applying Kirchhoff's rule and getting the correct sign for the potential difference across an inductor. The best approach is to assume a direction for I, assume dI/dt to be positive, and then write $L\, dI/dt$ for the potential difference across the inductor, obtaining the proper sign from Lenz's law. In Figure 28-21 there is a potential drop across the resistor from a to b of magnitude IR. If dI/dt is positive, the induced emf in the inductor is so directed from c to b as to decrease the current. Point b is therefore like the positive terminal of a battery. From b to c the potential drops by the amount $L\, dI/dt$. As we move from point c to point a, the potential increases by the amount of the emf of the battery, which is ε_0. Kirchhoff's first rule applied to this circuit therefore gives

Figure 28-21
Typical *LR* circuit.

$$\mathcal{E}_0 = IR + L\frac{dI}{dt} \qquad\qquad 28\text{-}26$$

This equation is similar to Equation 25-26, for an *RC* circuit. Again, we can understand the general behavior of the current *I* versus time without going into the mathematical details of the solution of the equation.

Just after the switch is closed, at *t* = 0, the current is zero and the rate of change of the current is, from Equation 28-26,

$$\left(\frac{dI}{dt}\right)_0 = \frac{\mathcal{E}_0}{L} \qquad\qquad 28\text{-}27$$

The current thus increases as expected. After a short time, the current has reached some positive value, and the rate of change is given by

$$\frac{dI}{dt} = \frac{\mathcal{E}_0}{L} - \frac{IR}{L}$$

At this time the current is still increasing, but its rate of increase is less than at *t* = 0. The final value of the current can be obtained by setting *dI/dt* equal to zero. From Equation 28-26 we see that the final value of the current is

$$I_f = \frac{\mathcal{E}_0}{R} \qquad \text{when } \frac{dI}{dt} = 0 \qquad\qquad 28\text{-}28$$

Figure 28-22 shows *I* versus *t*. The time for the current to reach an appreciable fraction of its final value depends on the resistance and inductance. The mathematical expression for the current as a function of time obtained by solving Equation 28-26 is

$$I = \frac{\mathcal{E}_0}{R}(1 - e^{-RT/L}) = I_f(1 - e^{-t/t_c}) \qquad\qquad 28\text{-}29$$

where

$$t_c = \frac{L}{R} \qquad\qquad 28\text{-}30$$

is the *time constant* of the circuit. The larger the self-inductance *L* or the smaller the resistance *R*, the longer it takes for the current to build up.

We note that the product of the initial slope $\mathcal{E}_0/L$ and the time constant $t_c = L/R$ equals the final current $\mathcal{E}_0/R$:

$$\left(\frac{dI}{dt}\right)_0 t_c = \frac{\mathcal{E}_0}{L}\frac{L}{R} = \frac{\mathcal{E}_0}{R} = I_f$$

If the current continued to increase at its initial rate, it would reach its final value after one time constant.

Figure 28-23 shows a slightly different arrangement, with an additional switch that allows us to remove the battery and an additional resistor *r* to protect the battery so that it is not shorted when both switches are momentarily closed. If both switches are open and we close switch S_1, the current builds up in the circuit as discussed above, except that the total resistance is now *r* + *R* and the final current is $\mathcal{E}/(r + R)$. Suppose that this switch has been closed for a time long compared with the time constant, which is now $L/(R + r)$, so that the current is approximately steady at its final value, which we shall call I_0.

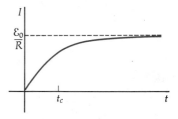

Figure 28-22
Graph of current *I* versus time *t* for the circuit in Figure 28-21, in which the switch is closed at *t* = 0. The current increases to its maximum value $\mathcal{E}_0/R$, reaching 63 percent of this value after time $t_c = L/R$.

Time constant

Figure 28-23
LR circuit with two switches so that the battery can be removed from the circuit. After the current in the inductor reaches its maximum value with S_1 closed, S_2 is closed and S_1 opened. The current then decreases in time as shown in Figure 28-24.

Switch S_2 is then closed and switch S_1 is opened, to remove the battery from consideration completely. Let us choose the time $t = 0$ when switch S_2 is closed. We now have a circuit with just a resistor and an inductor (loop *abcd*) with an initial current I_0. Again we analyze this circuit by applying Kirchhoff's rule. The potential drop from point *a* to *b* across the resistors is IR. Across the inductor, the potential drop from *b* to *c* is $L\, dI/dt$. (Since dI/dt will turn out to be negative, this is really a potential increase. Less confusion with signs arises if we always assume that the quantities I and dI/dt are positive when writing the equations. The solutions of the equations then give us the correct signs of the quantities.) We are now back at our original point. The sum of the potential drops must be zero. Thus

$$IR + L\frac{dI}{dt} = 0 \qquad\qquad 28\text{-}31$$

Since the current I is positive, the rate of change dI/dt is negative:

$$\frac{dI}{dt} = -\frac{R}{L}I \qquad\qquad 28\text{-}32$$

The induced emf is in the direction of the current and opposite to that indicated in Figure 28-23. In agreement with Lenz's law, the induced emf is in the direction that maintains the current and tends to prevent its decrease. Equation 28-32 is similar to Equation 25-21 for the discharge of a capacitor, with I replacing Q and L/R replacing the time constant RC.

Figure 28-24 shows a sketch of the current I versus time t. The original value of the current is I_0, and the original value of the slope is $dI/dt = -RI_0/L$. As the current decreases, the slope decreases in magnitude until both the current and its slope are zero. The mathematical expression for the current is

$$I = I_0 e^{-Rt/L} = I_0 e^{-t/t_c} \qquad\qquad 28\text{-}33$$

where $t_c = L/R$ is the time constant.

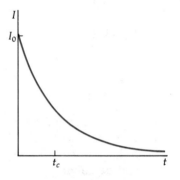

Figure 28-24
Current versus time after switch S_2 is closed in Figure 28-23. The current decreases as e^{-t/t_c}, where $t_c = L/R$ is the time constant.

28-8 Magnetic Energy

When a current is set up in a circuit like the one in Figure 28-21, only part of the energy supplied by the battery goes into Joule heat in the resistor; the rest of the energy is stored in the inductor. We can see this by multiplying each term in Equation 28-26 by the current. We have then

$$\mathcal{E}_0 I = I^2 R + LI\frac{dI}{dt} \qquad\qquad 28\text{-}34$$

The term on the left side of this equation represents the rate at which energy is supplied by the seat of emf. The first term on the right, $I^2 R$, is the rate at which energy is put into the resistor. This energy appears as heat. The second term, $LI\, dI/dt$, is the rate at which energy is put into the inductor and is the term we wish to discuss. Let U_m be the energy in the inductor. According to Equation 28-34, the rate of increase of this energy is

$$\frac{dU_m}{dt} = LI\frac{dI}{dt} \qquad\qquad 28\text{-}35$$

We can find the total energy in the inductor by integrating from the time $t = 0$, when the current is zero, to the time $t = \infty$, when the current has reached its final value I_f:

$$U_m = \int dU_m = \int LI \, dI = \tfrac{1}{2}LI_f^2 \qquad \text{28-36}$$

Energy in an inductor

This energy is stored in the self-inductance of the circuit, as we can see by considering what happens when the emf is removed from the circuit, as in Figure 28-23 with S_2 closed and S_1 open. Though there is now no power input into the circuit by the emf, the current does not immediately fall to zero, so there is still some power put into the resistor. The rate at which energy is put into the resistor in this case equals the rate at which the energy in the inductor is decreased. We can see this by multiplying each term in Equation 28-31 by the current I:

$$I^2R = -LI\frac{dI}{dt} \qquad \text{28-37}$$

Whenever a current is set up in a circuit, a magnetic field is set up. We can consider the energy stored in a circuit by virtue of its self-inductance and current as energy stored in the magnetic field. When the current is decreased, decreasing the energy $\tfrac{1}{2}LI^2$ in the inductor, the magnetic field is also decreased. The idea that energy is stored in a magnetic field is similar to the idea that energy is stored in an electric field when a capacitor is charged. In Chapter 23 we showed that the electrostatic energy stored in a parallel-plate capacitor can be written

$$U_e = \tfrac{1}{2}QV = \tfrac{1}{2}\epsilon_0 E^2 Ad$$

where the area of the plates A times their separation d is the volume in which there is an electrostatic field E. We then stated that this relation is more general than indicated by that example; namely, that whenever an electrostatic field is set up in some volume, there is energy stored $\tfrac{1}{2}\epsilon_0 E^2$ per unit volume. We now give an analogous argument for magnetic energy. Again, for simplicity, we consider a special case, that of a solenoid.

Let n be the number of turns per unit length of a solenoid and $A\ell$ be the volume, where A is the cross-sectional area and ℓ the length. The magnetic field inside the solenoid is

$$B = \mu_0 nI$$

and the self-inductance is

$$L = \mu_0 n^2 \ell A$$

Substituting this expression for L and $I = B/\mu_0 n$ for the current into Equation 28-36 for the energy of an inductor, we have

$$U_m = \tfrac{1}{2}LI^2 = \tfrac{1}{2}\mu_0 n^2 \ell A \left(\frac{B}{\mu_0 n}\right)^2 = \frac{B^2}{2\mu_0}\ell A$$

The energy stored in a solenoid can be written as the product of the volume ℓA and the term $\tfrac{1}{2}B^2/\mu_0$. Again, the result is more general than is indicated by this discussion. Whenever a magnetic field is set up in some volume $\mathcal{V}$ in space, energy is stored. The energy per unit volume η_m is

$$\eta_m = \frac{U_m}{\mathcal{V}} = \frac{B^2}{2\mu_0} \qquad \text{28-38}$$

Magnetic energy density

We see that there is energy stored in both the electric and magnetic fields. In general, if there is an electric field E and magnetic field B in some volume of space, the energy per unit volume associated with these fields is

$$\eta = \frac{U}{\mathcal{V}} = \tfrac{1}{2}\epsilon_0 E^2 + \frac{B^2}{2\mu_0} \qquad\qquad 28\text{-}39$$

28-9 *LC* and *LCR* Circuits

Figure 28-25 shows a capacitor connected to an inductor and a switch. Assume that the capacitor carries an initial charge Q_0 and the switch is initially open. At $t = 0$ we close the switch, and the charge flows through the inductor. For simplicity, we shall neglect any resistance in the circuit.

We have chosen the direction for the current in the circuit so that when the charge on the bottom capacitor plate is $+Q$, the current is

$$I = \frac{dQ}{dt}$$

This choice of direction is purely arbitrary. With this choice, the current will be negative just after the switch is closed. With this choice of positive sense for the current, the potential drop across the inductor from point a to b is $L\,dI/dt$. Across the capacitor from c to d there is a potential drop Q/C. Thus Kirchhoff's rule for this circuit gives

$$L\frac{dI}{dt} + \frac{Q}{C} = 0 \qquad\qquad 28\text{-}40$$

Substituting dQ/dt for I in this equation, we obtain for the charge Q in the capacitor

$$L\frac{d^2Q}{dt^2} + \frac{Q}{C} = 0$$

or

$$\frac{d^2Q}{dt^2} = -\frac{1}{LC}Q \qquad\qquad 28\text{-}41$$

Equation 28-41 is of the same form as the equation for the acceleration of a mass on a spring,

$$\frac{d^2x}{dt^2} = -\omega^2 x$$

where $\omega^2 = k/m$, k is the spring constant, and m is the mass.

In Chapter 11 we studied the solutions of this equation for simple harmonic motion and found that we could always write the solution in the form

$$x = A\cos(\omega t + \delta)$$

where $\omega = \sqrt{k/m}$ is the angular frequency, A is the maximum value of x (called the amplitude), and δ is the phase constant, which depends on the initial conditions. We can put Equation 28-41 in the same form by writing ω^2 for the quantity $1/LC$. Then

$$\frac{d^2Q}{dt^2} = -\omega^2 Q \qquad\qquad 28\text{-}42$$

Figure 28-25
LC circuit. The capacitor is initially charged. The charge on the plates and the current I are related by $I = dQ/dt$ for the choice of current direction indicated. Just after the switch is closed, I is negative.

$$\omega = \frac{1}{\sqrt{LC}} \qquad\qquad 28\text{-}43$$

The solution of Equation 28-42 is

$$Q = A \cos(\omega t + \delta) \qquad\qquad 28\text{-}44$$

The current is obtained by differentiating this solution:

$$I = \frac{dQ}{dt} = -\omega A \sin(\omega t + \delta) \qquad\qquad 28\text{-}45$$

For our initial conditions, the phase constant δ must be zero because the current is zero at time $t = 0$. Setting $t = 0$ in Equation 28-45, we have

$$I_0 = -\omega A \sin \delta = 0$$

and hence $\delta = 0$. Then our solution for the charge is

$$Q = A \cos \omega t$$

The constant A is just the value of the charge at $t = 0$ since $\cos 0 = 1$. Writing Q_0 for this initial charge, we have

$$Q = Q_0 \cos \omega t \qquad\qquad 28\text{-}46$$

and

$$I = -\omega Q_0 \sin \omega t \qquad\qquad 28\text{-}47$$

Figure 28-26 shows graphs of Q and I versus time. The charge oscillates between the values $+Q_0$ and $-Q_0$ with the angular frequency $\omega = 1/\sqrt{LC}$. The current also oscillates with this frequency and is 90° out of phase with the charge. The current is maximum when the charge is zero and zero when the charge is maximum.

Figure 28-26
Graphs of Q versus t and I versus t for the LC circuit in Figure 28-25.

In our study of the oscillation of a mass on a spring we found that the total energy is constant but oscillates between potential and kinetic energy. In our LC circuit we also have two kinds of energy, electrostatic energy and magnetic energy. The electrostatic energy of the capacitor is

$$U_e = \tfrac{1}{2}QV = \frac{Q^2}{2C}$$

where $V = Q/C$ is the voltage drop across the capacitor. Substituting $Q_0 \cos \omega t$ for Q in this equation, we have for the electrostatic energy of the capacitor

$$U_e = \frac{Q_0^2}{2C} \cos^2 \omega t \qquad\qquad 28\text{-}48$$

The electrostatic energy stored in the capacitor oscillates between its maximum value $Q_0^2/2C$ and zero. When the current in the circuit is I, the magnetic energy stored in the inductor is

$$U_m = \tfrac{1}{2}LI^2 \qquad\qquad 28\text{-}49$$

Substituting the value for the current from Equation 28-47 into this expression, we get

$$U_m = \tfrac{1}{2}L\omega^2 Q_0^2 \sin^2 \omega t = \frac{Q_0^2}{2C} \sin^2 \omega t \qquad \text{28-50}$$

where we have used the fact that ω^2 is equal to $1/LC$. The magnetic energy also oscillates between its maximum value of $Q_0^2/2C$ and zero. The sum of the electrostatic and magnetic energies is the total energy, which is constant in time:

$$U_{\text{total}} = U_e + U_m = \frac{Q_0^2}{2C} \cos^2 \omega t + \frac{Q_0^2}{2C} \sin^2 \omega t = \frac{Q_0^2}{2C}$$

Example 28-3 A 2-μF capacitor is initially charged to 20 V and then shorted across a 6-μH inductor. What are the frequency of oscillation and the maximum value of the current?

The frequency of oscillation is independent of the initial charge and depends only on the values of the capacitance and inductance. The frequency is

$$f = \frac{\omega}{2\pi} = \frac{1}{2\pi}\sqrt{\frac{1}{LC}} = \frac{1}{2\pi\sqrt{LC}}$$

$$= \frac{1}{2\pi\sqrt{(6 \times 10^{-6})(2 \times 10^{-6})}} = 4.59 \times 10^4 \text{ Hz}$$

According to Equation 28-47, the maximum value of the current is related to the maximum value of the charge by

$$I_m = \omega Q_0 = \frac{Q_0}{\sqrt{LC}}$$

The initial charge on the capacitor is

$$Q_0 = CV_0 = (2\ \mu F)(20\ V) = 40\ \mu C$$

Thus

$$I_m = \frac{40\ \mu C}{\sqrt{(6\ \mu H)(2\ \mu F)}} = 11.5\ A$$

In Figure 28-27 we include a resistor in series with the capacitor and inductor. Again we assume that the switch is initially open, with the capacitor carrying an initial charge Q_0, and we close the switch at $t = 0$. We need only modify Equation 28-40 by including the potential drop IR across the resistor. We then have from Kirchhoff's rule

$$L\frac{dI}{dt} + \frac{Q}{C} + IR = 0$$

or

$$L\frac{d^2Q}{dt^2} + \frac{Q}{C} + R\frac{dQ}{dt} = 0 \qquad \text{28-51}$$

using $I = dQ/dt$ as before. Equation 28-51 is analogous to Equation 11-39 for a damped harmonic oscillator. If the resistance is small, the charge and current still oscillate with very nearly the same frequency $1/\sqrt{LC}$ but the oscillations are damped; i.e., the maximum values of the charge and current decrease with each oscillation. We can under-

Figure 28-27
LCR circuit.

stand this qualitatively from energy considerations. If we multiply each term in Equation 28-51 by the current I, we have

$$IL\frac{dI}{dt} + I\frac{Q}{C} + I^2R = 0 \qquad\qquad 28\text{-}52$$

The first term in this equation is the current times the voltage across the inductor. This is the rate at which energy is put into the inductor or taken out of it, i.e., the rate of change of magnetic energy, which is positive or negative depending on whether I and dI/dt have the same or different signs. Similarly, the second term is the current times the voltage across the capacitor. This is the rate of change of the energy of the capacitor. Again, it may be positive or negative. The last term, I^2R, the rate at which energy is dissipated in the resistor as Joule heat, is always positive regardless of the sign of the current since it depends only on I^2. The sum of the electric and magnetic energies is not constant for this circuit because energy is continually dissipated in the resistor. Figure 28-28 shows graphs of Q versus t and I versus t for small resistance. If we increase R, the oscillations are more heavily damped, until a critical value of R is reached for which there is not even one oscillation. Figure

Figure 28-28
Graphs of Q versus t and I versus t for *LCR* circuit of Figure 28-27 for small resistance R.

28-29 shows the charge versus t for R greater than the critical damping value. The detailed solution of Equation 28-51 is exactly analogous to the solution of Equation 11-39 for the damped harmonic oscillator and will not be discussed here.

Figure 28-29
Graph of Q versus t for very large resistance R in *LCR* circuit of Figure 28-27.

Questions

8. It is not difficult to produce *LC* circuits with frequencies of oscillation of thousands of hertz or more, but it is difficult to produce *LC* circuits with small frequencies. Why?

9. How will the frequency of an *LCR* circuit be altered if the inductance in the circuit is doubled? The capacitance doubled? The voltage with which the capacitor is initially charged is doubled? The resistance in the circuit is doubled?

Electric Motors

Reuben E. Alley, Jr.
U.S. Naval Academy

In Thomas Edison's machine shop the various tools were driven from a single steam engine by means of shafts, pulleys, and belts, as shown in Figure 1. Such a power-distribution system was typical of most industrial establishments around 1900. The installation of electric-power distribution facilities and the development of practical electric motors made it possible to replace the complicated and hazardous system of belts and pulleys with individual energy sources for each machine. The growth of a nationwide electric-power network and the concomitant invention of several types of electric motors resulted in a revolution in our society as profound as the electronics revolution associated with the invention and application of transistors and integrated circuits.

Figure 1
Interior of Thomas Edison's second machine shop, Menlo Park, New Jersey, as it appeared during a New Year's Eve lighting demonstration, 1879. (*Collections of Greenfield Village and the Henry Ford Museum, Dearborn, Michigan.*)

A single electric motor may be supplied from a battery or from a small generator, but a nationwide power network is required to supply the millions of joules of electric energy that are converted into mechanical energy in such varied applications as steel rolling mills, paper mills, subway trains, electric locomotives, automobile assembly lines, and mining machinery. Although automobiles, ships, diesel locomotives, trucks, and airplanes do not depend upon the power network, they all use electric motors. In the home, electric motors are essential parts of furnaces, air conditioners, refrigerators, and washing machines. Imagine what your everyday life would be like if there were no electric motors.

Figure 2 shows a simple *dc motor.* Current from the battery magnetizes the soft-iron *armature,* which is free to rotate about axis AA' and turns to align itself with the *field* produced by the poles labeled N and S. As the armature rotates, it carries with it the *commutator,* whose two segments provide means for reversing current direction just as the armature reaches its equilibrium position. The mass of the armature ensures that it will rotate past the equilibrium position; then, because of its reversed polarity, there is a further rotation of a half revolution. Because the commutator reverses the current direction every $180°$, continuous rotation is achieved. Useful work can be done by the shaft of the motor.

The motor shown in Figure 2 has several disadvantages: (1) when power is disconnected, the motor tends to stop in the equilibrium position, and so there may be no starting torque, and (2) the torque is zero twice during each revolution. If the armature is provided with additional poles and windings, and if the commutator is divided into more segments, the torque is more nearly uniform and the motor always starts.

If the field is supplied from a permanent magnet, as in model trains and most battery-powered toys, the speed depends upon supply voltage and changes with load. For applications requiring appreciable amounts of power, electromagnets supply the field. Adjustment of field current permits control of speed independent of armature current. In general, if the load on a dc motor increases, speed decreases and more armature current is required. Field current can be adjusted to maintain constant speed. Automatic controls are frequently used with dc motors.

DC motors with field windings are very flexible. They may be connected in *series*, so that the armature current is also the field current; in *parallel* (or *shunt*), so that the field current is independent of armature current; or in a *compound* arrangement using two field windings, so that one is connected in series and the other is connected in parallel. DC motors are widely used for traction (subways and electric railways) and in applications where speed control is critical, e.g., steel rolling mills.

The dc motor in Figure 2 becomes a *synchronous motor* if the commutator is replaced by *slip rings* (Figure 3) and alternating current is supplied. Although this simple device has no starting torque, it will "lock in" if it is accelerated by external means to *synchronous speed* (determined by power-line frequency). As the load is increased, armature current increases but the speed stays constant unless the load is great enough to make the machine stall. Synchronous motors designed for use with three-phase power (see below) can be self-starting. When the horsepower rating is high, field current is much less than armature current. In order that the smaller current can be supplied through slip rings, most such machines have a stationary armature while the field rotates on the shaft that drives the load.

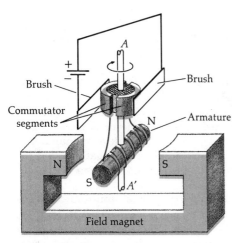

Figure 2
A simple dc motor. The field magnet may be a permanent magnet or a soft-iron core wrapped with many turns of wire, through which a direct current flows.

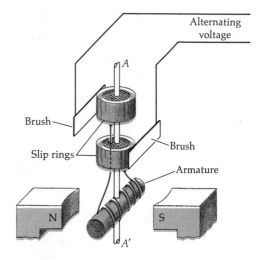

Figure 3
A simple synchronous motor. The field magnet is like that in Figure 2. Only the pole pieces are shown here.

The *universal series motor* has many low-power applications such as hand drills and small saws. A simple form of such a motor results if the current, before entering the commutator (Figure 2), first goes through the field winding. This motor will operate on either alternating or direct current. No-load speed depends upon the number of field poles, the number of armature poles, and the supply voltage. Many hand drills now are equipped with solid-state circuits that control speed by controlling applied voltage. These machines slow down under load, but they develop reasonably high torque and so are especially satisfactory for driving hand tools.

Practically all commercially generated electric power in the United States is three-phase, 60-Hz sinusoidal voltage. The term three-phase means that there are three sinusoidal voltages of equal amplitude and frequency whose successive peak values are separated in time by one-third of a cycle (see Figure 4). These three voltages are usually carried in the three-wire transmission lines that are a common sight around the countryside. One phase of the power appears between each pair of wires. Residences, almost without exception, are supplied with only a single phase, but all three phases are commonly supplied to industries.

Figure 4
Voltages in a three-phase power system. At any instant the sum of the three voltages is zero.

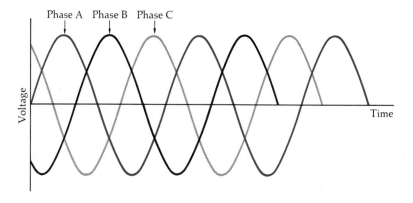

Figure 5 shows a simple three-phase *induction motor*. The winding of each of the three pairs of poles is connected to a different phase. The currents in the windings therefore reach their maximum values successively, and the result is a rotating magnetic field in the region between the poles. A stationary rotor of conducting material in this field will have currents set up in it. The magnetic fields of the *induced currents* interact with the rotating field to produce a torque. If there is no load on the motor, the torque accelerates the rotor to a speed almost equal to the synchronous speed of rotation of the field. (If the rotor actually attained synchronous speed, there would be no relative motion between field and rotor and so no induced currents or torque.) As the motor is loaded, rotor speed decreases; this results in larger induced currents and correspondingly greater torques. Speed may decrease to 75 percent of synchronous speed before the motor stalls. There are no moving electrical contacts in the three-phase induction motor, which can be built to develop hundreds of horsepower. The elimination of any possibility of electric sparks makes these motors especially attractive for applications in explosive atmospheres, e.g., mines or flour mills.

A practical induction motor usually has a cylindrical iron rotor with insulated copper bars embedded in its surface and interconnected to form good conducting paths for induced currents. In some motors the rotor windings, instead of being short-circuited, are connected to slip rings so that external resistors can be used for speed control, but the resulting moving electrical contacts remove one of the important advantages of induction motors.

If only one phase is connected in Figure 5, there is no rotating magnetic field and no starting torque. If a second winding in Figure 5 is connected in series with a large (hundreds of microfarads) capacitor and supplied from the same phase as the first winding, the currents in the two windings are suf-

Figure 5
Schematic cross section of a
simple three-phase induction
motor.

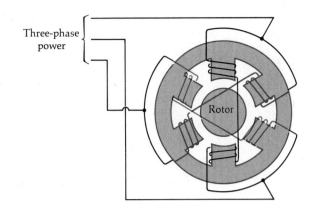

Three-phase power

Rotor

ficiently out of phase to provide a rotating field and therefore a starting torque. This fact is employed in the design of *single-phase* induction motors, used to drive refrigerators, washing machines, air conditioners, and furnaces.

Since 1965, especially in Great Britain, Germany, and Japan, serious efforts have been made to develop a high-speed mass-transit system based upon the *linear induction motor*. It is possible to achieve both acceleration and levitation (once speeds of about 30 km/h are reached) by electrical means. Such support for the vehicle eliminates most of the friction losses associated with the rolling of flanged wheels on steel rails.

Suppose the cylindrical shell that supports the poles of an induction motor is cut along an axial line and flattened (just as a tin can may be cut and unrolled). Then the poles will lie in a line and the magnetic field will move across the poles periodically. Now imagine a long strip of magnetic material with many poles equally spaced along its length. Let the first pole and every third succeeding pole be supplied from one phase of the power line; let the second pole and every third succeeding pole be supplied from the second phase; and let the third pole and every third succeeding pole be supplied from the third phase. The result will be a magnetic field that travels along the strip with a speed relative to the ground that depends upon the distance between poles and the frequency of the power. Now let a mass of conducting material be supported (by wheels running on rails, for example) so that it is free to move in the direction of the strip. When the mass is stationary, currents will be induced whose magnetic fields will interact with the moving field to produce an accelerating force along the direction of the strip (Figure 6). The mass will acquire a speed that is somewhat less than synchronous speed, just like the rotor of the induction motor. What has been described is a simple linear induction motor. Such a device was proposed in 1946 as a means of catapulting aircraft from a ship.

Figure 6
Schematic diagram of a
simple linear induction
motor.

Conducting material
on wheels

Direction of motion

Stationary
structure

Phase A
Phase B
Phase C

An alternative design of the linear induction motor involves mounting a number of poles on the vehicle and supplying them through flexible collectors in contact with a stationary three-phase distribution system. The interaction of the fields of the poles in the vehicle with currents induced in a stationary metal strip (aluminum backed by a soft magnetic material to provide a path for magnetic flux) produces support and accelerating force for the vehicle (Figure 7).

Figure 7
Linear induction motor with moving poles.

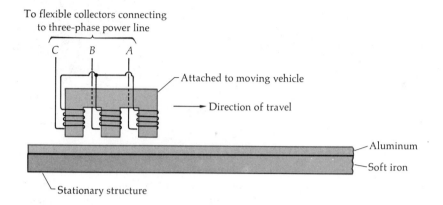

To flexible collectors connecting to three-phase power line

Attached to moving vehicle

Direction of travel

Aluminum

Soft iron

Stationary structure

In the United States some engineers are investigating the use of a single-phase *linear synchronous motor.* Speed variation is achieved through change in frequency of the alternating voltage supplied to the vehicle. The required steady magnetic field is stationary. One proposal is to operate the field windings at superconducting temperatures, thus greatly reducing the winding resistance and thereby reducing the power loss from the direct current supplying these windings.

Although no system has yet reached the commercial stage, some prototypes have been tested. There seems to be ample evidence that it will be possible to design and build a practical transportation system that will operate at speeds of 300 km/h or more based upon the concept of a linear motor.

Review

A. Define, explain, or otherwise identify:

Faraday's law, 767

Motional emf, 767

Lenz's law, 769

Self-induced emf, 772

Flip coil, 772

Ballistic galvanometer, 772

Eddy currents, 772

Betatron, 773

Self-inductance, 775

Mutual inductance, 775

Inductor, 778

Magnetic energy, 781

B. True or false:

1. The induced emf in a circuit is proportional to the magnetic flux through the circuit.

2. There can be an induced emf at an instant when the flux through the circuit is zero.

3. The induced emf in a circuit always decreases the magnetic flux through the circuit.

4. Faraday's law can be derived from the Biot-Savart law.

5. The energy in a magnetic field is proportional to B^2.

Exercises

Section 28-1, Motional EMF

1. A rod 30 cm long moves at 8 m/s in a plane perpendicular to a magnetic field of 500 G. Its velocity is perpendicular to the length of the rod. Find (a) the mag-

netic force on an electron in the rod, (b) the electrostatic field **E** in the rod, and (c) the potential difference V between the ends of the rod.

2. Find the speed of the rod in Exercise 1 if the potential difference between its ends is 6 V.

3. In Figure 28-5 let B be 0.8 T, $v = 10.0$ m/s, $\ell = 20$ cm, and $R = 2\ \Omega$. Find (a) the induced emf in the circuit, (b) the current in the circuit, and (c) the force needed to move the rod with constant velocity assuming negligible friction. (d) Find the power input by the force found in part (c) and the rate of heat production I^2R.

4. Work Exercise 3 for $B = 1.5$ T, $v = 6$ m/s, $\ell = 40$ cm, and $R = 1.2\ \Omega$.

Sections 28-2, Lenz's Law, and Section 28-3, Applications of Faraday's Law

5. The two loops in Figure 28-30 have their planes parallel to each other. As viewed from A toward B, there is a counterclockwise current in A. Give the direction of the current in loop B and state whether the loops attract or repel each other if the current in loop A is (a) increasing and (b) decreasing.

Figure 28-30
Exercise 5.

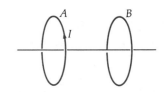

6. A bar magnet moves with constant velocity along the axis of a loop, as shown in Figure 28-31. (a) Make a qualitative sketch of the flux ϕ_m through the loop as a function of time t. Indicate the time t_1 when the magnet is halfway through the loop. (b) Sketch the current I in the loop versus time, choosing I positive when it is counterclockwise as viewed from the left.

7. Give the direction of the induced current in the circuit on the right in Figure 28-32 when the resistance in the circuit on the left is suddenly (a) increased; (b) decreased.

Figure 28-31
Exercise 6.

8. A uniform magnetic field **B** is established perpendicular to the plane of a loop of radius 5.0 cm, resistance 0.4 Ω, and negligible self-inductance. The magnitude of **B** is increasing at a rate of 40 mT/s. Find (a) the induced emf in the loop, (b) the induced current in the loop, and (c) the rate of Joule heating in the loop.

9. A 100-turn coil has radius 4.0 cm and resistance 25 Ω. At what rate must a perpendicular magnetic field change to produce a current of 4.0 A in the coil?

Figure 28-32
Exercise 7.

10. Show that if the flux through each turn of an N-turn coil of resistance R changes from ϕ_{m1} to ϕ_{m2} in any manner, the total charge that passes through the coil is given by $Q = N(\phi_{m2} - \phi_{m1})/R$.

11. The flux through a loop is given by $\phi_m = (t^2 - 4t) \times 10^{-1}$ T·m^2, where t is in seconds. (a) Find the induced emf ε as a function of time. (b) Find both ϕ_m and ε at $t = 0$, $t = 2$ s, $t = 4$ s, and $t = 6$ s.

12. For the flux given in Exercise 11, sketch ϕ_m versus t and ε versus t. (a) At what time is the flux maximum? What is the emf at this time? (b) At what times is the flux zero? What is the emf at these times?

13. A 100-turn circular coil has a diameter of 2.0 cm and resistance of 50 Ω. The plane of the coil is perpendicular to a uniform magnetic field of magnitude 1.0 T. The field is suddenly reversed in direction. (a) Find the total charge passing through the coil. If the reversal takes 0.1 s, find (b) the average current in the circuit and (c) the average emf in the circuit.

14. A 1000-turn coil of cross-sectional area 300 cm^2 and resistance 15.0 Ω is aligned with its plane perpendicular to the earth's magnetic field of 0.7 G (at the equator). If the coil is flipped over, how much charge flows through it?

15. A circular coil of 300 turns and radius 5.0 cm is connected to a ballistic galvanometer. The total resistance of the circuit is 20 Ω. The plane of the coil is originally aligned perpendicular to the earth's magnetic field at some point. When the coil is rotated through 90°, the charge passing through the galvanometer is measured to be 9.4 μC. Calculate the magnitude of the earth's magnetic field at that point.

Section 28-4, Eddy Currents

There are no exercises for this section.

Section 28-5, The Betatron

16. Consider a plane perpendicular to a magnetic field **B** which is uniform over a circular region of radius R and is essentially zero outside the circle ($r > R$). Show that if the magnetic field is changing at the rate dB/dt, the induced electric field at distance r from the center of the circle is tangent to the circle of radius r and has the magnitude

$$E = \begin{cases} \dfrac{r}{2}\dfrac{dB}{dt} & \text{for } r < R \\[2ex] \dfrac{R^2}{2r}\dfrac{dB}{dt} & \text{for } r > R \end{cases}$$

Section 28-6, Inductance

17. A coil with a self-inductance of 8.0 H carries a current of 3 A, which is changing at a rate of 200 A/s. Find (a) the magnetic flux through the coil and (b) the induced emf in the coil.

18. A coil with self-inductance L carries a current I, given by $I = I_0 \sin 2\pi ft$. Find and graph the flux ϕ_m and the self-induced emf as functions of time.

19. A solenoid has length 25 cm, radius 1 cm, and 400 turns, and carries a 3-A current. Find (a) B on the axis at the center, (b) the flux through the solenoid, assuming B to be uniform, (c) the self-inductance of the solenoid, and (d) the induced emf in the solenoid when the current changes at 150 A/s.

20. A circular coil of N_1 turns and area A_1 encircles a long tightly wound solenoid of N_2 turns, area A_2, and length ℓ_2, as shown in Figure 28-33. The solenoid carries a current I_2. (a) Show that the flux through the circular coil due to the current in the solenoid is given by $\phi_{m1} = \mu_0 N_1 (N_2/\ell_2) A_2 I_2$. (b) Explain why the flux depends on the area A_2 of the solenoid and not on that of the coil. (c) Find the mutual inductance of the solenoid and coil. (d) Explain why it is much more difficult to calculate the mutual inductance by assuming current I_1 in the coil and then finding the flux through the solenoid due to I_1.

Figure 28-33
Exercise 20.

21. A very small circular coil of N_1 turns and area A_1 is completely inside a long tightly wound solenoid with its plane perpendicular to the axis of the solenoid. The solenoid has N_2 turns, area A_2, and length ℓ_2 and carries current I_2. (a) Show that the flux through the circular coil due to the current in the solenoid is given by $\phi_{m1} = \mu_0 N_1 (N_2/\ell_2) A_1 I_2$. (b) Find the mutual inductance of the coil and solenoid. (c) Explain why it is much more difficult to calculate the mutual inductance by assuming current I_1 in the coil and then finding the flux through the solenoid due to I_1.

Section 28-7, *LR* Circuits

22. The current in an *LR* circuit is zero at time $t = 0$ and increases to half its final value in 4.0 s. (a) What is the time constant of this circuit? (b) If the total resistance is 5 Ω, what is the self-inductance?

23. A coil of resistance 8.0 Ω and self-inductance 4.0 H is suddenly connected across a constant potential difference of 100 V. Let $t = 0$ be the time of connection at which the current is zero. Find the current I and its rate of change dI/dt at times (a) $t = 0$, (b) $t = 0.1$ s, (c) $t = 0.5$ s, and (d) $t = 1.0$ s.

24. How many time constants must elapse before the current in an *LR* circuit which is originally zero reaches (a) 90 percent, (b) 99 percent, and (c) 99.9 percent of its final value?

25. The current in a coil of self-inductance 1 mH is 2.0 A at $t = 0$, when the coil is shorted through a resistor. The total resistance of the coil plus resistor is 10.0 Ω. Find the current after (a) 0.5 ms and (b) 10 ms.

26. Compute the initial slope dI/dt at $t = 0$ from Equation 28-33 and show that if the current decreased steadily at this rate, it would be zero after one time constant.

27. In Exercise 25, find the time for the current to decrease to one electron per second.

Section 28-8, Magnetic Energy

28. A solenoid of 2000 turns, area 4 cm², and length 30 cm carries a 4.0-A current. (a) Calculate the magnetic energy stored from $\frac{1}{2}LI^2$. (b) Divide your answer in part (a) by the volume of the solenoid to find the magnetic energy per unit volume in the solenoid. (c) Find B in the solenoid. (d) Compute the magnetic energy density from $\eta_m = B^2/2\mu_0$ and compare with part (b).

29. In the circuit in Figure 28-21 let $\varepsilon_0 = 12.0$ V, $R = 3.0\ \Omega$, and $L = 0.6$ H. The switch is closed at time $t = 0$. At time $t = 0.5$ s, find (a) the rate at which the battery supplies power, (b) the rate of Joule heating, and (c) the rate at which energy is being stored in the inductor.

30. Do Exercise 29 for the times $t = 1$ s and $t = 100$ s.

31. A coil with self-inductance 2.0 H and resistance 12.0 Ω is connected across a 24-V battery of negligible internal resistance. (a) What is the final current? (b) How much energy is stored in the inductor when the final current is attained?

32. In a plane electromagnetic wave such as a light wave, the magnitudes of the electric and magnetic fields are related by $E = cB$, where $c = 1/\sqrt{\epsilon_0\mu_0}$ is the speed of light. Show that in this case the electric and magnetic energy densities are equal.

33. Find (a) the magnetic energy, (b) the electric energy, and (c) the total energy in a volume of 1.0 m³ in which there is an electric field of 10^4 V/m and a magnetic field of 1 T.

Section 28-9, LC and LCR Circuits

34. Show from the definitions of the henry and farad that $1/\sqrt{LC}$ has units of s⁻¹.

35. What is the period of oscillation of an LC circuit consisting of a 2-mH coil and a 20-μF capacitor?

36. What inductance is needed with an 80-μF capacitor to construct an LC circuit which oscillates with frequency 60 Hz?

37. An LC circuit has capacitance C_1 and inductance L_1. A second circuit has $C_2 = \frac{1}{2}C_1$ and $L_2 = 2L_1$, and a third circuit has $C_3 = 2C_1$ and $L_3 = \frac{1}{2}L_1$. (a) Show that each of these circuits oscillates with the same frequency. (b) In which circuit would the maximum current be the greatest if in each case the capacitor were charged to potential V?

Problems

1. A conducting rod of length ℓ rotates at constant angular velocity ω about one end in a plane perpendicular to a uniform magnetic field **B** (Figure 28-34). (a) Show that the magnetic force on a charge q at distance r from the pivot is $Bqr\omega$. (b) Show that the potential difference between the ends of the rod is $V = \frac{1}{2}B\omega\ell^2$. (c) Draw any radial line in the plane from which to measure $\theta = \omega t$. Show that the area of the pie-shaped region between the reference line and the rod is $A = \frac{1}{2}\ell^2\theta$. Compute the flux through this area and show that $\varepsilon = \frac{1}{2}B\omega\ell^2$ follows from Faraday's law applied to this area.

Figure 28-34
Problem 1.

B in

2. A rod of length ℓ lies with its length perpendicular to a long wire carrying current I, as shown in Figure 28-35. The near end of the rod is a distance d away from the wire. The rod moves with speed v in the direction of the current I. (a) Show that the potential difference between the ends of the rod is given by

$$ V = \frac{\mu_0 I}{2\pi} v \ln \frac{d + \ell}{d} $$

(b) Use Faraday's law to obtain this result by considering the flux through a rectangular area $A = \ell vt$ swept out by the rod.

3. A rectangular loop 10 by 5.0 cm with resistance 1 Ω is pulled through a region of uniform magnetic field, $B = 1.0$ T (Figure 28-36), with constant speed $v = 2.0$ cm/s. The front end of the loop enters the region of magnetic field at time $t = 0$. (a) Find and graph the flux through the loop as a function of time. (b) Find and graph the induced emf and the current in the loop as functions of time. Neglect any self-inductance of the loop and extend your graphs from $t = 0$ to $t = 16$ s.

4. Show that for two inductors L_1 and L_2 connected in series such that none of the flux of either passes through the other, the effective inductance is given by $L = L_1 + L_2$.

5. Two concentric solenoids have the same length ℓ. The inner solenoid has n_1 turns per unit length and radius r_1, whereas the outer solenoid has n_2 turns per unit length and radius r_2 $(r_2 > r_1)$. Find the mutual inductance of these solenoids.

6. Show that for two inductors L_1 and L_2 connected in parallel such that none of the flux of one passes through the other, the effective inductance is given by $1/L = 1/L_1 + 1/L_2$.

7. When the current in a certain coil is 5.0 A and is increasing at the rate of 10.0 A/s, the potential difference across the coil is 140 V. When the current is 5.0 A and decreasing at the rate of 10.0 A/s, the potential difference is 60 V. Find the resistance and self-inductance of the coil.

8. A coaxial cable consists of two very thin-walled conducting cylinders of radii r_1 and r_2 (Figure 28-37). Current I goes in one direction down the inner cylinder and back up the outer cylinder. (a) Use Ampère's law to find B and show that $B = 0$ except in the region between the conductors. (b) Show that the magnetic energy density in the region between the cylinders is

$$ \eta_m = \frac{\mu_0 I^2}{8\pi^2 r^2} $$

(c) Find the magnetic energy in a cylindrical shell volume element of length ℓ and volume $d\mathcal{V} = \ell 2\pi r\, dr$ and integrate your result to show that the total magnetic energy in the volume of length ℓ between the cylinders is

$$ U_m = \frac{\mu_0}{4\pi} I^2 \ell \ln \frac{r_2}{r_1} $$

(d) Use the result of part (c) and $U_m = \frac{1}{2} L I^2$ to show that the self-inductance per unit length is

$$ \frac{L}{\ell} = \frac{\mu_0}{2\pi} \ln \frac{r_2}{r_1} $$

9. In the circuit of Figure 28-21 let $\varepsilon = 12.0$ V, $R = 3.0$ Ω, and $L = 0.6$ H (as in Exercise 29). After a time $t = t_c$, find (a) the total energy supplied by the battery and (b) the energy stored in the inductor. Hint: Find the rates as functions of time and integrate from $t = 0$ to $t = t_c = L/R$.

10. In Figure 28-37 compute the flux through a rectangular area of sides ℓ and $r_2 - r_1$ between the conductors. Show that the self-inductance per unit length can be found from $\phi_m = LI$ [see part (d) of Problem 8].

Figure 28-35
Problem 2.

Figure 28-36
Problem 3.

Figure 28-37
Problems 8 and 10.

11. An ac generator rectangular loop of dimensions a and b has N turns. The loop is connected to slip rings (Figure 28-38) and rotates with angular velocity ω in a uniform magnetic field **B**. (a) Show that the potential difference between the two slip rings is $\varepsilon = NBab\omega \sin \omega t$. ($b$) If $a = 1.0$ cm, $b = 2.0$ cm, $N = 1000$, and $B = 2$ T, at what angular frequency ω must the coil be rotated to generate an emf whose maximum value is 110 V?

Figure 28-38
Problem 11.

12. A conducting rod of length ℓ, resistance R, and mass m rides on a pair of horizontal frictionless rails of negligible resistance (Figure 28-39). A uniform magnetic field **B** is perpendicular to the plane of the rails. The rod is moving with initial speed v_0 when points a and b are connected with a conductor of negligible resistance. (a) Find the retarding force on the rod. (b) Show that the speed is given by $v = v_0 e^{-t/T}$, where $T = mR/(B\ell)^2$, assuming the connection across ab is made at $t = 0$. (c) Find the rate of Joule heat production I^2R and show that the total heat produced equals $\frac{1}{2}mv_0^2$.

Figure 28-39
Problems 12 and 14.

13. In the circuit in Figure 28-23 the current is $I_0 = 4.0$ A when switch S_2 is closed and S_1 opened. The inductance is 50 mH, and the resistance is 150 Ω. (a) Find the rate of Joule heat production I^2R as a function of time and (b) calculate the total heat produced. (c) Calculate the initial energy stored in the inductor and show that this equals your answer to part (b).

14. A battery of emf ε and negligible internal resistance is connected across ab in Figure 28-39, and the rod is placed at rest across the rails at $t = 0$. (a) Find the force on the rod as a function of speed v and write Newton's second law for the rod when it has speed v. (b) Show that the rod reaches a terminal velocity and find an expression for it. (c) What is the current when the rod reaches its terminal velocity?

15. Show that the electric field in a charged parallel-plate capacitor cannot go abruptly to zero at the edge by evaluating $\oint \mathbf{E} \cdot d\ell$ for the rectangular curve shown in Figure 28-40 and applying Faraday's law to this curve.

Figure 28-40
Problem 15.

16. Show by direct substitution that Equation 28-51 is satisfied by $Q = Q_0 e^{-Rt/2L} \cos \omega' t$, where $\omega' = \sqrt{(1/LC) - (R/2L)^2}$ and Q_0 is the charge on the capacitor at $t = 0$.

17. (a) Compute the current $I = dQ/dt$ from the solution of Equation 28-51 given in Problem 16 and show that

$$I = -I_0 \left(\sin \omega' t + \frac{R}{2L\omega'} \cos \omega' t \right) e^{-Rt/2L}$$

where $I_0 = \omega' Q_0$. (b) Show that this can be written

$$I = -\frac{I_0}{\cos \delta} (\cos \delta \sin \omega' t + \sin \delta \cos \omega' t) e^{-Rt/2L}$$

$$= -\frac{I_0}{\cos \delta} \sin (\omega' t + \delta) e^{-Rt/2L}$$

where $\tan \delta = R/2L\omega'$. When $R/2L\omega'$ is small, $\cos \delta \approx 1$ and $I \approx -I_0 \sin (\omega' t + \delta) e^{-Rt/2L}$.

CHAPTER 29   Magnetism in Matter

Objectives   After studying this chapter, you should:

1. Be able to list the three types of magnetic materials and discuss the origins, directions, and strengths of the magnetic effects in each.

2. Be able to state the definition of the magnetic intensity **H**.

3. Be able to state the analogous expressions of Ampère's law, Gauss' law, and Coulomb's law for the magnetic intensity **H**.

4. Be able to relate the magnetic vectors **B, H,** and **M** using the definitions of magnetic susceptibility and permeability.

5. Be able to derive the relation between magnetic moment and angular momentum for a charged particle moving in a circle.

6. Be able to sketch the general shape of the B-versus-H curve for a ferromagnetic material.

7. Be able to state Curie's law for paramagnetic materials.

8. Be able to describe the general temperature dependence of the magnetization in paramagnetic materials and explain its origin.

In studying electric fields in matter we found that the electric field is affected by the presence of electric dipoles. For polar molecules, which have a permanent electric dipole moment, the dipoles are aligned by the electric field; whereas for nonpolar molecules, electric dipoles are induced by the external field. In both cases, the dipoles are aligned parallel to the external electric field, and the alignment tends to weaken the external field.

Similar but more complicated effects occur in magnetism. Atoms have magnetic moments due to the motion of their electrons. In addition, each electron has an intrinsic magnetic moment associated with its spin. The net magnetic moment of an atom depends on the arrangement of the electrons in the atom. Unlike the situation with electric dipoles, the alignment of magnetic dipoles parallel to an external magnetic field tends to increase the field. We can see this difference by comparing the lines of **E** for an electric dipole with the lines of **B** for a magnetic dipole, e.g., a small current loop as shown in Figure 29-1. Far from the dipoles, the lines are identical, but in the region inside the dipole the lines of **B** and **E** are oppositely directed. Thus in an electrically polarized material, the dipoles create an electric field *antiparallel* to their dipole-moment vector, whereas in a magnetically polarized material, the dipoles create a magnetic field *parallel* to the magnetic-dipole-moment vectors.

We can classify materials into three categories, paramagnetic, diamagnetic, and ferromagnetic. Paramagnetic and ferromagnetic materials have molecules with permanent magnetic dipole moments. In *paramagnetism*, these moments do not interact strongly with each other and are normally randomly oriented. In the presence of an external magnetic field, the dipoles are partially aligned in the direction of the field, thereby increasing the field. However, at ordinary temperatures and ordinary external fields, only a very small fraction of the molecules are aligned because thermal motion tends to randomize their orientation. The increase in the total magnetic field is therefore very small. *Ferromagnetism* is much more complicated. Because of a strong interaction between neighboring magnetic dipoles, a high degree of alignment can be achieved even with weak external magnetic fields, thereby causing a very large increase in the total field. Even when there is no external magnetic field, ferromagnetic material may have magnetic dipoles aligned, as in permanent magnets. *Diamagnetism* is the result of an induced magnetic moment opposite in direction to the external field. The induced dipoles thus weaken the resultant magnetic field. This effect occurs in all materials but is very small and often masked by the paramagnetic or ferromagnetic effects if the individual molecules have permanent magnetic dipole moments.

In Section 27-5 we discussed the magnetic field produced by a permanent magnet, characterizing the magnet by the magnetic moment

Three types of magnetic materials

(a)

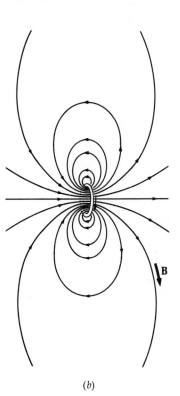

(b)

Figure 29-1
(*a*) Electric field lines of an electric dipole. (*b*) Magnetic field lines of a magnetic dipole. Far from the dipoles, the field patterns are identical. In the region inside the dipoles, the lines of **B** and **E** are oppositely directed.

per unit volume **M**, known as the magnetization vector. We found that the field due to the magnetized material is the same as that due to a current per unit length M on the surface of the magnet, and we argued that this amperian current can be understood in terms of atomic current loops which do not completely cancel on the surface. The discussion and results of that section apply to any magnetized material, whether it is paramagnetic, diamagnetic, or ferromagnetic. We describe the magnetization of any such material by the magnetic dipole moment per unit volume **M**. Except for permanent magnets, this magnetization is caused by some external magnetic field and disappears when the external field is removed. The magnetization vector **M** is parallel to the external field that causes it for paramagnetism and ferromagnetism, and antiparallel for diamagnetism. The magnetic induction **B** due to the magnetization of the material is very small compared with the external field for paramagnetism and diamagnetism, and very large for ferromagnetism.

The calculation of the resultant magnetic-induction field **B** is complicated when magnetized material is in the presence of an external field. Although the field due to the material can be calculated from Ampère's law or the Biot-Savart law using the amperian current associated with the magnetization vector **M**, the magnetization itself depends on the resultant magnetic-induction field **B**. Because of this complication, a new vector, the magnetic intensity **H**, is defined in an attempt to separate out the external field from that due to the material. In many cases of interest, **H** can be found from the external conduction currents in wires alone without reference to any material present. The magnetization **M** can then be related to **H**, and the resultant magnetic induction field **B** can be found from **H** and **M**. We define the magnetic intensity **H** in Section 29-1 and discuss some of its properties. We then estimate the order of magnitude of the magnetic moment of an atom and use it to estimate the possible magnitude of the magnetization **M** for paramagnetic and ferromagnetic materials. After a brief look at paramagnetism and diamagnetism we shall discuss qualitatively the complicated phenomenon of ferromagnetism, which has many important practical applications.

29-1 Magnetic Intensity **H**

Let us consider a long solenoid with n turns per unit length carrying a current I. The magnetic induction at the center of the solenoid is

$$B_0 = \mu_0 n I \qquad \text{29-1}$$

This expression is good if we are far from the ends of the solenoid. Since we wish to avoid any complications from end effects, we shall assume that the solenoid is very long. (End effects can be completely eliminated in practice, using a doughnut-shaped solenoid called a *toroid*, illustrated in Figure 29-2, but we continue to use a solenoid because we have already studied it in some detail.) If we now insert some material inside the solenoid, there will be an additional contribution to the magnetic-induction field due to the magnetization of the material. If **M** is the magnetic moment per unit volume, the induction field $\mathbf{B}_m$ produced by this magnetization is given by an expression similar to Equation 29-1, with the amperian current per unit length M replacing the current per unit length in the solenoid nI:

$$\mathbf{B}_m = \mu_0 \mathbf{M} \qquad \text{29-2}$$

Figure 29-2
Toroid. The magnetic field is similar to that of a solenoid, but there are no end effects.

The total magnetic induction at the center of the solenoid is the vector sum of these separate fields:

$$\mathbf{B} = \mathbf{B}_0 + \mu_0\mathbf{M} \qquad\qquad 29\text{-}3$$

(We have written this as a vector sum because **M** is in the opposite direction to $\mathbf{B}_0$ for diamagnetic materials.) As mentioned above, it is convenient to separate out the external magnetic field from that due to the material. The *magnetic-intensity* vector **H** is defined by

$$\mathbf{H} = \frac{\mathbf{B}}{\mu_0} - \mathbf{M} \qquad\qquad 29\text{-}4a$$

Magnetic intensity **H** *defined*

Then

$$\mathbf{B} = \mu_0\mathbf{H} + \mu_0\mathbf{M} \qquad\qquad 29\text{-}4b$$

For a toroid or very long solenoid, for which we can neglect the effects at the ends of the material, **H** is simply the original external magnetic induction $\mathbf{B}_0$ divided by μ_0. The magnitude of **H** at the center of a long solenoid is

$$H = nI \qquad\qquad 29\text{-}5$$

From Equation 29-4 we see that the dimensions of **H** are the same as those for **M**, which are amperes per metre. (Because the number of turns of wire often comes into the equation for **H**, as in Equation 29-5, the units of **H** are often stated as ampere-turns per metre.)

In situations like this where we can neglect end effects of the magnetized material, the magnetic intensity **H** is determined by the external conduction current, whereas in the equations for **B** the current I repre-cases **H** can be obtained from Ampère's law or the Biot-Savart law in the forms

$$\oint_C \mathbf{H} \cdot d\boldsymbol{\ell} = I \qquad\qquad 29\text{-}6$$

and

$$d\mathbf{H} = \frac{1}{4\pi}\frac{I\,d\boldsymbol{\ell} \times \hat{\mathbf{r}}}{r^2} \qquad\qquad 29\text{-}7$$

It is important to realize that these equations differ from the corresponding equations for the magnetic induction **B** not only in the absence of the constant μ_0 but also because the current I is the macroscopic conduction current, whereas in the equations for **B** the current I represents any kind of current, including the atomic or amperian current associated with the magnetization of the material.

Example 29-1 Use Equations 29-6 and 27-13 to find **B** and **H** for a long cylinder which has a magnetization **M** and is wound with n turns per unit length carrying a current I.

Figure 29-3 shows a rectangular path for computing the line integrals in these equations. As in Section 27-4, we assume that the tangential component of the magnetic field is constant along path ab and zero along the other legs of the path. The conduction current through the rectangle is $nI\ell$, where ℓ is the length from a to b. Thus Equation 29-6 gives

$$\oint \mathbf{H} \cdot d\boldsymbol{\ell} = H\ell = nI\ell \qquad \text{or} \qquad H = nI$$

Figure 29-3
Path for calculation of $\oint \mathbf{H} \cdot d\boldsymbol{\ell}$ and $\oint \mathbf{B} \cdot d\boldsymbol{\ell}$ in Example 29-1. The only contribution is from a to b.

To calculate B we need to include the atomic current, which is $M\ell$ since M is the atomic current per unit length. Thus Equation 27-13 gives

$$\oint \mathbf{B} \cdot d\boldsymbol{\ell} = \mu_0 I_{\text{total}} = \mu_0 (I_{\text{con}} + I_{\text{mag}})$$

or

$$B\ell = \mu_0 (nI\ell + M\ell)$$

$$B = \mu_0 (nI + M)$$

At points near the ends, or poles, of a magnetized material, the magnetic intensity $\mathbf{H}$ is not completely determined by the external conduction currents because of a contribution from the poles of the material. We can see this by considering the flux of $\mathbf{H}$ through any closed surface. Using the defining equation 29-4, we have

$$\oint_S \mathbf{H} \cdot \hat{\mathbf{n}} \, dA = \frac{1}{\mu_0} \oint_S \mathbf{B} \cdot \hat{\mathbf{n}} \, dA - \oint_S \mathbf{M} \cdot \hat{\mathbf{n}} \, dA$$

Since the net flux of the magnetic induction $\mathbf{B}$ is necessarily zero, we have

$$\oint_S \mathbf{H} \cdot \hat{\mathbf{n}} \, dA = - \oint_S \mathbf{M} \cdot \hat{\mathbf{n}} \, dA \qquad\qquad 29\text{-}8$$

In a homogeneous material the magnetization $\mathbf{M}$ is uniform. The flux of $\mathbf{M}$ through any closed surface within the material is then zero. There is a net flux of $\mathbf{M}$ only at the ends of the material. Let us consider the uniformly magnetized cylinder shown in Figure 29-4. The flux of $\mathbf{M}$ through the surface S_1 enclosing the left end of the cylinder is positive and has the magnitude MA, where A is the cross-sectional area of the cylinder. Similarly, the net flux of M through surface S_2 enclosing the right end is $-MA$. From our discussion in Section 27-5 we note that MA is just the pole strength of the material and that the pole at the left is negative and the pole at the right is positive. In general, we can define the pole strength of a magnetized material q_m by

$$q_m = - \oint_S \mathbf{M} \cdot \hat{\mathbf{n}} \, dA \qquad\qquad 29\text{-}9$$

Then the flux of the magnetization vector $\mathbf{H}$ is

$$\oint_S \mathbf{H} \cdot \hat{\mathbf{n}} \, dA = q_m \qquad\qquad 29\text{-}10$$

This equation is similar to Gauss' law for the flux of the electric field $\mathbf{E}$, with the magnetic pole strength q_m replacing the electric charge q. It follows from this equation that $\mathbf{H}$ varies inversely as the square of the distance from a magnetic pole according to

$$\mathbf{H} = \frac{1}{4\pi} \frac{q_m}{r^2} \hat{\mathbf{r}} \qquad\qquad 29\text{-}11$$

If we wish to calculate $\mathbf{H}$ at a point near a magnetic pole, we must add the contribution due to the pole given by Equation 29-11 to that due to the conduction currents, which can be calculated from Ampère's law or the Biot-Savart law. In many cases we are interested in the magnetic intensity $\mathbf{H}$ only in regions far from any poles, e.g., at the center of a long solenoid, and we can ignore the contribution given by Equation 29-11.

Figure 29-4
Surfaces S_1 and S_2 enclosing the ends of a uniformly magnetized cylinder. The net flux of $\mathbf{M}$ out of S_1 is $+MA$, and that out of S_2 is $-MA$, where A is the area of the cylinder.

Gauss' law for $\mathbf{H}$

Coulomb's law for $\mathbf{H}$

Example 29-2 A permanent cylindrical bar magnet of radius a and length $L = 10a$ has uniform magnetization **M** to the right parallel to its axis. Find **H** and **B** on the axis at the center of the magnet and just outside the right end.

Since there are no conduction currents in this example, **H** is due solely to the magnetic poles. If we consider the center of the magnet to be far from the ends so that the poles can be neglected, **H** = 0 and **B** = μ_0**M** at that point. We can find a correction of this approximation by using Coulomb's law for **H** at the center due to the two poles. At the center, both the positive pole on the right and the negative pole on the left contribute equally to give **H** to the left in the direction opposite **M**. Using $q_m = \pi a^2 M$ and $r = \frac{1}{2}L$ in Equation 29-11, we have for the magnitude of H due to either pole,

$$H = \frac{1}{4\pi}\frac{q_m}{r^2} = \frac{1}{4\pi}\frac{\pi a^2 M}{(\frac{1}{2}L)^2} = \frac{a^2}{L^2}M$$

Adding the contributions due to each pole, we obtain for H at the center

$$\mathbf{H} = -\frac{2a^2}{L^2}\mathbf{M}$$

The magnetic induction **B** at the center is found from Equation 29-4b:

$$\mathbf{B} = \mu_0(\mathbf{H} + \mathbf{M}) = \mu_0\mathbf{M}\left(1 - \frac{2a^2}{L^2}\right)$$

We note that for $a = L/10$ the correction due to the poles is 2 percent.

Just outside the right end, the near pole can be treated as a disk of pole density $\sigma_m = M$ (Equation 27-21) and the far pole can be treated as a point pole a distance L away. We found in Chapter 20 that a disk charge produces an electric field at a nearby point given by $\sigma/2\epsilon_0$. Since Equations 29-10 and 29-11 for **H** are the same as the corresponding equations for **E** except for the absence of ϵ_0, the magnetic intensity **H** produced by a disk of pole density σ_m is $\sigma_m/2$. The total **H** just outside the right end of the magnet is then

$$H = +\frac{\sigma_m}{2} = \frac{1}{4\pi}\frac{q_m}{L^2} = +\frac{M}{2} - \frac{1}{4\pi}\frac{\pi a^2 M}{L^2}$$

or

$$\mathbf{H} = \frac{\mathbf{M}}{2}\left(1 - \frac{a^2}{2L^2}\right)$$

Since there is no magnetization **M** outside the magnet, **B** is just μ_0**H**,

$$\mathbf{B} = \frac{\mu_0\mathbf{M}}{2}\left(1 - \frac{a^2}{2L^2}\right)$$

If we neglect the term $a^2/2L^2$ due to the far pole, the magnetic induction B is $\frac{1}{2}\mu_0 M$, approximately half its value at the center of the magnet, a result which also holds for a long solenoid.

Questions

1. How are **B** and **H** related in a vacuum?

2. Must **B** and **H** necessarily be parallel?

29-2 Magnetic Susceptibility and Permeability

For paramagnetic and diamagnetic materials the magnetization **M** is proportional to the magnetic intensity **H**:

$$\mathbf{M} = \chi_m \mathbf{H} \qquad\qquad 29\text{-}12$$

The proportionality factor χ_m is called the *magnetic susceptibility*. Since **M** and **H** have the same dimensions, the susceptibility is dimensionless. It is positive for paramagnetic materials and negative for diamagnetic materials. For diamagnetic materials the susceptibility is essentially independent of temperature, but for paramagnetic materials the susceptibility decreases as the temperature increases. This temperature dependence for paramagnetic materials is the result of the tendency for thermal vibrations to reduce the alignment of the permanent dipole moments, discussed in more detail in Section 29-4.

Magnetic susceptibility

From Table 29-1 listing typical values of the susceptibility, we see that χ_m is much less than 1. If **M** is proportional to **H**, then **B** is also proportional to **H**. Substituting $\chi_m \mathbf{H}$ for **M** in Equation 29-4*b*, we can write

$$\mathbf{B} = \mu_0(\mathbf{H} + \mathbf{M}) = \mu_0(\mathbf{H} + \chi_m\mathbf{H}) = \mu_0(1 + \chi_m)\mathbf{H} \qquad 29\text{-}13$$

or

$$\mathbf{B} = \mu\mathbf{H} \qquad\qquad 29\text{-}14$$

where

$$\mu = (1 + \chi_m)\mu_0 \qquad\qquad 29\text{-}15$$

is called the *permeability* of the material. Because the susceptibility is so small, the permeability of all paramagnetic and diamagnetic material is very nearly equal to the permeability of free space μ_0.

Magnetic permeability

Table 29-1
Magnetic susceptibility of various materials at 20°C

Material	χ_m
Aluminum	2.3×10^{-5}
Bismuth	-1.66×10^{-5}
Copper	-0.98×10^{-5}
Diamond	-2.2×10^{-5}
Gold	-3.6×10^{-5}
Magnesium	1.2×10^{-5}
Mercury	-3.2×10^{-5}
Silver	-2.6×10^{-5}
Sodium	-0.24×10^{-5}
Titanium	7.06×10^{-5}
Tungsten	6.8×10^{-5}
Hydrogen (1 atm)	-9.9×10^{-9}
Carbon dioxide (1 atm)	-2.3×10^{-9}
Nitrogen (1 atm)	-5.0×10^{-9}
Oxygen (1 atm)	2090×10^{-9}

Equation 29-14 is also written for ferromagnetic materials, but it is difficult to interpret because the magnetization **M** is not a linear function of **H**. The permeability defined by Equation 29-14 depends not only on the value of **H** but also on the previous state of magnetization of a ferromagnetic material. In such cases, **B** and **M** are multivalued functions of **H**. The maximum value of μ for ferromagnetic materials is typically several thousand times greater than μ_0. We shall discuss the relation between **B**, **H**, and **M** for ferromagnetic materials in more detail in Section 29-6.

Questions

3. Why are some values of χ_m in Table 29-1 positive and others negative?

4. Why is it more convenient to list χ_m than μ for the materials in Table 29-1?

29-3 Atomic Magnetic Moments

The magnetization **M** of a paramagnetic or ferromagnetic material can be related to the permanent magnetic moments of the individual atoms of the material. In this section we calculate the order of magnitude of these magnetic moments from the Bohr theory of the atom, in which the electrons move in circular orbits around the nucleus. Although this theory has severe defects, its results are in agreement with those from the correct quantum-mechanical theory of the atom in many cases and insight can often be obtained from the simpler Bohr picture. We can use the simple circular motion of a charged particle to show that there is a general connection between the angular momentum of the particle and the magnetic moment (Figure 29-5). Consider a charge moving in a circle of radius r with speed v. The time for each revolution, the period T, is related to the speed and radius by

$$vT = 2\pi r$$

This motion results in a current loop and thus produces a magnetic moment. The current is the charge q divided by the period T:

$$I = \frac{q}{T} = \frac{qv}{2\pi r}$$

The magnetic moment is the product of the current and the area πr^2:

$$m = IA = \frac{qv}{2\pi r}\,\pi r^2 = \tfrac{1}{2}qvr \qquad\qquad 29\text{-}16$$

If the mass of the particle is M (we use M here for the mass to avoid confusion with m for the magnetic moment), the angular momentum is

$$L = Mvr = M\frac{2m}{q}$$

If the charge q is positive, the angular momentum and magnetic moment are in the same direction. We can therefore write

$$\mathbf{m} = \frac{q}{2M}\,\mathbf{L} \qquad\qquad 29\text{-}17$$

Figure 29-5
Particle of charge q and mass M moving in a circle of radius r. The angular momentum is into the paper with magnitude Mvr, and the magnetic moment is in (if q is positive) with magnitude $\tfrac{1}{2}qvr$.

804 CHAPTER 29 Magnetism in Matter

For the electron, with negative charge, the magnetic moment points in the direction opposite to the angular momentum. Equation 29-17 is the general classical relation between magnetic moment and angular momentum. It also holds in the quantum-mechanical theory of the atom for orbital angular momentum but not for the intrinsic spin angular momentum of the electron. For electron spin, the magnetic moment is twice that predicted by this equation. This extra factor of 2 is a quantum-mechanical result which has no analog in classical mechanics.

In both the Bohr theory of the atom and in the rigorous quantum-mechanical theory, the orbital angular momentum is quantized and takes on only values which are integral multiples $h/2\pi$, where h is a fundamental constant called Planck's constant,* and has the value

$$h = 6.63 \times 10^{-34} \text{ J·s}$$

The combination $h/2\pi$ occurs often and is designated by $\hbar$, where

$$\hbar = \frac{h}{2\pi} = \frac{6.63 \times 10^{-34}}{2\pi} \text{J·s} = 1.05 \times 10^{-34} \text{ J·s}$$

The orbital angular momentum L can therefore be written

$$L = N\hbar \qquad\qquad 29\text{-}18$$

where N is an integer. The spin angular momentum of the electron is also quantized and has magnitude $\frac{1}{2}\hbar$. It is therefore convenient to write the magnetic moment as

$$\mathbf{m} = \frac{q\hbar}{2M}\frac{\mathbf{L}}{\hbar} \qquad\qquad 29\text{-}19a$$

For an electron, $q = -e$ and $M - m_e$. The magnetic moment and angular momentum are then oppositely directed, and their magnitudes are related by

$$m = \frac{e\hbar}{2m_e}\frac{L}{\hbar} = m_B\frac{L}{\hbar} \qquad\qquad 29\text{-}19b$$

where

$$m_B = \frac{e\hbar}{2m_e} \qquad\qquad 29\text{-}20 \qquad \textit{Bohr magneton}$$

is called a *Bohr magneton*. It is a convenient unit with which to measure atomic magnetic moments. For example, the magnetic moment associated with the spin of an electron is 1 Bohr magneton. Although the calculation of the magnetic moment of any atom is a complicated quantum-mechanical problem, the result of theory and experiment for all atoms is that the magnetic moment is of the order of a few Bohr magnetons (or zero for atoms with closed-shell electron structure). The magnitude of the Bohr magneton in SI units is

$$m_B = \frac{(1.60 \times 10^{-19} \text{ C})(1.05 \times 10^{-34} \text{ J·s})}{2(9.11 \times 10^{-31} \text{ kg})}$$

$$= 9.27 \times 10^{-24} \text{ A·m}^2$$

The SI unit for magnetic moment, the ampere-metre², can also be expressed as a joule per tesla (see Exercise 10 in Chapter 26). The value of the Bohr magneton is therefore

$$m_B = 9.27 \times 10^{-24} \text{ A·m}^2 = 9.27 \times 10^{-24} \text{ J/T} \qquad\qquad 29\text{-}21$$

* Quantization and Planck's constant will be discussed in Chapter 36.

If all the atoms or molecules in some material have their magnetic moments aligned, the magnetic moment per unit volume of the material is just the product of the number of molecules per unit volume n and the magnetic moment of each molecule. For this extreme case, the saturation magnetization vector M_s is

$$M_s = nm \qquad\qquad\qquad 29\text{-}22$$

The number of molecules per unit volume is related to the molecular weight, the density, and Avogadro's number. For example, the number of iron molecules (atoms) per unit volume is

$$n = \frac{6.02 \times 10^{23} \text{ atoms/mol}}{55.8 \text{ g/mol}} \frac{7.9 \text{ g}}{1 \text{ cm}^3} \frac{10^6 \text{ cm}^3}{1 \text{ m}^3}$$

$$= 8.52 \times 10^{28} \text{ atoms/m}^3$$

If we assume that the magnetic moment of each iron atom is 1 Bohr magneton, the maximum value of the magnetization vector in iron has the magnitude

$$M_s = (8.52 \times 10^{28} \text{ atoms/m}^3) \, (9.27 \times 10^{-24} \text{ A·m}^2/\text{atom})$$
$$= 7.90 \times 10^5 \text{ A/m}$$

The magnetic induction on the axis inside a long iron cylinder with maximum magnetization is then

$$B = \mu_0 M_s = (4\pi \times 10^{-7} \text{ T·m/A}) \, (7.90 \times 10^5 \text{ A/m})$$
$$= 0.993 \text{ T} \approx 1 \text{ T}$$

The measured saturation magnetic induction of annealed iron is about 2.16 T, indicating that the magnetic moment of an iron atom is slightly greater than 2 Bohr magnetons. This magnetic moment is due mainly to the spins of two unpaired electrons in the iron atom.

29-4 Paramagnetism

Paramagnetism occurs in materials whose atoms have permanent magnetic moments interacting with each other only very weakly. When there is no external magnetic field, these magnetic moments are randomly oriented. In the presence of an external magnetic field they tend to line up parallel to the field, but this is counteracted by the tendency for the moments to be randomly oriented due to thermal motion. The fraction of moments that line up with the field depends on the strength of the field and the temperature. At very low temperatures and high external fields nearly all the moments are aligned with the field. In this situation the contribution to the total magnetic field due to the material is very large, as indicated in the numerical estimates in the previous section. Even with the largest magnetic field obtainable in the laboratory, the temperature must be as low as a few degrees absolute in order to obtain a high degree of alignment. At higher temperatures and weaker external fields only a small fraction of the moments are aligned with the field, and the contribution of the material to the total magnetic field is very small. We can state this more quantitatively by comparing the energy of a magnetic moment in an external magnetic field with the thermal energy, which is of the order of kT, where k is Boltzmann's constant and T is the absolute temperature. The potential energy of a magnetic moment in an external magnetic field is least when the moment is parallel to the field and greatest when it is antiparallel to the field. We can find the expression for this potential energy by computing the work that must be done to rotate a magnetic dipole in an external field. Let a dipole of moment **m** make an angle θ with the magnetic induction **B** (Figure 29-6a). The torque exerted by the field on the dipole has the magnitude

$$\tau = mB \sin \theta$$

If we wish to rotate the dipole to increase the angle by $d\theta$, the work we must do is

$$dW = \tau \, d\theta = mB \sin \theta \, d\theta$$

The work we must do to rotate the dipole from $\theta = 0$ to $\theta = 180°$ (Figure 29-6b) equals the change in potential energy:

$$W = \int_0^{180°} mB \sin \theta \, d\theta = -mB \cos \theta \Big]_0^{180°} = 2mB = \Delta U \qquad 29\text{-}23$$

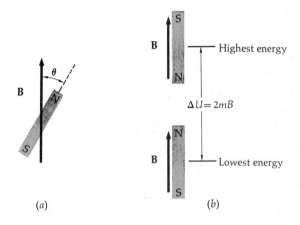

(a)   (b)

Figure 29-6
(a) Magnetic dipole whose moment makes an angle θ with a magnetic field **B**. (b) When the dipole is antiparallel to **B**, the energy is greater by the amount $2mB$ than when it is parallel to **B**.

The potential energy when $\theta = 180°$ is thus greater than at $\theta = 0°$ by the amount $2mB$. For a typical magnetic moment of 1 Bohr magneton and a typical strong field of 1 T, this difference in potential energy is

$$2mB = 2(9.27 \times 10^{-24} \text{ J/T}) \, (1 \text{ T}) = 1.85 \times 10^{-23} \text{ J}$$

At a normal temperature of $T = 300$ K, the typical thermal energy kT is 4.14×10^{-21} J, about 200 times greater than the difference in potential energy when the dipole is aligned parallel or antiparallel with the field. Thus even in a strong field of 1 T, most of the moments will be essentially randomly oriented because of thermal motions.

Figure 29-7 shows a plot of the ratio of the magnetization M to its saturation value M_s versus a dimensionless parameter $x = mB/kT$. At $x = 0$, corresponding to zero external field, the magnetization is zero because the dipoles are randomly aligned. At very large values of x, which can occur only for high fields and very low temperatures, nearly all the dipoles are aligned with the field and $M/M_s = 1$. In the region of small x the curve is approximately linear, with a slope of $\frac{1}{3}$. In this region we have

$$M = \frac{1}{3} \frac{mB}{kT} M_s \qquad 29\text{-}24$$

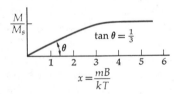

Figure 29-7
M/M_s versus mB/kT. At weak external fields or high temperatures, M/M_s increases linearly with mB/kT. When mB is much greater than kT, M reaches its saturation value M_s.

The result that the magnetization varies inversely with the absolute temperature was discovered experimentally by Pierre Curie and is known as *Curie's law*.

Curie's law

29-5 Diamagnetism

Diamagnetism was discovered by Faraday in 1846, when he found that a piece of bismuth is repelled by either pole of a magnet, indicating that the external field of the magnet induces a magnetic dipole in the bismuth in the direction opposite the field. We can understand this effect qualitatively using Lenz's law. Figure 29-8 shows two positive charges moving in circular orbits with the same speed but in opposite directions. Their magnetic moments are in opposite directions and cancel. (It is simpler to consider positive charges even though it is the negatively charged electrons which provide the magnetic moments in matter.) Consider now what happens when an external magnetic field **B** is turned on in the direction into the page. According to Lenz's law, currents will be induced to oppose the change in flux. If we assume that the radius of the circle is not changed, the charge on the left will be speeded up to increase its flux out of the page and the charge on the right will be slowed down to decrease its flux into the page. In each case, the *change* in magnetic moment of the charges will be in the direction out of the page, opposite that of the external field.

We can estimate the magnitude of this effect by relating the change in the speeds of the particles to the change in the centripetal force due to the external magnetic field. We assume that the radius of the orbit does not change and that the change in speed is small compared with the original speed. Both these assumptions can be justified. The original centripetal force is provided by the electrostatic force of attraction of the electron to the nucleus. Calling this force F and setting it equal to the mass times the acceleration, we have

$$F = \frac{m_e v^2}{r} \qquad\qquad 29\text{-}25$$

where m_e is the electron mass. In the presence of an external magnetic field there is an additional force on each particle, $q\mathbf{v} \times \mathbf{B}$. Note that on the particle on the left in Figure 29-8 this force is inward, thus increasing the centripetal force. This is necessary because, according to our argument from Lenz's law, this particle must speed up to oppose the change in flux. Similarly, the force is outward on the particle on the right, decreasing the centripetal force. Again, this is in the correct direction since this particle slows down to oppose the change in flux when the field is turned on. In each case the magnitude of the change in force is qvB since **v** and **B** are perpendicular. Since this change is very small, we use the differential approximation. From Equation 29-25 we have

$$\Delta F \approx dF = \frac{2m_e v}{r}\, dv \approx \frac{2m_e v}{r}\, \Delta v$$

Substituting qvB for ΔF and solving for Δv gives

$$\Delta v = \frac{qrB}{2m_e} \qquad\qquad 29\text{-}26$$

This change in speed causes a change in the magnetic moment in the outward direction for both particles. The magnetic moment of a charge moving in a circle is related to its speed by Equation 29-16,

$$m = \tfrac{1}{2}qvr$$

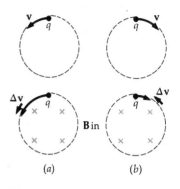

Figure 29-8
(*a*) Charge moving in a circle with magnetic moment out (assuming q positive). In the presence of an external magnetic field the speed is increased. (*b*) Charge moving in a circle with its magnetic moment in. An external magnetic field causes a decrease in its speed. In both cases the *change* in the magnetic moment is out.

Then

$$\Delta m = \tfrac{1}{2}qr\,\Delta v = \tfrac{1}{2}qr\,\frac{qrB}{2m_e} = \frac{q^2r^2}{4m_e}B \qquad\qquad 29\text{-}27$$

When we use $R = 10^{-10}$ m for a typical radius, the values of q and m_e for the electron, and a strong field of $B = 1$ T, the order of magnitude of this induced magnetic moment is

$$\Delta m = \frac{(1.60\times 10^{-19}\ \text{C})^2(10^{-10}\ \text{m})^2}{4(9.11\times 10^{-31}\ \text{kg})}\,(1\ \text{T}) \approx 7\times 10^{-29}\ \text{A·m}^2$$

which is about 10^5 times smaller than a Bohr magneton. Since m is of the order of a Bohr magneton, $\Delta m/m$ is about 10^{-5}. This justifies our assumption that the change in speed is very small.

Since the induced magnetic moment is antiparallel to the external field whether the original moment is parallel or antiparallel to the field, there is a contribution from each electron in the atom to the induced magnetic moment of the atom. The magnetization vector **M** is the product of the number of *electrons* per unit volume times the induced magnetic moment of each electron. If n is the number of atoms per unit volume and Z is the atomic number, the number of electrons per unit volume is nZ. (Of course, an accurate calculation of the induced dipole moment of an atom must take into account the different values of r^2 for the different electrons and the fact that the orbits are not perpendicular to the external field.)

Atoms with closed-shell electron structures have zero angular momentum and therefore no permanent magnetic dipole moment. They are diamagnetic. Nonferromagnetic atoms with permanent magnetic dipole moments are either paramagnetic or diamagnetic, depending on which effect is stronger. Since the diamagnetic effect does not depend on temperature, and alignment of permanent moments decreases with temperature for both paramagnetic and ferromagnetic substances, all materials are diamagnetic at sufficiently high temperatures.

29-6 Ferromagnetism

Ferromagnetism occurs in pure iron, cobalt, and nickel, in alloys of these metals with each other and with some other elements, and in a few other substances (gadolinium, dysprosium, and a few compounds). In these substances a small external magnetic field can produce a very large degree of alignment of the atomic magnetic dipole moments, which, in some cases, can persist even when there is no external magnetizing field. This occurs because the magnetic dipole moments of atoms of these substances exert strong forces on their neighbors so that over a small region of space the moments are aligned with each other even with no external field. These dipole forces are predicted for these substances by quantum mechanics but cannot be explained with classical physics. At temperatures above a critical temperature, called the *Curie temperature*, these forces disappear and ferromagnetic materials become paramagnetic.

Curie temperature

The region of space over which the magnetic dipole moments are aligned is called a *domain*. The size of a domain is usually microscopic. Within the domain, all the magnetic moments are aligned, but the

Domain

direction of alignment varies from domain to domain so that the net magnetic moment of a macroscopic piece of material is zero in the normal state. Figure 29-9 illustrates this situation. When an external magnetic field is applied, the boundaries of the domains (called *domain walls*) shift and the direction of alignment within a domain may change so that there is a net magnetic moment in the direction of the applied field. Since the degree of alignment is great even for a small external field, the magnetic field produced in the material by the dipoles is often much greater than the external field.

  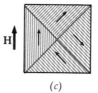

(a)   (b)   (c)

Figure 29-9
Schematic illustration of magnetization by domain changes: (a) unmagnetized; (b) magnetization by domain boundary changes; (c) magnetization by rotation of domains.

Consider magnetizing a long iron rod inside a solenoid by gradually increasing the current in the solenoid windings. Assume that the rod and solenoid are long enough to permit us to neglect end effects. The magnetic intensity **H** at the center of the solenoid is then simply related to the current by

$$H = nI = \frac{N}{\ell} I \qquad\qquad 29\text{-}28$$

where N is the number of turns and ℓ is the length. The magnetic induction **B** is given by

$$\mathbf{B} = \mu_0\mathbf{H} + \mu_0\mathbf{M} \qquad\qquad 29\text{-}29$$

where **M** is the magnetization. For iron and other ferromagnetic materials, the magnetization **M** is often much greater than the magnetic intensity **H** by a factor of several thousand or more.

William Vandivert

Photomicrograph showing magnetic domains in a thin single crystal of yttrium iron garnet magnified 330 times. The photograph was made with transmitted light through the crystal between crossed Polaroids. The different directions of magnetization of the domains give rise to different colors in the transmitted light. The magnetization is pointing down into the paper in the light striped regions and out of the paper in the dark striped regions. In the large gray areas, the magnetization lies in the plane of the paper.

Figure 29-10 shows a plot of B versus H. As H is gradually increased from zero, B increases from zero along the part of the curve from the origin O to point P_1. The flattening of this curve near point P_1 indicates that the magnetization M is approaching its saturation value M_s when all the atomic dipoles are aligned. The external field needed to produce saturation in a given ferromagnetic material is called the saturation intensity H_s. A further increase in H above H_s increases B only through the term $\mu_0 H$ in Equation 29-29. When H is gradually decreased from point P_1, there is no corresponding decrease in the magnetization. The shift of the domains in a ferromagnetic material is not completely reversible, and some magnetization remains even when H is reduced to zero, as indicated in Figure 29-10. This effect is called *hysteresis*, from a word which means to lag. The value of the magnetic induction at point r when H is zero is called the *remanent field* B_r. If the current in the solenoid is now reversed so that **H** is in the opposite direction, the magnetic induction B is gradually brought to zero at point c. The value of H needed to reduce B to zero is called the *coercive force* H_c. The remaining part of the hysteresis curve is obtained by further increasing the current in the opposite direction until point P_2 is reached, corresponding to saturation in the opposite direction; decreasing the current to zero at point P_3; and increasing the current again to produce a magnetic intensity H in the original direction. We can define the magnetic susceptibility χ_m formally just as we did for paramagnetic or diamagnetic materials:

$$\mathbf{M} = \chi_m \mathbf{H}$$

Then

$$\mathbf{B} = \mu_0(\mathbf{H} + \mathbf{M}) = \mu_0(1 + \chi_m)\mathbf{H} = K_m \mu_0 \mathbf{H} = \mu \mathbf{H} \qquad \text{29-30}$$

where

$$K_m = 1 + \chi_m = \frac{\mu}{\mu_0} \qquad \text{29-31}$$

is called the *relative permeability*. Since, according to Figure 29-10, **B** and **H** are sometimes parallel and sometimes antiparallel and either can be zero when the other is not, μ and K_m may have any value from zero to infinity and may be positive or negative depending on **H** and the history of the material. If we confine the definitions of μ and K_m to that part of the magnetization curve from the origin O to point P_1, these quantities are always positive and have maximum values characteristic of the material. Table 29-2 lists the saturation magnetization M_s and the maximum value of K_m for some ferromagnetic materials. Note that the maximum values of K_m are much greater than 1. In a ferromagnetic material the quantity $\mu_0 \mathbf{H}$ can usually be neglected compared with $\mu_0 \mathbf{M}$.

Hysteresis

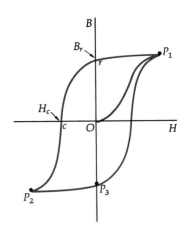

Figure 29-10
B versus H for hard ferromagnetic material. The outer curve is called a hysteresis curve.

Table 29-2
Maximum values of $\mu_0 M$ and K_m for some ferromagnetic materials

Material	$\mu_0 M_s$, T	K_m
Iron (annealed)	2.16	5,500
Iron-silicon (96% Fe, 4% Si)	1.95	7,000
Permalloy (55% Fe, 45% Ni)	1.60	25,000
Mu-metal (77% Ni, 16% Fe, 5% Cu, 2% Cr)	0.65	100,000

The area enclosed by the hysteresis curve represents loss in energy because of the irreversibility of the process. The energy appears in the material as heat. We can see this as follows. The emf in the solenoid due to the changing flux is given by Faraday's law:

$$\varepsilon = \frac{d\phi}{dt} = NA\frac{dB}{dt} \qquad\qquad 29\text{-}32$$

where A is the cross-sectional area of the coil and rod. (We neglect the minus sign in Faraday's law because we are interested in magnitudes only.) The rate at which work is done against this emf is

$$\frac{dW}{dt} = \varepsilon I \qquad\qquad 29\text{-}33$$

Using Equation 29-32 for the emf and Equation 29-28 for the current, we have

$$\frac{dW}{dt} = \left(NA\frac{dB}{dt}\right)\frac{H\ell}{N} = \ell AH\frac{dB}{dt}$$

or

$$dW = \ell AH\,dB \qquad\qquad 29\text{-}34$$

The sum of this work for a complete hysteresis cycle is thus the area enclosed in the B-versus-H curve times the volume ℓA of the material. If the hysteresis effect is small, so that the area is small, indicating a small energy loss, the material is called *magnetically soft* (soft iron is an example). The hysteresis curve for a magnetically soft material is shown in Figure 29-11. Here the remanent field B_r and the coercive force H_c are nearly zero, and the energy loss per cycle is small. Magnetically soft materials are used for transformer cores to increase the induction B without incurring a large energy loss when the field alternates many times per second.

On the other hand, it is desirable to have a large remanent field B_r and a large coercive force H_c in a permanent magnet (the large coercive force is important so that the magnetization will not be destroyed by small stray fields). Magnetically hard materials, e.g., carbon steel and the alloy Alnico 5, are used for permanent magnets.

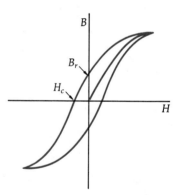

Figure 29-11
Hysteresis curve for magnetically soft material. The remanent field B_r and the coercive force H_c are small compared with those for a magnetically hard material (Figure 29-10).

Review

A. Define, explain, or otherwise identify:

Paramagnetism, 797	Permeability of free space, 802
Diamagnetism, 797	Bohr magneton, 804
Ferromagnetism, 797	Curie's law, 806
Magnetization, 798	Curie temperature, 808
Magnetic intensity, 799	Magnetic domain, 808
Magnetic pole strength, 800	Hysteresis, 810
Gauss' law for **H**, 800	Remanent field, 810
Coulomb's law for **H**, 800	Coercive force, 810
Magnetic susceptibility, 802	Magnetically soft, 811
Permeability, 802	Magnetically hard, 811

B. True or false:

1. Diamagnetism occurs in all materials.

2. Diamagnetism is the result of induced magnetic dipole moments.

3. In free space, **H** is always zero.

4. Paramagnetism is the result of partial alignment of permanent magnetic dipole moments.

5. Hysteresis is associated with a loss in electromagnetic energy.

6. Magnetically hard materials are good for transformer cores.

Exercises

Section 29-1, Magnetic Intensity H

1. A tightly wound solenoid 20 cm long has 400 turns carrying a current of 4 A such that its axial field is in the z direction. Neglecting end effects, find **H** and **B** at the center when (a) there is no core and (b) there is an iron core in the solenoid with magnetization $\mathbf{M} = 1.2 \times 10^6 \mathbf{k}$ A/m.

2. Consider a single magnetic pole of strength q_m far away from any other pole. Use Gauss' law for **H** (Equation 29-10) to derive Coulomb's law for **H** (Equation 29-11).

3. In Exercise 1 the cross-sectional area of the solenoid core is 4 cm². Assuming the same current and magnetization as in Exercise 1, find (a) the magnetic pole strength q_m of the core, (b) **H** at the center of the solenoid due to the magnetic poles, (c) the resultant **H** at the center due to both the magnetic poles and the current in the windings, and (d) **B** at the center.

4. For the permanent magnet in Example 29-2, show that just inside the right end of the magnet (a) $\mathbf{H} = -\frac{1}{2}\mathbf{M}(1 + a^2/2L^2)$ and (b) $\mathbf{B} = \frac{1}{2}\mu_0\mathbf{M}(1 - a^2/2L^2)$. (c) How do these results compare with those in Example 29-2 for the point just outside the magnet?

Section 29-2, Magnetic Susceptibility and Permeability

5. Which of the four gases listed in Table 29-1 are diamagnetic and which are paramagnetic?

6. If the solenoid in Exercise 1 has an aluminum core, find **H**, **M**, and **B** at the center, neglecting end effects.

7. Repeat Exercise 6 for a tungsten core.

Section 29-3, Atomic Magnetic Moments

8. Nickel has density 8.7 g/cm³ and molecular mass 58.7 g/mol. Its saturation magnetization is $\mu_0 M_s = 0.61$ T. Calculate the magnetic moment in Bohr magnetons of a nickel atom.

9. Repeat Exercise 8 for cobalt, which has density 8.9 g/cm³, molecular mass 58.9 g/mol, and saturation magnetization $\mu_0 M_s = 1.79$ T.

Section 29-4, Paramagnetism

10. Show that Curie's law predicts that the magnetic susceptibility for a paramagnetic substance is given by $\chi_m = m\mu_0 M_s/3kT$.

11. Assume that the magnetic moment of an aluminum atom is 1 Bohr magneton. The density of aluminum is 2.7 g/cm³, and its molecular mass is 27 g/mol. (a) Calculate M_s and $\mu_0 M_s$ for aluminum. (b) Use the result of Exercise 10 to calculate χ_m at $T = 300$ K. (c) Explain why you expect this result to be larger than that listed in Table 29-1.

12. In a simple model of paramagnetism we can consider that some fraction f of the molecules have their magnetic moment aligned with the external magnetic

field and the rest of the molecules are randomly oriented to produce no contribution to the magnetic moment. (a) Use this model and Curie's law to show that at temperature T and external field B this fraction aligned is $f = mB/3kT$. (b) Calculate this fraction for $T = 300$ K, $B = 1$ T, assuming m to be 1 Bohr magneton.

Section 29-5, Diamagnetism

There are no exercises for this section.

Section 29-6, Ferromagnetism

13. The saturation magnetic intensity for annealed iron is $H_s = 1.6 \times 10^5$ A/m. Find the permeability μ and the relative permeability K_m at saturation (see Table 29-2).

14. For annealed iron the relative permeability K_m has its maximum value of about 5500 at $H = 125$ A/m. (This is well below the saturation intensity.) Find M and B when K_m is maximum.

15. The coercive force for a certain permanent bar magnet is $H_c = 4.4 \times 10^4$ A/m. The bar magnet is to be demagnetized by placing it inside a long solenoid 15 cm long with 600 turns. What is the minimum current needed in the solenoid to demagnetize the magnet?

16. A long solenoid has 50 turns/cm and carries a current of 2 A. The solenoid is filled with iron, and B is measured to be 1.72 T. (a) What is H (neglecting end effects)? (b) What is M? (c) What is the relative permeability K_m for this case?

17. When the current in the solenoid in Exercise 16 is 0.2 A, the magnetic induction is measured to be 1.58 T. (a) Neglecting end effects, what is H? (b) What is M? (c) What is the relative permeability K_m?

Problems

1. A toroid has mean radius R and cross-sectional radius r (Figure 29-2), where $r < R$. (a) Use Equation 29-6 to show that $H = nI = IN/2\pi R$, where n is the number of turns per unit length and I is the current. (b) When the toroid is filled with material, it is called a *Rowland ring*. Find $\mathbf{H}$ and $\mathbf{B}$ in such a ring. Assume a magnetization $\mathbf{M}$ everywhere parallel to $\mathbf{H}$.

2. Show that the normal component of the magnetic induction $\mathbf{B}$ is continuous across any surface. Do this by applying Gauss' law for $\mathbf{B}$ ($\oint \mathbf{B} \cdot \hat{n}\, dA = 0$) to a pillbox gaussian surface which has a face on each side of the surface.

3. Use Equation 29-10 to show that the normal component of $\mathbf{H}$ on one side of a surface differs from that on the other side of the surface by the amount σ_m, where σ_m is the magnetic pole density on the surface (see Problem 2).

4. A Rowland ring has windings with 5 turns/cm carrying current $I = 1$ A. The magnetic induction in the ring is 1.40 T. (a) Find $\mathbf{H}$ and $\mathbf{M}$ in the ring. (b) What is K_m? (c) A small gap is cut in the ring. Find $\mathbf{B}$ and $\mathbf{H}$ in the gap assuming that they are unchanged in the bulk of the material (see Problems 2 and 3).

5. In our derivation of the magnetic moment induced in an atom we used the Bohr model and assumed that the radius of the orbit did not change in the presence of an external magnetic field. Show that the assumption of constant radius is justified by showing that when $\mathbf{B}$ is applied, there is an impulse which increases or decreases the speed of the electron by just the correct amount given by Equation 29-26. (a) Use Faraday's law and the expression $\varepsilon = \oint \mathbf{E} \cdot d\ell$ to show that the induced electric field is related to the rate of change of the magnetic field by $E = \frac{1}{2}r\, dB/dt$, assuming r to be constant. (b) Use Newton's second law to show that the change in speed of the electron dv is related to the change in B by $dv = (qr/2m_e)\, dB$. Integrate to obtain Equation 29-26.

6. Equation 29-27 gives the induced magnetic moment for a single electron in an orbit which has its plane perpendicular to **B**. If an atom has Z electrons, a reasonable simplifying assumption is that on the average one-third have their planes perpendicular to **B**. Show that the diamagnetic susceptibility obtained from Equation 29-27 is then

$$\chi_m = \frac{-nZq^2r^2}{12m_e}\mu_0$$

where n is the number of atoms per unit volume. Use $n \approx 6 \times 10^{28}$ atoms/m³ and $r \approx 5 \times 10^{-11}$ m to estimate χ_m for $Z \approx 50$.

7. Use the values in Table 29-3 to plot B versus H and K_m versus H.

8. A long solenoid of length ℓ and cross-sectional area A has n turns per unit length and carries a current I. It is filled with iron of relative permeability K_m. Find the self-inductance of the solenoid.

Table 29-3
For Problem 7

H, A/m	B, T
0	0
50	0.04
100	0.67
150	1.00
200	1.2
500	1.4
1,000	1.6
10,000	1.7

CHAPTER 30   Alternating-Current Circuits

Objectives   After studying this chapter, you should:

1. Be able to state the definition of rms current and relate it to the maximum current in an ac circuit.

2. Be able to state the definition of capacitive reactance, inductive reactance, and impedance.

3. Be able to draw a phasor diagram for a series *LRC* circuit and from it relate the phase angle ϕ to the capacitive reactance, inductive reactance, and resistance.

4. Be able to state the definition of the Q value and discuss its significance.

5. Be able to state the resonance condition for an *LRC* circuit with generator and sketch the power versus ω for both a high-Q and low-Q circuit.

6. Be able to describe a step-up and a step-down transformer.

In Chapter 28 we looked briefly at circuits with inductance, capacitance, and resistance and saw that in an *LC* circuit with no resistance, the energy oscillates between electric energy in the capacitor and magnetic energy in the inductor. With resistance in the circuit, the oscillations are damped as energy is dissipated as heat. In this chapter we look at similar circuits driven by a generator of sinusoidal alternating current. Much of the mathematics involved in the detailed analysis of such circuits is studied in a course in differential equations. As in the last sections of Chapter 28 and in the study of forced oscillators in Chapter 11, we shall content ourselves with a qualitative analysis of these circuits, to bring out the main features without relying on differential-equation theory.

We begin by studying a simple ac generator which puts out a sinusoidal emf, and then we consider the results of placing a resistor, capacitor, or inductor separately across its terminals. We then look at a combination of these circuit elements in series with such a generator. There are several reasons for studying sinusoidal currents and voltages.

Although practical generators are considerably different from the simple device we shall study in Section 30-1, they are designed to put out a sinusoidal emf. We shall see that when the generator output is sinusoidal, the current in an inductor, capacitor, or resistor is also sinusoidal, though not necessarily in phase with the generator emf. When the emf and current are both sinusoidal, their maximum values can be simply related. Finally, an important reason for studying sinusoidal currents is that even when the current in some circuit is not sinusoidal, it can be analyzed in terms of sinusoidal components using Fourier analysis. The results for sinusoidal currents can therefore be applied to any kind of alternating current.

30-1   An AC Generator

A simple generator of alternating current is a coil rotating in a uniform magnetic field, as shown in Figure 30-1, where the unit vector $\hat{n}$ normal to the plane of the coil makes an angle θ with a uniform magnetic field **B**. The magnetic flux through the coil is

$$\phi_m = NBA \cos \theta \qquad\qquad 30\text{-}1$$

where N is the number of turns and A the area of the coil. If the coil is mechanically rotated, the flux through the coil will change and, according to Faraday's law, there will be an emf in the coil. Let ω be the angular velocity of the coil. Then

$$\theta = \omega t \qquad\qquad 30\text{-}2$$

where we have chosen $\theta = 0$ at time $t = 0$ for convenience. Substituting this expression for θ into Equation 30-1, we have for the flux

$$\phi_m = NBA \cos \omega t \qquad\qquad 30\text{-}3$$

The emf in the coil will then be

$$\mathcal{E} = -\frac{d\phi_m}{dt} = -NBA(-\omega \sin \omega t) = +NBA\omega \sin \omega t$$

or

$$\mathcal{E} = \mathcal{E}_{max} \sin \omega t \qquad\qquad 30\text{-}4$$

where

$$\mathcal{E}_{max} = NBA\omega \qquad\qquad 30\text{-}5$$

is the maximum value of the emf. We can thus produce a sinusoidal emf in a coil by rotating it with constant angular velocity in a magnetic field. In this type of emf source, mechanical energy is converted into electric energy. In circuit diagrams, an ac generator is represented by the symbol ⊝ . Although practical generators are considerably more complicated, they work on the principle that there is an alternating emf in a coil rotating in a magnetic field, and they are designed so that the emf produced is sinusoidal.

Questions

1. Does the sinusoidal nature of the emf depend on the size or shape of the coil?

Figure 30-1
A coil rotating in a magnetic field with constant angular velocity generates a sinusoidal emf.

2. How could such a coil be used to generate a nonsinusoidal emf?

3. If such a generator delivers electric energy to a circuit, where does the energy come from?

30-2   Alternating Current in a Resistor

The simplest ac circuit consists of a generator and a resistor, as shown in Figure 30-2. We find the current in this circuit as usual by Kirchhoff's rule. Let the emf of the generator be given by

$$\mathcal{E} = \mathcal{E}_{max} \sin \omega t \qquad\qquad 30\text{-}6$$

Then Kirchhoff's rule gives

$$\mathcal{E} = IR = \mathcal{E}_{max} \sin \omega t \qquad\qquad 30\text{-}7$$

The current is proportional to the generator voltage and is given by

$$I = \frac{\mathcal{E}_{max}}{R} \sin \omega t = I_{max} \sin \omega t \qquad\qquad 30\text{-}8$$

where the maximum current is

$$I_{max} = \frac{\mathcal{E}_{max}}{R} \qquad\qquad 30\text{-}9$$

The instantaneous power dissipated in the resistor is

$$I^2 R = (I_{max} \sin \omega t)^2 R = I^2_{max} R \sin^2 \omega t$$

This power varies from zero to its maximum value $I^2_{max} R$, as shown in Figure 30-3. We are usually interested in the average power over one or more cycles. The average value of $\sin^2 \omega t$ over one or more cycles is $\frac{1}{2}$.* The average power dissipated in the resistor is thus

$$(I^2 R)_{av} = \tfrac{1}{2} I^2_{max} R \qquad\qquad 30\text{-}10$$

Figure 30-2
An ac generator in series with a resistor R.

Figure 30-3
Plot of $\sin^2 \omega t$ versus ωt. The average value over a cycle is $\frac{1}{2}$. This result can be obtained from the fact that $\cos^2 \omega t$ has the same average value over a cycle, and from $\sin^2 \omega t + \cos^2 \omega t = 1$.

Most ac ammeters and voltmeters measure root-mean-square (rms) values of current and voltage rather than the maximum or peak values. The rms value of the current, I_{rms}, is defined by

$$I_{rms} = \sqrt{(I^2)_{av}} \qquad\qquad 30\text{-}11$$

rms current defined

The average value of I^2 is

$$(I^2)_{av} = (I_{max} \sin \omega t)^2_{av} = \tfrac{1}{2} I^2_{max}$$

---

* This can be seen from the identity $\sin^2 \omega t + \cos^2 \omega t = 1$, and the fact that the average values of $\sin^2 \omega t$ and $\cos^2 \omega t$ are equal.

The rms value of the current is thus

$$I_{rms} = \frac{1}{\sqrt{2}} I_{max} \qquad \text{30-12}$$

The rms value of any quantity that varies sinusoidally equals the maximum value of that quantity divided by $\sqrt{2}$. In terms of the rms current, the average power dissipated in the resistor is

$$(I^2R)_{av} = I_{rms}^2 R \qquad \text{30-13}$$

The average power delivered by the generator is of course equal to that dissipated in the resistor for this simple circuit:

$$P_{av} = (\mathcal{E}I)_{av} = \mathcal{E}_{max}I_{max}(\sin^2 \omega t)_{av}$$

or

$$P_{av} = \tfrac{1}{2}\mathcal{E}_{max}I_{max} = \mathcal{E}_{rms}I_{rms} \qquad \text{30-14}$$

where $\mathcal{E}_{rms} = (1/\sqrt{2})\,\mathcal{E}_{max}$. Since Equation 30-9 holds for the rms values of emf and current as well as the maximum values,

$$I_{rms} = \frac{\mathcal{E}_{rms}}{R} \qquad \text{30-15}$$

The power input of the emf (Equation 30-14) equals the power dissipated in the resistor (Equation 30-13).

30-3 Alternating Current in a Capacitor

Figure 30-4 shows a capacitor connected across the terminals of a generator. Assuming I to be positive in the clockwise direction as indicated, we have

$$I = \frac{dQ}{dt}$$

where Q is the charge on the capacitor. Kirchhoff's rule for this circuit gives

$$\mathcal{E} = \frac{Q}{C} = \mathcal{E}_{max} \sin \omega t \qquad \text{or} \qquad Q = \mathcal{E}_{max} C \sin \omega t$$

The current is

$$I = \frac{dQ}{dt} = \omega\mathcal{E}_{max}C \cos \omega t = I_{max} \cos \omega t \qquad \text{30-16}$$

where

$$I_{max} = \mathcal{E}_{max}\omega C \qquad \text{30-17}$$

The current in this circuit is not in phase with the generator voltage, which is also the potential drop across the capacitor. Using the trigonometric identity

$$\cos \omega t = \sin (\omega t + 90°)$$

we have

$$I = I_{max} \sin (\omega t + 90°)$$

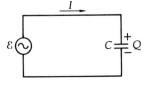

Figure 30-4
An ac generator in series with a capacitor C.

The current and voltage across the capacitor are plotted in Figure 30-5, from which we see that the maximum value of the voltage occurs 90° or one-fourth period after the maximum value of the current. The voltage drop across the capacitor is said to *lag* the current by 90°, or one-fourth period.

The relation between the maximum (or rms) current and the maximum (or rms) voltage can be written in a form similar to Ohm's law:

$$I_{max} = \omega C \varepsilon_{max} = \frac{\varepsilon_{max}}{1/\omega C} = \frac{\varepsilon_{max}}{X_C} \qquad 30\text{-}18a$$

or

$$I_{rms} = \frac{\varepsilon_{rms}}{X_C} \qquad 30\text{-}18b$$

where

$$X_C = \frac{1}{\omega C} \qquad 30\text{-}19$$

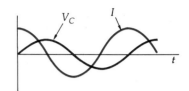

Figure 30-5
Plot of current I and voltage V_C across the capacitor versus time. The maximum in V_C occurs one-fourth period after the maximum in the current. The voltage V_C thus lags the current by one-fourth period, or 90°.

Capacitive reactance

is called the *capacitive reactance* of the circuit. Like resistance, capacitive reactance has units of ohms; the larger the reactance for a given emf, the smaller the maximum value of the current. Unlike resistance, the capacitive reactance depends on the frequency of the current: the larger the frequency, the smaller the reactance. The instantaneous power input by the generator is

$$P = \varepsilon I = (\varepsilon_{max} \sin \omega t)(I_{max} \cos \omega t)$$
$$= \varepsilon_{max} I_{max} \sin \omega t \cos \omega t$$

The average power input by the generator is zero. We can see this by writing $\sin \omega t \cos \omega t = \frac{1}{2} \cos 2\omega t$. This term oscillates twice during each cycle and is negative as often as it is positive.

Example 30-1 A 20-μF capacitor is placed across a generator which has a maximum emf of 100 V. Find the reactance and maximum current when the frequency is 60 Hz and when it is 5000 Hz.

These frequencies correspond to angular frequencies ω_1 and ω_2, given by $\omega_1 = 2\pi f_1 = 2\pi(60 \text{ s}^{-1}) = 337 \text{ s}^{-1}$ and $\omega_2 = 2\pi(5000 \text{ s}^{-1}) = 3.14 \times 10^4 \text{ s}^{-1}$. The reactances at these frequencies are

$$X_1 = \frac{1}{\omega_1 C} = [377(20 \times 10^{-6})]^{-1} = 133 \ \Omega$$

and

$$X_2 = \frac{1}{\omega_2 C} = [(3.14 \times 10^4)(20 \times 10^{-6})]^{-1} = 1.59 \ \Omega$$

The maximum currents are then

$$I_{1,max} = \frac{\varepsilon_{max}}{X_1} = \frac{100 \text{ V}}{133 \ \Omega} = 0.754 \text{ A}$$

and

$$I_{2,max} = \frac{100 \text{ V}}{1.59 \ \Omega} = 62.8 \text{ A}$$

30-4 Alternating Current in an Inductor

Figure 30-6 shows an inductor across the terminals of an ac generator. Kirchhoff's rule for this circuit gives

$$\mathcal{E} = L\frac{dI}{dt} = \mathcal{E}_{max} \sin \omega t \qquad 30\text{-}20$$

Solving for the current I gives

$$I = -\frac{\mathcal{E}_{max}}{\omega L} \cos \omega t = \frac{\mathcal{E}_{max}}{\omega L} \sin(\omega t - 90°) \qquad 30\text{-}21$$

(We have neglected the constant of the integration because it depends on the initial conditions, which are of no interest for this discussion.) From the plot in Figure 30-7 of the current and voltage across the inductor we see that the maximum value of the voltage occurs 90° or one-fourth period before the corresponding maximum value of the current. The voltage across an inductor is said to *lead* the current by 90° or one-fourth period.

The maximum (or rms) current is related to the maximum (or rms) voltage across the inductor by

$$I_{max} = \frac{\mathcal{E}_{max}}{\omega L} = \frac{\mathcal{E}_{max}}{X_L} \qquad 30\text{-}22a$$

or

$$I_{rms} = \frac{\mathcal{E}_{rms}}{X_L} \qquad 30\text{-}22b$$

where

$$X_L = \omega L \qquad 30\text{-}23$$

The quantity X_L is called the *inductive reactance*. As the frequency increases, the inductive reactance increases and the maximum current decreases. As it was for an ac generator and capacitor, the average power input of the generator into the inductor is zero. This is because the emf is proportional to $\sin \omega t$ and the current to $\cos \omega t$, and $(\sin \omega t \cos \omega t)_{av} = 0$.

30-5 *LCR* Circuit with Generator

An important circuit containing many of the features of most ac circuits is a series *LCR* circuit with a generator (Figure 30-8). We discussed a similar circuit without the generator in Chapter 28, finding that the current oscillates with angular frequency nearly equal to $\omega_0 = 1/\sqrt{LC}$ (assuming that the resistance is small) and that the maximum value of the current decreases exponentially with time. Kirchhoff's rule applied to this circuit with the generator gives

$$L\frac{dI}{dt} + IR + \frac{Q}{C} = \mathcal{E}_{max} \sin \omega t \qquad 30\text{-}24$$

with $I = dQ/dt$. This equation is analogous to Equation 11-53 for the driven oscillator. The first term $L\,dI/dt = L\,d^2Q/dt^2$ is analogous to $m\,d^2x/dt^2$; the second term $IR = (dQ/dt)R$ is the damping term analogous to the term $bv = b\,dx/dt$ in Equation 11-53; the third term Q/C is

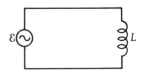

Figure 30-6
An ac generator in series with an inductor L.

Figure 30-7
Plot of current I and voltage V_L across the inductor as functions of time. The maximum in V_L occurs one-fourth period before the maximum current. The voltage V_L thus leads the current by one-fourth period, or 90°.

Inductive reactance

Figure 30-8
A series *LRC* circuit with an ac generator.

analogous to the restoring-force term kx. The solution of this equation for the current I contains a part called the transient current, which decreases exponentially with time and is identical to the solution of the equation without the driving term on the right side, plus a steady-state current which does not decrease exponentially in time. At a sufficient time after the switch is closed, the transient current is negligible compared with the steady-state current. We shall ignore the transient current in our discussion of this circuit and concentrate on the steady-state solution. The current obtained by solving Equation 30-24 is

$$I = I_{max} \sin (\omega t - \phi) \qquad \text{30-25}$$

where

$$\tan \phi = \frac{X_L - X_C}{R} \qquad \text{30-26}$$

and

$$I_{max} = \frac{\mathcal{E}_{max}}{\sqrt{(X_L - X_C)^2 + R^2}} = \frac{\mathcal{E}_{max}}{Z} \qquad \text{30-27}$$

The quantity

$$Z = \sqrt{(X_L - X_C)^2 + R^2} \qquad \text{30-28}$$

is called the *impedance*. Combining these results and converting to the rms values $I_{rms} = I_{max}/\sqrt{2}$ and $\mathcal{E}_{rms} = \mathcal{E}_{max}/\sqrt{2}$, we have

$$I_{rms} = \frac{\mathcal{E}_{rms}}{Z} \sin (\omega t - \phi) \qquad \text{30-29}$$

Impedance

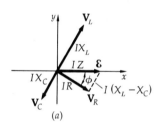

(a)

We can obtain these results from a simple vector diagram called a *phasor diagram*. At any instant, the current is the same in each element of the series circuit. We represent this current by a vector (called a phasor) which has magnitude equal to I_{max} and which makes an angle $\omega t - \phi$ with the *x* axis. This vector rotates counterclockwise with angular frequency ω. The instantaneous current is the *y* component of this vector. If we multiply this vector by the resistance R, we obtain a vector $\mathbf{V}_R$ representing the voltage drop across the resistor. These vectors are parallel because the voltage drop across the resistor is in phase with the current. The voltage drop across the capacitor lags the current by 90°. This voltage is therefore represented by a vector $\mathbf{V}_C$ which has magnitude $I_{max}X_C$ and lags the vector $\mathbf{V}_R$ by 90°, as shown in Figure 30-9. The *y* component of this vector equals the instantaneous voltage drop across the capacitor. Similarly, we represent the voltage drop across the inductor by a vector $\mathbf{V}_L$ which has magnitude $I_{max}X_L$ and which leads $\mathbf{V}_R$ by 90°. The sum of the *y* components of these three vectors equals the instantaneous sum of the voltage drops across the resistor, capacitor, and inductor. By Kirchhoff's law, this sum equals the instantaneous emf. The sum of these *y* components is the *y* component of the resultant vector:

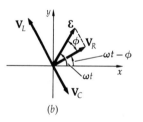

(b)

Figure 30-9
Phasor diagram for an *LRC* circuit with generator. The emf $\mathcal{E}$ and the voltage drops V_R, V_C, and V_L are the *y* components of the vectors or phasors representing these quantities. In (*a*) these vectors are shown at $t = 0$. The vector $\mathbf{V}_R$ has magnitude IR and is in phase with the current. $\mathbf{V}_C$ lags the current by 90° and $\mathbf{V}_L$ leads the current by 90°. The current and $\mathbf{V}_R$ lag the emf by ϕ, where $\tan \phi = (X_L - X_C)/R$. (The current I can be either I_{max} or I_{rms}.) In (*b*) each of the vectors is rotated through the angle ωt.

$$\mathcal{E} = \mathbf{V}_R + \mathbf{V}_L + \mathbf{V}_C \qquad \text{30-30}$$

The vector $\mathcal{E}$ representing the emf makes an angle ωt with the *x* axis and makes an angle ϕ with the current and resistance vector $\mathbf{V}_R$ (Figure 30-9*b*). From the right triangle in Figure 30-9*a* we obtain for the magnitude of the resultant vector

$$\mathcal{E}_{max} = |\mathbf{V}_R + \mathbf{V}_L + \mathbf{V}_C| = \sqrt{V_R^2 + (V_L - V_C)^2}$$

$$= I_{max} \sqrt{R^2 + (X_L - X_C)^2} = I_{max} Z$$

This result is Equation 30-27. We see also from the diagram that the phase angle ϕ is given by Equation 30-26.

Resonance

Since both the inductive reactance $X_L = \omega L$ and the capacitive reactance $X_C = 1/\omega C$ depend on the angular frequency ω of the applied emf, the impedance Z and the rms current I_{rms} also depend on ω. As we increase ω, the inductive reactance increases and the capacitive reactance decreases. When X_L and X_C are equal, the impedance Z has its minimum value $Z_{min} = R$ and I_{rms} has its greatest value. The value of ω for which X_L and X_C are equal is obtained from

$$\omega L = \frac{1}{\omega C}$$

or

$$\omega = \frac{1}{\sqrt{LC}} = \omega_0 \qquad \text{30-31}$$

Resonance condition

The impedance is minimum and the rms current maximum when the angular frequency of the emf equals the natural angular frequency $\omega_0 = 1/\sqrt{LC}$. At this frequency the circuit is said to be at *resonance*. This resonance condition in a driven LCR circuit is exactly the same as that in a driven simple harmonic oscillator. We note from Equation 30-26 that when $X_L = X_C$, the phase angle ϕ is zero; i.e., at resonance the current is in phase with the applied emf.

The instantaneous power supplied by the generator to the circuit is

$$P = \mathcal{E}I = \mathcal{E}_{max} \sin \omega t \, I_{max} \sin (\omega t - \phi)$$

The average power is most easily computed by first expanding the function $\sin (\omega t - \phi)$ using the trigonometric identity for the sine of the sum of two angles:

$$\sin (\omega t - \phi) = \sin \omega t \cos \phi - \cos \omega t \sin \phi$$

Then

$$P = \mathcal{E}_{max}I_{max} \cos \phi \sin^2 \omega t - \sin \phi \sin \omega t \cos \omega t$$

The time average of $\sin^2 \omega t$ is $\frac{1}{2}$, and that of $\sin \omega t \cos \omega t$ is zero, as we have seen before. The average power is therefore

$$P_{av} = \tfrac{1}{2}\mathcal{E}_{max}I_{max} \cos \phi = \mathcal{E}_{rms}I_{rms} \cos \phi \qquad \text{30-32}$$

As in a dc circuit, the power supplied depends on the product of the emf and the current. However, Equation 30-32 also contains the factor $\cos \phi$, called the *power factor*. The power supplied is maximum at resonance as can be seen from this equation, since $\phi = 0$ at resonance and I_{rms} has its maximum value.

The power can be expressed as a function of the angular frequency ω. From the triangle in Figure 30-9a we have

$$\cos \phi = \frac{R}{Z} \qquad \text{30-33}$$

When we use this result and write $I_{rms} = \mathcal{E}_{rms}/Z$, the average power is

$$P_{av} = \mathcal{E}_{rms}^2 \frac{R}{Z^2} \qquad \text{30-34}$$

From the definition of the impedance Z we have

$$Z^2 = (X_L - X_C)^2 + R^2 = \left(\omega L - \frac{1}{\omega C}\right)^2 + R^2$$

$$= \frac{L^2}{\omega^2}\left(\omega^2 - \frac{1}{LC}\right)^2 + R^2 = \frac{L^2}{\omega^2}(\omega^2 - \omega_0^2)^2 + R^2$$

where we have used $\omega_0^2 = 1/LC$. Using this expression for Z^2, we obtain for the average power as a function of ω

$$P_{av} = \frac{\mathcal{E}_{rms}^2 R\omega^2}{L^2(\omega^2 - \omega_0^2)^2 + \omega^2 R^2} \qquad \text{30-35}$$

Equation 30-35 is the same as Equation 11-62 for the average power input in a driven oscillator with R replacing the damping constant b, L replacing the mass m, and $\mathcal{E}$ replacing the driving force F_0 (the factor of $\frac{1}{2}$ in Equation 11-62 is the result of using the maximum driving force F_0 rather than the rms force).

Figure 30-10 shows a plot of the average power versus generator frequency ω for two values of the resistance R. The smaller the resistance, the more sharply peaked the resonance. Let $\Delta\omega = \omega_2 - \omega_1$ be the width of the resonance curve, where ω_1 and ω_2 are the two values of ω for which P_{av} is half its maximum value. For a sharply peaked resonance we can find a useful approximation for the width as follows. At resonance, the denominator of the expression on the right in Equation 30-35 is just $\omega^2 R^2$. The power will be half its maximum value when this denominator is approximately twice this value (neglecting the variation of the numerator with ω). Then at the half-power points

$$L^2(\omega^2 - \omega_0^2)^2 \approx \omega^2 R^2$$

or

$$L(\omega^2 - \omega_0^2) = L(\omega - \omega_0)(\omega + \omega_0) \approx \pm\omega R$$

Since ω and ω_0 are nearly equal for a sharply peaked resonance, we can replace ωR with $\omega_0 R$ and $\omega + \omega_0$ with $2\omega_0$. Then we have approximately

$$\omega - \omega_0 \approx \pm\frac{R}{2L}$$

$$\omega \approx \omega_0 \pm \frac{R}{2L}$$

That is, $\omega_2 \approx \omega_0 + R/2L$ and $\omega_1 \approx \omega_0 - R/2L$. The width is then

$$\Delta\omega \approx \frac{R}{L} \qquad \text{30-36}$$

As we did for the resonance curve for a mechanical driven oscillator, we describe the width of the resonance by a dimensionless parameter, the Q value or quality factor, which in general is defined by $Q = L\omega_0/R$. When Q is greater than about 2 or 3, the resonance is sharply peaked and our approximations for the width are valid. Then Q is simply the ratio of the resonance frequency to the width:

$$Q = \frac{L\omega_0}{R} = \frac{\omega_0}{\Delta\omega} \qquad \text{30-37} \qquad \textit{Q value}$$

A common application of series resonance circuits is a radio receiver, where the resonance frequency of the circuit is varied by varying the capacitance. Resonance occurs when the natural frequency of the circuit

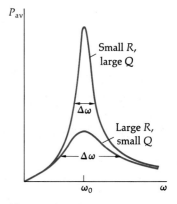

Figure 30-10
Plot of average power P_{av} given by Equation 30-35 versus ω for small and large R. The Q value is the ratio of the resonance frequency ω_0 to the resonance width $\Delta\omega$. A large Q value indicates a narrow resonance.

equals the frequency of the radio waves picked up at the antenna. At resonance, there is a relatively large current in the antenna circuit. If the Q value of the circuit is sufficiently high, currents due to other stations off resonance will be negligible compared with those due to the station to which the circuit is tuned.

Example 30-2 A series LCR circuit with $L = 2$ H, $C = 2\ \mu$F, and $R = 20\ \Omega$ is driven by a generator of maximum emf 100 V and variable frequency. Find the resonance frequency ω_0, the phase ϕ, and maximum current I_{max} when the generator angular frequency is $\omega = 400$ rad/s.

The resonance frequency is $\omega_0 = 1/\sqrt{LC} = 1/\sqrt{(2\ \text{H})(2 \times 10^{-6}\ \text{F})} = 500$ rad/s. When the generator frequency is 400 rad/s, it is well below the resonance frequency. The capacitive and inductive reactances at 400 rad/s are

$$X_C = \frac{1}{\omega C} = \frac{1}{(400)(2 \times 10^{-6})} = 1250\ \Omega$$

and

$$X_L = \omega L = (400)(2) = 800\ \Omega$$

The total reactance is $X_L - X_C = 800\ \Omega - 1250\ \Omega = -450\ \Omega$. This is of much greater magnitude than the resistance, a result which always holds far from resonance. The total impedance is

$$Z = \sqrt{(X_L - X_C)^2 + R^2} \approx 450\ \Omega$$

since 20^2 is negligible compared with 450^2. The maximum current is then

$$I_{max} = \frac{\mathcal{E}_{max}}{Z} = \frac{100\ \text{V}}{450\ \Omega} = 0.222\ \text{A}$$

This is small compared with I_{max} at resonance, which is $(100\ \text{V})/(20\ \Omega) = 5$ A. The phase angle ϕ is given by

$$\tan \phi = \frac{X_L - X_C}{R} = \frac{-450\ \Omega}{20\ \Omega} = -22.5 \qquad \phi = -87°$$

From Equation 30-25 we see that a negative phase angle means that the current leads the generator voltage.

Example 30-3 Find the resonance width and Q value of the circuit of Example 30-2.

In that example we found the resonance frequency to be $\omega_0 = 500$ rad/s. The Q value is then

$$Q = \frac{L\omega_0}{R} = \frac{2(500)}{20} = 50$$

and the width of the resonance is

$$\Delta\omega = \frac{\omega_0}{Q} = \frac{500}{50} = 10\ \text{rad/s}$$

Questions

4. In general, does the power factor depend on frequency?

5. Can the instantaneous power delivered by the generator ever be negative?

6. What is the power factor for a circuit which has inductance or capacitance or both but zero resistance?

7. What is the power factor for a circuit with resistance but no inductance or capacitance? Does this power factor depend on frequency?

30-6 The Transformer

In order to transport power with a minimum I^2R heat loss in transmission lines, it is economical to use a high voltage and low current. On the other hand, safety and other considerations, e.g., insulation, make it convenient to use power to run motors and other electrical appliances at lower voltage and higher current. This is accomplished by using a transformer, a device for changing ac voltage and current without appreciable loss in power. If V is the voltage and I the current, the instantaneous power is VI. If the voltage is to be changed without change in power, the current must also be changed.

Figure 30-11 diagrams a simple transformer consisting of two coils of wire around a common core of soft iron. The coil carrying the input power is called the *primary*, and the other coil is called the *secondary*. Either coil of a transformer can be used for the primary or secondary. The function of the iron core is to increase the flux greatly for a given current and to guide it so that nearly all the flux through one turn of one coil goes through each turn of both coils. The iron core is laminated to reduce eddy-current losses. Other possible losses are the I^2R losses in the coils, which can be reduced by using a low-resistance wire for the coils, and hysteresis losses in the core, which can be reduced by using soft iron. It is relatively easy to design a transformer for which power is transferred from the primary to the secondary with efficiency of 90 to 99 percent. We shall discuss only an ideal transformer, which has no losses.

Consider an ac generator of emf ε across the primary of N_1 turns with the secondary coil of N_2 turns open. Because of the iron core, there is a large flux through each coil even with a very small magnetizing current I_m in the primary circuit. We can neglect the resistance of the coil compared with the inductive reactance. The primary is then a simple circuit consisting of an ac generator and a pure inductance, as discussed in Section 30-4. The current and voltage in the primary are out of phase by 90°, and the average power in the primary circuit is zero. We can see from Faraday's law that the voltage across the secondary is just the turns ratio N_2/N_1 times the voltage across the primary. The induced emf V_1 in the primary circuit is

$$V_1 = -\frac{d\phi}{dt}$$

The total flux ϕ is the flux through each turn ϕ_{turn} times the number of turns N_1:

$$V_1 = -N_1 \frac{d\phi_{\text{turn}}}{dt} \qquad\qquad 30\text{-}38$$

Assuming no flux leakage out of the iron core, the flux through each

Figure 30-11
Transformer with N_1 turns in the primary and N_2 turns in the secondary.

turn is the same for both coils. Thus the total flux through the secondary coil is $N_2\phi_{turn}$, and the voltage across the secondary coil is

$$V_2 = -N_2 \frac{d\phi_{turn}}{dt} \qquad\qquad 30\text{-}39$$

Comparing these equations, we see that

$$V_2 = \frac{N_2}{N_1} V_1 \qquad\qquad 30\text{-}40$$

According to Kirchhoff's law for the primary circuit, the applied emf $\mathcal{E}$ is the negative of the induced emf V_1:

$$\mathcal{E} - \frac{d\phi}{dt} = 0$$

$$\mathcal{E} = \frac{d\phi}{dt} = -V_1 \qquad\qquad 30\text{-}41$$

In terms of the applied emf, the voltage across the secondary coil is

---

$$V_2 = -\frac{N_2}{N_1} \mathcal{E} \qquad\qquad 30\text{-}42$$

The transformer is called a *step-up transformer* if N_2 is greater than N_1 so that the output voltage is greater than the input voltage. If N_2 is less than N_1, it is a *step-down transformer*. There is no current in the secondary because that circuit is open. The very small current I_m in the primary coil is 90° out of phase with the emf in that coil.

Consider now what happens when we put a resistance R, called a *load resistance*, across the secondary coil. There will then be a current I_2 in the secondary circuit which is in phase with the voltage V_2 across the resistance. This current will set up an additional flux through each turn ϕ'_{turn} proportional to $N_2 I_2$. This flux adds to the original flux ϕ_{turn} set up by the original magnetizing current in the primary I_m. However, the voltage across the primary coil is determined by the generator emf, which is unaffected by the secondary circuit. According to Equation 30-41, the flux in the iron core must therefore change at the original rate; i.e., the total flux in the iron core must be the same as with no load across the secondary. The primary coil thus draws an additional current I_1 to maintain the original flux ϕ_{turn}. The flux through each turn produced by this additional current is proportional to $N_1 I_1$. Thus the additional current I_1 in the primary is related to the current I_2 in the secondary by

---

$$N_1 I_1 = -N_2 I_2 \qquad\qquad 30\text{-}43$$

The negative sign indicates that these currents are 180° out of phase because they produce counteracting fluxes. Since I_2 is in phase with V_2, the additional current in the primary I_1 is in phase with the applied emf.

Figure 30-12 is a phasor diagram of the phase relationships between the voltages and currents. The total current in the primary I is the "vector" sum of the original magnetizing current I_m and the additional current I_1, which is usually very much greater than I_m. The power delivered by the generator is the product of the rms applied emf, the rms total current in the primary I_{rms}, and the power factor $\cos\phi$, where ϕ is the phase angle between the applied emf and the total current I. Since I_1 is in phase with the applied emf, ϕ is the angle between I_1 and I shown in Figure 30-12.

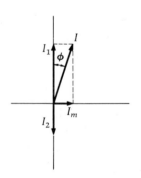

Figure 30-12
Phase relations between the magnetizing current I_m, the secondary current I_2, and the additional primary current I_1. The total primary current I is nearly equal to I_1 because I_m is usually small. Then the power factor, $\cos\phi$, is nearly 1.

We note from this figure that $I \cos \phi$ is just the additional current I_1, and so the power input by the primary is

$$P = \mathcal{E}_{rms} I_{rms} \cos \phi = \mathcal{E}_{rms} I_{1,rms}$$

Using Equations 30-42 and 30-43 to relate the applied emf and the additional current I_1 in the primary to the voltage V_2 and the current I_2 in the secondary, we have

$$\mathcal{E} I_1 = \left(-\frac{N_1}{N_2} V_2\right)\left(-\frac{N_2}{N_1} I_1\right) = V_2 I_2 \qquad 30\text{-}44$$

Thus

$$\mathcal{E}_{rms} I_{1,rms} = V_{2,rms} I_{2,rms} \qquad 30\text{-}45$$

The power input to the primary equals the power output in the secondary, as assumed for an ideal transformer with no losses.

In most cases the additional current in the primary I_1 is much greater than the original magnetizing current I_m with no load. This can be demonstrated by putting a light bulb in series with the primary coil: it is much brighter when there is a load across the secondary than when the secondary circuit is open. If I_m can be neglected, Equation 30-43 relates the total currents in the primary and secondary circuit.

The currents I_1 and I_2 can be related to the resistance R across the secondary. For the secondary current we have simply

$$I_2 = \frac{V_2}{R}$$

Using Equations 30-42 and 30-43 to write I_2 and V_2 in terms of the primary current I_1 and the primary emf $\mathcal{E}$, we have

$$I_1 = \frac{\mathcal{E}}{(N_1/N_2)^2 R} \qquad 30\text{-}46$$

The current I_1 is the same as if we connected a resistance $(N_1/N_2)^2 R$ across the generator. This effect is called *impedance transformation* since in general some combination of capacitance, inductance, and resistance with impedance Z is connected across the secondary coil as the load.

Review

A. Define, explain, or otherwise identify:

rms current, 817
Capacitive reactance, 819
Inductive reactance, 820
Impedance, 821
Phasor, 821
Resonance condition, 822
Power factor, 822
Resonance width, 823

Q value, 823
Transformer, 825
Primary, 825
Secondary, 825
Step-up transformer, 826
Step-down transformer, 826
Impedance transformation, 827

B. True or false:

1. At very high frequencies a capacitor acts like a short circuit.

2. A circuit with a high Q value has a narrow resonance curve.

3. At resonance the impedance equals the resistance.

4. Off resonance, the current always lags the generator voltage.

5. At resonance, the current and generator voltage are in phase.

6. If a transformer increases the current, it must decrease the voltage.

Exercises

Section 30-1, An AC Generator

1. A coil has an area of 4 cm², has 200 turns, and rotates in a magnetic field of 0.5 T. (*a*) What must its angular frequency be to generate an emf of 10 V maximum? (*b*) If it rotates at 60 Hz, what is ε_{max}?

2. In what magnetic field should the coil of Exercise 1 rotate to generate an emf of 10 V maximum at 60 Hz?

3. (*a*) Show that the rms emf of a generator is related to the maximum emf by $\varepsilon_{rms} = 0.707\varepsilon_{max}$. (*b*) The rms value of standard house voltage is 120 V. What is the maximum voltage?

Section 30-2, Alternating Current in a Resistor

4. A 100-W light bulb is plugged into a standard 120-V rms outlet. Find (*a*) I_{rms}, (*b*) I_{max}, and (*c*) the maximum power.

5. A 3-Ω resistor is placed in series with a 12.0-V (maximum) generator of frequency 60 Hz. (*a*) What is the angular frequency of the current? (*b*) Find I_{max} and I_{rms}. What is (*c*) the maximum power into the resistor, (*d*) the minimum power, and (*e*) the average power?

6. A 5.0-kW rms electric clothes dryer runs on 220 V rms. Find (*a*) I_{rms} and (*b*) I_{max}. (*c*) Find these same quantities for a dryer of the same power that operates at 120 V rms.

Section 30-3, Alternating Current in a Capacitor

7. What is the reactance of a 1.0-nF capacitor at (*a*) 60 Hz, (*b*) 6 kHz, and (*c*) 6 MHz?

8. Find the reactance of a 10.0-μF capacitor at (*a*) 60 Hz, (*b*) 600 Hz, and (*c*) 6 kHz.

9. Sketch a graph of X_C versus frequency f for $C = 100\ \mu$F.

10. An emf of 10.0 V maximum and frequency 20 Hz is applied to a 20-μF capacitor. Find (*a*) I_{max} and (*b*) I_{rms}.

11. At what frequency is the reactance of a 10-μF capacitor (*a*) 1 Ω, (*b*) 100 Ω, and (*c*) 0.01 Ω?

12. Show that the maximum power delivered by a generator to a capacitor is $P_{max} = \varepsilon_{rms}I_{rms}$. What is the average power?

Section 30-4, Alternating Current in an Inductor

13. What is the reactance of a 1.0-mH inductor at (*a*) 60 Hz, (*b*) 600 Hz, and (*c*) 6 kHz?

14. An inductor has a reactance of 100 Ω at 80 Hz. (*a*) What is its inductance? (*b*) What is its reactance at 160 Hz?

15. At what frequency would the reactance of a 10.0-μF capacitor equal that of a 1.0-mH inductor?

Section 30-5, *LCR* Circuit with Generator

16. Show that Equation 30-27 can be written

$$I_{max} = \frac{\omega\varepsilon_{max}}{\sqrt{L^2(\omega^2 - \omega_0^2)^2 + \omega^2 R^2}}$$

17. A series LCR circuit with $L = 10$ mH, $C = 2$ μF, and $R = 5$ Ω is driven by a generator of maximum emf 100 V and variable angular frequency ω. Find (a) the resonance frequency ω_0 and (b) I_{rms} at resonance. When $\omega = 8000$ rad/s, find (c) X_C and X_L, (d) Z and I_{rms}, and (e) the phase angle ϕ.

18. (a) Show that Equation 30-26 can be written $\tan \phi = L(\omega^2 - \omega_0^2)/\omega R$. Find ϕ approximately (b) at very low frequencies and (c) at very high frequencies.

19. For the circuit of Exercise 17, let the generator frequency be $f = \omega/2\pi = 1$ kHz. Find (a) the resonance frequency $f_0 = \omega_0/2\pi$, (b) X_C and X_L, (c) the total impedance Z and I_{rms}, and (d) the phase angle ϕ.

20. A series LCR circuit in a radio receiver is tuned by a variable capacitor so that it can resonate at frequencies from 500 to 1600 kHz. If $L = 1.0$ μH, find the range of C necessary to cover this range of frequencies.

21. FM radio stations have carrier frequencies which are separated by 0.20 MHz. When the radio is tuned to a station such as 100.1 MHz, the resonance width of the receiver circuit should be much smaller than 0.2 MHz so that adjacent stations are not received. If $f_0 = 100.1$ MHz and $\Delta f = 0.05$ MHz, what is the Q value of the circuit?

22. (a) Find the power factor for the circuit in Example 30-2 when $\omega = 400$ rad/s. (b) At what angular frequency is the power factor $\frac{1}{2}$?

23. Find (a) the Q value and (b) the resonance width for the circuit of Exercise 17. (c) What is the power factor when $\omega = 8000$ rad/s?

24. Show that when there is resistance and capacitance but no inductance, the power factor is given by $\cos \phi = RC\omega/\sqrt{1 + (RC\omega)^2}$. Sketch the power factor versus ω.

25. An ac generator of maximum emf 20 V is connected in series with a 20-μF capacitor and an 80-Ω resistor. There is no inductance in the circuit. Find (a) the power factor, (b) the rms current, and (c) the average power if the angular frequency of the generator is 400 rad/s.

26. Express the power factor as a function of ω and sketch for a circuit with resistance and inductance in series with a generator.

27. A coil can be considered to be a resistance and an inductance in series. Assume that $R = 100$ Ω and $L = 0.4$ H. The coil is connected across a 120-V 60-Hz line. Find (a) the power factor, (b) the rms current, and (c) the average power supplied.

28. Find the power factor and the phase ϕ for the circuit of Exercise 17 when the generator frequency is (a) 900 Hz, (b) 1.1 kHz, and (c) 1.3 kHz.

Section 30-6, The Transformer

29. A transformer has 400 turns on the primary and 8 turns on the secondary. (a) Is this a step-up or step-down transformer? (b) If the primary is connected across 120 V rms, what is the open-circuit voltage across the secondary? (c) If the primary current is 0.1 A, what is the secondary current, assuming negligible magnetizing current and no power loss?

30. The primary of a step-down transformer has 250 turns and is connected to a 120-V rms line. The secondary is to supply 20 A at 9 V. Find (a) the current in the primary and (b) the number of turns in the secondary, assuming 100 percent efficiency.

31. A transformer has 500 turns in its primary, which is connected to 120 V rms. Its secondary coil is tapped at three places to give an output of 2.5, 7.5, and 9 V. How many turns are needed for each part of the secondary coil?

Problems

1. A coil with resistance and inductance is connected to a 120-V rms 60-Hz line. The average power supplied to the coil is 60 W, and the rms current is 1.5 A. Find (a) the power factor, (b) the resistance of the coil, and (c) the inductance of the coil. (d) Does the current lead or lag the voltage? What is the phase angle ϕ?

2. Compute by direct integration the area under the curve $\sin^2 \omega t$ from $t = 0$ to $t = T = 2\pi/\omega$ and show that it equals $\frac{1}{2}T$.

3. In a certain LCR circuit $X_C = 16\ \Omega$ and $X_L = 4\ \Omega$ at some frequency. The resonance frequency is $\omega_0 = 10^4$ rad/s. (a) Find L and C. If $R = 5\ \Omega$ and $\mathcal{E}_{max} = 26$ V, find (b) the Q value and (c) the maximum current.

4. Show that the formula $P_{av} = \mathcal{E}_{rms}^2 R/Z^2$ gives the correct result for a circuit containing only a generator and (a) a resistor, (b) a capacitor, (c) an inductor.

5. In a series LCR circuit connected to an ac generator whose maximum emf is 200 V, the resistance is 60 Ω and the capacitance is 8.0 μF. The inductor can be varied from 8.0 to 40.0 mH by insertion of an iron core into a solenoid. The angular frequency is 2500 rad/s. If the capacitor voltage is not to exceed 150 V, find (a) the maximum current and (b) the range of L that is safe to use.

6. Sketch the impedance Z versus ω for (a) a series LR circuit, (b) a series RC circuit, and (c) a series LCR circuit.

7. When an LCR series circuit is connected to a 120-V rms 60-Hz line, the current is $I_{rms} = 11.0$ A and the current leads the emf by 45°. (a) Find the power supplied to the circuit. (b) What is the resistance? (c) If the inductance $L = 0.05$ H, find the capacitance C. (d) What capacitance or inductance would you add to make the power factor 1?

8. (a) Show that Equation 30-26 can be written $\tan \phi = Q(\omega^2 - \omega_0^2)/\omega\omega_0$. (b) Show that near resonance, $\tan \phi \approx 2Q(\omega - \omega_0)/\omega$. (c) Sketch ϕ versus x where $x = \omega/\omega_0$, for a circuit with high Q and for one with low Q.

9. Show by direct substitution that the current given by Equation 30-25 with I_{max} and ϕ given by Equations 30-26 and 30-27 satisfies Equation 30-24. *Hint:* Use trigonometric identities for the sine and cosine of the sum of two angles and write the equation in the form $A \sin \omega t + B \cos \omega t = 0$. Since this equation must hold for all time, $A = 0$ and $B = 0$.

10. A resistor and inductor are *in parallel* across an emf $\mathcal{E} = \mathcal{E}_{max} \sin \omega t$ as in Figure 30-13. Show that (a) $I_R = (\mathcal{E}_{max}/R) \sin \omega t$, (b) $I_L = (\mathcal{E}_{max}/X_L) \sin (\omega t - 90°)$, and (c) $I = I_R + I_L = I_{max} \sin (\omega t - \phi)$, where $\tan \phi = R/X_L$ and $I_{max} = \mathcal{E}_{max}/Z$ with $Z^{-2} = R^{-2} + X_L^{-2}$.

Figure 30-13
Parallel RL circuit and ac generator for Problem 10.

11. A certain electrical device draws 10 A rms and has an average power of 720 W when connected to a 120-V rms 60-Hz power line. (a) What is the impedance of the device? (b) What series combination of resistance and reactance is this device equivalent to? (c) If the current leads the emf, is the reactance inductive or capacitive?

12. An ac generator is in series with a capacitor and an inductor in a circuit with negligible resistance. (a) Show that the charge on the capacitor obeys the equation

$$L\frac{d^2Q}{dt^2} + \frac{Q}{C} = \mathcal{E}_{max} \sin \omega t$$

(b) Show by direct substitution that this equation is satisfied by $Q = Q_{max} \sin \omega t$ if

$$Q_{max} = -\frac{\mathcal{E}_{max}}{L(\omega^2 - \omega_0^2)}$$

(c) Show that the current can be written $I = I_{max} \sin(\omega t - \phi)$, where

$$I_{max} = \frac{\mathcal{E}_{max}}{|\omega^2 - \omega_0^2|} = \frac{\mathcal{E}_{max}}{|X_L - X_C|}$$

and $\phi = -90°$ for $\omega < \omega_0$ and $\phi = +90°$ for $\omega > \omega_0$.

13. A method for measuring inductance is to connect the inductor in series with a known capacitance, a resistor, an ac ammeter, and a variable-frequency signal generator. The frequency of the signal generator is varied and the emf kept constant until the current is maximum. If $C = 10~\mu F$, $\mathcal{E}_{max} = 10$ V, $R = 100~\Omega$, and I is maximum at $\omega = 5000$ rad/s, what is L? What is I_{max}?

14. A resistor, inductor, and capacitor are *in parallel* across an ac generator $\mathcal{E} = \mathcal{E}_{max} \sin \omega t$, as shown in Figure 30-14. Show that the total current in the generator is given by $I = I_{max} \sin(\omega t - \phi)$, where $I_{max} = \mathcal{E}_{max}/Z$ with $Z^{-2} = R^{-2} + (1/X_L - 1/X_C)^2$ and $\tan \phi = R(1/X_L - 1/X_C)$.

Figure 30-14
Parallel LCR circuit and ac generator for Problem 14.

15. Find the power factor for the parallel circuit of Problem 14 and show that $\phi = 0$ when $X_C = X_L$.

16. A resistor and capacitor are connected *in parallel* across a sinusoidal emf $\mathcal{E} = \mathcal{E}_{max} \sin \omega t$, as in Figure 30-15. (a) Show that the current in the resistor is $I_R = (\mathcal{E}_{max}/R) \sin \omega t$. (b) Show that the current through the capacitor branch is $I_C = (\mathcal{E}_{max}/X_C) \sin(\omega t + 90°)$. (c) Show that the total current is given by $I = I_R + I_C = I_{max} \sin(\omega t + \phi)$, where $\tan \phi = R/X_C$ and $I_{max} = \mathcal{E}_{max}/Z$ with $Z^{-2} = R^{-2} + X_C^{-2}$.

Figure 30-15
Parallel RC circuit and ac generator for Problem 16.

CHAPTER 31   Maxwell's Equations and
Electromagnetic Waves

Objectives   After studying this chapter, you should:

1. Be able to write down Maxwell's equations and discuss the experimental basis of each.

2. Be able to state the expression for the speed of an electromagnetic wave in terms of the fundamental constants μ_0 and ϵ_0.

3. Be able to state the expression for the Poynting vector and discuss its significance.

4. Be able to state the relationships between the Poynting vector, the intensity of an electromagnetic wave, and radiation pressure.

5. Be able to calculate the radiation pressure and the maximum values of E and B from the intensity of an electromagnetic wave.

About 1860 the great English physicist James Clerk Maxwell found that the experimental laws of electricity and magnetism, which we have studied in Chapters 20 to 30, could be conveniently summarized in a set of equations now known as *Maxwell's equations*. They relate the electric and magnetic field vectors **E** and **B** to their sources, which are electric charges, currents, and changing fields. Maxwell's equations express in a concise mathematical form the laws of Coulomb, Gauss, Biot-Savart, and Ampère (with the important addition of Maxwell's displacement current). They play a role in classical electromagnetism analogous to that of Newton's laws in classical mechanics. In principle, all problems in electricity and magnetism can be solved from Maxwell's equations, just as all problems in classical mechanics can be solved from Newton's laws. Maxwell's equations are considerably more complicated than Newton's laws, however, and their application to the solution of most problems involves mathematics beyond the scope of this book. Nevertheless, Maxwell's equations are of great theoretical importance. Maxwell showed that these equations could be combined to yield a wave equation for the electric and magnetic field vectors **E** and **B**. Such *elec-*

tromagnetic waves are caused by oscillating charges and currents. They were first produced in the laboratory by Heinrich Hertz in 1887. Maxwell showed that the speed of electromagnetic waves is predicted to be

$$c = \frac{1}{\sqrt{\mu_0 \epsilon_0}}$$ 31-1

where ϵ_0 is the constant appearing in Coulomb's and Gauss' laws and μ_0 is that in the Biot-Savart law and Ampère's law. When the measured values of these fundamental constants are put into Equation 31-1, the speed of electromagnetic waves is found to be about 3×10^8 m/s, the same as the measured speed of light. Maxwell noted this "coincidence" with great excitement and correctly surmised that light itself is an electromagnetic wave.

   In this chapter we state Maxwell's equations and relate each one to the laws of electricity and magnetism already studied. We then show that Maxwell's equations imply that the electric and magnetic field vectors obey a wave equation which describes waves propagating through free space with speed $c = 1/\sqrt{\mu_0 \epsilon_0}$.

31-1   Maxwell's Equations

Maxwell's equations are

$$\oint_S \mathbf{E} \cdot \hat{\mathbf{n}}\, dA = \frac{1}{\epsilon_0} Q$$ 31-2

$$\oint_S \mathbf{B} \cdot \hat{\mathbf{n}}\, dA = 0$$ 31-3

$$\oint_C \mathbf{E} \cdot d\boldsymbol{\ell} = -\frac{d}{dt} \int_S \mathbf{B} \cdot \hat{\mathbf{n}}\, dA$$ 31-4

$$\oint_C \mathbf{B} \cdot d\boldsymbol{\ell} = \mu_0 I + \mu_0 \epsilon_0 \frac{d}{dt} \int_S \mathbf{E} \cdot \hat{\mathbf{n}}\, dA$$ 31-5

Equation 31-2 is Gauss' law; it states that the flux of the electric field through any closed surface equals $1/\epsilon_0$ times the net charge inside the surface. As discussed in Chapter 21, Gauss' law implies that the electric field due to a point charge varies inversely as the square of the distance from the charge. This law describes how the lines of $\mathbf{E}$ diverge from a positive charge and converge on a negative charge. Its experimental basis is Coulomb's law.

   Equation 31-3, sometimes called Gauss' law for magnetism, states that the flux of the magnetic induction vector $\mathbf{B}$ is zero through any closed surface. This equation describes the experimental observation that the lines of $\mathbf{B}$ do not diverge from any point in space or converge to any point; i.e., isolated magnetic poles do not exist.

   Equation 31-4 is Faraday's law. The integral of the electric field around any closed curve C is the emf. This emf equals the (negative) rate of change of the magnetic flux through any surface S bounded by the curve. (This is not a closed surface, and so the magnetic flux through S is not necessarily zero.) Faraday's law describes how the lines of $\mathbf{E}$ encircle an area through which the magnetic flux is changing. Faraday's law relates the electric field vector $\mathbf{E}$ to the magnetic field vector $\mathbf{B}$.

Equation 31-5, Ampère's law with Maxwell's displacement-current modification, states that the line integral of the magnetic induction **B** around any closed curve C equals μ_0 times the current through any surface bounded by the curve plus $\mu_0\epsilon_0$ times the rate of change of the electric flux through such a surface. This law describes how the magnetic-induction lines encircle an area through which a current is passing or the electric flux is changing.

The charges and currents in these equations include the bound charges which occur on dielectrics and the magnetization or atomic currents which occur in magnetized materials. To be useful when such materials are present, Equations 31-5 and 31-2 are rewritten in terms of free charge and conduction currents (plus displacement currents) only. This is done by substituting **H** for **B** in Equation 31-5 and omitting the μ_0. Similarly, Equation 31-2 can be written in terms of free charge only by defining a new vector **D**, called the *displacement vector*, which is related to **E** and the electric polarization of dielectrics analogously to the way **B**, **H**, and the magnetic polarization **M** are related. Since we shall not consider the complications introduced by polarized materials, we need not be concerned with such refinements.

31-2 The Wave Equation for Electromagnetic Waves

In Section 14-8 we showed that the harmonic wave functions for waves on a string and for sound waves obey a partial differential equation called the *wave equation:*

$$\frac{\partial^2 y(x,\ t)}{\partial x^2} = \frac{1}{v^2}\frac{\partial^2 y(x,\ t)}{\partial t^2} \qquad \text{31-6}$$

In this equation, $y(x,\ t)$ is the wave function, which for string waves is the displacement of the string and for sound waves can be either the pressure change or the displacement of air particles from equilibrium. The derivatives are partial derivatives because the wave function depends on both x and t. The quantity v is the velocity of the wave, which depends on the medium and on the frequency if the medium is dispersive. We also showed in Section 14-8 that the wave equation for string waves can be derived by applying Newton's laws of motion to a string under tension, and we found that the velocity of the wave is $\sqrt{T/\mu}$, where T is the tension and μ the linear mass density.

The solutions of this equation, studied in Chapter 14, were harmonic wave functions of the form

$$y = y_0 \sin (kx - \omega t)$$

where $k = 2\pi/\lambda$ is the wave number and $\omega = 2\pi f$ is the angular frequency.

In this section we shall use Maxwell's equations to derive the wave equation for electromagnetic waves. We shall not consider how such waves arise from the motion of charges but merely show that the laws of electricity and magnetism imply a wave equation, which in turn implies the existence of electric and magnetic fields **E** and **B** propagating through space with the velocity of light c. We shall consider only free space, in which there are no charges or currents. We shall also assume that the electric and magnetic fields **E** and **B** are functions of time and one space coordinate only, which we shall take to be the x coordinate. This assumption is equivalent to the assumption of plane waves.

We can use Equations 31-2 and 31-3 to show that the electric and magnetic fields associated with electromagnetic waves must be transverse to the direction of propagation. For the situation we are considering, i.e., free space with no charge density and no current, Equations 31-2 and 31-3 state that the net flux of either **E** or **B** through any closed surface is zero. Let us consider the flux of **E** through a cube of sides Δx, Δy, and Δz, as shown in Figure 31-1. We are assuming that **E** depends only on x, the direction of propagation. Consider first the cube's top and bottom faces; their area is $\Delta x \, \Delta z$. Flux through these faces depends on E_y. Since, by our assumption, E_y does not depend on y, the flux of **E** into the cube through the bottom face must equal the flux out through the top face. Similar reasoning leads to the result that there can be no contribution to the net flux through the faces of area $\Delta x \, \Delta y$ because E_z does not depend on z. The only possible contribution is through the faces of area $\Delta y \, \Delta z$. The flux out of the right face is $E_{xR} \, \Delta y \, \Delta z$, where E_{xR} is the value of E_x at the right face. The flux in through the left face is $E_{xL} \, \Delta y \, \Delta z$, where E_{xL} is E_x at the left face. Since the net flux out of the cube must be zero, E_{xR} must equal E_{xL}; that is, E_x must be the same at the left and right sides of the cube. Thus E_x does not depend on x. If the electric field has any component at all in the x direction, it must be constant in space. Such a constant field is unrelated to electromagnetic waves and can be ignored for this discussion. Similar reasoning applied to the flux of **B** through such a cube leads to the conclusion that B_x must also be constant in space. Thus the part of the electric or magnetic fields which varies in time and space must be perpendicular to the direction of propagation.

To obtain the wave equation relating the time and space derivatives of either the electric field **E** or the magnetic field **B**, we first relate the time derivative of one of the field vectors to the space derivative of the other. We do this by applying the other two Maxwell equations (31-4 and 31-5) to appropriately chosen curves in space. Let us assume for simplicity that the electric field is in the y direction. We can relate the spatial dependence of E_y to the time dependence of B_z by applying Equation 31-4 (which is Faraday's law) to the rectangular curve of sides Δx and Δy lying in the xy plane, as shown in Figure 31-2. Assuming that Δx and Δy are very small, the line integral of **E** around this curve is approximately

$$\oint \mathbf{E} \cdot d\boldsymbol{\ell} = E_y(x_2) \, \Delta y - E_y(x_1) \, \Delta y = [E_y(x_2) - E_y(x_1)] \, \Delta y$$

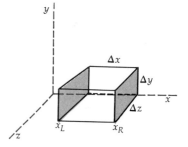

Figure 31-1
Volume element with sides parallel to the coordinate planes. If **E** depends only on x and t, there is no contribution to the net flux of **E** from the top and bottom or from the front and back sides. If the net flux of **E** out of the volume is zero, E_x must have the same value at x_R as at x_L.

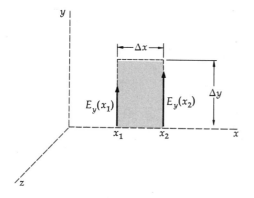

Figure 31-2
Rectangular curve in the xy plane for derivation of Equation 31-7.

The contributions of the type $E_x \Delta x$ from the top and bottom of this curve cancel because we have assumed that **E** does not depend on y (or z). Assuming that Δx is very small, we can approximate the difference in E_y at the points x_1 and x_2 by

$$E_y(x_2) - E_y(x_1) = \Delta E_y \approx \frac{\partial E_y}{\partial x} \Delta x$$

Then

$$\oint \mathbf{E} \cdot d\boldsymbol{\ell} \approx \frac{\partial E_y}{\partial x} \Delta x\, \Delta y$$

The flux of the magnetic induction through this curve is approximately

$$\int \mathbf{B} \cdot \hat{\mathbf{n}}\, dA = B_z \Delta x\, \Delta y$$

Faraday's law then gives

$$\frac{\partial E_y}{\partial x} \Delta x\, \Delta y = -\frac{\partial B_z}{\partial t} \Delta x\, \Delta y$$

or

$$\frac{\partial E_y}{\partial x} = -\frac{\partial B_z}{\partial t} \qquad\qquad 31\text{-}7$$

Equation 31-7 implies that if there is a component of electric field E_y that depends on x, there must be a component of magnetic induction B_z that depends on time; or, conversely, if there exists a magnetic induction field B_z that depends on time, there must be an electric field E_y that depends on x. We can get a similar equation relating E_y to B_z by applying Equation 31-5 to the curve of sides Δx and Δz in the xz plane, as shown in Figure 31-3. For the case with no currents Equation 31-5 is

$$\oint_C \mathbf{B} \cdot d\boldsymbol{\ell} = \mu_0 \epsilon_0 \frac{d}{dt} \int_S \mathbf{E} \cdot \hat{\mathbf{n}}\, dA$$

We omit the details of this calculation, which is similar to that already done; the result is

$$\frac{\partial B_z}{\partial x} = -\mu_0 \epsilon_0 \frac{\partial E_y}{\partial t} \qquad\qquad 31\text{-}8$$

which relates the space variation of the magnetic induction B_z and the time variation of the electric field E_y.

We can eliminate either B_z or E_y from Equations 31-7 and 31-8 by differentiating either equation with respect to x or t. If we differentiate both sides of Equation 31-7 with respect to x, we obtain

$$\frac{\partial}{\partial x}\left(\frac{\partial E_y}{\partial x}\right) = -\frac{\partial}{\partial x}\left(\frac{\partial B_z}{\partial t}\right) \qquad \text{or} \qquad \frac{\partial^2 E_y}{\partial x^2} = -\frac{\partial}{\partial t}\left(\frac{\partial B_z}{\partial x}\right)$$

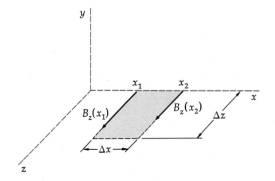

Figure 31-3
Rectangular curve in the xz plane for the derivation of Equation 31-8.

where we have interchanged the order of the time and space derivatives on the right side. We now use Equation 31-8 for $\partial B_z/\partial x$:

$$\frac{\partial^2 E_y}{\partial x^2} = -\frac{\partial}{\partial t}\left(-\mu_0\epsilon_0 \frac{\partial E_y}{\partial t}\right)$$

or

$$\frac{\partial^2 E_y}{\partial x^2} = \mu_0\epsilon_0 \frac{\partial^2 E_y}{\partial t^2} \qquad\qquad 31\text{-}9 \qquad \textit{Wave equation for } E_y$$

Comparing this equation with Equation 31-6, we see that E_y obeys a wave equation for waves with speed

$$v = \frac{1}{\sqrt{\mu_0\epsilon_0}}$$

which is Equation 31-1. If we had instead chosen to eliminate E_y from Equations 31-7 and 31-8 (by differentiating Equation 31-7 with respect to t, for example), we would have obtained an equation identical to Equation 31-9 with B_z replacing E_y. We have thus shown that both the electric field E_y and the magnetic field B_z obey a wave equation for waves traveling with the velocity $1/\sqrt{\mu_0\epsilon_0}$, which is the velocity of light. As we noted in discussing harmonic waves, a particularly important solution to Equation 31-9 is the harmonic wave function of the form

$$E_y = E_{y0} \sin (kx - \omega t) \qquad\qquad 31\text{-}10$$

If we substitute this solution into either Equation 31-7 or 31-8, we can show that the magnetic induction B_z is in phase with the electric field E_y. From Equation 31-7 we have

$$\frac{\partial E_y}{\partial x} = kE_{y0} \cos (kx - \omega t) = -\frac{\partial B_z}{\partial t}$$

Solving for B_z gives

$$B_z = \frac{k}{\omega} E_{y0} \sin (kx - \omega t) = B_{z0} \sin (kx - \omega t) \qquad 31\text{-}11 \qquad \textit{E and B are in phase}$$

where

$$B_{z0} = \frac{k}{\omega} E_{y0} = \frac{E_{y0}}{c} \qquad\qquad 31\text{-}12$$

and $c = \omega/k$ is the velocity of the wave.*

We have shown that Maxwell's equations imply the wave equation 31-9 for the electric field component E_y and the magnetic induction component B_z, and that if E_y varies harmonically, as in Equation 31-10, the magnetic induction B_z is in phase with E_y and has an amplitude related to the amplitude of E_y by Equation 31-12. The electric and magnetic fields are perpendicular to each other and to the direction of the wave propagation, the x direction here.

From Figure 31-4 we note that the direction of propagation (the x direction in this case) is parallel to the vector $\mathbf{E} \times \mathbf{B}$. The vector $(\mathbf{E} \times \mathbf{B})/\mu_0 = \mathbf{E} \times \mathbf{H}$ is known as the *Poynting vector* (after John H. Poynting):

$$\mathbf{S} = \frac{\mathbf{E} \times \mathbf{B}}{\mu_0} = \mathbf{E} \times \mathbf{H} \qquad\qquad 31\text{-}13 \qquad \textit{Poynting vector}$$

* In obtaining Equation 31-11 by integration from the previous equation an arbitrary constant of integration arises, but we have omitted this constant magnetic induction field from Equation 31-11 because it plays no part in the electromagnetic waves we are interested in. Note that if any constant electric field is added to Equation 31-10, the new function still satisfies the wave equation.

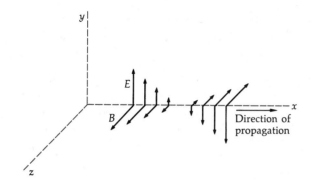

Figure 31-4
Electric and magnetic field
vectors in a plane electromagnetic wave at one instant in
time. **E** and **B** are in phase.
The direction of propagation
of the wave is the direction of
E × **B**, which in this case is
the x direction.

We shall now show that its average magnitude equals the intensity of
the electromagnetic wave. In our discussion of the transport of energy
by waves of any kind, we showed that the intensity (average energy per
unit time per unit area) is in general equal to the product of the average
energy density (energy per unit volume) and the speed of the wave. The
energy per unit volume associated with an electric field **E** is

$$\eta_e = \tfrac{1}{2}\epsilon_0 E^2$$

For the harmonic wave given by Equation 31-10 we have

$$\eta_e = \tfrac{1}{2}\epsilon_0 E_0^2 \sin^2 (kx - \omega t)$$

For simplicity in notation, we shall drop the y- and z-component subscripts. The energy per unit volume associated with a magnetic induction field **B** is

$$\eta_m = \frac{B^2}{2\mu_0}$$

From Equations 31-11 and 31-12 we have

$$\eta_m = \frac{B_0^2}{2\mu_0} \sin^2 (kx - \omega t) = \frac{E_0^2}{2\mu_0 c^2} \sin^2 (kx - \omega t)$$

But since $c^2 = 1/\epsilon_0\mu_0$,

$$\eta_m = \tfrac{1}{2}\epsilon_0 E_0^2 \sin^2 (kx - \omega t)$$

and we see that the magnetic and electric energy densities are equal.
The total electromagnetic energy per unit volume is

$$\eta = \eta_e + \eta_m = \epsilon_0 E_0^2 \sin^2 (kx - \omega t)$$

$$= \frac{E_0^2}{\mu_0 c^2} \sin^2 (kx - \omega t) \qquad\qquad 31\text{-}14$$

The energy per unit time per unit area is the product of this energy density and the speed c:

$$\eta c = \frac{E_0^2}{\mu_0 c} \sin^2 (kx - \omega t)$$

$$= [E_0 \sin (kx - \omega t)] \left[\frac{E_0}{\mu_0 c} \sin (kx - \omega t) \right]$$

$$= E \frac{B}{\mu_0} = EH = |\mathbf{S}|$$

The intensity at any point x is the time average of the energy per unit
time per unit area. From Equation 31-14 we see that the energy density
is proportional to $\sin^2 (kx - \omega t)$. The average of this quantity over one

or more cycles is $\frac{1}{2}$. The average energy density is then

$$\eta_{\text{av}} = \frac{E_0^2}{2\mu_0 c^2}$$

and the intensity is

$$I = c\eta_{\text{av}} = \frac{E_0^2}{2\mu_0 c} = \frac{E_0}{\sqrt{2}} \frac{B_0}{\mu_0\sqrt{2}} = E_{\text{rms}} H_{\text{rms}} = |\mathbf{S}|_{\text{av}} \qquad 31\text{-}15$$

where $E_{\text{rms}} = E_0/\sqrt{2}$ is the rms value of E and $H_{\text{rms}} = B_{\text{rms}}/\mu_0 = B_0/(\mu_0\sqrt{2})$ is the rms value of H.

The magnitude of the Poynting vector is the instantaneous power per unit area. The rate at which electromagnetic energy flows through any area is the flux of the Poynting vector through that area:

$$P = \int \mathbf{S} \cdot \hat{\mathbf{n}} \, dA = \int (\mathbf{E} \times \mathbf{H}) \cdot \hat{\mathbf{n}} \, dA \qquad 31\text{-}16$$

where P is the power through the area and $\mathbf{S} = \mathbf{E} \times \mathbf{H}$ is the Poynting vector.

In Chapter 14 we pointed out that traveling waves carry momentum as well as energy. The momentum carried by an electromagnetic wave equals the energy carried divided by the speed c. Since the intensity of a wave is the energy per unit time per unit area, the intensity divided by c is the momentum carried by the wave per unit time per unit area. The SI unit of momentum per unit time is the unit of force, the newton. The intensity divided by c is therefore a force per unit area, or pressure, with SI units of $N/m^2 = Pa$. This pressure is called radiation pressure p_r:

Radiation pressure

$$p_r = \frac{I}{c} \qquad 31\text{-}17$$

We can relate the radiation pressure to the electric or magnetic fields using Equations 31-15 and 31-12, giving

$$p_r = \frac{I}{c} = \frac{E_0^2}{2\mu_0 c^2} = \frac{B_0^2}{2\mu_0} = \frac{1}{c} |\mathbf{S}|_{\text{av}} \qquad 31\text{-}18$$

Example 31-1 A 100-W light bulb emits electromagnetic waves uniformly in all directions. Find the intensity, the radiation pressure, and the electric and magnetic fields at a distance of 3 m from the bulb, assuming that 50 W goes into electromagnetic radiation.

At a distance r from the bulb, the energy is spread uniformly over an area $4\pi r^2$. The intensity is therefore

$$I = \frac{50 \text{ W}}{4\pi r^2} = 0.442 \text{ W/m}^2 \qquad \text{at } r = 3 \text{ m}$$

The radiation pressure is the intensity divided by the speed of light: $p_r = I/c = (0.442 \text{ W/m}^2)/(3 \times 10^8 \text{ m/s}) = 1.47 \times 10^{-9}$ Pa. This is a very small pressure compared with atmospheric pressure of the order of 10^5 Pa. The maximum value of the magnetic field is, from Equation 31-18,

$$B_0 = (2\mu_0 p_r)^{1/2} = [2(4\pi \times 10^{-7})(1.47 \times 10^{-9})]^{1/2} = 6.08 \times 10^{-8} \text{ T}$$

The maximum value of the electric field is the speed of light times B_0:

$$E_0 = cB_0 = 18.2 \text{ V/m}$$

The electric and magnetic fields are of the form $E = E_0 \sin (kx - \omega t)$ and $B = B_0 \sin (kx - \omega t)$ with $E_0 = 18.2$ V/m and $B_0 = 6.08 \times 10^{-8}$ T.

James Clerk Maxwell (1831–1879)

C. W. F. Everitt
Stanford University

Maxwell aged about 32.
(*Courtesy of the Master and Fellows of Peterhouse, Cambridge, England.*)

One day in 1877 a young Scottish undergraduate named Donald MacAlister, afterwards a distinguished physician and academic statesman, wrote home from Cambridge University that he had just had dinner with a professor who was "one of the best of our men, and a thorough old Scotch laird in ways and speech." This description of James Clerk Maxwell was accurate. He was the proprietor of an estate of 2000 acres in the southwest of Scotland and a man with all the qualities of the better kind of Victorian country gentleman: cultivated, considerate of his tenants, taking his due part in local affairs, an expert swimmer and horseman. He was a descendant through one line (the Maxwells) of a family long famous in Scottish Border history, and through another (the Clerks) of one of the leading families in Edinburgh society. His uncle, Sir George Clerk, had been a prominent Conservative politician. Few would suspect that the "Scotch laird" who seemed to MacAlister charmingly old-fashioned in 1877 was also a scientist whose writings remain astonishingly up to date in the 1980s; that he was the greatest mathematical physicist since Newton; that he had created the electromagnetic theory of light and predicted the existence of radio waves; that he had written the first significant paper on control theory; that he was just then writing a profound article on statistical mechanics, a science he and Ludwig Boltzmann jointly invented; that he had performed with his wife's aid a brilliant series of experiments on color vision and had taken the first color photograph; and that in the remaining 2 years until his death in 1879, at the age of 48, he would lay the foundations of another new subject that was to reach fruition in the twentieth century, rarefied gas dynamics.

Maxwell owed his interest in science to his father, John Clerk Maxwell, who before marrying in 1826 and moving to his estate, had lived in Edinburgh and regularly attended meetings of the Royal Society of Edinburgh. His great interest was in mechanical contrivances; he published one scientific paper, a proposal for an automatic-feed printing press, in 1831, the year Maxwell was born. The neat ingenuity and carefully thought-out details of Maxwell's own scientific apparatus bear the stamp of his father's mind. Maxwell's first 10 years were spent in the country, remote from any school, and his education was in the hands of his mother, Frances Cay, a woman of intelligence and great force of character. From her he imbibed the love of English literature that left its mark in the graceful literary style of his writings. She died when he was 8 years old, and after a disastrous 2-year experiment with a tutor, he was sent to Edinburgh to attend Edinburgh Academy, then and now one of the leading schools in Britain. There, after a slow start, where his back-country accent and odd dress earned him the nickname "Dafty," his intellect began to blossom, and through family connections he was gradually introduced to the magnificent intellectual and scientific culture then flourishing in Edinburgh.

One friend of the Clerks and Cays was D. R. Hay, a decorative artist interested in principles of design. He happened in 1845 to be seeking, not too successfully, a geometrical method of drawing oval curves, similar to the string property of the ellipse. Maxwell at age 14 discovered that where the string for the ellipse is folded back on itself n times toward one focus and m times towards the other, like the block and tackle of a crane, a true oval is formed, of the kind first studied in the 1680s by René Descartes in connection with the refraction of light. Figures 1a and 1b reproduce four ovals from Maxwell's paper, three with $m = 1$ and $n = 2$, the other $m = 2$ and $n = 3$. (Figure 1c illustrates one of the interesting optical properties of these ovals.) Simple as the idea was, it had not been thought of before. Maxwell's father showed it to James David Forbes, Professor of Natural Philosophy at Edinburgh University, and between them the two men secured its publication in the *Proceedings of the Royal Society of Edinburgh*. Thereafter Maxwell often went to meetings of the Society and soon came

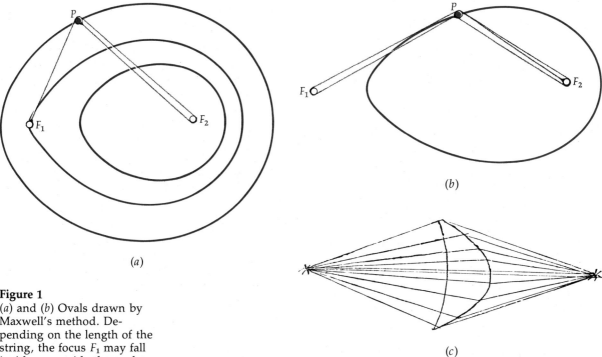

(a)

(b)

(c)

Figure 1

(a) and (b) Ovals drawn by Maxwell's method. Depending on the length of the string, the focus F_1 may fall inside or outside the oval. The ovals in (a) are generated by attaching the string at the focus F_1, passing it around the pencil P and the other focus F_2, and fixing it at P. The oval in (b) is generated by attaching the string at F_2, passing it around P and F_1, passing it back around P and F_2, and finally fixing it at P. (*Redrawn from* The Scientific Papers of James Clerk Maxwell, *Vol. I, p. 1.*) (c) Optical properties of a Cartesian oval. This figure reproduced from an early manuscript of Maxwell's shows a lens with a convex oval surface to the right and a concave spherical surface to the left, the center of the sphere being on the inner focus of the oval. A lens of this kind has perfect imaging properties for the two foci: a point source of light at either focus forms a perfect point image at the other.

to know many of the leading scientists in Edinburgh. One encounter, which he afterwards described as a turning point in his life, was when his uncle John Cay took him and a friend to see the private laboratory of William Nicol, inventor of the Nicol polarizing prism. Nicol later presented him with a pair of prisms, with which over the next few years he made many experiments on polarized light, discovering among other things the *Maxwell spot,* a curious visual phenomenon arising from a previously unknown polarizing structure in the foveal region of the human eye. The prisms are now on exhibit in the Cavendish Laboratory Museum at Cambridge.

Maxwell's undergraduate career was unusually protracted. He spent 3 years at Edinburgh University and another $3\frac{1}{4}$ at Cambridge. Unlike Einstein, he enjoyed student life and was fortunate to gain the attention of some outstanding teachers. At Edinburgh he was influenced by two powerful and sharply contrasted men, Forbes and Sir William Hamilton, the metaphysician. Forbes was an experimentalist, who had invented the seismometer and done important work on the polarization of infrared radiation, besides achieving fame as one of the earliest British Alpinists. He gave Maxwell the run of his laboratory and with his help began the experiments on color vision that led eventually to Maxwell's own work on the subject. Hamilton, who had a genius for inspiring youth, imparted to Maxwell the ranging philosophic vision that can be seen in the many interesting metaphysical asides in his papers.

In 1850 Maxwell went up to Cambridge. By then his mathematical bent was clear and the Cambridge Mathematical Tripos offered the finest training in applied mathematics along with the most grueling examination system the wit of man has devised. Like many another clever undergraduate before and since, Maxwell worked hard while pretending not to; but he just missed the coveted position of Senior Wrangler, having to be content with second place. His private tutor was William Hopkins, the founder of modern geophysics and arguably the greatest teacher Cambridge has ever produced. Others who influenced him were G. G. Stokes, the mathematical physicist who held the chair Newton had occupied, and William Whewell, the mathematician-philosopher of whom it was remarked that "science was his forte and omniscience his foible." Maxwell's student friendships were more with classical scholars than scientists. He too affected omniscience. After a discussion on New Year's Eve 1854 one friend wrote of

> Maxwell as usual showing himself acquainted with every subject upon which the conversation turned. I never met a man like him. I do believe there is not a single subject on which he cannot talk, and talk well too, displaying always the most curious and out of the way information.

Maxwell's electromagnetic theory of light is rooted in the work of two men, Michael Faraday and William Thomson. Faraday's invention of the electric motor and his researches on electromagnetic induction, electrochemistry, dielectric and diamagnetic action, and magneto-optical rotation made him in Maxwell's words "the nucleus of everything electric since 1830." His contributions to theory lay in his progressively advancing ideas about lines of electric and magnetic force, in particular the geometrical relations governing electromagnetic phenomena and the idea that magnetic forces might be accounted for not by direct attractions and repulsions between elements of current but by attributing to lines of force the property of shortening themselves and repelling each other sideways (Figure 2). Thomson's role was to relate lines of force to existing theories in electrostatics and magnetostatics, to invent a number of highly ingenious analytical techniques for solving electrical problems, and to emphasize the cardinal importance of energy principles in electromagnetism. Maxwell then introduced a series of new concepts: the *electrotonic function* (vector potential), the energy density of the field, the displacement current, and the significance of the operation *curl* in the field equations; he organized the subject into a coherent structure and made in 1861 the momentous discovery of the equivalence between light and electromagnetic waves.

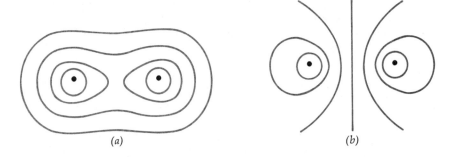

Figure 2
Faraday's explanation of forces between current-carrying wires. The two diagrams show the lines of force observed when currents are flowing in parallel wires. Faraday assumed that the lines of force tend to shorten and repel each other sideways. (*a*) For wires with currents flowing in the same direction, the lines of force pull the two wires together. (*b*) For wires with currents flowing in opposite directions, the lines of force push the wires apart.

(a)   (b)

The first part of Maxwell's paper "On Faraday's Lines of Force" (1855–1856) developed an analogy, due in essence to Thomson, between lines of electric and magnetic force and stream lines in a moving incompressible fluid. Maxwell applied this to interpret many of Faraday's observations, prefacing his paper with a luminous discussion of the significance of analogies in physics, a subject he enlarged on soon afterwards in a philosophical essay read in April 1856 to the Apostles, a famous Cambridge quasi-secret intellectual society of which he was a member. Next, still building on Faraday and Thomson, Maxwell extended the discussion to electromagnetism. He formulated a group of equations summarizing the relations of the electric and magnetic fields to the charges and currents producing them—the beginnings of what we now call Maxwell's equations. They described the phenomena with elegant precision from a point of view completely different from the then-popular action-at-a-distance theories of André-Marie Ampère and Wilhelm Weber. A nice feature was that the central theorem in all this work was one which, following Maxwell, we now call *Stokes' theorem*. It had been published in January 1854 by Stokes *as an examination question* in the Smith's Prize Examination at Cambridge, which Maxwell sat for immediately after completing his bachelor's degree.

After such a brilliant start one might have expected a rush of papers following up the new ideas. But other physicists ignored them, and Maxwell had the habit of investigating different subjects in turn, often with long intervals between successive papers in the same field. Six years elapsed before the appearance of his next paper, "On Physical Lines of Force," published in four parts in 1861–1862. During the interval Maxwell made brilliant contributions to three distinct subjects before returning to electromagnetism: color vision, the theory of Saturn's rings, and the kinetic theory of gases. He left Cambridge, became a professor at Marischal College in Aberdeen, married the daughter of the Principal of the College, and then found himself in the odd position of being forced to retire at age 29 with a life pension after his chair had been abolished when the two universities in Aberdeen were united by Act of Parliament. Fortunately the chair at King's College in London, had just fallen vacant, and so he went there.

"On Physical Lines of Force" contained Maxwell's extraordinary molecular-vortex model of the electromagnetic field. In order to account for the pattern of stresses associated with lines of force by Faraday, Maxwell investigated the properties of a medium occupying all space in which tiny molecular vortices rotate with their axes parallel to the lines of force. The closer together the lines are, the faster the rotation of the vortices. In a medium of this kind the lines of force do tend to shorten themselves and repel each other sideways, yielding the right forces between currents and magnets; the question is: what makes the vortices rotate? Here Maxwell put forward an idea as ingenious as it was weird. He postulated that an electric current consists in the motion of tiny particles that mesh like gear wheels with the vortices, and that the medium is filled with similar particles between the vortices. Figure 3 gives the picture. Maxwell remarks:

> I do not bring [this hypothesis] forward as a mode of connexion existing in nature . . . [but] I venture to say that anyone who understands [its] provisional and temporary character . . . will find himself helped rather than hindered by it in his search for the true interpretation of [electromagnetic] phenomena.

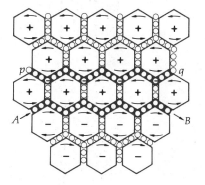

Figure 3
Maxwell's vortex model of the magnetic field. The rotating vortices represent lines of magnetic force. They mesh with small particles which act like gear wheels. In free space the particles are restrained from moving, except for a small elastic reaction (the displacement current), but in a conducting wire they are free to move. Their motion constitutes an electric current, which in turn sets the vortices in rotation, creating the magnetic field around the wire. A and B represent current through a wire, and p and q represent an induced current in an adjacent wire. (*Redrawn from* The Scientific Papers of James Clerk Maxwell, *Vol. 1, fig. 2 after p. 488.*)

The question then was how to fit electrostatic phenomena into the model. Maxwell made the medium an elastic one. Thus magnetic forces were accounted for by rotations in the medium, and electric forces by its elastic distortion. Any elastic medium will transmit waves. In Maxwell's medium the velocity of the waves turned out to be related to the ratio of electric to magnetic forces. Putting in numbers from an experiment of 1856 by G. Kohlrausch and W. Weber, Maxwell found to his astonishment that the propagation velocity was equal to the velocity of light. With excitement manifested in italics he wrote "we can scarcely avoid the inference that *light consists in the transverse undulation of the same medium which is the cause of electric and magnetic phenomena.*"

Having made the great discovery, Maxwell promptly jettisoned his model. Instead of attempting a more refined mechanical explanation of the phenomena, he formulated a system of electromagnetic equations from which he deduced

Balance arm of apparatus with which Maxwell and Charles Hockin compared the ratio of electric to magnetic forces to determine the characteristic velocity, showing that it was equal to the velocity of light (1868).

that waves of electric and magnetic force would propagate through space with the velocity of light. That is why his is called an *electromagnetic* theory of light, in contrast to the theories of the mechanical ether that preceded it. The theory appeared in two papers of 1865 and 1868, and in its most general form in the great *Treatise on Electricity and Magnetism*, published in 1873, a work of such scope that Robert Andrews Millikan, author of the famous oil-drop experiment to measure the charge on the electron, ranked it with Newton's *Principia* in considering them the two most influential books in the history of physics, "the one creating our modern mechanical world and the other our modern electrical world." Maxwell wrote it during a period of retirement from academic life after he had resigned his chair at King's College in 1865, finishing the work off somewhat hastily after his appointment in 1871 as Cavendish Professor of Experimental Physics at Cambridge.

Equally profound were Maxwell's contributions to statistical and molecular physics. They began with a paper in 1859 on the kinetic theory of gases, in which Maxwell introduced the velocity distribution function and enunciated the *equipartition theorem*, which in its original form stated that the average translational and rotational energies of large numbers of colliding molecules, whether of the same or different species, are equal. One surprising conclusion, afterwards confirmed experimentally by Maxwell and his wife, was that the viscosity of a gas should over a wide range be independent of its pressure. Another result was Maxwell's estimate of the mean free path of a gas molecule, which Loschmidt in 1865 applied to make the first serious estimates of the diameters of molecules. Later, Maxwell developed the general theory of transport phenomena, from which the Boltzmann equation is derived; invented the concept of ensemble averaging; created rarefied-gas dynamics; and conceived that "very small BUT lively being" the Maxwell demon. Brilliant as the successes of classical statistical mechanics were, its failures, as Maxwell saw, were in some ways even more striking. The equipartition theorem gave an answer for the ratio of the specific heats of gases that disagreed with experiment, while some of Boltzmann's theorems "proved too much" because they would apply to the properties of solids and liquids as well as gases. These questions remained shrouded in mystery until the emergence in 1900 of Planck's quantum hypothesis. Writing about them in 1877, Maxwell confessed his bewilderment and stated that nothing remained but to adopt the attitude of "thoroughly conscious ignorance that is the prelude to every real advance in knowledge."

Maxwell was an unusually sensitive man, with strong religious feeling and a fascinating and astonishing sense of humor.* Many of his letters reveal a delightfully sly irony. He also had some talent for writing poetry, usually light, but occasionally touching a deeper note. The last stanza of one poem to his wife, written in 1867, was

> All powers of mind, all force of will
> May lie in dust when we are dead,
> But love is ours, and shall be still
> When earth and seas are fled.

Apparatus with which Maxwell and his wife measured the viscosity of gases as a function of pressure and temperature (1863–1865).

* For more details of Maxwell's life and work see L. Campbell and W. Garnett, *The Life of James Clerk Maxwell*, Johnson Reprint Corp., Harcourt, Brace & Jovanovich, New York, 1970 (reprint of Oxford 1882 edition); and C. W. F. Everitt, *James Clerk Maxwell: Physicist and Natural Philosopher*, Scribner, New York, 1975.

Review

A. Define, explain, or otherwise identify the following:

Maxwell's equations, 833
Wave equation, 834

Poynting vector, 837
Radiation pressure, 839

B. True or false:

1. Maxwell's equations apply only to fields that are constant in time.

2. The wave equation can be derived from Maxwell's equations.

3. Electromagnetic waves are transverse.

4. In an electromagnetic wave, the electric and magnetic fields are in phase.

5. In an electromagnetic wave, the electric and magnetic field vectors **E** and **B** are equal in magnitude.

6. In an electromagnetic wave, the electric and magnetic energy densities are equal.

Exercises

Section 31-1, Maxwell's Equations

1. Write the Maxwell equation which corresponds most closely to the law stated: (a) Faraday's law; (b) Ampère's law; (c) Gauss' law; (d) Gauss' law for magnetism.

Section 31-2, The Wave Equation for Electromagnetic Waves

2. Show by direct substitution that the wave function $E_y = E_0 \sin (kx - \omega t) = E_0 \sin k(x - ct)$, where $c = \omega/k$ satisfies Equation 31-9.

3. An electromagnetic wave has intensity 100 W/m². Find (a) the radiation pressure p_r, (b) E_{rms}, and (c) B_{rms}.

4. Use the known values of μ_0 and ϵ_0 in SI units to compute $c = 1/\sqrt{\mu_0\epsilon_0}$ and show that its value is approximately 3×10^8 m/s.

5. The amplitude of an electromagnetic wave is $E_0 = 400$ V/m. Find (a) E_{rms}, (b) B_{rms}, (c) the intensity I, and (d) the radiation pressure p_r.

6. (a) Show that if E is in volts per metre and B in teslas, the units of the Poynting vector $(\mathbf{E} \times \mathbf{B})/\mu_0$ are watts per square metre. (b) Show that if the intensity I is in watts per square metre, the units of radiation pressure $p_r = I/c$ are newtons per square metre.

7. (a) An electromagnetic wave of intensity 200 W/m² is incident normally on a black card of sides 20 by 30 cm which absorbs all the radiation. Find the force exerted on the card by the radiation. (b) Find the force exerted by the same wave if the card reflects all the radiation incident on it.

8. Find the force exerted by the electromagnetic wave on the reflecting card in part (b) of Exercise 7 if the radiation is incident at an angle of 30° to the normal.

Problems

1. A laser beam has a diameter of 1.0 mm and average power of 1.5 mW. Find the intensity of the beam, E_{rms}, B_{rms}, and the radiation pressure.

2. The intensity of sunlight striking the upper atmosphere is 1.4 kW/m². (a) Find E_{rms} and B_{rms} due to the sun at the upper atmosphere of the earth. (b) Find the average power output of the sun. (c) Find the intensity and the radiation pressure at the surface of the sun.

3. (*a*) Find the intensity and radiation pressure 6 m from a 500-W light bulb, assuming it to be a point source. (*b*) Find E_{rms} and B_{rms} at this distance from the bulb. (Assume half the power goes into electromagnetic radiation.)

4. A 10- by 15-cm card has a mass of 2 g and is perfectly reflecting. The card hangs in a vertical plane and is free to rotate about a horizontal axis through one edge. The card is illuminated uniformly by an intense light so that the card makes an angle of 1° with the vertical. Find the intensity of the light.

5. A long cylindrical conductor of radius *a* and resistivity ρ carries a steady current *I* uniformly distributed over its cross-sectional area. (*a*) Use Ohm's law to relate the electric field **E** in the conductor to *I*, ρ, and *a*. (*b*) Find the magnetic field **B** just outside the conductor. (*c*) At $r = a$ (the edge of the conductor) use the results of parts (*a*) and (*b*) to compute the Poynting vector $\mathbf{S} = \mathbf{E} \times \mathbf{H} = (1/\mu_0)\mathbf{E} \times \mathbf{B}$. In what direction is **S**? (*d*) Find the flux of **S** through the surface of the wire of length *L* and area $2\pi aL$ and show that the rate of energy flow into the wire equals I^2R, where *R* is the resistance.

6. (*a*) Using arguments similar to those given in the text, show that

$$\frac{\partial E_z}{\partial x} = \frac{\partial B_y}{\partial t} \quad \text{and} \quad \frac{\partial B_y}{\partial x} = \mu_0\epsilon_0\frac{\partial E_z}{\partial t}$$

(*b*) Show that E_z and B_y also satisfy the wave equation.

7. Discuss how Maxwell's equations might have to be modified in a region of space containing magnetic monopoles. (A magnetic monopole is a particle carrying a single "magnetic charge" qm, that is, an isolated magnetic pole. Its existence was predicted by P. A. M. Dirac in 1931. The discovery of one such particle was reported in 1975, but the evidence is not conclusive.)

CHAPTER 32   Light

Electromagnetic radiation with wavelengths in the range of about 400 to 700 nm, to which the eye is sensitive, is called light. This phenomenon has intrigued mankind for centuries. Despite the many advances in recent times, much important research in both the theory of light and its applications is still being carried out today. To gain historical perspective, we first look briefly at some early theories and experiments which help us understand light as a wave motion. We then discuss the basic phenomena of reflection, refraction, and polarization. They can all be adequately understood by using rays to describe the straight-line propagation of light and neglecting interference and diffraction effects. As discussed in Chapter 15, this ray approximation is valid for the propagation of any wave motion if the wavelength is small compared with any apertures or obstacles—often the situation in optics because of the very small wavelengths of light. In the next chapter we shall apply our knowledge of reflection and refraction to the study of image formation by mirrors and lenses, a subject known as ray optics or geometrical optics. The fascinating study of the interference and diffraction of light will be taken up in Chapter 34.

32-1 Waves or Particles?

The controversy over the nature of light is one of the most interesting in the history of science. Early theories considered light to be a stream of particles which emanated from a source and caused the sensation of vision upon entering the eye. The most influential proponent of this particle theory of light was Newton. Using it, he was able to explain the laws of reflection and refraction.

When light strikes a boundary between two transparent media, e.g., air and water or air and glass, part of the light energy is reflected and part transmitted. The reflected ray makes an angle with the normal to the surface equal to that of the incident ray, whereas the transmitted ray is bent toward the normal (Figure 32-1). (If the light is leaving the water or glass and entering air, the transmitted ray is bent away from the normal.) This bending of the ray is called *refraction*.

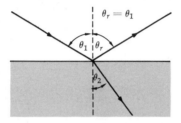

Figure 32-1
When a light ray strikes the boundary between two different media, e.g., air and glass, part of the energy is reflected and part is transmitted. The angle of reflection equals the angle of incidence. The angle of refraction θ_2 is less than the angle of incidence for air-to-glass refraction.

The law of reflection of light from a plane boundary is easily explained by the particle theory. Figure 32-2 shows a particle bouncing off a hard plane surface. If there is no friction, the component of particle momentum parallel to the surface is not changed by the collision but the component perpendicular to the wall is reversed (assuming that the mass of the wall is much greater than that of the particle and that the collision is elastic). Thus the angle of reflection equals the angle of incidence, as is observed for light rays reflecting from a plane mirror.

Figure 32-2
Reflection of a particle. If the surface is frictionless, the angle of reflection θ_r equals the angle of incidence θ_1.

Figure 32-3 illustrates Newton's explanation of refraction at an air-glass or air-water surface. Newton assumed that the light particles are strongly attracted to the glass or water so that when they approach the surface, they receive a momentary impulse which increases the component of momentum perpendicular to the surface. Thus the direction of the momentum of the light particle changes, the light beam being bent toward the normal to the surface, in agreement with observation. An important feature of this theory is that the velocity of light must be greater in water or glass than in air to account for the bending of the beam toward the normal when it enters the water or glass and away from the normal when it enters air from either of those media. Experimental determination of the velocity of light in water did not come until about 1850, nearly 200 years later.

Figure 32-3
Refraction of a particle from air into water or glass. According to Newton's particle theory of light, the light particle is attracted by the water or glass, and its component of momentum normal to the surface is increased, bending the path toward the normal to the surface.

The chief proponents of the wave theory of light propagation were Christian Huygens and Robert Hooke. Huygens, using his wavelet construction (Section 15-3), was able to explain reflection and refraction, assuming that light travels more slowly in a glass or water medium than in air. Newton saw the virtues of the wave theory of light, particularly as it explained the colors formed by thin films, which Newton studied extensively. However, he rejected the wave theory because of the observed straight-line propagation of light:

> To me the fundamental supposition itself seems impossible, namely, that the waves or vibrations of any fluid can, like the rays of light, be propagated in straight lines, without a continual and very extravagant spreading and bending every way into the quiescent medium, where they are terminated by it. I mistake if there be not both experiment and demonstration to the contrary.*

Because of Newton's great reputation and authority, this reluctant rejection of the wave theory of light, based on lack of evidence of diffraction, was strictly adhered to by Newton's followers. Even after evidence of diffraction was available, they sought to explain it as scattering of light particles from the edges of slits. Newton's particle theory of light was accepted for more than a century.

In 1801 Thomas Young revived the wave theory of light. He was one of the first to introduce the idea of interference as a wave phenomenon in both light and sound. Figure 32-4 shows his drawing of the combination of waves from two sources. His observation of interference with light was a clear demonstration of the wave nature of light. Young's work went unnoticed by the scientific community for more than a decade, however. Perhaps the greatest advance in the general acceptance of the wave theory of light was due to the French physicist Augustin Fresnel (1788–1827), who performed extensive experiments on interference and diffraction and put the wave theory on a mathematical basis.

* Newton, *Opticks*, 1704. Reprint, Dover, New York, 1952.

The Science Museum, London

Portrait of Thomas Young, British physicist, physician, and egyptologist (1773–1829).

Figure 32-4
Thomas Young's drawing of a two-source interference pattern. (*Courtesy of the Niels Bohr Library, American Institute of Physics.*)

He showed, for example, that the observed rectilinear propagation of light is a result of the very short wavelength of visible light. An interesting triumph of the wave theory was an experiment suggested to Fresnel by S. Poisson, who sought to discredit the wave theory. Poisson noted that if an opaque disk is illuminated by light from a source on its axis, the Fresnel wave theory predicts that light waves bending around the edge of the disk should meet and interfere constructively on the axis, producing a bright spot in the center of the shadow of the disk. Poisson considered this to be a ridiculous contradiction of fact, but Fresnel's immediate demonstration that such a spot does in fact exist convinced many doubters that the wave theory of light is valid (see Figure 32-5).

In 1850 Jean Foucault measured the speed of light in water and showed that it is less than that in air, thus ruling out Newton's particle theory.

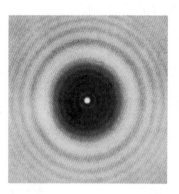

Figure 32-5
Diffraction of light by an opaque disk, showing the bright spot in the center of the shadow. (*From M. Cagnet, M. Françon, and J. C. Thrierr,* Atlas of Optical Phenomena, *plate 33, Springer-Verlag, Berlin, 1962.*)

In 1860 James Clerk Maxwell published his mathematical theory of electromagnetism, which predicted the existence of electromagnetic waves propagating with the speed $c = 1/\sqrt{\epsilon_0\mu_0} \approx 3 \times 10^8$ m/s, the speed of light, and correctly suggested that this agreement is not accidental but indicates that light is an electromagnetic wave. Maxwell's theory was confirmed in 1886 by Hertz, who produced and detected waves in the laboratory by strictly electrical means. Hertz used a spark gap in a tuned circuit to generate the waves and another similar circuit to detect them. In the latter half of the nineteenth century, Kirchhoff and others applied Maxwell's equations to explain the interference and diffraction of light and other electromagnetic waves and put Huygens' empirical methods of construction on a firm mathematical basis.

This discussion of the wave-particle controversy of light would not be complete if we did not mention the discovery in the twentieth century that although the wave theory is generally correct in describing light and other electromagnetic waves, it fails to account for all their observed properties. One of the ironies of the history of science is that in his famous experiment of 1887, which confirmed Maxwell's wave theory, Hertz also discovered the *photoelectric effect*, which can be explained only by a particle model of light, as Einstein showed only a few years later. Hertz noticed that the passage of sparks in the receiving gap of his apparatus was facilitated by light from the generating gap. Investigating this effect in 1900, Lenard found that light falling on a metal surface ejects electrons and that the energies of these electrons do *not* depend on the intensity of the light. This result was quite surprising since the intensity is the energy per second per unit area falling on the metal surface. In 1905 Einstein demonstrated that this result can be explained by the assumption that the energy of a light wave is *quantized* into small bundles, called *photons*. The energy of a photon is propor-

The Bettmann Archive

Nineteenth-century engraving of Augustin Jean Fresnel (1788–1827).

Photons

tional to the frequency of the wave. According to Einstein, the energy of a photon is

$$E = hf \qquad\qquad 32\text{-}1$$

where $h = 6.63 \times 10^{-34}$ J·s is Planck's constant, and f is the frequency of the light wave.

An electron ejected from a metal surface exposed to light receives its energy from a single photon. When the intensity of light of a given frequency is increased, more photons fall on the surface, ejecting more electrons, but the energy of each electron is not increased. Using this model, Einstein predicted that the maximum energy of an electron ejected in the photoelectric effect and escaping from the metal would increase linearly with the frequency of incident light, and that if the frequency of the incident light were below a certain threshold frequency f_t (which depends on the kind of metal), the photoelectric effect would not occur, no matter what the intensity, because no single photon would have enough energy to eject an electron. These predictions were accurately confirmed about 10 years later in difficult experiments performed by the American physicist R. A. Millikan. Thus a particle model of light was reintroduced.

Complete understanding of this dual nature of light did not come until the 1920s, when experiments by C. J. Davisson and L. Germer and by G. P. Thompson showed that electrons (and other "particles") also have a dual nature and exhibit the wave properties of interference and diffraction besides their well-known particle properties. The behavior of fundamental quantities such as light, electrons, and other subatomic particles is correctly described by the theory of quantum mechanics worked out by E. Schrödinger, W. Heisenberg, P. A. M. Dirac, and others. Although this theory differs from both a classical wave theory and a classical particle theory, in some circumstances the quantum theory resembles a classical wave theory and in others it resembles a classical particle theory. For example, the propagation of these fundamental quantities can always be described as a wave propagation exhibiting the usual wave effects of interference and diffraction. On the other hand, the exchange of energy between these fundamental quantities, as in the photoelectric effect, is usually best described in terms of particle mechanics. As discussed in Chapter 15, when the wavelength of any wave is very small compared with the various obstacles and apertures it encounters, the ray approximation is valid and the propagation is in straight lines. The wave nature of electrons and other so-called "particles" is not easily observed because of their extremely short wavelengths—the same difficulty which prevented Newton from observing the wave character of light. On the other hand, the particle nature of the energy exchanges involving light is often not noticed because the number of photons is so enormous and the energy of each photon is so small. This difficulty is not unlike that of observing the particle nature of gas molecules which exert pressure on the walls of a container.

Questions

1. The spreading of a beam of light after passing through a very small opening was observed before Newton's time, but it was argued that this was to be expected if the beam consisted of particles moving in not quite perfectly parallel paths. Explain.

2. Supporters of the particle theory of light were able to explain the slight diffraction of light observed, but they could not hope to explain

Young's observation of interference on the basis of the particle theory. Why?

3. According to Newton's theory of light, reflection and refraction are the result of attractive forces between the particles of light and the atoms of the refracting material. Explain why this attraction affects the path of a light beam only as it enters or leaves the refracting material and not in the center of the material.

4. Newton concluded that the force between particles of light and atoms of matter must be an attractive force. Can you see why? How would absorption of light by matter be explained in Newton's theory?

32-2   Electromagnetic Waves

Electromagnetic waves include light, radio, x-rays, gamma rays, microwaves, and others, all involving the propagation of electric and magnetic fields through space with speed $c = 3 \times 10^8$ m/s (in vacuum). All electromagnetic waves are generated when electric charge is accelerated. The differences between the various types of electromagnetic waves are in their frequency and wavelength. Table 32-1 shows the

Table 32-1

Electromagnetic spectrum

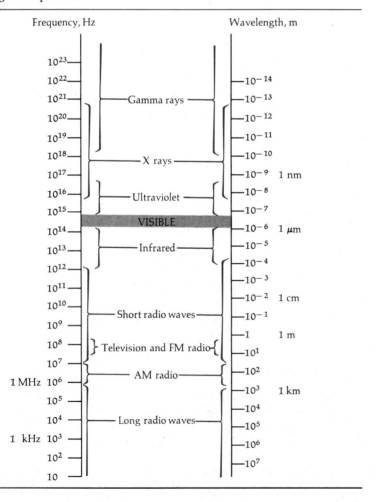

electromagnetic spectrum and the names usually associated with the various frequency and wavelength ranges. The ranges often are not well defined and sometimes overlap. For example, electromagnetic waves with wavelength of about 0.1 nm are called x-rays if their origin is atomic and gamma rays if their origin is nuclear. The human eye is sensitive to electromagnetic radiation of wavelengths from about 400 to 700 nm, the range called *visible light*, but the word light is also used for wavelengths just beyond the visible range. Ultraviolet light is electromagnetic radiation just on the shorter-wavelength side of the visible range, and infrared light is that just on the longer-wavelength side. There are no limits on the wavelengths of electromagnetic radiation: all frequencies and wavelengths are theoretically possible.

A complete mathematical description of electromagnetic waves can be derived from Maxwell's equations (Chapter 31), but such a description is much too difficult for an introductory course. As shown in Chapter 31, the oscillating electric and magnetic fields associated with an electromagnetic wave are mutually perpendicular and perpendicular to the direction of propagation of the wave, as illustrated in Figure 32-6.

Figure 32-6
The electric and magnetic field vectors are in phase, perpendicular to each other, and perpendicular to the direction of propagation of the wave.

These fields are produced by charges which accelerate. In particular, according to classical electromagnetic theory, if a charge oscillates with simple harmonic motion of frequency f, it radiates energy in the form of electromagnetic waves of the same frequency. Figure 32-7 shows a simple and common type of radiating charge system, two equal but opposite charges whose separation varies harmonically with time. The radiation from such an oscillating electric dipole is called *electric-dipole radiation*. Some characteristics of electric-dipole radiation are important for our understanding of light:

Electric-dipole radiation

1. The magnitude of the electric field at large distances varies inversely as the distance and is proportional to $\sin \theta$, where θ is the angle between the dipole axis and the point of observation as shown in Figure 32-7. Since the energy density and therefore the intensity is proportional to E^2, the intensity of electric-dipole radiation varies as $1/r^2$ and is proportional to $\sin^2 \theta$.

2. The electric field in an electromagnetic wave from an electric dipole is in the plane formed by the dipole axis and the direction of propagation. This is the plane of the paper in Figure 32-7.

Figure 32-7
Electric dipole. The intensity of radiation is proportional to $\sin^2 \theta$. It is zero along the direction of the dipole and maximum perpendicular to this direction.

Many electromagnetic waves exhibit the characteristics of electric-dipole radiation. For example, radio waves are produced by currents oscillating in a straight antenna, an arrangement equivalent to an electric dipole. Light waves come from charge oscillations in atoms. Although the classical theory of electromagnetic radiation has been modified by the theory of quantum mechanics, many of the features of light and other electromagnetic waves can be understood in terms of the classical theory.

All electromagnetic waves are similar except for their wavelength and frequency, but this difference is very important. For example, it is not possible to generate light waves using ordinary circuits to produce the acceleration of the charges because the eye is sensitive only to electromagnetic waves with frequency of the order of 10^{14} Hz, much greater than can be attained with ordinary circuits. As we know, the behavior of waves closely depends on the relative sizes of the wavelength and of the physical objects and apertures the waves encounter. Since the wavelength of light is in the rather narrow range of about 400 to 700 nm, it is much smaller than most obstacles and apertures and the ray approximation often is valid.

Questions

5. What sort of radiation is an electromagnetic wave whose frequency is 10^5 Hz? 10^{10} Hz? 10^{15} Hz? 10^{20} Hz?

6. A dipole antenna on top of a high broadcasting tower has a vertical axis. Describe qualitatively the variation in intensity you would observe if you walked on the ground along a straight path which passes directly under the antenna, starting a great distance from the tower and continuing to a great distance on the other side.

7. If you want to broadcast radio waves from a dipole antenna, would you place the axis of the antenna horizontally or vertically? Why?

32-3   The Speed of Light

The first effort to measure the speed of propagation of light was made by Galileo. He and a partner situated themselves on hilltops about a kilometre apart, each with a lantern and a shutter to cover it. Galileo proposed to measure the time for light to traverse twice the distance between the experimenters. A would uncover his lantern, and when B saw the light, B would uncover his. The time between A's uncovering his lantern and seeing the light from B's would be the time for light to travel back and forth between the experimenters. Though this method is sound in principle, the speed of light is so great that the time interval to be measured is much smaller than fluctuations in the human response time, and Galileo was unable to obtain any value for the speed of light.

The first indication of the true magnitude of the speed of light came from astronomical observations of the period of one of the moons of Jupiter. This period is determined by measuring the time between eclipses (when the moon disappears behind Jupiter). The eclipse period is about 42.5 h, but measurements made when the earth is moving away from Jupiter, as from point A to C in Figure 32-8, give a greater time for this period than measurements made when the earth is moving toward Jupiter, as from point C to A in the figure. Since these measurements differ by only about 15 s from the average value, the discrepancies were difficult to measure accurately. In 1675 the astronomer Ole Roemer attributed these discrepancies to the fact that the velocity of light is not infinite. During the 42.5 h between eclipses of Jupiter's moon, the distance between the earth and Jupiter changed, making the path for the light longer or shorter. Roemer devised the following method for measuring the cumulative effect of these discrepancies. We neglect the motion of Jupiter (it is much slower than that of the earth). When the

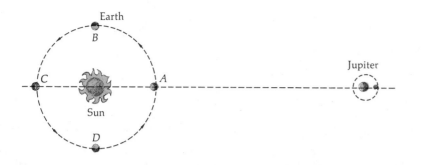

Figure 32-8
Roemer's method of mea-
suring the speed of light. The
time between eclipses of Jupiter's moon appears greater
when the earth is moving
from A to C than when it is
moving from C to A. The difference is due to the time it
takes light to travel the distance traveled by the earth
along the line of sight
during one period of the
moon.

earth is at the nearest point A, the period of the moon is measured. The time when an eclipse should occur $\frac{1}{2}$ y later, when the earth is at point C, is computed. The observed eclipse time is about 16 min later than predicted. This is the time it takes light to travel a distance equal to the diameter of the earth's orbit.

The first nonastronomical measurement of the velocity of light was made by the French physicist A. H. L. Fizeau in 1849. His method (Figure 32-9) was greatly improved by Foucault, who replaced the toothed wheel by a rotating mirror. In about 1850, Foucault measured the speed of light in air and in water, showing that it is less in water. Precise measurements of the speed of light using essentially the same method were performed by the American physicist A. A. Michelson from 1880 to 1930.

Another method does not involve light directly but is based on Maxwell's electromagnetic theory, which predicts the speed of light to be $c = 1/\sqrt{\epsilon_0\mu_0}$. The electrical constant ϵ_0 can be determined by accurately measuring the capacitance of a parallel-plate capacitor, and the magnetic constant μ_0 can be determined by measuring the force between two current-carrying wires. (This is essentially the calibration experiment to determine the SI unit of current, the ampere.) Accurate electrodynamic measurements of the speed of light were first made by Rosa and Dorsey of the U.S. Bureau of Standards in 1906.

Figure 32-9
Fizeau's method of measuring
the speed of light. Light from
the source is reflected by
mirror B and transmitted
through a gap in the toothed
wheel to mirror A. The speed
of light is determined by
measuring what angular
speed of the wheel will
permit the reflected light to
be transmitted by the next
gap in the toothed wheel so
that an image of the source is
observed. (*Redrawn from Bernard Jaffe, Michelson and the
Speed of Light. Copyright* ©
*1960 by Doubleday & Company,
Inc. Used by permission of the
publisher.*)

There is general agreement between the results of these various methods for determining the speed of light and many others which we shall not discuss. The accepted value for the speed of light is now

$$c = 2.997925 \pm 0.000003 \times 10^8 \text{ m/s}$$

The value 3×10^8 m/s is accurate enough for nearly all calculations.

Question

8. Estimate the time required for light to make the round trip in Galileo's experiment to determine c.

32-4　Reflection

When waves of any type strike a plane barrier, new waves are generated which move away from the barrier. Experimentally it is found that the rays corresponding to the incident and reflected waves make equal angles with the normal to the barrier and that the reflected ray lies in the *plane of incidence* formed by the incident ray and the normal. This *law of reflection* can be stated by saying that *the reflected ray lies in the plane of incidence such that the angle of reflection equals the angle of incidence.* The law of reflection holds for all waves whether they are water waves, sound waves, electromagnetic waves, or others. Figure 32-10 shows ultrasonic plane waves in water reflecting from a steel plate. The law of reflection can be derived in several ways. According to Huygens' principle (Section 15-3), each point on a given wavefront can be considered to be a point source of secondary wavelets. Figure 32-11 shows a plane wavefront AA' striking a barrier at point A. The angle θ_i between the ray corresponding to this wavefront and the normal to the barrier is called the *angle of incidence*. As can be seen from the figure, the angle of incidence θ_i equals the angle ϕ_i between the incident wavefront and the barrier. The position of the wavefront after a time Δt is found by constructing wavelets of radius $v\,\Delta t$ with centers on the wavefront AA'. Wavelets which do not strike the barrier form the portion of the new wavefront BB'. Wavelets which do strike the barrier are reflected and form the portion of the new wavefront BB''. By a similar construction,

Law of reflection

Figure 32-10
Ultrasonic plane waves in water reflecting from a steel plate. (*Battelle-Northwest Photography.*)

Figure 32-11
Plane wave reflected at a plane barrier. The angle θ_i between the incident ray and the normal to the barrier is the angle of incidence. It is the same as the angle ϕ_i between the incident wavefront and the barrier.

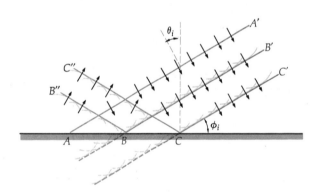

the wavefront $C''CC'$ is obtained from the Huygens' wavelets originating on the wavefront $B''BB'$. Figure 32-12 is an enlargement of a portion of Figure 32-11 showing a part of the original wavefront AP which strikes the barrier in time Δt. In this time the wavelet from point P reaches the barrier at point B, and the wavelet from point A reaches point B''. The reflected wavefront BB'' makes an angle ϕ_r with the barrier which is equal to the angle of reflection θ_r between the reflected ray and the normal to the barrier. The triangles ABP and BAB'' are both right triangles with a common side AB and equal sides $AB'' = BP =$

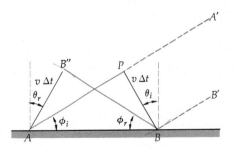

Figure 32-12
Geometry of Huygens' construction for the calculation of the law of reflection. The wavefront AP originally intersects the barrier at point A. After a time Δt, the Huygens' wavelet from P strikes the barrier at point B, and the one from A reaches point B''. Since the triangles ABP and BAB'' are congruent, the angles θ_i and θ_r are equal.

$v\,\Delta t$. Hence these triangles are congruent, and the angles ϕ_i and ϕ_r are equal, implying that the angle of reflection θ_r equals the angle of incidence θ_i, which is the law of reflection. Having derived this law from consideration of the propagation of wavefronts and Huygens' principle, it is now much easier to consider only the rays which are perpendicular to the wavefronts.

A second derivation of the law of reflection uses an interesting principle enunciated in the seventeenth century by the French mathematician Pierre de Fermat, stated for our purposes as follows:

The path taken by light in traveling from one point to another is such that the time of travel is a minimum when compared with nearby paths.

Fermat's principle

In Figure 32-13 we assume that light leaves point A, strikes the plane surface, which we can consider to be a mirror, and travels to point B. We wish to find the path taken by light. The problem of reflection stated for the application of Fermat's principle is: At what point P in Figure 32-13 must light strike the mirror so that it will travel from point A to point B in the least time? Since the light always travels in the same medium for this problem, the time will be minimum when the distance is minimum. In Figure 32-13 the distance APB is the same as the distance $A'PB$, where A' is the image point of the source A. Point A' lies along the perpendicular from A to the mirror and is equidistant behind the mirror. Obviously, as we vary point P, the distance $A'PB$ is least when the points A', P, and B lie on a straight line. We can see from the figure that when this occurs, the angle of incidence equals the angle of reflection.

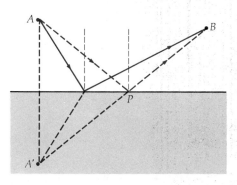

Figure 32-13
Geometry for deriving the law of reflection by Fermat's principle.

Although we have derived the law of reflection in several different ways, we have not discussed the detailed mechanism by which light is reflected by atoms, nor have we developed expressions for the relative intensities of the reflected and transmitted light. A complete discussion of the reflection of light taking into account the fact that light is an electromagnetic wave of very great frequency and applying the laws of electricity and magnetism to such a wave cannot be given in an elementary text. The results of such a treatment are in full accord with experiment, and we shall mention a few.

The fraction of energy reflected at a boundary between two media depends on the angle of incidence, the indexes of refraction of the two media, and the polarization of the incident light wave. The *index of refraction* n of a medium is related to the speed v of waves in that medium and is defined by

$$n = \frac{c}{v}$$

32-2   *Index of refraction*

where c is the speed of light in vacuum. The general expression for the fraction of energy reflected is quite complicated. When light is incident normally, the reflected intensity is

$$I = \left(\frac{n_2 - n_1}{n_2 + n_1}\right)^2 I_0 \qquad\qquad 32\text{-}3$$

where I_0 is the incident intensity and n_1 and n_2 are the indexes of refraction of the two media. For a typical case of reflection from an air-glass surface for which $n_1 = 1$ and $n_2 = 1.5$, Equation 32-3 gives $I = \frac{1}{25}I_0$; only about 4 percent of the energy is reflected, the rest being transmitted.

For light reflected from an air-glass surface, the reflected wave is 180° out of phase with the incident wave when the incident wave is in air, but the two waves are in phase when the incident wave is in glass. We can understand this from our discussion of the reflection of pulse waves on a string. In Chapter 14 we found that when a pulse is incident on a boundary beyond which the velocity of the pulse is slower, the pulse is inverted; but it is not inverted if the velocity of the pulse is greater beyond the boundary. If we consider a harmonic wave to be a series of pulses, an inversion of the pulse is equivalent to a 180° phase difference between the incident and reflected wave. This phase difference can be observed experimentally (see Chapter 34).

32-5 Refraction

When light or any other wave is incident on a boundary surface separating two media, some of the energy is reflected and some transmitted. The angle between the transmitted ray and the normal to the surface θ_2, called the angle of refraction, is related to the angle of incidence θ_1 by

$$n_1 \sin \theta_1 = n_2 \sin \theta_2 \qquad\qquad 32\text{-}4 \qquad \textit{Snell's law}$$

where n_1 and n_2 are the indexes of refraction of the first and second media, respectively, as defined by Equation 32-2. This result was discovered experimentally in 1621 by W. Snell and is known as Snell's law or the law of refraction. Like the law of reflection, this result can be derived either from Huygens' or Fermat's principle. In Figure 32-14 we apply Huygens' construction to find the wavefront of the transmitted wave, in much the same way as we did for reflection. AP is a portion of a wavefront in medium 1 at angle of incidence θ_1 which intersects the line dividing the media at point A.

In time Δt the wavelet from P travels the distance $v_1 \Delta t$ and reaches the point B on the line AB separating the two media, while the wavelet from point A travels the distance $v_2 \Delta t$ into the second medium. The new wavefront BB' is not parallel to the original wavefront AP because the speeds v_1 and v_2 are different. From the triangle APB

$$\sin \phi_1 = \frac{v_1 \Delta t}{AB} \qquad \text{or} \qquad AB = \frac{v_1 \Delta t}{\sin \phi_1} = \frac{v_1 \Delta t}{\sin \theta_1}$$

using the fact that the angle ϕ_1 equals the angle of incidence θ_1. Similarly, from triangle $AB'B$,

$$\sin \phi_2 = \frac{v_2 \Delta t}{AB} \qquad \text{or} \qquad AB = \frac{v_2 \Delta t}{\sin \phi_2} = \frac{v_2 \Delta t}{\sin \theta_2}$$

Figure 32-14
Application of Huygens' principle to the refraction of plane waves at the surface separating a medium in which the wave speed is v_1 from a medium in which the wave speed is v_2. For the case shown, in which v_2 is less than v_1, the rays are bent toward the normal to the surface.

where $\theta_2 = \phi_2$ is the angle between the refracted ray and the normal to the line separating the media. θ_2 is the angle of refraction. Equating the two values of AB gives

$$\frac{\sin\theta_1}{v_1} = \frac{\sin\theta_2}{v_2} \qquad\qquad 32\text{-}5$$

If we write the wave speeds v_1 and v_2 in terms of the indexes of refraction $n = c/v$, Equation 32-5 becomes Equation 32-4, Snell's law. Note that if v_2 is less than v_1, n_2 is greater than n_1 and the angle of refraction θ_2 is less than the angle of incidence θ_1. The ray is then bent toward the normal. For example, when light passes from air to glass, the ray is bent toward the normal; when it passes from glass to air, the ray is bent away from the normal.

The frequency of the transmitted (and reflected) wave is the same as that of the incident wave. This can be understood in terms of Huygens' principle where we think of the incident wave at the barrier as the source of the transmitted wave. Since the speed of the transmitted wave differs from that of the incident wave, their wavelengths also differ. This can be seen in Figure 32-15, which shows plane waves in a ripple tank incident on a line barrier where the wave speed changes abruptly because the water depth differs on each side of the barrier.

Figure 32-15
Refraction of plane waves in a ripple tank at a boundary at which the wave speed changes because the depth of water changes. Note that reflection also occurs at the boundary. (*Courtesy of Film Studio, Education Development Center, Newton, Mass.*)

Derivation of the law of refraction from Fermat's principle is more complicated than that of the law of reflection. Figure 32-16 shows possible paths for light traveling from point A in air to point B in glass. Point P_1 is on the straight line between A and B, but this path is not the one for the shortest travel time because light travels with a smaller velocity in the glass. If we move slightly to the right of P_1, the total path length is greater but the distance traveled in the slower medium is less than for the path through P_1. It is not apparent from the figure which path is that of least time, but it is not surprising that a path slightly to the right of the straight-line path takes less time because the time gained by traveling a shorter distance in the glass more than makes up for the time lost traveling a longer distance in air. As we move the point of intersection of the possible path to the right of point P_1, the total time for the travel from A to B decreases until we reach a minimum point $P_{\min}$. Beyond this point, the time saved by traveling a shorter distance in the glass does not compensate for the greater time required for the greater distance traveled in air. The details of the derivation of Equation 32-4 from Fermat's principle are given at the end of this section.

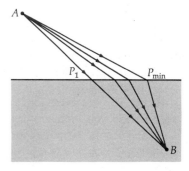

Figure 32-16
Geometry for deriving Snell's law of refraction from Fermat's principle. The point $P_{\min}$ is the point at which light must strike the glass for travel time from A to B to be a minimum.

Solving Equation 32-4 for sin θ_2 gives

$$\sin \theta_2 = \frac{n_1}{n_2} \sin \theta_1 \qquad\qquad 32\text{-}6$$

If n_2 is greater than n_1 (as in the case of refraction of light entering a glass or water medium from air), this equation can be solved for the angle of refraction θ_2 for any angle of incidence θ_1. However, if n_2 is less than n_1 (as in the case of refraction of light entering air from glass or water), Equation 32-6 gives values for sin θ_2 greater than 1 if the angle of incidence is greater than some critical angle θ_c defined by

$$\sin \theta_c = \frac{n_2}{n_1} \qquad\qquad 32\text{-}7$$

Since the sine of any angle must be less than 1, Equation 32-6 cannot be solved for the angle of refraction for angles of incidence greater than θ_c. This situation is illustrated in Figure 32-17 for light incident on a glass-to-air surface, where $n_1 = 1.5$ and $n_2 = 1$. The critical angle is then given by sin $\theta_c = \frac{2}{3}$, $\theta_c = 41.8°$. When the angle of incidence is greater than this critical angle, there is no refracted wave in the second medium. All the energy is reflected. This phenomenon is called *total internal reflection*. An interesting application of total internal reflection is the transmission of a beam of light down a long narrow transparent glass fiber. If the beam begins approximately parallel to the axis of the fiber, it will always strike the walls of the fiber at angles greater than the critical angle, and no light energy will be lost through the walls of the fiber.

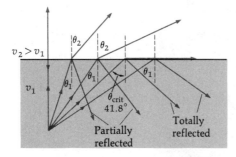

Figure 32-17
Total internal reflection. At the critical andlgeθ_{crit}, given by sin $\theta_{crit} = v_1/v_2$, the angle of refraction is 90°, and the refracted ray emerges parallel to the boundary surface. At angles of incidence greater than the critical angle, there is no refracted ray. If the first medium is glass and the second air, v_1/v_2 is approximately $\frac{2}{3}$ and the critical angle is 41.8°.

A complete discussion of refraction of light based on the laws of electricity and magnetism is beyond the scope of this text. The intensity of the transmitted light can always be found by conservation of energy if the intensity of the reflected light is known. The sum of the energy transmitted and reflected must equal the energy incident.

Understanding the mechanism of refraction by atoms is also more difficult than for reflection. We can consider the transmitted wave as the resultant of the interference of the incident wave and a wave radiated from the atoms, which absorb and reradiate some of the light. For light entering glass from air, there is a phase lag between the reradiated wave and the incident wave. Thus the resultant wave also lags in phase behind the incident wave. This phase lag means that the position of a wave crest of the transmitted wave is retarded relative to the position of a wave crest of the incident wave in the medium. Therefore in a given time, the transmitted wave does not travel as far in the medium as the original incident wave; i.e., the phase velocity of the transmitted wave is less than that of the incident waves. Thus the index of refraction, defined to be the ratio of the velocity of light in vacuum to that in glass, is greater than 1.

Figure 32-18 shows the index of refraction of a glass as a function of wavelength. The slight decrease in n as λ increases means that the wave speed in a medium increases as λ increases; i.e., the wave speed in a medium depends on the wavelength and frequency of the light. This is the phenomenon of *dispersion,* discussed for sound waves and waves on strings in Section 15-7. When a beam of light is incident at some angle on a glass surface, the angle of refraction of shorter wavelengths

Dispersion

Figure 32-18
Typical dispersion curve for light in glass. The index of refraction decreases slightly as the wavelength increases, indicating that the speed of light in the glass depends on wavelength and frequency.

(toward the blue end of the visible spectrum) is slightly smaller than that of the longer wavelengths (toward the red end of the spectrum). A beam of white light is therefore spread out or dispersed into its component colors or wavelengths (Figure 32-19). The formation of a rainbow is a familiar example of dispersion of sunlight by refraction in water droplets.

Figure 32-19
A beam of white light incident on a glass prism is dispersed into its component colors. The index of refraction decreases as the wavelength increases as shown in Figure 32-18, so the longer wavelengths (red) are bent less than the shorter wavelengths (blue).

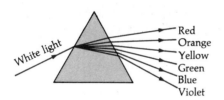

Optional

Derivation of Snell's Law from Fermat's Principle

Figure 32-20 shows the geometry for finding the path of least time. If L_1 is the distance traveled in medium 1 with index of refraction n_1 and L_2 the distance in medium 2 with index of refraction n_2, the time for light to go along this path is

$$t = \frac{L_1}{v_1} + \frac{L_2}{v_2} = \frac{L_1}{c/n_1} + \frac{L_2}{c/n_2} = \frac{n_1 L_1}{c} + \frac{n_2 L_2}{c}$$

32-8

Figure 32-20
Geometry for deriving Snell's law from Fermat's principle.

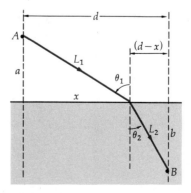

The quantity nL is called the *optical-path length*. According to Fermat's principle, light will travel along the path for which the time is minimum. From Equation 32-8 this is the path for which the optical-path length is a minimum. We can derive Snell's law of refraction by consid-

Newton demonstrating the
dispersion of sunlight
with a glass prism.

ering the time as a function of a single parameter indicating the position of point P_{min}. In terms of the distance x in Figure 32-20 we have

$$L_1^2 = a^2 + x^2 \qquad \text{and} \qquad L_2^2 = b^2 + (d - x)^2 \qquad\qquad 32\text{-}9$$

Figure 32-21 shows the time t as a function of x. At the value of x for
which the time is a minimum, the slope of this graph is zero, or

$$\frac{dt}{dx} = 0$$

Differentiating Equation 32-8 gives

$$\frac{dt}{dx} = \frac{1}{c}\left(n_1 \frac{dL_1}{dx} + n_2 \frac{dL_2}{dx}\right)$$

For the path of minimum time we have

$$n_1 \frac{dL_1}{dx} + n_2 \frac{dL_2}{dx} = 0 \qquad\qquad 32\text{-}10$$

We can compute these derivatives from Equations 32-9. We have

$$2L_1 \frac{dL_1}{dx} = 2x \qquad \text{or} \qquad \frac{dL_1}{dx} = \frac{x}{L_1}$$

Figure 32-21
Sketch of time for light to
travel from point A to B in
Figure 32-16 versus distance x
measured along the refracting
surface. The time is minimum
at the point for which the
angles of incidence and refraction obey Snell's law.

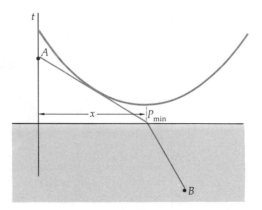

But x/L_1 is just $\sin \theta_1$, where θ_1 is the angle of incidence. Thus

$$\frac{dL_1}{dx} = \sin \theta_1$$

Similarly

$$2L_2 \frac{dL_2}{dx} = 2(d - x)(-1)$$

or

$$\frac{dL_2}{dx} = -\frac{d - x}{L_2} = -\sin \theta_2$$

where θ_2 is the angle of refraction. Hence Equation 32-10 is

$$n_1 \sin \theta_1 + n_2(-\sin \theta_2) = 0$$

or

$$n_1 \sin \theta_1 = n_2 \sin \theta_2$$

which we recognize as Snell's law.

32-6 Polarization

In any transverse wave, the vibration is perpendicular to the direction of propagation and can therefore be resolved into rectangular components in the plane perpendicular to that direction. For example, in an electromagnetic wave moving in the x direction, the electric and magnetic fields lie in a plane parallel to the yz plane and can be resolved into y and z components. If the electric field vector remains parallel to a line in space, the wave is said to be linearly polarized. The magnetic field then also remains parallel to a line in space.* Linear polarization is also called *plane polarization*. Electric-dipole radiation is linearly polarized.

Transverse waves can also be circularly polarized or elliptically polarized. In circularly or elliptically polarized electromagnetic waves, the electric field vector rotates in a circle or an ellipse. For example, for a wave moving in the x direction, the y and z components of the electric field for such a wave are given by

$$E_y = E_{y0} \sin (kx - \omega t)$$

and

$$E_z = E_{z0} \sin \left(kx - \omega t + \frac{\pi}{2} \right) = E_{z0} \cos (kx - \omega t) \qquad \text{32-11}$$

If the amplitudes E_{y0} and E_{z0} are equal, the wave is circularly polarized. Otherwise the polarization is elliptical.

The various kinds of polarization of transverse waves can be visualized most easily by considering mechanical waves on a string. If one end of a long string is moved back and forth along a line, the resulting waves on the string are linearly polarized, each element of the string vibrating along a line. On the other hand, if the end of the string is moved with constant speed in a circle, a circularly polarized wave will be propagated along the string with each element of the string moving in a

---

* In discussing electromagnetic waves it is simpler to consider only one of the fields, such as the electric field, since the direction of the magnetic field is then determined by the fact that $\mathbf{E} \times \mathbf{B}$ is the direction of propagation. The electric field is usually chosen because most detectors, including the human eye, are sensitive to $\mathbf{E}$ rather than to $\mathbf{B}$.

circle. Similarly, if the string is moved in an elliptical path, the resulting wave is elliptically polarized.

Most waves produced by a single source are polarized; e.g., string waves produced by the vibration of one end of a string, or electromagnetic waves produced by a single atom or a single electric dipole antenna are polarized. Waves produced by many sources are usually unpolarized. A typical light wave, for example, is produced by millions of atoms acting independently. The electric field for a light wave propagating in the x direction can be resolved into y and z components at any instant, but these components have a phase difference which varies randomly in time because there is no correlation between the electric field vectors produced by the different atoms. Such a light wave is unpolarized.

Four phenomena produce polarized light from unpolarized light:

1. Absorption

Means of polarization

2. Reflection

3. Scattering

4. Birefringence

We look briefly at each in this section.

Absorption

A common method of polarization is absorption in a sheet of commercial material called Polaroid, invented by E. H. Land in 1938. This material contains long-chain hydrocarbon molecules which are aligned when the sheet is stretched in one direction during the manufacturing process. These chains become conducting (at optical frequencies) when the sheet is dipped in a solution containing iodine. When light is incident with its electric field vector **E** parallel to the chains, electric currents are set up along the chains and the light energy is absorbed. If the electric field **E** is perpendicular to the chains, the light is transmitted. The direction perpendicular to the chains is called the *transmission axis*. We shall make the simplifying assumption that all the light is transmitted when **E** is parallel to the transmission axis and all the light is absorbed when **E** is perpendicular to the transmission axis.

Consider a light beam in the z direction incident on a Polaroid which has its transmission axis in the y direction. On the average, half of the incident light has its **E** vector in the y direction and half in the x direction. Thus half the intensity is transmitted, and the transmitted light is linearly polarized with its **E** vector in the y direction.

Suppose we have a second piece of Polaroid whose transmission axis makes an angle θ with that of the first, as in Figure 32-22. The **E** vector of the light between the Polaroids can be resolved into two components, one parallel and one perpendicular to the transmission axis of the second Polaroid. If we call the direction of the transmission axis of the second Polaroid y',

$$E_{y'} = E \cos \theta \quad \text{and} \quad E_{x'} = E \sin \theta$$

Only the component $E_{y'}$ is transmitted by the second Polaroid. The transmitted intensity is proportional to the square of the transmitted amplitude. Thus, if I_1 is the intensity between the two Polaroids, the intensity transmitted by both Polaroids is

$$I_{\text{total}} = I_1 \cos^2 \theta \qquad \qquad \text{32-12}$$

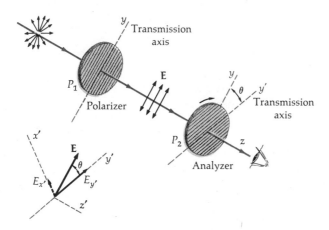

Figure 32-22
Two Polaroids with their transmission directions making an angle θ with each other. Only the component $E \cos \theta$ is transmitted through the second Polaroid. If the original intensity is I_0, the intensity between the Polaroids is $\frac{1}{2}I_0$ and that transmitted by both Polaroids is $\frac{1}{2}I_0 \cos^2 \theta$.

(The intensity incident on the second Polaroid is of course half that incident on the first.) When two polarizing elements are placed in succession in a beam of light as described here, the first is called the *polarizer* and the second is called the *analyzer*. If the polarizer and analyzer are crossed, i.e., have their axes perpendicular to each other, no light gets through when the absorption is complete. In practice, usually some weak red or blue light can be observed through crossed Polaroids, indicating that the absorption is not complete over the entire visible spectrum. Equation 32-12 is known as *Malus' law* after its discoverer, E. L. Malus (1775–1812). It applies to any two polarizing elements whose transmission directions make an angle θ with each other.

Malus' law

Polarization of electromagnetic waves by absorption can be demonstrated with microwaves (electromagnetic waves with wavelengths on the order of centimetres). In a typical microwave generator, polarized waves are radiated by a dipole antenna. An absorber can be made of a screen of parallel straight wires (Figure 32-23). An electric field parallel to the wires sets up currents in the wires, which absorb energy. If the waves are perpendicular to the wires, no currents are set up and the wave is transmitted. The transmission axis of this polarizer is perpendicular to the direction of the wires. When the axis is parallel to the dipole antenna (the wires are perpendicular to the dipole), the microwaves are transmitted. If the absorber is rotated 90° from this position, the waves are absorbed.

(a)

(b)

Figure 32-23
Polarization of microwaves. The electric field of the microwaves is vertical, parallel to the vertical dipole radiator. (a) When the metal wires are horizontal (transmission axis vertical), the waves are transmitted, as indicated by the high reading on the detector. (b) When the wires are vertical (transmission axis horizontal), the waves are absorbed, as indicated by the low reading on the detector. (*Courtesy of Larry Langrill, Oakland University.*)

Reflection

When unpolarized light is reflected from a plane surface, e.g., that separating air and glass or air and water, the reflected light is partially polarized. The degree of polarization depends on the angle of incidence and the indexes of refraction of the two media. When the angle of incidence is such that the reflected and refracted rays are perpendicular to each other, the reflected light is completely polarized. This result was discovered experimentally by Sir David Brewster in 1812.

Figure 32-24 shows light incident at the polarizing angle θ_p for which the reflected light is completely polarized. The electric field vector **E** of the incident light can be resolved into components parallel and perpendicular to the plane of incidence. The reflected light is completely polarized with its electric field vector perpendicular to the plane of incidence. We can relate the polarizing angle θ_p to the indexes of refraction of the media using Snell's law. If n_1 is the index of refraction of the first medium and n_2 that of the second medium, we have

$$n_1 \sin \theta_p = n_2 \sin \theta_2 \qquad\qquad 32\text{-}13$$

where θ_2 is the angle of refraction. From Figure 32-24 we see that the sum of the angle of reflection and the angle of refraction is 90°. Since the angle of reflection equals the angle of incidence, we have

$$\theta_2 = 90° - \theta_p$$

Then

$$n_1 \sin \theta_p = n_2 \sin (90° - \theta_p) = n_2 \cos \theta_p$$

or

$$\tan \theta_p = \frac{n_2}{n_1} \qquad\qquad 32\text{-}14 \qquad \textit{Brewster's law}$$

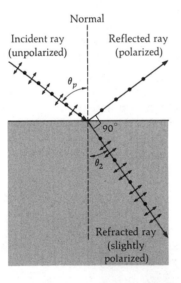

Figure 32-24
Polarization by reflection. The incident wave is unpolarized and has components of **E** (indicated by the arrows) parallel to the plane of incidence and components (indicated by the dots) perpendicular to the plane of incidence. For incidence at the polarizing angle θ_p shown, the reflected light is completely polarized with **E** perpendicular to the plane of incidence as indicated by the dots. Then the transmitted beam is partially polarized because only a small fraction of the light is reflected. At other angles of incidence, the reflected light is partially polarized.

Equation 32-14 is known as *Brewster's law*. Although the reflected light is completely polarized when the incident angle is θ_p, the transmitted light is only partially polarized because only a small fraction of the incident light is reflected. If the incident light itself is polarized with its electric field vector in the plane of incidence, there is no reflected light when the angle of incidence is θ_p (Figure 32-25). There is no simple method of deriving Brewster's law; it can be derived from the electromagnetic theory of light. We can understand the result qualitatively from Figure 32-25. If we consider the molecules of the second medium to be oscillating in the direction of the electric field of the refracted ray, they cannot radiate energy along the direction of oscillation, which would be the direction of the reflected ray.

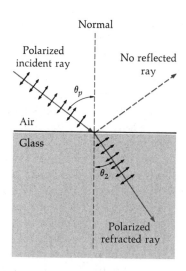

Figure 32-25
Polarized light incident at the polarizing angle. If the incident light has no component of **E** perpendicular to the plane of incidence, there is no reflected light.

Because of the polarization of reflected light, sunglasses made of polarizing material can be very effective in cutting out glare. If the light is reflected from a horizontal surface such as a lake or snow on the ground, the plane of incidence will be vertical and the electric field vector of the reflected light will be predominantly horizontal. Sunglasses with their transmission axis vertical will then reduce the glare by absorbing much of the reflected light.

Scattering

The phenomenon of absorption and reradiation is called *scattering*. Reflection of light is actually a scattering of light by a large number of scattering centers closely spaced compared with the wavelength. Refraction is a similar phenomenon in which the scattered light interferes with the incident light. The term scattering, however, usually refers to the situation in which the scattering centers are separated by distances not small compared with the wavelength of light. A familiar example of light scattering is that from clusters of air molecules (due to random fluctuations in the density of air) which tend to scatter short wavelengths more than long wavelengths, giving the sky its blue color.

Scattering can be demonstrated by adding a small amount of powdered milk to a container of water. The milk particles absorb light and reradiate it as dipole radiators. Consider a beam of light in the z direction (Figure 32-26). The **E** vector is therefore in the x and y directions. The scattering particles oscillate parallel to the **E** vector of the incident wave, i.e., in the x and y directions but not in the z direction. If we look perpendicular to the beam, say in the x direction, we see light radiated by charges in the scattering centers oscillating in the y direction but we do not see radiation due to the oscillations in the x direction because no light is radiated in the direction along the line of the dipoles. We also see no radiation with **E** in the z direction because the scattering particles do not oscillate in that direction. The light radiated in the x direction is thus polarized with its **E** vector in the y direction. This is easily demonstrated by polarizing the incident beam with a Polaroid. If the incident light contains only **E** vectors in the x direction, no scattered light will be observed in the x direction, whereas light will be observed in the y direction. Rotating the Polaroid 90° will reverse the situation: no scattered light in the y direction and scattered light in the x direction.

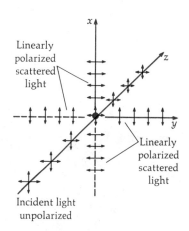

Figure 32-26
Polarization by scattering. Unpolarized light propagating in the z direction is incident on a scattering center at the origin. The light scattered in the x direction is polarized with its **E** vector in the y direction, while the light scattered in the y direction is polarized in the x direction.

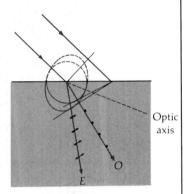

Figure 32-27
Huygens wavelets for the *O* and *E* rays when light is incident on a birefringent crystal whose optic axis is in the plane of incidence. The *E* ray is polarized with its electric field in the plane of incidence, whereas the *O* ray has its electric field perpendicular to the plane of incidence. The *O* ray obeys Snell's law of refraction, but the *E* ray does not.

Figure 32-28
Huygens wavelets for the *O* and *E* rays when unpolarized light is incident on a birefringent crystal whose optic axis is perpendicular to the plane of incidence. In this case both rays obey Snell's law of refraction.

Birefringence

Birefringence is a complicated phenomenon occurring in materials, e.g., calcite, which are anisotropic, i.e., have a preferred direction. Although most materials are isotropic (all directions are equivalent), some crystals, because of their atomic structure, have different optical properties for light traveling in different directions through the crystal. The velocity of light in calcite and some other materials is not the same in all directions. Such materials are called *double-refracting* or *birefringent*. When a light ray is incident on a calcite crystal, it may be separated into two rays which travel with different velocities and are polarized in mutually perpendicular directions. The Huygens wavelets corresponding to one ray are spherical, as in propagation through ordinary isotropic materials, and this ray is called the *ordinary (O) ray*. In the *extraordinary (E) ray* the Huygens wavelets are ellipsoids of revolution, indicating that the velocity of propagation depends on direction. Along the axis of revolution, the two rays propagate with the same speed. The Huygens wavelets are tangent to each other in this direction, called the *optic axis*. Perpendicular to the optic axis, the difference in speed of the two rays is maximum. In calcite, the extraordinary ray travels faster than the ordinary ray perpendicular to the optic axis. The index of refraction for the ordinary ray in calcite is about $n_O = 1.66$. For the extraordinary ray n_E varies from 1.66 along the optic axis to 1.49 perpendicular to it.

The propagation of both the ordinary ray and extraordinary ray through a birefringent crystal can be traced by construction of the Huygens wavelets for the rays. For a general orientation of the crystal this is extremely complex. We shall restrict our discussion to the cases in which the optic axis of the crystal is either in the plane of incidence or perpendicular to it. Then both the *O* and *E* rays are in the plane of incidence, and the polarization of the rays is relatively simple. The ordinary ray is polarized with its electric field perpendicular to the plane formed by the optic axis and the direction of propagation, whereas the extraordinary ray has its electric field in the plane containing the optic axis and the direction of propagation. The Huygens wavelets and polarizations are shown in Figures 32-27 and 32-28 for the optic axis in the plane of incidence and perpendicular to the plane of incidence, respectively. Note the spatial separation of the rays.

In Figure 32-29 a light beam is incident normally on the face of a crystal whose optic axis is perpendicular to the surface. The two rays propagate along the optic axis with the same speed, and no separation of the rays is observed. In Figure 32-30 the light is incident normally on a crystal whose optic axis is not perpendicular to the surface. In this case, even for normal incidence the extraordinary ray is deflected. The two rays are separated in space and emerge from the crystal as parallel rays separated by an amount that depends on the thickness of the crystal (Figure 32-31). This effect can be demonstrated by placing a calcite crystal over a small source and observing the two images of the source through the crystal. A Polaroid analyzer can be used to show that the two rays are polarized in mutually perpendicular directions. If the crystal is rotated about the line of the incident ray, the extraordinary ray is rotated through space.

There are several interesting applications of double refraction. One is the Nicol prism, which is used as a polarizer. Figure 32-32 shows the geometry of such a prism. A crystal is cut so that the two rays are separated in space. The crystal is cut along the diagonal and cemented together with a transparent cement which has an index of refraction en-

Figure 32-29
Huygens wavelets for light incident normally on a birefringent crystal whose optic axis is perpendicular to the surface. Both rays propagate with the same speed and in the same direction, and so there is no separation of the rays.

Figure 32-30
Huygens wavelets for light incident normally on a birefringent crystal whose optic axis is not perpendicular to the surface. The E ray is deflected even at normal incidence.

Figure 32-31
A narrow beam of unpolarized light incident normally on a birefringent crystal oriented as in Figure 32-30 is split into two beams (exaggerated in this figure). When the crystal is rotated in space, the E ray rotates in space.

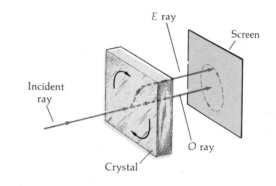

Figure 32-32
A Nicol prism made by cutting a birefringent crystal and gluing it back together. The O ray is totally reflected at the surface of the crystal and glue, but the E ray is not. The emerging beam is polarized.

suring that one ray is totally reflected whereas the other is transmitted. The advantage of a Nicol prism over Polaroid is that the emerging light is completely polarized over the entire visible spectrum.

Double refraction can also be used to produce a rotation of the plane of polarization of linearly polarized light and to produce circularly polarized light from linearly polarized light. Figure 32-33 shows a light beam incident normally on the surface of a crystal whose optic axis is parallel to the surface. The Huygens wavelets for the O and E rays have circular cross sections but of different radii. The two rays travel in the same direction as the incident ray but with different speeds and different polarizations. In calcite, for example, the extraordinary ray travels faster than the ordinary ray so that its wavefront gets ahead of that for the O ray.

Figure 32-33
Huygens wavelets for unpolarized light incident normally on a birefringent crystal whose optic axis is parallel to the surface. The O and E rays propagate in the same direction but with different speeds. The two beams emerge from the crystal with a phase difference which depends on the indexes of refraction for the two rays and on the thickness of the crystal.

The beams emerge with a phase difference which depends on the thickness of the plate. In a *half-wave plate* the thickness is such that the number of wavelengths of the ordinary ray is greater than the number of wavelengths of the extraordinary ray by an odd number of half wavelengths. The phase difference is thus 180°. (Since the ordinary ray travels slower, it has a smaller wavelength, $\lambda = v/f$, than the extraordinary ray. A given thickness thus has more ordinary waves in it than extraordinary waves.) If the incident light is polarized in the direction at 45° to the polarizations of the two rays, the emerging light will have its polarization rotated as shown in Figure 32-34. In a *quarter-wave plate* the thickness is such that there is a 90° phase difference between the rays when they emerge. The resulting electric field is of the form $E_x = E_0 \times \sin \omega t$, $E_y = E_0 \cos \omega t$. The **E** vector thus rotates in a circle, and the wave is circularly polarized.

Many common substances, e.g., cellophane and transparent tape, are birefringent. Interesting and beautiful patterns can be observed by placing such materials between crossed Polaroids. With no substance between the Polaroids, no light is transmitted, but a sheet of cellophane acts as a half-wave plate for light of a certain color, depending on the thickness of the cellophane. It rotates the plane of polarization of the light between the Polaroids, and some light of that color is transmitted.

Figure 32-34
Rotation of the direction of polarization by a half-wave plate birefringent crystal. Light from the Polaroid is linearly polarized with its **E** vector making an angle of 45° with the optic axis, which is parallel to the surface of the crystal, as in Figure 32-33. The thickness of the crystal is such that the E and O rays differ in phase by 180° upon emergence from the crystal. The effect is to rotate the direction of polarization by 90°.

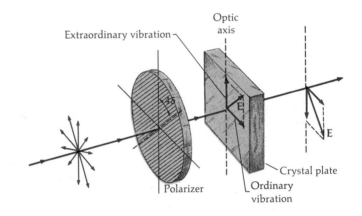

Double image produced by birefringent calcite crystal. (*From Richard T. Weidner and Robert L. Sells,* Elementary Classical Physics, *vol. II, 2d ed., fig. 36.17, p. 743. Copyright © 1973 by Allyn and Bacon, Inc. Used by permission.*)

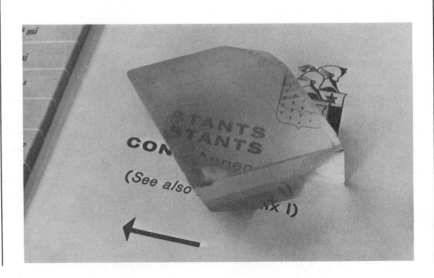

Radar Astronomy

G. H. Pettengill
Massachusetts Institute of Technology

Figure 1
Steerable parabolic reflector, 84 ft in diameter, used for radar astronomy at 23-cm wavelength at M.I.T.'s Lincoln Laboratory, Westford, Massachusetts. [*From M. I. Skolnik (ed.), Radar Handbook, p. 33-18, McGraw-Hill Book Company, New York, 1970, by permission of the publishers.*]

Essentially all we know about celestial objects has come from measuring the electromagnetic radiation they emit or reflect. In passing through the earth's atmosphere, however, much of this radiation is absorbed. In fact, we can observe celestial objects from the ground only at wavelengths lying within two major spectral windows, where our atmosphere is relatively transparent. The shorter-wavelength, or optical, window shows us the splendor of the night sky and has made the development of optical astronomy possible. The longer-wavelength, or radio, window has been increasingly exploited in recent years and provides the basis for radio astronomy.

Radio waves can be collected and focused by antennas using large parabolic reflecting surfaces, as shown in Figure 1. This particular antenna uses an additional reflector to redirect the incident energy collected by the paraboloid onto a feed horn, visible near the center of the primary reflector. The same antenna can be used for radiating energy by reversing the receiving process. In this case, a transmitter is attached to the feed horn, and its radiation is modified by reflection into a plane wave extending across the aperture of the large paraboloid. At great distances from the antenna, diffraction transforms this circular section of a plane wave into a spherical wavefront with a cone angle (beam width) proportional to the ratio of the radio wavelength to the reflector's diameter.

If the radio transmitter is sufficiently powerful (typically several hundred kilowatts) and the antenna sufficiently large (30 m or so in diameter), it is possible to detect radar echoes from the moon and the nearer planets. Since the time, frequency, and polarization characteristics of the coherent transmitted waveform are precisely known, the alteration of these characteristics as seen in the echo can be related to properties of the distant scattering surface and the propagating medium en route. The study of celestial objects in this way is called *radar astronomy*.

How does the distant surface scatter the incident radar energy? If the surface contains regions, or facets, which are many wavelengths across and flat

Figure 2
Schematic representation of radar scattering and the dispersion of echo power in delay and frequency introduced by finite target size and rotation. [*After M. I. Skolnik (ed.), Radar Handbook, p. 33-3, McGraw-Hill Book Company, New York, 1970, by permission of the publishers.*]

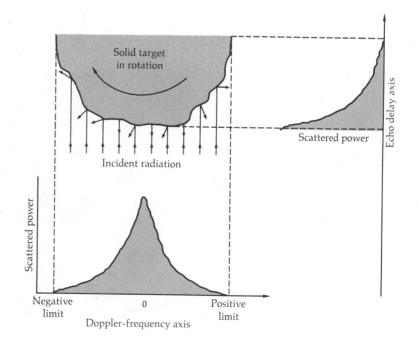

within plus or minus the wavelength, the reflection is coherent, or specular; i.e., it emerges at an angle with respect to the facet normal which is equal to the angle of incidence. Since the radar observes only backscattered radiation, only surface facets oriented at right angles to the line of sight are seen, as diagrammed schematically in Figure 2. If the surface is rough, so that its height varies rapidly by a significant fraction of the wavelength as one moves across it, the scattering is diffuse and depends only slightly on the angle of incidence to the surface.

Rotation of the target introduces a doppler frequency shift which varies linearly across the target as shown in Figure 2. Therefore, if one analyzes the echo power in frequency, one can establish a direct correspondence between the echo intensity observed in a given frequency interval and the scattering efficiency of a calculable location on the surface which imparts those particular values of frequency shift. Similarly, the distribution of echo power in time delay can be related to scattering from identifiable regions of the surface lying at different distances from the radar.

By analyzing the echo power simultaneously as a function of delay and frequency, radar scattering maps of the surface can be obtained, as shown in Figure 3. These radar maps clearly show the existence of large-scale surface features, such as craters and mountain ranges, and are of particular interest in the study of the cloud-shrouded surface of Venus, which is not otherwise accessible to study from afar. This technique can also be turned around and used to determine the planet's rotation. In this way it was discovered by radar that Mercury does not always keep one face turned to the sun, as was formerly thought, but rotates 50 percent faster than its average orbital rate.

By timing the total round-trip echo delay, the distance between the radar and the surface of the distant celestial object is obtained, since the velocity of radio waves is known. In many current experiments, the timing accuracy is better than one-millionth of a second, while the total round-trip echo delay exceeds 1000 s. Thus, we measure distance with a fractional accuracy of 1×10^{-9}, or a thousand times more accurately than is possible using optical observations alone. Not many measurements in physics achieve this level of accuracy, which is sufficient to make observable results of many extremely small perturbing effects on the orbits of the inner planets. Several of these effects are identifiable with relativistic "corrections" and have been used successfully to confirm the accuracy of Einstein's formulation of the theory of general relativity.

Figure 3
Radar map of the moon at 70-cm wavelength, made by T. W. Thompson using the Arecibo radar in Puerto Rico. Note the unusually strong diffuse scattering by the crater Tycho in the bottom center of the map, indicating that it has an exceptionally rough surface. (*From T. W. Thompson, Atlas of Lunar Radar Maps at 70-cm Wavelength. The Moon, vol. 10, no. 1, p. 54, May 1974.*)

Review

A. Identify the contributions the following people made to the understanding of light: Young, Fresnell, Foucault, Maxwell, Hertz, Einstein, Fizeau, Roemer, Fermat, Brewster.

B. Define, explain, or otherwise identify:

Photoelectric effect, 850
Photon, 850
Electric-dipole radiation, 853
Plane of incidence, 856
Law of reflection, 856
Fermat's principle, 857
Index of refraction, 857

Snell's law, 858
Polarization, 863
Malus' law, 865
Brewster's law, 866
Scattering, 867
Birefringence, 868

Exercises

Section 32-1, Waves or Particles?

1. Show that if the wavelength of an electromagnetic wave is given in nanometres, the photon energy in electronvolts is given by $E = (1240 \text{ eV·nm})/\lambda$.

2. (a) Use the result of Exercise 1 to calculate the range of photon energies in the visible spectrum of wavelengths from about 400 to 700 nm. (b) Molecular transitions involve the absorption or emission of photons of energy less than 1 eV. Is this radiation in the infrared (long-wavelength) or ultraviolet (short-wavelength) part of the electromagnetic spectrum?

3. What is the energy of an x-ray photon whose wavelength is 0.1 nm?

4. A prominent radiation from sodium has a wavelength of 589 nm. What is the energy of the photon of this wavelength?

Section 32-2, Electromagnetic Waves

5. What is the frequency of a 3-cm microwave?

6. What is the frequency of an x-ray of wavelength 0.1 nm?

7. Find the wavelength for a typical AM radio frequency and for a typical FM radio frequency.

8. What is the wavelength of a photon whose energy is 5 eV? In what part of the electromagnetic spectrum is it?

9. A radiating electric dipole lies along the z axis. Let I_1 be the intensity of the radiation at a distance $r = 10$ m and at angle $\theta = 90°$. Find the intensity (in terms of I_1) at (a) $r = 30$ m, $\theta = 90°$; (b) $r = 10$ m, $\theta = 45°$; (c) $r = 20$ m, $\theta = 30°$.

10. (a) For the situation described in Exercise 9, at what angle is the intensity at $r = 5$ m also equal to I_1? (b) At what distance is the intensity equal to I_1 at $\theta = 45°$?

Section 32-3, The Speed of Light

11. A light-year is the distance light travels in 1 year. (a) Find the number of kilometres in a light-year. (b) The spiral galaxy in the Andromeda constellation is about 2×10^{19} km away from us. How many light-years is this?

12. How long does it take light to travel from the sun to the earth, a distance of about 1.5×10^{11} m? What is the distance to the sun in light-minutes?

13. On a rocket sent to Mars to take pictures, the camera is triggered by radio waves which (like all electromagnetic waves) travel with the speed of light. What is the time delay between sending and receiving the signal from the earth to Mars? (Take the distance to Mars to be 9.7×10^{10} m.)

Section 32-4, Reflection

14. Calculate the fraction of light energy reflected from water at normal incidence ($n = 1.33$ for water).

15. Light is incident normally on a slab of glass of index of refraction $n = 1.5$. Reflection occurs at both surfaces of the slab. About what percentage of the incident light energy is transmitted by the slab?

16. A physics student playing pocket billiards wants to strike his cue ball so that it hits a cushion and then hits the eight ball squarely. He chooses several points on the cushion and for each point he measures the distance from it to the cue ball and to the eight ball. He aims at the point for which the sum of these distances is least. Will his cue ball hit the eight ball? How is this method related to Fermat's principle?

17. A point source of light is 5 cm above a plane reflecting surface (such as a mirror). Draw a ray from the source which strikes the surface at an angle of incidence of about 45°; draw two more at angles slightly less than 45°; and draw the reflected ray for each. The reflected rays appear to diverge from a point called the image of the light source. Draw dotted lines extending the reflected rays back until they meet at a point behind the surface, to locate the image point.

Section 32-5, Refraction

18. The index of refraction of water is 1.33. Find the angle of refraction of a beam of light in air hitting the water surface at an angle of (a) 20°, (b) 30°, (c) 45°, and (d) 60° with the normal, and indicate these rays on a diagram.

19. Repeat Exercise 18 for a beam of light initially in water and incident on a water-air surface.

20. What is the critical angle for total internal reflection for light traveling from water ($n = 1.33$) to air ($n = 1.00$)?

21. Find the speed of light in water ($n = 1.33$) and in glass ($n = 1.5$).

22. A beam of monochromatic red light of wavelength 700 nm in air travels in water. What is the wavelength in water? Does a swimmer under water observe the same color or a different color for this light?

23. The critical angle for total internal reflection in diamond is about 24°. What is the index of refraction of diamond?

24. A beam of light strikes a plane glass surface at an angle of incidence of 45°. The index of refraction of the glass varies with wavelength according to the graph in Figure 32-18. How much smaller is the angle of refraction for blue light of wavelength 400 nm than for red light of wavelength 700 nm?

25. Use Figure 32-18 to calculate the critical angles for total internal reflection from a glass-air surface for blue light of wavelength 400 nm and red light of wavelength 700 nm.

26. A swimmer at S in Figure 32-35 develops a leg cramp while swimming near the shore of a calm lake and calls for help. A lifeguard at L hears the call. The lifeguard can run 9 m/s and swim 3 m/s. He knows physics and chooses a path that will take the least time to reach the swimmer. Which of the paths shown in Figure 32-35 does he take? Explain.

Section 32-6, Polarization

27. Two Polaroid sheets have their transmission directions crossed so that no light gets through. A third sheet is inserted between the two so that its transmission direction makes an angle θ with the first sheet. Unpolarized light of intensity I_0 is incident on the first sheet. Find the intensity transmitted through all three sheets if (a) $\theta = 45°$; (b) $\theta = 30°$.

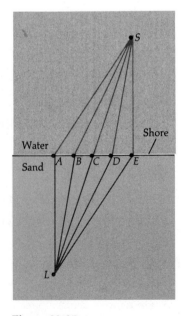

Figure 32-35
Swimmer at S and lifeguard at L for Exercise 26.

28. Two Polaroid sheets are inserted between two other Polaroid sheets which have their transmission directions crossed, so that the angle between each successive pair of sheets is 30°. Find the transmitted intensity if the original light is unpolarized with intensity I.

29. The polarizing angle for a certain substance is 60°. (a) What is the angle of refraction of light incident at this angle? (b) What is the index of refraction of this substance?

30. The critical angle for total internal reflection for a substance is 45°. What is the polarizing angle for this substance?

31. What is the polarizing angle for glass with $n = 1.5$?

32. A beam of linearly polarized light strikes a calcite crystal such that its electric vector makes an angle of 30° with the optic axis (Figure 32-36). What is the ratio of the intensities of the ordinary and extraordinary rays?

33. Light of wavelength λ in air is incident on a slab of calcite so that the ordinary and extraordinary rays travel in the same direction as in Figure 32-33. Show that the phase difference between these rays after traversing a thickness t is

$$\delta = \frac{2\pi}{\lambda}(n_O - n_E)t$$

Figure 32-36
Exercise 32.

Problems

1. Show that the transmitted intensity through a glass slab of index of refraction n for normally incident light is approximately given by

$$I_T = I_0 \left[\frac{4n}{(n+1)^2}\right]^2$$

2. (a) Use the result of Problem 1 to find the ratio of the transmitted intensity to the incident intensity through N parallel slabs of glass for light of normal incidence. (b) Find this ratio for three slabs of glass with $n = 1.5$. (c) How many slabs of glass with $n = 1.5$ will reduce the intensity to 10 percent of the incident intensity?

3. This problem is a refraction analogy. A band is marching down a football field with a constant speed v_1. About midfield, the band comes to a section of muddy ground, which has a sharp boundary making an angle of 30° with the 50-yd line, as shown in Figure 32-37. In the mud, the marchers move with speed $v_2 = \frac{1}{2}v_1$. Diagram how each line of marchers is bent as it encounters the muddy section of the field so that eventually the band is marching in a different direction. Indicate the original direction by a ray and the final direction by a second ray, and find the angles between these rays and the line perpendicular to the boundary line. Is their direction of motion bent toward the perpendicular to the boundary line or away from it?

Figure 32-37
Problem 3.

Marching band   50-yd line

4. Light is incident normally upon one face of a prism of glass where the index of refraction is n (Figure 32-38). The light is totally reflected at the right side. (a) What is the minimum value n can have? (b) When this prism is immersed in a liquid whose index of refraction is 1.15, there is still total reflection, but in water, whose index is 1.33, there no longer is total reflection. Use this information to establish limits for possible values of n.

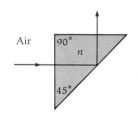

Figure 32-38
Problem 4.

5. Investigate the effect on the critical angle of a thin film of water on a glass surface for rays which originate in the glass. Take $n = 1.5$ for glass and $n = 1.33$ for water. (a) What is the critical angle for total internal reflection at the glass-water surface? (b) Is there any range of incident angles that are greater than θ_c for glass-to-air refraction and for which light rays will leave the glass *and* the water and pass into the air medium?

6. Show that a light ray transmitted through a glass slab emerges parallel to the incident ray but displaced from it. For an incident angle of 60°, index of refraction of glass $n = 1.5$, and slab thickness of 10 cm, find the displacement measured along the glass surface from the point at which the incident ray would emerge if there were no slab.

7. A prism is made of calcite cut so that its upper face contains the optic axis. The incident ray is unpolarized and normally incident upon the surface (Figure 32-39). (a) What must the angle of the prism be for the ordinary ray to be totally reflected internally? (b) Show that the extraordinary ray will not then be internally reflected, so that the emerging light will be plane-polarized.

Figure 32-39
Problem 7.

8. The indexes for ordinary and extraordinary waves given for calcite are for green light ($\lambda = 540$ nm). (a) What would be the minimum thickness for a quarter-wave plate? (b) What would be the minimum thickness for a half-wave plate? (c) Would you expect these to be quarter- and half-wave plates for, say, red light? Why or why not?

9. In Exercise 26, let the shoreline be the x axis and the swimmer be at point $x = 20$ m, $y = 30$ m. The lifeguard is at $y = -30$ m on the y axis (Figure 32-40). (a) Find the point x_0 at which the lifeguard enters the water if he can run 3 times as fast as he can swim, using the approximation $\sin \theta = \tan \theta$ for the angles made by his path and the normal to the shoreline. (b) Find x_0 to three significant figures by trial and error, using your answer from (a) as a first guess. (c) Find the time for the lifeguard to reach the swimmer if he runs at 9 m/s and swims at 3 m/s and compare with the time it would take if the guard followed a straight-line path from his initial position to that of the swimmer.

Figure 32-40
Problem 9.

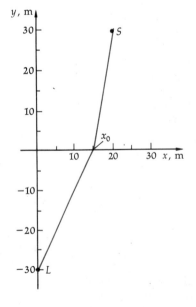

CHAPTER 33   Geometric Optics

Objectives   After studying this chapter you should:

1. Be able to draw simple ray diagrams to locate images and determine whether they are real or virtual, erect or inverted, and enlarged or reduced for spherical mirrors and thin lenses.

2. Be able to locate the image formed by a spherical mirror or by a thin lens using the mirror or thin-lens equation and calculate the magnification of the image.

3. Be able to locate the image formed by a single refracting surface.

4. Be able to discuss the origin of the various aberrations that occur with mirrors and lenses.

5. Be able to discuss how the eye works.

6. Be able to show with a simple diagram why close objects appear larger than far objects.

7. Be able to describe how a simple magnifier works and calculate the angular magnification of a simple magnifier.

8. Be able to describe with diagrams and equations how a microscope and telescope work.

The wavelength of light is often very small compared with the size of any obstacles or apertures it encounters, and diffraction effects can be neglected. The study of such situations, for which the ray approximation is valid and light propagates in straight lines, is known as *geometric optics*. In this chapter, we shall apply the laws of reflection and refraction discussed in Chapter 32 to study the formation of images by mirrors and lenses.

33-1 Plane Mirrors

Figure 33-1 shows a narrow bundle of light rays from a point source P reflected from a plane mirror. After reflection, the rays diverge exactly as if they came from a point P' behind the plane of the mirror. The point P' is called the *image* of the object P. When these rays enter the eye, they cannot be distinguished from rays diverging from a source at P' with no mirror. The image is called a *virtual image* because the light does not actually emanate from the image but only appears to. Geometric construction using the law of reflection shows that the image point lies on the line through the object perpendicular to the plane of the mirror and

Virtual image

Figure 33-1
Image formed by a plane mirror. The rays from point P, which strike the mirror and enter the eye, appear to come from the image point P' behind the mirror. The image can be seen by the eye anywhere in the region indicated.

at a distance behind the plane equal to that from the plane to the object. The image can be seen by an eye anywhere in the region indicated, in which a line from the image to the eye passes through the mirror. We note that the object need not be directly in front of the mirror. An image can be seen as long as the object is above the plane of the mirror.

Figure 33-2 shows an image of an extended object formed by a plane mirror. The size of the image is the same as that of the object.

Figure 33-3 illustrates the formation of multiple images by two plane mirrors making an angle with each other. Light reflected from mirror 1 strikes mirror 2 just as if it came from the image point P_1. The image P_1 is called the *object point* for mirror 2. Its image is at point P_{12}. This image will be formed whenever the image point P_1 is in front of the plane of mirror 2. The image at point P_2 is due to rays from the object which re-

Figure 33-2
Image of an extended object in a plane mirror. The image is the same size as the object.

$x' = x$

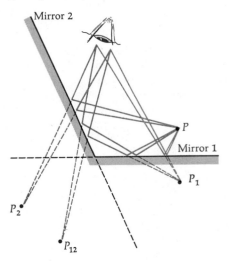

Figure 33-3
Images formed by two plane mirrors. P_1 is the image of the object P in mirror 1, and P_2 is the image of the object in mirror 2. Point P_{12} is the image of P_1 in mirror 2 seen when light rays from the object reflect first from mirror 1 and then from mirror 2, as shown. The image P_2 does not have an image in mirror 1 because it is behind that mirror.

flect directly from mirror 2. Since P_2 is behind the plane of mirror 1, it cannot serve as an object point for a further image in mirror 1. The number of multiple images formed by two mirrors depends on the angle between the mirrors and the position of the object.

Question

1. How tall must a mirror be for a standing person to see his entire reflection in it?

33-2 Spherical Mirrors

Figure 33-4 shows a bundle of rays from a point on the axis of a concave spherical mirror reflecting from the mirror and converging at point P'. The rays then diverge from this point just as if there were an object at that point. This image is called a *real image* because the light actually does emanate from the image point. It can be seen by an eye at the left of the image looking into the mirror. It could also be observed on a ground-glass viewing screen or photographic film placed at the image point. A virtual image cannot be observed on a screen at the image point because there is no light there. Despite this distinction between real and virtual images, the light rays diverging from a real image and those appearing to diverge from a virtual image are identical, so that no distinction is made by the eye when viewing a real or a virtual image.

From Figure 33-4 we see that only rays which strike the mirror at points near the axis AO are reflected through the image point. Such rays are called *paraxial rays*. Because other rays converge to different points near the image point, the image appears blurred, an effect called *spherical aberration*. The image can be sharpened by reducing the size of the mirror so that nonparaxial rays do not strike it. Although the image is then sharper, its brightness is reduced because less light is reflected.

Using the law of reflection and elementary geometry, we can relate the image distance s' to the object distance s and the radius of curvature r. The geometry is shown in Figure 33-5. The result is

$$\frac{1}{s} + \frac{1}{s'} = \frac{2}{r} \qquad \text{33-1}$$

H. M. Null/Rapho Guillumette Pictures

The lake serves as a plane mirror producing virtual images of the trees. The images are inverted and the same size as the trees.

Spherical aberration

Figure 33-4
Rays from a point object P on the axis AO of a concave spherical mirror form an image at P' if the rays strike the mirror near the axis. Nonparaxial rays striking the mirror at points far from O are not reflected through the image point P'.

Figure 33-5
Geometry for calculating the image distance s' from the object distance s and the radius of curvature r. The angle β is an exterior angle to the triangle PAC and is therefore equal to $\alpha + \theta$. Similarly, from triangle PAP', $\gamma = \alpha + 2\theta$. Eliminating θ from these equations gives $2\beta = \gamma + \alpha$. Equation 33-1 follows from the small-angle approximations $\alpha \approx l/s$, $\beta \approx l/r$, and $\gamma \approx l/s'$.

The derivation of this equation assumes that angles made by the incident and reflected rays with the axis are small, an assumption equivalent to that of paraxial rays.

When the object distance is much greater than the radius of curvature of the mirror, the term $1/s$ in Equation 33-1 can be neglected, resulting in $s' = \frac{1}{2}r$ for the image distance. This distance is called the *focal length f* of the mirror, and the image point is called the *focal point F*.

$$f = \tfrac{1}{2}r \qquad\qquad 33\text{-}2$$

In terms of the focal length f the mirror equation is

$$\frac{1}{s} + \frac{1}{s'} = \frac{1}{f} \qquad\qquad 33\text{-}3$$

The focal point is the point at which parallel rays corresponding to plane waves from infinity are focused, as illustrated in Figure 33-6.

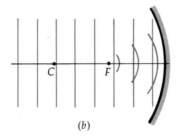

(a)   (b)

Figure 33-6
(a) Parallel rays strike a concave mirror and are reflected through the focal point at a distance $\frac{1}{2}r$. (b) The incoming wavefronts are plane waves; upon reflection they become spherical waves, which converge at the focal point.

(Again, only paraxial rays are focused at a single point.) Figure 33-7 shows rays from a point source at the focal point which strike the mirror and are reflected parallel to the axis. This illustrates a property of waves called *reversibility*. If we reverse the direction of a reflected ray, the law of reflection assures that the reflected ray will be along the original incoming ray but in the opposite direction. Reversibility holds also for refracted rays. If we have a real image of a source formed by a reflecting or refracting surface, we can place a source at the image point and the new image will be formed at the position of the original source.

A useful method of locating images is by geometric construction of a ray diagram, as illustrated in Figure 33-8, where the object is a human figure perpendicular to the axis a distance s from the mirror. By a judicious choice of rays from the head of the figure we can quickly locate the image. A ray from the head parallel to the axis is reflected through the focal point a distance $\frac{1}{2}r$ from the mirror, as shown. Another ray, through the center of curvature of the mirror, strikes the mirror perpen-

Figure 33-7
Illustration of reversibility. Rays diverging from a point source at the focal point of a concave mirror are reflected from the mirror as parallel rays. The rays are the same as in Figure 33-6a but in the reverse direction.

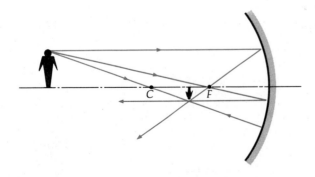

Figure 33-8
Ray diagram for location of image by geometric construction.

dicular to the surface and is reflected back along its original path. The intersection of these two rays locates the image point of the head. This can be checked by a third ray through the focal point, which is reflected back parallel to the axis.

We see from the figure that the image is inverted and is not the same size as the object. The magnification of the optical system (the spherical mirror in this case) is defined to be the ratio of the image size to the object size. Comparison of the triangles in Figure 33-9 shows that the magnification is just the ratio of the distances s' and s.

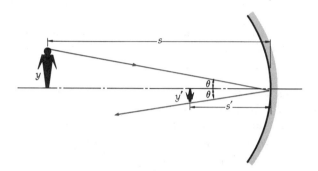

Figure 33-9
Geometry for finding the magnification of a spherical mirror. From the upper triangle, $\tan \theta = y/s$, and from the lower triangle, $\tan \theta = -y'/s'$, where the negative sign is introduced because y' is negative. Then the magnification is $m = y'/y = -s'/s$.

When the object is between the mirror and its focal point, the rays reflected from the mirror do not converge but appear to diverge from a point behind the mirror, as illustrated in Figure 33-10. In this case the image is virtual and erect. If s is less than $\frac{1}{2}r$ in Equation 33-1, the image distance s' turns out to be negative. We can apply Equations 33-1 and 33-3 to this case and to convex mirrors if we adopt a convenient sign convention. Whether the mirror is convex or concave, real images can be formed only on the same side of the mirror as the object; virtual images are formed on the opposite side, where there are no actual light rays. Distances to points on the real side are taken to be positive; distances to points on the virtual side are taken to be negative. Thus for the

Sign convention

concave mirror, s and r are positive, and s' is positive or negative depending on whether the image is real or virtual. For a convex mirror (Figure 33-11) the center of curvature is on the virtual side, and so r is taken to be negative. The focal length is also negative. For either case, Equation 33-1 gives the image distance s' in terms of the object distance and radius of curvature. The lateral magnification of the image is given by

$$m = \frac{y'}{y} = -\frac{s'}{s} \qquad \qquad 33\text{-}4$$

Figure 33-10
When the object is between the focal point and the mirror, the image is virtual and behind the mirror. Here it is located by a ray along the line from the focal point through the head of the figure and a second ray from the center of curvature through the head. The first ray is reflected parallel to the axis and the second is reflected back on itself. These rays diverge from a point behind the mirror. A third ray (not shown) could be drawn from the head parallel to the axis. It is reflected through the focal point, and its extension behind the mirror intersects the other rays at the image point.

Figure 33-11
Ray diagram for a convex mirror. The ray parallel to the axis is reflected as if it came from the focal point to the right of the mirror, and the ray toward the center of curvature is reflected back on itself. A third ray (not shown) could be drawn toward the focal point. It would be reflected parallel to the axis, and its extension behind the mirror would intersect the other two rays of the image.

A negative magnification, which occurs when both s and s' are positive, indicates that the image is inverted. Although these equations with this sign convention are relatively easy to use, practical work in optics often requires only knowledge of whether the image is real or virtual and an approximate knowledge of its location. This is easiest to obtain by constructing a ray diagram. A similar method for locating images produced by a lens and various lens combinations (Section 33-4) is often of more practical use than the equations relating the image position to the object position.

Example 33-1 An object 2 cm high is 10 cm from a convex mirror with a radius of curvature of 10 cm. Locate the image and find its height.

Since the center of curvature of a convex mirror is on the virtual-image side of the mirror, the focal length is negative:

$$f = \tfrac{1}{2}r = \tfrac{1}{2}(-10 \text{ cm}) = -5 \text{ cm}$$

Using Equation 33-3 to find the image distance gives

$$\frac{1}{10 \text{ cm}} + \frac{1}{s'} = \frac{1}{f} = -\frac{1}{5 \text{ cm}}$$

$$\frac{1}{s'} = -\frac{2}{10 \text{ cm}} - \frac{1}{10 \text{ cm}} = -\frac{3}{10 \text{ cm}}$$

$$s' = 3\tfrac{1}{3} \text{ cm}$$

Convex mirror resting on paper with equally spaced parallel stripes. The image of each point in front of the mirror is virtual and behind the mirror. Note the reduction in size and distortion in the shape of the image and the large number of stripes that are seen.

Fundamental Photographs

The image distance is negative, indicating a virtual image behind the mirror. The magnification is

$$m = -\frac{s'}{s} = -\frac{-3\frac{1}{3}}{10} = +\frac{1}{3}$$

The image is erect and one-third the size of the object. Its size is

$$y' = my = \tfrac{1}{3}(2 \text{ cm}) = \tfrac{2}{3} \text{ cm}$$

The ray diagram for this example is similar to Figure 33-11.

Edward Lettau, Photo Researchers

Questions

2. Under what conditions will a concave mirror produce an erect image? A virtual image? An image smaller than the object? An image larger than the object?

3. Answer Question 2 for a convex mirror.

Convex mirrors are useful for wide-angle viewing when the distortion in shape is not important.

33-3 Images Formed by Refraction

Figure 33-12 illustrates the formation of an image by refraction from a spherical surface separating two media with indexes of refraction n_1 and n_2. Again, only paraxial rays converge to one point. We can derive an equation relating the image distance to the object distance, the radius of

Figure 33-12
Image formed by refraction at a spherical surface.

curvature, and the indexes of refraction by applying Snell's law of refraction to these rays and using the small-angle approximation. This derivation is given in Figure 33-13. The resulting equation is

$$\frac{n_1}{s} + \frac{n_2}{s'} = \frac{n_2 - n_1}{r} \qquad 33\text{-}5$$

We can use the same sign convention for this equation as we used for mirrors, but we must note that for refraction real images are formed to the right of the surface (if the object is to the left) and virtual images to the left. Thus s' and r are taken to be positive if the image and center of curvature lie to the right of the surface.

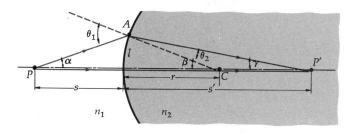

Figure 33-13
Geometry for the derivation of Equation 33-5. The angles θ_1 and θ_2 are related by Snell's law, which for small angles can be written $n_1\theta_1 = n_2\theta_2$. From the triangle ACP' we obtain

$$\beta = \theta_2 + \gamma = \frac{n_1}{n_2}\theta_1 + \gamma$$

or

$$n_1\theta_1 = n_2\beta - n_2\gamma$$

From triangle PAC we obtain $\theta_1 = \alpha + \beta$. Eliminating θ_1 gives $n_1\alpha + n_2\gamma = (n_2 - n_1)\beta$. Equation 33-5 follows from substituting the small-angle approximations $\alpha \approx l/s$, $\gamma \approx l/s'$, and $\beta \approx l/r$.

We can apply Equation 33-5 to find the apparent depth of an object under water when viewed from directly overhead. For this case the surface is a plane surface, the radius of curvature is infinite, and the image and object distances are related by

$$\frac{n_1}{s} + \frac{n_2}{s'} = 0$$

$$s' = -\frac{n_2}{n_1} s \qquad\qquad 33\text{-}6$$

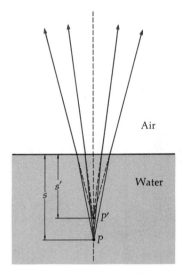

Apparent depth

The negative sign indicates that the image is virtual and on the same side of the refracting surface as the object, as shown in the ray diagram in Figure 33-14.

Figure 33-14
The image P' appears at the depth s', which is less than the actual depth s of the object P. When viewed in air from directly overhead, the apparent depth equals the real depth divided by the index of refraction of water.

Example 33-2 Find the apparent depth of a fish resting 1 m below the surface of water which has an index of refraction $n = \frac{4}{3}$.

Using $n_1 = \frac{4}{3}$ and $n_2 = 1$ in Equation 33-6, we obtain

$$s' = -\frac{3}{4}(1 \text{ m}) = -75 \text{ cm}$$

The apparent depth is three-fourths the actual depth. Note that this result holds only when the object is viewed from directly overhead so that the rays are paraxial.

33-4 Lenses

The most important application of Equation 33-5 is in finding the position of the image formed by a lens. We do this by considering the refraction at each surface separately. Consider a glass lens of index of refraction n with air on both sides. Let the radii of curvature of the surfaces of the lens be r_1 and r_2. If an object is at a distance s from the first surface, application of Equation 33-5 gives for the distance of the image due to the refraction at the first surface

$$\frac{1}{s} + \frac{n}{s'} = \frac{n-1}{r_1} \qquad\qquad 33\text{-}7$$

This image is usually not formed (unless the lens is extremely thick) because the light is again refracted at the second surface. Assume, for example, that the image distance s_1' is negative, indicating a virtual image to the left of the first surface. The light leaving this surface strikes the second surface *as if* it came from an object at the image position. If the thickness of the lens is t, the distance of this point from the second surface is $-s_1' + t$. (For this case $-s_1'$ is positive because s_1' is negative.) We can find the final image position due to both refractions by using this distance for the object distance for the second surface. If s_1' turns out to be positive, indicating a real image to the right of the first surface, the image point is to the right of the second surface if s_1' is greater than the thickness t and to the left of the surface if t is greater than s_1'. In either case the distance from the surface is $-s_1' + t$. Even if this image is to the right of the second surface, we can use this distance in Equation 33-5 to find the image distance of the final image. In this case the object distance for the second refraction is negative, indicating a virtual object on the right side of the surface. Thus for all possible values of the first image distance s_1', the image formed by refraction at the second surface is at a distance s' from this surface given by

$$\frac{n}{-s_1' + t} + \frac{1}{s'} = \frac{1 - n}{r_2} = -\frac{n - 1}{r_2}$$ 33-8

For this refraction the light is traveling from the glass medium of index n into the air medium of index 1.

Bausch & Lomb

For a general lens of thickness t, it is usually easier to find the distance s_1' numerically from Equation 33-7 and use this result in Equation 33-8 to find s' than to eliminate s_1' from these two equations. However, in many cases, the thickness t is much smaller than any of the other distances involved. For such a *thin lens* we can neglect t in Equation 33-8 and easily eliminate s_1' from these equations. Solving for n/s_1' in each equation, we obtain

$$\frac{n}{s_1'} = \frac{n - 1}{r_1} - \frac{1}{s} = \frac{1}{s'} + \frac{n - 1}{r_2}$$

or

$$\frac{1}{s} + \frac{1}{s'} = (n - 1)\left(\frac{1}{r_1} - \frac{1}{r_2}\right)$$ 33-9

Equation 33-9 gives the image distance s' in terms of the object distance s and the properties of the thin lens—r_1, r_2, and the index of refraction n. As with mirrors, the focal length of a thin lens is defined to be the image distance when the object distance is very large. Setting s equal to infinity and writing f for the image distance s', we obtain

$$\frac{1}{f} = (n - 1)\left(\frac{1}{r_1} - \frac{1}{r_2}\right)$$ 33-10

and

$$\frac{1}{s} + \frac{1}{s'} = \frac{1}{f}$$ 33-11   *Thin-lens equation*

Equation 33-10 is sometimes called the *lens-maker's equation* because it gives the focal length of a thin lens in terms of the properties of the lens. Equation 13-11 is the same as that for a spherical mirror. Note that one or both of the radii of curvature may be negative according to our sign convention. For example, in the double convex lens shown in Figure 33-15, the first radius r_1 is positive because the center of curvature lies

Parallel light rays incident on a double convex lens. The refracted rays converge at the second focal point of the lens. Reflected rays from each surface of the lens can also be seen. The first surface acts as a convex mirror producing diverging reflected rays, whereas the second surface acts as a concave mirror producing converging reflected rays.

Figure 33-15
Double convex lens. Both surfaces bend the light rays toward the axis. For this lens, r_1 is positive and r_2 is negative in Equation 33-10, resulting in a positive focal length.

on the right side of the surface, which is the real side, but r_2 is negative because the center of curvature of the second surface lies on the left or virtual side of the surface. In this case, both surfaces tend to bend a light ray toward the axis of the lens. This lens is a *converging lens*. Since both $1/r_1$ and $-1/r_2$ are positive and $n-1$ is positive, the focal length f given in Equation 33-10 must be positive, indicating that parallel light from the left is focused on the right, or real, side of the lens. A converging lens has a positive focal length and is therefore called a *positive lens*.

Bausch & Lomb

Parallel light rays incident on a double concave lens. The refracted rays diverge as if they came from the first focal point to the left of the lens. As in the previous photograph, reflected rays from each surface of the lens can also be seen. Here, the first surface acts as a concave mirror producing converging reflected rays, whereas the second surface acts as a convex mirror producing diverging reflected rays.

Figure 33-16 shows a double concave lens. Both surfaces tend to diverge light rays away from the lens axis. Since both $1/r_1$ and $-1/r_2$ are negative, the focal length is negative, indicating that parallel light from the left is diverged as if it came from a point on the left of the lens. A *diverging lens* has a negative focal length and is called a *negative lens*. For any other lens in which r_1 and r_2 are both positive or both negative, the lens is converging or diverging depending on which radius of curvature has the greater magnitude. If we turn any lens around, we interchange r_1 and r_2 and change their signs; i.e., the new radii of curvature for a lens which is turned around are related to the old by $r_1' = -r_2$ and $r_2' = -r_1$. We see from Equation 33-10 that the new focal length f' has the same magnitude and sign as the old. Thus, for example, if parallel light strikes a double convex lens from the *right*, it is focused at a point on the left a distance f from the lens. The two points on the left and right of any thin lens a distance f from the lens are the first and second focal points of the lens, designated by F and F'. Using the reversibility property of light rays, we can see that light diverging from either of the focal points and striking the lens will leave the lens as a parallel beam. If we set the object distance s in Equation 33-10 equal to the focal distance f, we obtain $s' = \infty$, indicating that the light is not focused but emerges as a parallel beam.

Figure 33-16
Double concave lens. Both surfaces bend the light rays away from the axis. Here r_1 is negative and r_2 is positive in Equation 33-10, resulting in a negative focal length.

If we have two or more thin lenses, we can find the final image produced by the system by finding the first image distance and using it along with the distance between lenses to find the object distance for the second lens. That is, we consider each image, whether it is formed or not, as the object point for the next lens.

As with images formed by mirrors, it is convenient to locate the image by graphical methods. Figure 33-17 illustrates the graphical method for a converging lens. Three convenient rays from the head of the figure are (1) a ray parallel to the axis of the lens which is bent through the focal point on the right of the lens, (2) a ray through the center of the lens (vertex) which is undeflected, and (3) a ray through the first focal point of the lens which emerges parallel to the axis. These three rays converge to the image point, as indicated. In this case the image is real and inverted. From Figure 33-17 we have $\tan \theta = y/s = -y'/s'$. The lateral magnification is then

$$m = \frac{y'}{y} = -\frac{s'}{s} \qquad\qquad 33\text{-}12$$

Figure 33-17
Ray diagram for a thin converging lens. For simplicity we assume all the bending to take place at a line. The ray through the center of the lens is undeflected because the lens surfaces there are nearly parallel and close together.

This expression is the same as that for mirrors. Again a negative magnification indicates that the image is inverted.

Example 33-3 A double convex thin lens made of glass of index of refraction $n = 1.5$ has both radii of curvature of magnitude 20 cm. An object 2 cm high is placed 10 cm from the lens. Find the focal length of the lens, locate the image, and find its size.

We first calculate the focal length from Equation 33-10, noting that $r_1 = 20$ cm and $r_2 = -20$ cm by our sign convention. The focal length is thus

$$\frac{1}{f} = (1.5 - 1) \left(\frac{1}{20 \text{ cm}} - \frac{1}{-20 \text{ cm}} \right) = \frac{1}{20 \text{ cm}} \qquad \text{or} \qquad f = 20 \text{ cm}$$

Figure 33-18 shows a ray diagram for a small object 10 cm from the lens. The ray parallel to the axis is bent through the focal point, as shown. The ray through the vertex is undeflected. These rays are diverging on the right of the lens. The image is thus located by extending the rays back until they meet. These two rays are sufficient to locate the image.

Figure 33-18
Ray diagram for Example 33-3. When the object is between the first focal point and a converging lens, the image is virtual and erect.

(As a check, we could draw a third ray from the object along a line through the focal point in front of the lens. This ray would then leave the lens parallel to the axis.) We can see immediately from this drawing

Bausch & Lomb

Parallel light rays incident on a diverging lens followed by a converging lens. The net effect of such a lens system can be converging rays, diverging rays, or parallel rays, depending on the lens separation and the relative size of the focal lengths of the two lenses. Here the net effect is to produce converging rays from parallel rays.

that the image is virtual, erect, and enlarged. It is on the same side of the lens as the object and is about twice as far away from the lens. Since it is quite easy to make an error in the calculation of the image distance from Equation 33-11, it is always a good idea to check your result with a ray diagram.

The image distance is found algebraically from Equation 33-11:

$$\frac{1}{10 \text{ cm}} + \frac{1}{s'} = \frac{1}{20 \text{ cm}}$$

$$\frac{1}{s'} = \frac{1}{20 \text{ cm}} - \frac{1}{10 \text{ cm}} = -\frac{1}{20 \text{ cm}} \quad \text{or} \quad s' = -20 \text{ cm}$$

The image distance is negative, indicating that the image is virtual and to the left of the lens. The magnification is $m = -s'/s = -(-20 \text{ cm})/(10 \text{ cm}) = +2$. The image is thus twice as large as the object, and it is erect. Since the size of the object is given as 2 cm, the size of the image is 4 cm.

Example 33-4 A second lens of focal length +10 cm is placed 20 cm to the right of the lens in Example 33-3. Locate the final image.

Figure 33-19 shows a ray diagram for this example. The rays used to locate the image of the first lens are not necessarily convenient rays for the second lens. In this case one of the rays is, the one through the focal point of the first lens, which coincides with the center of the second lens. We see that the final image is real, inverted, and just outside the focal point of the second lens. We locate its position algebraically by noting that the virtual image of the first lens is 20 cm to the left of that lens and 40 cm to the left of the second lens. Using $s = 40$ cm and $f = 10$ cm, we have

$$\frac{1}{40} + \frac{1}{s'} = \frac{1}{10}$$

giving

$$s' = \tfrac{40}{3} \text{ cm} = 13.3 \text{ cm}$$

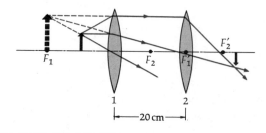

Figure 33-19
Ray diagram for Example 33-4. The image due to the first lens serves as the object for the second lens. The final image is located by drawing two rays from the first image. In this case one of the original rays used to locate the first image is undeflected by the second lens. A second ray from the first image parallel to the axis locates the final image.

Example 33-5 Two lenses of focal length 10 cm each are 15 cm apart. Find the final image of an object 15 cm from one of the lenses.

In the ray diagram of Fig. 33-20 the image of the first lens would be 30 cm from the lens if the second lens were not there. We calculate this using $s = 15$ cm and $f = 10$ cm in the thin-lens equation:

$$\frac{1}{15} + \frac{1}{s'} = \frac{1}{10}$$

$$\frac{1}{s'} = \frac{1}{10} - \frac{1}{15} = \frac{1}{30} \quad \text{or} \quad s' = 30 \text{ cm}$$

Figure 33-20
Ray diagram for Example 33-5. The image due to the first lens is to the right of the second lens. The rays do not actually converge to that image because they are refracted by the second lens. That image therefore serves as a virtual object for the second lens. The final image is found by drawing rays toward the first image, as shown. Here, a ray parallel to the axis and one through the center of the second lens are used.

This image is not formed because the light rays strike the second lens before they reach the image position. We can locate the final image graphically by choosing rays which are heading toward the unformed image when they strike the lens. A ray parallel to the axis and one through the center of the second lens are sufficient. We see that the final image is between the second lens and its focal point. The final image can be located algebraically by using the first image as the object for the second lens. Since this image is not formed because it is to the right of the second lens, it is a *virtual object*. Since it is 15 cm to the right of the second lens, the object distance is $s = -15$ cm. Then

$$\frac{1}{-15} + \frac{1}{s'} = \frac{1}{f} = \frac{1}{10}$$

$$\frac{1}{s'} = \frac{1}{10} + \frac{1}{15} = \frac{5}{30} = \frac{1}{6} \quad \text{or} \quad s' = 6 \text{ cm}$$

Questions

4. A fish under water is viewed from a point not directly overhead. Is its apparent depth greater or less than three-fourths its actual depth?

5. Under what conditions will the focal length of a thin lens be positive? Negative?

6. Under what conditions will the image formed by a positive thin lens be erect? Inverted? Real? Virtual? Larger than the object? Smaller than the object?

7. The focal length of a lens is different for different colors of light. Why?

33-5 Aberrations

When all the rays from a point object are not focused at a single image point, the resulting blurring of the images is called an *aberration*. The blurring of the image of an object point on the axis due to rays which strike a mirror or lens at points far from the axis is called *spherical aberration*. The similar blurring of the image of a point off the axis due to the different magnification of different parts of the lens or mirror is called *coma*, after the comet-shaped image observed. The imaging of a point source off axis into two mutually perpendicular lines at different locations is called *astigmatism*. The distortion in shape of the image of an extended object due to the fact that the magnification depends on the distance of the object point from the axis is called *distortion*. These aberrations, common to both mirrors and lenses, are the result of the laws of reflection and refraction applied to spherical surfaces. They are not evident in our simple equations for mirrors and lenses because we have used small-angle approximations in the derivation of these equations. Some of the aberrations can be eliminated or partially corrected by using nonspherical surfaces for the mirror or lens, but this is usually difficult and costly compared with the production of spherical surfaces. One example of a nonspherical reflecting surface is the parabolic mirror illustrated in Figure 33-21. Parallel rays incident on the parabolic surface are reflected and focused at a common point no matter how far they are from the axis. Parabolic reflecting surfaces are important in large astronomical telescopes, in which a large reflecting surface is needed to make the intensity of the image as great as possible. A parabolic surface can also be used in a searchlight to produce a parallel beam of light from a small source placed at the focus of the surface.

An important aberration of lenses not found in mirrors is *chromatic aberration* due to the variation in the index of refraction with wavelength. From Equation 33-10 we see that even for paraxial rays, the focal length of a lens depends on the index of refraction and is therefore slightly different for different wavelengths. Chromatic and other aberrations can be partially corrected by using combinations of lenses instead of a single lens. For example, a positive lens and a negative lens of greater focal length can be used together to produce a converging lens system which has much less chromatic aberration than a single lens of the same focal length.

Figure 33-21
Spherical aberration can be eliminated by using a parabolic surface, which reflects all parallel rays through a common focal point.

33-6   The Eye

The optical system of prime importance is the eye, shown in Figure 33-22. Light enters the eye through a variable aperture, the pupil, and is focused by the cornea-lens system on the retina, a film of nerve fibers covering the back surface. The retina contains tiny sensing structures called *rods* and *cones*, which receive the image and transmit the information along the optic nerve to the brain. The shape of the crystalline

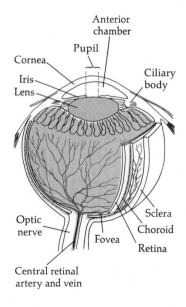

Figure 33-22
Cutaway view of the human eye. The amount of light entering the eye is controlled by the iris, which regulates the size of the pupil. Because the lens is surrounded by material of nearly equal index of refraction, most of the refraction occurs at the cornea. The lens thickness is controlled by the ciliary muscle. The cornea and lens together focus the image on the retina, which contains about 125 million receptors called rods and cones and about 1 million optic-nerve fibers. (*From H. Curtis,* Biology, *2d ed., p. 696. Worth Publishers, Inc., New York, 1975.*)

lens can be slightly altered by the action of the ciliary muscle. When the eye is focused on an object far away, the muscle is relaxed and the cornea-lens system has its maximum focal length, about 2.5 cm, the distance from the cornea to the retina. When the object is brought closer to the eye, the ciliary muscle tenses, increasing the curvature of the lens slightly and decreasing its focal length, so that the image is again focused on the retina. This process is called *accommodation*. If the object is too close to the eye, the lens cannot focus the light on the retina and the image is blurred. The closest point for which the lens can focus the image on the retina is called the *near point*. The distance from the eye to the near point varies greatly from one person to another and with age. At the age of 10 years, the near point may be as close as 7 cm, whereas at 60 years it may recede to 200 cm because of the loss of flexibility of the lens. The standard value taken for the near point is 25 cm.

If the relaxed eye images distant objects behind the retina, the person is said to be farsighted. Farsightedness is corrected with a converging lens (Figure 33-23). On the other hand, if light from a distant object is focused in front of the retina by the relaxed eye, the person is nearsighted. Nearsightedness is corrected with a diverging lens (Figure 33-24).

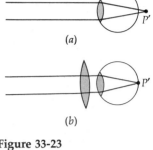

Figure 33-23
(a) The lens of a farsighted eye focuses parallel rays to form an image behind the retina. (b) A converging lens corrects this defect by bringing the image onto the retina.

(a)   (b)

Figure 33-24
(a) The lens of a nearsighted eye focuses parallel rays to form an image in front of the retina. (b) A diverging lens corrects this defect.

Example 33-6 By how much must the focal length of the cornea-lens system of the eye change when the object is moved from infinity to the near point at 25 cm? Assume that the distance from the cornea to the retina is 2.5 cm.

When the object is at infinity, the focal length of the cornea-lens system is 2.5 cm. When the object is at 25 cm, the focal length f must be such that the image distance is 2.5 cm. From Equation 33-11 we have

$$\frac{1}{25 \text{ cm}} + \frac{1}{2.5 \text{ cm}} = \frac{1}{f}$$

$$\frac{1}{f} = \frac{1}{25 \text{ cm}} + \frac{10}{25 \text{ cm}} = \frac{11}{25 \text{ cm}}$$

$$f = \frac{25 \text{ cm}}{11} = 2.27 \text{ cm}$$

The focal length must therefore decrease by 0.23 cm.

(*Below, left*) Section of the retina of the human eye, magnified 500 times. (*From F. W. Sears and M. W. Zemansky*, University Physics, *4th ed., p. 582. Addison-Wesley Publishing Company, Inc., Reading, Mass., 1970.*) (*Below*) Rods (cylindrical structures extending into lower foreground) and cones (upper middle and left) as seen through a scanning electron microscope, magnified 1600 times. (*Courtesy of E. R. Lewis.*)

Nerve fibers

Receptor layer

Rods   Cones

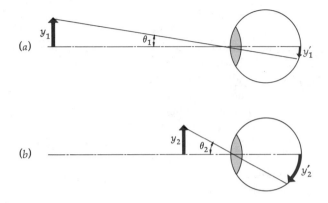

(a)

(b)

Figure 33-25
(a) A distant object of height y_1 looks small because the size of the image on the retina is small. (b) When the same object is closer, it looks larger because the retinal image is larger. The size of the image on the retina is approximately proportional to the angle θ subtended by the object.

The apparent size of an object is determined by the size of the image on the retina. The larger the image on the retina, the greater the number of rods and cones activated. From Figure 33-25 we see that the size of the image on the retina is greater when the object is close than it is when the object is far away. Even though the actual size of the object does not change, its apparent size is greater when it is brought closer to the eye. Since the near point is the closest point to the eye for which a sharp image can be formed on the retina, the distance to the near point is called the *distance of most distinct vision*. A convenient measure of the size of the image on the retina is the angle subtended by the object at the eye (Figure 33-25). When an object of height y is at the near point, the angle subtended is

$$\theta_0 = \frac{y}{25 \text{ cm}} \qquad \qquad 33\text{-}13$$

where we have taken 25 cm to be the distance to the near point and have used the small-angle approximation $\theta_0 \approx \tan \theta_0$.

33-7 The Simple Magnifier

The apparent size of an object can be increased by using a converging lens called a *simple magnifier*. In Figure 33-26b a small object of height y is at the first focal point of a thin converging lens of focal length f, which is less than the distance of most distinct vision (taken to be 25 cm). The image formed by the converging lens is virtual, erect, and at infinity. It subtends an angle θ given by

$$\theta = \frac{y}{f}$$

Viewed through the lens, the object looks larger because the image on the retina is larger by the factor θ/θ_0, where θ_0, given by Equation 33-13, is the angle subtended by the object at the near point. The ratio θ/θ_0 is called the angular magnification or magnifying power M of the lens:

$$M = \frac{\theta}{\theta_0} = \frac{25 \text{ cm}}{f} \qquad \qquad 33\text{-}14$$

The angular magnification of a simple magnifier can be increased if the object is moved closer to the magnifier, since the image moves in from infinity and the angle subtended increases. The largest usable magnifi-

25 cm

(a)

f

(b)

Figure 33-26
(a) An object at the near point subtends an angle θ_0 at the eye. (b) When the object is at the focal point of the converging lens, the image (not shown) is at infinity and can be viewed by the relaxed eye. When f is less than 25 cm, the converging lens acts as a simple magnifier by allowing the object to be brought closer to the eye, increasing the angle subtended by the object to θ and thereby increasing the image size on the retina.

cation occurs when the image is at the near point of the eye, 25 cm. The angle subtended is then y/s, where s is the object distance for the magnifier. We can calculate s from the thin-lens equation 33-11. Using $s' = -25$ cm, we have

$$\frac{1}{s} + \frac{1}{-25} = \frac{1}{f}$$

$$\frac{1}{s} = \frac{1}{f} + \frac{1}{25} = \frac{25 + f}{25f} \qquad s = \frac{25f}{25 + f}$$

Then

$$\theta = \frac{y}{s} = \frac{25 + f}{25f}\, y \qquad \text{and} \qquad \frac{\theta}{\theta_0} = \frac{25 + f}{f} = 1 + \frac{25}{f}$$

When the image of the magnifier is viewed at the near point, the magnification is

$$M = 1 + \frac{25 \text{ cm}}{f} \tag{33-15}$$

It is left as an exercise to show that in this case the angular magnification is equal to the linear magnification $m = y/y'$.

Simple magnifiers are used as eyepieces or oculars in compound microscopes and telescopes to view the image formed by another lens or lens system. To correct aberrations, combinations of lenses with a resulting short positive focal length may be used as a simple magnifier in place of a single lens, but the principle is the same. For convenience we shall use $(25 \text{ cm})/f$ for the magnification of a simple magnifier, though the choice between Equations 33-14 and 33-15 depends on whether the image is viewed at infinity or the near point. Of course, if the distance to the near point is much greater than 25 cm, the effective magnification of a simple magnifier is greater. For example, if the near point for a person is 50 cm from the eye, a simple magnifier of focal length f gives an angular magnification of $(50 \text{ cm})/f$ when the image is viewed at infinity.

33-8   The Compound Microscope and the Telescope

The compound microscope and the telescope are interesting applications of the principles of geometric optics. In their simplest forms each consists of two converging lenses. The lens nearest the object, called the *objective*, forms a real image of the object. The lens nearest the eye, called the *ocular* or *eyepiece*, is used as a simple magnifier to view the image formed by the objective.

The compound microscope (Figure 33-27) is used to look at very small objects at short distances. The distance between the second focal point of the objective and the first focal point of the ocular is called the *tube length l* and is usually fixed. The object is placed just outside the focal point of the objective so that an enlarged image is formed at the first focal point of the ocular a distance $l + f_o$ from the objective, where f_o is the focal length of the objective. From Figure 33-27, $\tan \beta = y/f_o = -y'/l$. The lateral magnification of the objective is therefore

$$m_o = \frac{y'}{y} = -\frac{l}{f_o} \tag{33-16}$$

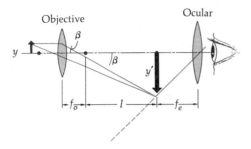

Objective   Ocular

Figure 33-27
Schematic diagram of a compound microscope, consisting of two positive lenses, the objective of focal length f_o, and the ocular, or eyepiece, of focal length f_e, separated by a distance $l + f_o + f_e$. The real image of the object formed by the objective is viewed by the ocular, which acts as a simple magnifier.

The overall magnification of the compound microscope is the product of the lateral magnification of the objective and the angular magnification of the ocular $M_e = (25 \text{ cm})/f_e$, where f_e is the focal length of the ocular, or eyepiece:

$$m = m_o M_e = - \frac{l}{f_o} \frac{25 \text{ cm}}{f_e} \qquad \text{33-17}$$

The astronomical telescope, illustrated schematically in Figure 33-28, is used to view objects which are far away and often large. The purpose of the objective is to form a real image at the first focal point of the ocular. Since the object is usually a great distance from the telescope, the image is at the second focal point of the objective. The objective and ocular are therefore separated by a distance $f_o + f_e$, where f_o and f_e are the focal lengths of the objective and ocular, respectively.

Let θ_o be the angle subtended by the object when it is viewed directly by the unaided eye. This is the same as the angle subtended by the object at the objective shown in the figure. The angle θ_e in the figure is that subtended by the final image. The angular magnification of the telescope is θ_e/θ_o. Using the small-angle approximation, we have

$$\tan \theta_o = - \frac{y'}{f_o} \approx \theta_o$$

where we have introduced the negative sign to make θ_o positive when y' is negative. Similarly,

$$\tan \theta_e = \frac{y'}{f_e} \approx \theta_e$$

Since y' is negative, θ_e is negative, indicating that the image is inverted. The angular magnification of the telescope is then

$$M = \frac{\theta_e}{\theta_o} = - \frac{f_o}{f_e} \qquad \text{33-18}$$

Seventeenth-century telescope used by Galileo.

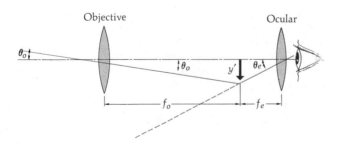

Objective   Ocular

Figure 33-28
Schematic diagram of an astronomical telescope. The objective forms a real image of a distant object near its second focal point, which coincides with the first focal point of the ocular. The ocular serves as a simple magnifier which views this image.

Review

A. Define, explain, or otherwise identify:

Virtual image, 878
Real image, 879
Paraxial ray, 879
Spherical aberration, 879
Focal length, 880
Focal point, 880
Reversibility, 880
Lateral magnification, 881
Apparent depth, 884
Converging lens, 886

Positive lens, 886
Diverging lens, 886
Negative lens, 886
Virtual object, 889
Chromatic aberration, 890
Accommodation, 891
Near point, 891
Angular magnification, 892
Objective, 893
Ocular, 893

B. True or false:

1. Virtual and real images look the same to the eye.

2. A virtual image cannot be displayed on a screen.

3. Aberrations occur only for real images.

4. A negative image distance implies that the image is virtual.

5. All rays parallel to the axis of a spherical mirror are reflected through a single point.

6. A diverging lens cannot form a real image from a real object.

7. The image distance for a positive lens is always positive.

8. Chromatic aberration does not occur with mirrors.

9. The image formed by the objective of a telescope is inverted and larger than the object.

10. The final image of a telescope is virtual.

Exercises

Section 33-1, Plane Mirrors

1. The image of the point object shown in Figure 33-29 is viewed by an eye as shown. Draw a bundle of rays from the object which reflect from the mirror and enter the eye. For this object position and mirror, indicate the region of space in which the eye can be placed to see the image.

Figure 33-29
Exercise 1.

2. A point object is placed between parallel mirrors separated by 30 cm. The object is 10 cm from the lower mirror and 20 cm from the upper mirror. (a) Find the distances from the upper mirror to the first four images which occur above the mirror. (b) Find the distances from the lower mirror to the first four images which occur below the lower mirror.

3. Two plane mirrors make an angle of 90°. Show that there are three images for any position of the object. Draw appropriate bundles of rays from the object to the eye for viewing each image.

4. Two plane mirrors make an angle of 60° with each other. Consider a point object on the bisector of the angle between the mirrors. Show on a sketch the location of all the images formed.

5. Repeat Exercise 4 for two mirrors at an angle of 120°.

Section 33-2, Spherical Mirrors

6. A concave spherical mirror has a radius of curvature of 50 cm. Draw ray diagrams to locate the image (if one is formed) for an object at a distance of (a) 100 cm, (b) 50 cm, (c) 25 cm, and (d) 10 cm from the mirror. For each case state whether the image is real or virtual; erect or inverted; and enlarged, reduced, or the same size as the object.

7. Use the mirror equations 33-2 to 33-4 to locate and describe the images for the object distances and mirror of Exercise 6.

8. Repeat Exercise 6 with a convex mirror of the same radius of curvature.

9. Repeat Exercise 7 with a convex mirror of the same radius of curvature.

10. Show that a convex mirror cannot form a real image of a real object no matter where the object is placed by showing that s' is always negative for positive s.

11. Show that a concave mirror forms a virtual image of a real object when $s < f$ and a real image when $s > f$ (see Exercise 10).

12. A certain telescope uses a concave spherical mirror of radius 8 m as its objective. Find the location and diameter of the image of the moon, which has a diameter of 3.5×10^6 m and is 3.8×10^8 m from the mirror.

Section 33-3, Images Formed by Refraction

13. A sheet of paper with writing on it is protected by a thick glass plate having an index of refraction of 1.5. If the plate is 2 cm thick, at what distance beneath the top of the plate does the writing appear when viewed from directly overhead?

14. A very long glass rod has one end ground to a convex hemispherical surface of radius 5 cm. Its index of refraction is 1.5. (a) A point object is on the axis in air a distance 20 cm from the surface. Find the image and state whether it is real or virtual. (b) Repeat for an object at 5 cm from the surface. (c) Repeat for an object very far from the surface. Draw a ray diagram for this case.

15. At what distance from the rod of Exercise 14 should an object be placed so that the light rays in the rod are parallel to the axis? Draw a ray diagram for this situation.

16. Repeat Exercise 14 for a glass rod with a concave hemispherical surface of radius (−) 5 cm.

17. Repeat Exercise 14 for the glass rod immersed in water of index of refraction 1.33; this means that the objects are in water.

18. Repeat Exercise 14 for a glass rod with a concave surface ($r = -5$ cm) immersed in water (index of refraction 1.33).

Section 33-4, Lenses

19. The following thin lenses are made of glass with an index of refraction of 1.5. Make a sketch and find the focal length in air for each: (a) double convex: $r_1 = 10$ cm, $r_2 = -21$ cm; (b) plano-convex: $r_1 = \infty$, $r_2 = -10$ cm; (c) double concave: $r_1 = -10$ cm, $r_2 = +10$ cm; (d) plano-concave: $r_1 = \infty$, $r_2 = +20$ cm.

20. Glass with an index of refraction of 1.6 is used to make a thin lens which has radii of equal magnitude. Find the radii of curvature and make a sketch if the focal length in air is to be (a) +5 cm, (b) −5 cm.

21. The following thin lenses are made of glass of index of refraction 1.5. Make a sketch and find the focal length in air for each: (a) $r_1 = 20$ cm, $r_2 = 10$ cm; (b) $r_1 = 10$ cm, $r_2 = 20$ cm; (c) $r_1 = -10$ cm, $r_2 = -20$ cm.

22. For the following object distances and focal lengths of thin lenses in air find the image distance and the magnification and state whether the image is real or virtual, erect or inverted. (a) $s = 40$ cm, $f = 20$ cm; (b) $s = 10$ cm, $f = 20$ cm; (c) $s = 40$ cm, $f = -30$ cm; (d) $s = 10$ cm, $f = -30$ cm.

23. What is meant by a negative object distance? How can it occur? Find the image distance and magnification and state whether the image is virtual or real, erect or inverted, for a thin lens in air when (a) $s = -20$ cm, $f = +20$ cm; (b) $s = -10$ cm, $f = -30$ cm.

24. An object 3.0 cm high is placed 20 cm from a thin lens of focal length 5.0 cm. Draw a careful ray diagram to find the position and size of the image and check your result using the thin-lens equation.

25. Repeat Exercise 24 for an object 1.0 cm high placed 10 cm from a thin lens of focal length 5.0 cm.

26. Repeat Exercise 24 for an object 1.0 cm high placed 10 cm from a thin lens whose focal length is -5.0 cm.

27. A thin converging lens of focal length 10 cm is used to obtain an image which is twice as large as a small object. (a) Where should the object be placed? (b) Where is the image? (c) Is the image real or virtual? Erect or inverted?

28. Show that a diverging lens can never form a real image from a real object. *Hint:* Show that s' is always negative.

29. Show that a converging lens forms a real image from a real object when s is greater than f. *Hint:* Show that s' is then positive.

30. Show that if two thin lenses of focal lengths f_1 and f_2 are placed in contact, they are equivalent to a single thin lens of focal length $f = f_1 f_2/(f_1 + f_2)$.

31. Two converging lenses each of focal length 10 cm are separated by 35 cm. An object is 20 cm to the left of the first lens. Find the final image using both ray diagrams and the lens equation. What is the overall lateral magnification?

Section 33-5, Aberrations

32. A concave spherical mirror has a radius of curvature 6.0 cm. A point object is on the axis at 9.0 cm from the mirror. Construct a careful ray diagram showing rays from the object which make angles of 5, 10, 30, and 60° with the axis, strike the mirror, and are reflected back across the axis. (Use a compass to draw the mirror, and use a protractor to measure the angles needed to find the reflected rays.) What is the spread along the axis of the image for these rays?

33. For the mirror of Exercise 32, draw rays parallel to the axis at distances 0.5, 1, 2, and 4 cm and find the points at which the reflected rays cross the axis.

34. A double convex lens of radii $r_1 = +10$ cm and $r_2 = -10$ cm is made from a glass with an index of refraction of 1.53 for blue light and 1.47 for red light. Find the focal lengths of this lens for red and blue light.

Section 33-6, The Eye

In the following exercises take the distance from the cornea-lens system of the eye to the retina to be 2.5 cm.

35. Suppose the eye were designed like a camera with a lens of fixed focal length $f = 2.5$ cm which could move toward or away from the retina. Approximately how far would the lens have to move to focus the image of an object 25 cm from the eye on the retina? *Hint:* Find the distance from the retina to the image behind it for an object at 25 cm.

36. Since the index of refraction of the lens of the eye is not very different from that of the surrounding material, most of the refraction takes place at the

cornea, where n changes abruptly from 1.0 in air to about 1.4. Assuming the cornea to be homogeneous with an index of refraction of 1.4, calculate its radius if it focuses parallel light on the retina a distance 2.5 cm away.

37. If two point objects close together are to be seen as two distinct objects, the images must fall on the retina on two different cones that are not adjacent; i.e., there must be an unactivated cone between them. The separation of the cones is about 1 μm. What is the smallest angle the two points can subtend (see Figure 33-30)? How close can two points be if they are 20 m from the eye?

Figure 33-30
Exercise 37.

Section 33-7, The Simple Magnifier

38. Show that when the image is viewed at the near point, the lateral and angular magnifications of a simple magnifier are equal.

39. What is the magnifying power of a lens of focal length 5 cm when (a) the image is viewed at infinity and (b) the image is viewed at the near point 25 cm from the eye?

40. A lens of focal length 10 cm is used as a simple magnifier with the image at infinity by one person whose near point is at 25 cm and by another whose near point is at 50 cm from the eye. What is the effective magnifying power of the lens for each person? Compare the size of the image on the retina when each looks at the same object with the magnifier.

Section 33-8, The Compound Microscope and the Telescope

41. A crude symmetric hand-held microscope consists of two converging lenses each of focal length 5.0 cm fastened in the ends of a tube 30 cm long. (a) What is the "tube length"? (b) What is the lateral magnification of the objective? (c) What is the overall magnification of the microscope? (d) How far from the objective should the object be placed?

42. Repeat Exercise 41 for the same two lenses separated by 40 cm.

43. A simple telescope has an objective with a focal length of 100 cm and an ocular with a focal length of 5 cm. It is used to look at the moon, which subtends an angle of about 0.009 rad. (a) What is the size of the image formed by the objective? (b) What angle is subtended by the final image at infinity?

44. The objective lens of the refracting telescope at the Yerkes Observatory has a focal length of about 19.5 m. What is the diameter of the image of the moon formed by the objective? (The angle subtended by the moon is about 0.009 rad.)

Problems

1. A slide projector is to form a real magnified image 4 m from its lens. The width of the slide (object) is 3.5 cm, and that of the image is to be 105 cm. What focal length should the projector lens have?

2. (a) Show that the focal length f' for a thin lens of index of refraction n_2 in a medium of index n_1 is given by

$$\frac{1}{f'} = \frac{n_2 - n_1}{n_1}\left(\frac{1}{r_1} - \frac{1}{r_2}\right)$$

(b) Show that if f is the focal length of a thin lens in air, its focal length in water is f', given by

$$f' = \frac{n_1(n_2 - 1)}{n_2 - n_1}f$$

where n_1 is the index of refraction of water and n_2 is that of the lens material.

3. You wish to see an image of your face for makeup or shaving. If you want the image to be upright, virtual, and magnified 1.5 times when your face is 30 cm from the mirror, what kind of mirror should you use, convex or concave, and what should its focal length be?

4. A glass rod 96 cm long with an index of refraction of 1.6 has its ends ground to convex spherical shapes of radii 8 cm and 16 cm. A point object is in air on the axis 20 cm from the 8-cm-radius end. (*a*) Find the image distance due to refraction at the first surface. (*b*) Find the final image due to refraction at both surfaces. (*c*) Is the final image real or virtual?

5. Repeat Problem 4 for a point object in air 20 cm from the 16-cm-radius end.

6. Show that as an object moves away from a spherical mirror, its image always moves in the opposite direction. *Hint:* Compute ds' in terms of ds from the mirror equation.

7. Find the focal length of a thick double convex lens with an index of refraction of 1.5, thickness 4 cm, and radii +20 and −20 cm.

8. A small object is 20 cm from a thin positive lens of focal length 10 cm. To the right of the lens is a plane mirror which crosses the axis at the focal point of the lens and is tilted so that the reflected rays do not go back through the lens (see Figure 33-31). Sketch a ray diagram showing the final image. Is this image real or virtual?

Figure 33-31
Problem 8.

9. Find the final image for the situation in Problem 8 except that the mirror is not tilted. Assume that the image is viewed by an eye to the left of the object, looking through the lens into the mirror.

10. A horizontal concave mirror with a radius of curvature of 50 cm holds a layer of water of refractive index 1.33 and maximum depth 1 cm. At what height above the mirror must an object be placed so that its image is at the same position as the object?

11. When a bright light source is placed 30 cm in front of a lens, there is an erect image 7.5 cm from the lens. There is also a faint inverted image 6 cm in front of the lens due to reflection from the front surface of the lens. When the lens is turned around, this weaker inverted image is 10 cm in front of the lens. Find the index of refraction of the lens.

12. An object is 15 cm in front of a positive lens of focal length 15 cm. A second positive lens of focal length 15 cm is 20 cm from the first lens. Find the final image and draw a ray diagram.

13. Work Problem 12 if the second lens is a diverging lens of focal length −15 cm.

14. A glass ball of radius 10 cm has an index of refraction 1.5. The back half of the ball is silvered so that it acts as a concave mirror (Figure 33-32). Find the position of the final image seen by an eye to the left of the object and ball for an object at (*a*) 30 cm and (*b*) 20 cm to the left of the front surface of the ball.

Figure 33-32
Problem 14.

15. (a) Show that to obtain magnification of magnitude m with a thin positive lens of focal length f, the object distance must be given by

$$s = \frac{m+1}{m} f$$

(b) A 35-mm camera has a lens with a 50-mm focal length. It is used to take a picture of a person 175 cm tall so that the image size on the film is 35 mm. How far from the camera should the person stand?

16. A convenient form of the thin-lens equation due to Newton measures the object and image distances from the focal points. Show that if $x = s - f$ and $x' = s' - f$, the thin-lens equation can be written $xx' = f^2$ and the lateral magnification is given by $m = -x'/f = -f/x$. Indicate x and x' on a sketch.

17. Show that a small change dn in the index of refraction of a lens material produces a small change in the focal length df given approximately by

$$\frac{df}{f} = -\frac{dn}{n-1}$$

Use this to find the focal length of a thin lens for blue light, for which $n = 1.53$, if the focal length for red light, for which $n = 1.47$, is 20 cm.

18. The lateral magnification of a spherical mirror or a thin lens is given by $m = -s'/s$. Show that for objects of small horizontal extent, the longitudinal magnification is approximately $-m^2$. Do this by showing that $ds'/ds = -s'^2/s^2$.

19. A disadvantage of the astronomical telescope for terrestrial use, e.g., at a football game, is that the image is inverted. A galilean telescope uses a converging lens as its objective but a diverging lens as its eyepiece. The image formed by the objective is beyond the diverging lens at its second focal point so that the final image is virtual and at infinity. Show with a ray diagram that the final image is erect. Show that the angular magnification is $M = -f_o/f_e$, where f_o is the focal length of the objective and f_e is the focal length of the ocular; f_e is negative, making M positive.

20. A galilean telescope is designed so that the final image is at the near point, 25 cm from the eye. The focal length of the objective is 100 cm and that of the ocular is -5 cm. (a) If the object distance is 30 m, where is the image of the objective? (b) What is the object distance for the ocular so that the final image is at the near point? (c) How far apart are the lenses? (d) If the object height is 1 m, what is the height of the final image? What is the angular magnification? (See Problem 19.)

CHAPTER 34   Physical Optics: Interference and Diffraction

Objectives   After studying this chapter you should:

1. Be able to work problems involving interference in thin films.

2. Be able to describe the Michelson interferometer.

3. Be able to sketch the two-slit-interference intensity pattern and locate the interference maxima and minima.

4. Be able to use the vector model to find the resultant of several harmonic waves.

5. Be able to sketch the interference pattern due to three or more equally spaced slits.

6. Be able to sketch the single-slit diffraction pattern and locate the first diffraction minimum.

7. Be able to sketch the combined interference-diffraction pattern for several slits.

8. Be able to state the Rayleigh criterion for resolution and use it to investigate the conditions for resolution of two objects.

9. Be able to discuss the use of diffraction gratings and find the resolving power of a grating.

Physical optics is the study of the interference and diffraction of light. These are the important phenomena that distinguish waves from particles. In Chapter 15 we discussed the interference of sound waves from two point sources and gave a brief qualitative discussion of the diffraction of sound. Since the analytical treatment of interference and diffraction is the same for all waves whether they are sound waves, waves on strings, water waves, or electromagnetic waves, you should review Chapter 15 before you begin this chapter.

When harmonic waves of the same frequency and wavelength but different phase come from two sources and combine, the resultant wave is a harmonic wave whose amplitude depends on the phase difference.

If the phase difference is 0 or an integer times 2π, the waves are in phase and the interference is constructive. The resultant amplitude equals the sum of the individual amplitudes, and the intensity (which is proportional to the square of the amplitude) is maximum. If the phase difference is $\frac{1}{2}\pi$ or an odd integer times $\frac{1}{2}\pi$, the waves are out of phase and the interference is destructive. The resultant amplitude is then the difference between the individual amplitudes, and the intensity is a minimum. If the amplitudes are equal, the maximum intensity is 4 times that of either source and the minimum intensity is zero. A common cause of phase difference between two waves is a difference in path length traveled by the two waves. A path difference of Δx contributes a phase difference δ given by

$$\delta = \frac{\Delta x}{\lambda} 2\pi \qquad\qquad 34\text{-}1$$

Another cause of phase difference is the $180° = \frac{1}{2}\pi$ phase change a wave undergoes upon reflection from a boundary surface when the wave speed is greater in the first (incident) medium than in the second.

As mentioned in Chapter 15, interference of waves from two sources is not observed unless the sources are coherent, i.e., the phase difference between the waves from the sources is constant in time. Because a light beam is usually the result of millions of atoms radiating independently, two different light sources are usually not coherent; the phase difference between the waves from such sources fluctuates randomly many times per second. Coherence in optics is usually achieved by splitting the light beam from a single source into two or more beams, which then combine to produce an interference pattern. This splitting can be caused by (1) reflection from two nearby surfaces of a thin film (Section 34-1), (2) reflection from a mirror with the beams combined by reflection from other mirrors, as in the Michelson interferometer (Section 34-2), or (3) diffraction of the beam from two small openings or slits in an opaque barrier (Section 34-3). Coherent sources can also be obtained by using a single point source and its image in a plane mirror for the two sources, an arrangement called Lloyd's mirror.

34-1 Interference in Thin Films

Perhaps the most commonly observed case of interference of light is that due to reflection of light from the two surfaces of a thin film of water, air, or oil. Consider viewing, at small angles with the normal, a thin film of water (such as a soap bubble), as in Figure 34-1. Part of the light is reflected from the upper surface. Since light travels more slowly in water than in air, there is a 180° phase change in this reflection. Part of the light enters the film and is partially reflected by the bottom water-air surface. There is no phase change in this reflection. If the light is nearly perpendicular to the surfaces, both the ray reflected from the top surface and the one reflected from the bottom surface can enter the eye at point P in the figure. The path difference between these two rays is $2t$, where t is the thickness of the film. This path difference produces a phase difference of $2\pi(2t/\lambda')$, where λ' is the wavelength of the light in the film; λ' is related to the wavelength λ_0 in air by

$$\lambda' = \frac{v}{f} = \frac{c/n}{f} = \frac{\lambda_0}{n} \qquad\qquad 34\text{-}2$$

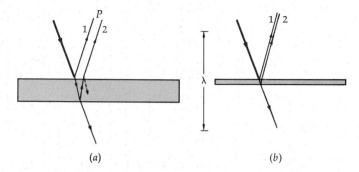

Figure 34-1
(*a*) Light rays reflected from
the top and bottom surfaces
of a thin film are coherent
and produce interference. (*b*)
If the film thickness is much
smaller than the wavelength
of light, the interference will
be destructive because one of
the rays undergoes a change
in phase of 180° upon re-
flection.

where n is the index of refraction. The phase difference between these
two rays is thus 180° plus that due to the path difference. Destructive
interference occurs when the path difference $2t$ is zero or a whole
number of wavelengths (in the film). Constructive interference occurs if
the path difference is an odd number of half wavelengths. We can
express these conditions mathematically as

$$\frac{2t}{\lambda'} = \begin{cases} m & \text{destructive} \qquad 34\text{-}3 \\ \\ m + \tfrac{1}{2} & \text{constructive} \qquad 34\text{-}4 \end{cases} \quad m = 0, 1, 2, 3, \ldots$$

Figure 34-2
Interference of light reflected
from a thin film of water
resting on a glass surface. In
this case, both rays undergo a
change in phase of 180° upon
reflection.

When a thin water film lies on a glass surface (Figure 34-2), the ray
that reflects from the lower water-glass surface also undergoes a 180°
phase change because the index of refraction of glass (about 1.5) is
greater than that of water (about 1.33). Both the rays shown in the figure
undergo a 180° phase change upon reflection. The phase difference
between these rays is due solely to the path difference and is given by
$\delta = 2\pi(2t/\lambda')$.

A common example of this type of interference occurs when there is a
thin film of oil on a water-covered surface or street. Because the
thickness of the film varies from point to point, constructive interference occurs for different wavelengths at different points, giving rise to
colored bands.

Figure 34-3 illustrates the interference pattern observed when light is
reflected from an air film between a spherical glass surface and a plane
glass surface in contact. These circular interference fringes are known as
Newton's rings. Near the point of contact, where the path difference
between the ray reflected from the upper glass-air surface and the lower
air-glass surface is essentially zero or at least small compared with the
wavelength of light, the interference is perfectly destructive because of
the 180° phase shift of the ray reflected from the lower air-glass surface.
This region is therefore dark. The first bright fringe occurs at a radius
such that the path difference contributes a phase difference of 180°,
which adds to that due to the phase shift upon reflection and produces a
total phase difference of 360°, or zero. The computation of the fringe
spacing in terms of the radius of curvature of the spherical piece of glass
is left as a problem.

Figure 34-3
Newton's rings observed
with light reflected from a
thin film of air between a
plane glass and convex glass
surface. At the center the
thickness of the air film is
negligible and the interference is destructive because of
the phase change of one of
the rays.

Example 34-1 A wedge-shaped film of air is made by placing a small
slip of paper between the edges of two flat pieces of glass. Light of
wavelength 500 nm is incident normally on the glass plates, and interference fringes are observed by reflection. If the angle made by the
plates is 3×10^{-4} rad, how many interference fringes per unit length
are observed?

Because of the 180° phase change of the ray reflected from the bottom plate (Figure 34-4), the first fringe near the point of contact (where the path difference is zero) will be dark. Let x be the horizontal distance to the mth dark fringe where the plate separation is t (Figure 34-4). Since the angle θ is very small, it is given approximately by

$$\theta = \frac{t}{x}$$

Using Equation 34-3 for m, we have

$$m = \frac{2t}{\lambda'} = \frac{2t}{\lambda}$$

since the film is an air film. Substituting $t = x\theta$ gives

$$m = \frac{2x\theta}{\lambda} \quad \text{or} \quad \frac{m}{x} = \frac{2\theta}{\lambda} = \frac{2(3 \times 10^{-4})}{5 \times 10^{-5} \text{ cm}} = 12 \text{ cm}^{-1}$$

where we have used $\lambda = 5 \times 10^{-7}$ m $= 5 \times 10^{-5}$ cm. We therefore observe 12 dark fringes per centimetre. In practice, the number of fringes per centimetre, which is easy to count, can be used to determine the angle.

Figure 34-4
Light incident normally on a wedge of air between two glass plates. The path difference $2t$ is proportional to x. Alternate bright and dark bands are seen when viewed from above.

Figure 34-5 shows interference fringes produced by a wedge-shaped air film between two flat glass plates, as in Example 34-1. The straightness of the fringes indicates the flatness of the glass plates. Such plates are called *optically flat*. A similar air wedge formed by two ordinary glass plates yields the irregular fringe pattern in Figure 34-6, indicating that these plates are not optically flat.

Figure 34-5
Straight-line fringes from a wedge of air like Figure 34-4. The straightness of the fringes indicates the flatness of the glass plates. (*Courtesy of T. A. Wiggins.*)

Figure 34-6
Fringes from a wedge of air between glass plates that are not optically flat. (*Courtesy of T. A. Wiggins.*)

Questions

1. Why must a film used to observe interference colors be thin?

2. Nonreflecting coatings for lenses are made by covering the lens with an extremely thin layer of material having an index of refraction greater than that of the glass. Explain how this works. Why does the coating have to be thin compared with the wavelength of visible light?

3. What effect does increasing the wedge angle in Example 34-1 have on the spacing of the interference fringes? If the angle is too large, fringes are not observed. Why?

4. The spacing of Newton's rings decreases rapidly as the diameter of the rings increases. Explain qualitatively.

34-2   The Michelson Interferometer

An interferometer is a device using interference fringes to make precise measurements. Figure 34-7 is a schematic diagram of a Michelson interferometer. Light from a broad source strikes a beam-splitter plate A, which is partially silvered. The light is partially reflected and partially transmitted by this plate. The reflected beam travels to mirror M_2 and is reflected back toward the eye at O. The transmitted beam travels through a compensating plate B of the same thickness as A to mirror M_1 and is reflected back to the eye at O. The purpose of the compensating plate is to make both beams pass through the same thickness of glass.

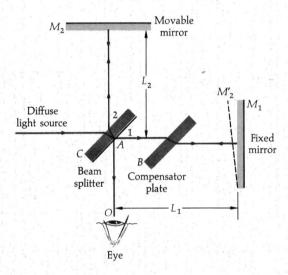

Figure 34-7
Michelson interferometer. The dashed line M_2' is the image of mirror M_2 in the mirror A. The interference fringes observed are those of a small wedge formed by the sources M_1 and M_2'. As M_2 (and therefore M_2') is moved, the fringes move past the field of view.

Mirror M_1 is fixed, but mirror M_2 can be moved back and forth with a fine and accurately calibrated screw adjustment. The two beams combine at O and form an interference pattern. This pattern is most easily understood by considering the light sources to be mirror M_1 and the image of mirror M_2 labeled M_2' in the diagram. This image is formed by the beam-splitting mirror. If the mirrors M_1 and M_2 are exactly perpendicular to each other and equidistant from the beam splitter, the image M_2' will coincide with M_1. If not, M_2' will be slightly displaced and make a small angle with M_1, as shown in the diagram. The interference pattern at O will then be that of a thin wedge-shaped film of air between M_1 and M_2', as discussed in Example 34-1. If mirror M_2 is now moved, the fringe pattern will shift. Suppose, for example, that the mirror M_2 is moved back $\frac{1}{4}\lambda$. The image M_2' will then move an additional distance $\frac{1}{4}\lambda$ away from M_1, increasing the thickness of the wedge $\frac{1}{4}\lambda$ at each point. This will introduce an additional path difference of $\frac{1}{2}\lambda$ everywhere in the wedge, and the fringe pattern will move over by one-half fringe; i.e., a previously dark fringe will now be a bright fringe, etc. As mirror M_2 is moved, the displacement of the fringe pattern is observed. If the

Albert A. Michelson in his laboratory. (*Courtesy of the Niels Bohr Library, American Institute of Physics.*)

distance the mirror is moved is known, the wavelength of the light can be determined. Michelson used such an interferometer to measure the wavelength of light emitted by krypton 86 in terms of the standard metre bar. This measurement was then used to redefine the standard metre in terms of this wavelength.

Another use for the Michelson interferometer is to measure the index of refraction of air (or of some other gas). One of the beams is enclosed in a container which can be evacuated. The wavelength in air λ' is related to that in vacuum by $\lambda' = \lambda_0/n$, where n is the index of refraction of air (about 1.0003). When the container is evacuated, the wavelength increases so that there are fewer waves in this distance, causing a shift in the fringe pattern. By measuring the shift the index of refraction can be determined (see Exercise 7). We shall discuss another example of the use of the Michelson interferometer in the next chapter when we discuss the famous Michelson-Morley experiment to measure the absolute velocity of the earth through space.

A student-type Michelson interferometer. The fringes are produced on a ground-glass screen by light from a laser. (*Courtesy of Dr. L. Velinsky.*)

34-3   The Two-Slit Interference Pattern

Interference patterns of light from two or more sources can be observed only if the sources are coherent, i.e., are in phase or have a phase difference that is constant in time. We have mentioned that the randomness of the radiation process of atoms means that two different light sources are generally incoherent. The interference of thin films discussed previously can be observed because the two beams come from the same light source but are separated by reflection. In Young's famous two-slit experiment used to demonstrate the wave nature of light, two coherent sources are produced by illuminating two parallel slits with a single source. We assume here that the width of each slit is small compared with the wavelength. (We shall treat the general case in a later section.) Then, because of diffraction, each slit acts as a line source (equivalent to a point source in two dimensions). The interference pattern is observed on a screen far from the slits. This pattern was discussed in Section 15-2. At very large distances from the slits, the lines from the two slits to some point P on the screen are approximately parallel, and the path difference is approximately $d \sin \theta$, as shown in Figure 34-8. We thus have interference maxima at an angle given by

$$d \sin \theta = m\lambda \qquad \text{34-5}$$

Two-slit interference maxima

and minima at

$$d \sin \theta = (m + \tfrac{1}{2})\lambda \qquad \text{34-6}$$

where $m = 0, 1, 2, 3, \ldots$. The phase difference at a point P is $2\pi/\lambda$ times the path difference $d \sin \theta$:

$$\delta = \frac{2\pi}{\lambda} d \sin \theta \qquad \text{34-7}$$

Figure 34-8
Two slits as coherent sources of light for the observation of interference in Young's experiment. Since the screen is very far away compared with the separation of the slits, the rays from the slits to a point on the screen are approximately parallel. The path difference between the two rays is $d \sin \theta$.

The distance x measured along the screen from the central point to point P is related to θ by $y = L \tan \theta$, where L is the distance from the slits to the screen. For small θ we have $\sin \theta \approx \tan \theta \approx \theta$. Then $\sin \theta \approx y/L$, and Equation 34-7 becomes

$$\delta \approx \frac{2\pi}{\lambda} \frac{yd}{L} \qquad \text{34-8}$$

To calculate the intensity on the screen at a general point P we need to add two harmonic wave functions that differ in phase, as we did in Chapter 15 when we discussed the interference of sound waves. The wave functions for electromagnetic waves are the electric field vectors. Let E_1 be the electric field at some point P on the screen due to the waves from slit 1, and let E_2 be the electric field due to waves from slit 2. Since the angles θ of interest are small, we can assume these fields to be parallel and consider only their magnitudes. Both electric fields oscillate with the same frequency (since they result from a single source that illuminates both slits), and they have the same amplitude, assuming that the slits are the same size and are approximately equidistant from the screen. (The path difference is only of the order of a few wavelengths of light at most.) They have a phase difference δ given by Equations 34-7 and 34-8. If we represent these wave functions by

$$E_1 = E_0 \sin \omega t \qquad \text{and} \qquad E_2 = E_0 \sin (\omega t + \delta)$$

the resultant wave function is

$$E = E_1 + E_2 = E_0 \sin \omega t + E_0 \sin (\omega t + \delta)$$
$$= 2E_0 \cos \tfrac{1}{2}\delta \sin (\omega t + \tfrac{1}{2}\delta) \qquad \text{34-9}$$

where we have used a trigonometric identity for the sum of two sine functions:

$$\sin A + \sin B = 2 \cos \tfrac{1}{2}(A - B) \sin \tfrac{1}{2}(A + B) \qquad \text{34-10}$$

(This is the same result as Equation 15-1.) The amplitude of the resultant wave is $2E_0 \cos \tfrac{1}{2}\delta$. It has its maximum value of $2E_0$ when the waves are in phase ($\delta = 0$, or an integer times 2π), and is zero when they are 180° out of phase ($\delta = \tfrac{1}{2}, \tfrac{3}{2}, \ldots$). Since the intensity is proportional to the square of the amplitude, the intensity at any point P is

$$I = 4I_0 \cos^2 \tfrac{1}{2}\delta \qquad \text{34-11}$$

Two-slit interference intensity

where I_0 is the intensity on the screen due to either slit acting separately. This intensity pattern is shown in Figure 34-9 as a function of $\sin \theta$, which is equivalent to a plot of intensity versus y for small θ. If we average over many interference maxima and minima, we obtain $2I_0$ for the average intensity since the average value of $\cos^2 \delta$ is $\tfrac{1}{2}$. This is the intensity that would arise from the two sources if they acted independently without interference. It is the intensity we would observe if the sources were incoherent, because then there would be an additional phase difference between them that fluctuates randomly so that only the average intensity could be observed.

Figure 34-9
Plot of relative intensity versus $\sin \theta$ observed on a screen far away from two slits as in Figure 34-8. The maximum intensity is proportional to $4A_0^2$ at $\sin \theta = m\lambda/d$, where m is an integer and A_0 is the amplitude from each source separately. This is 4 times the intensity of either source acting separately. The average relative intensity $2A_0^2$, indicated by the dashed line, equals twice that of each source acting separately. Since $\sin \theta \approx y/L$ for small θ, this is also a plot of relative intensity versus y.

Another method of producing the two-slit interference pattern is an arrangement shown in Figure 34-10 and known as Lloyd's mirror. A single slit is placed at a distance $\tfrac{1}{2}d$ above the plane of a mirror. Light striking the screen directly from the source interferes with that reflected from the mirror. The reflected light can be considered to come from the virtual image of the slit formed by the mirror. Because of the 180° change in phase on reflection at the mirror, the interference pattern is that of two coherent line sources which differ in phase by 180°. The pattern is the same as that shown in Figure 34-9 for two slits except that the maxima and minima are interchanged. The central fringe just above the mirror at a point equidistant from the sources is dark. Constructive interference occurs at points for which the path difference is $\tfrac{1}{2}$ wavelength or any odd number of half wavelengths. At these points, the 180° phase difference due to the path difference combines with the 180° phase difference of the sources to produce constructive interference.

Lloyd's mirror

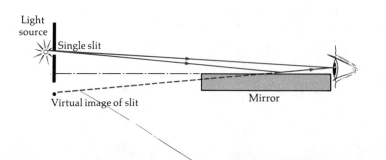

Figure 34-10
Lloyd's mirror for observation of double source interference with light. The two sources (light source and its image) are coherent and out of phase by 180° because of the phase change on reflection. The central interference band at points equidistant from the sources is dark.

34-4 The Vector Model for Addition of Harmonic Waves

To calculate the interference pattern produced by three, four, or more sources and to calculate the diffraction pattern of a single slit we need to combine several harmonic waves of the same frequency but differing in phase. A simple geometric interpretation of harmonic wave functions leads to a method of addition of harmonic waves of the same frequency by geometric construction. This method lets us find the sum of two or more harmonic waves geometrically without having to remember the trigonometric identity of Equation 34-10. It is useful even if the amplitudes of the waves are different or there are more than two waves. The method is based on the fact that the y (or x) component of the resultant of two vectors equals the sum of the y (or x) components of the vectors. Let

$$E_1 = A_1 \sin (kx - \omega t) \quad \text{and} \quad E_2 = A_2 \sin (kx - \omega t + \delta)$$

be the wave functions of the two waves. We wish to add these waves at some point x and some time t. We can simplify our notation by writing α for the quantity $kx - \omega t$. Our problem is then to find the sum

$$E_1 + E_2 = A_1 \sin \alpha + A_2 \sin (\alpha + \delta)$$

Consider a vector of magnitude A_1 making an angle α with the x axis (Figure 34-11). The y component of this vector is $A_1 \sin \alpha$, which is the wave function E_1. Similarly, the wave function $E_2 = A_2 \sin (\alpha + \delta)$ is the y component of a vector of magnitude A_2 making an angle $\alpha + \delta$ with the x axis. By the laws of vector addition, the sum of these components equals the y component of the resultant vector, as shown in Figure 34-11. The y component of the resultant vector, $A' \sin (\alpha + \delta')$, is a harmonic wave function which is the sum of the two original wave functions,

$$A_1 \sin \alpha + A_2 \sin (\alpha + \delta) = A' \sin (\alpha + \delta') \qquad 34\text{-}12$$

where $\alpha = kx - \omega t$ and A' (the amplitude of the resultant wave) and δ' (the phase of the resultant wave relative to the first wave) are found by adding the vectors representing the waves, as in Figure 34-11. As time

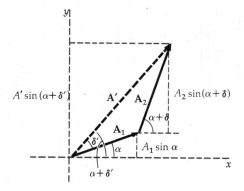

Figure 34-11
The wave function $A_1 \sin \theta$ is the y component of the vector $\mathbf{A}_1$ making angle α with the x axis. The wave function $A_2 \sin (\alpha + \delta)$ is the y component of the vector $\mathbf{A}_2$ making angle $\alpha + \delta$ with the x axis. The sum of these wave functions is $A' \sin (\alpha + \delta')$, which is the y component of the resultant vector $\mathbf{A}' = \mathbf{A}_1 + \mathbf{A}_2$.

varies, α varies. The vectors representing the two wave functions and the resultant vector representing the resultant wave function rotate in space, but their relative positions do not change because all the vectors rotate with the same angular velocity ω.

Example 34-2 Use the vector method of addition to derive Equation 34-9 for the superposition of two waves of the same amplitude.

Figure 34-12 shows the vectors representing two waves of equal amplitude A and the resultant wave of amplitude A'. These three vectors form an isosceles triangle in which the two equal angles are δ'. Since the sum of these angles equals the exterior angle δ, we have

$$\delta' = \tfrac{1}{2}\delta$$

The amplitude A' can be found from the right triangles formed by bisecting the resultant vector, as shown in Figure 34-12b. From these triangles we have

$$\cos \tfrac{1}{2}\delta = \frac{\tfrac{1}{2}A'}{A}$$

Therefore the amplitude is given by $A' = 2A \cos \tfrac{1}{2}\delta$, and the resultant wave is

$$A' \sin(\alpha + \delta') = 2A \cos \tfrac{1}{2}\delta \sin(\alpha + \tfrac{1}{2}\delta)$$

in agreement with Equation 34-9, with $\alpha = \omega t$.

Example 34-3 Find the resultant of the two waves

$$E_1 = 4 \sin(kx - \omega t) \quad \text{and} \quad E_2 = 3 \sin(kx - \omega t + 90°)$$

Figure 34-13 shows the vector diagram for this addition. The vectors make an angle of 90° with each other. The resultant of these two vectors has the magnitude 5 and makes an angle of 37° with the first vector, as shown. The sum of these two waves is

$$E_1 + E_2 = 5 \sin(kx - \omega t + 37°)$$

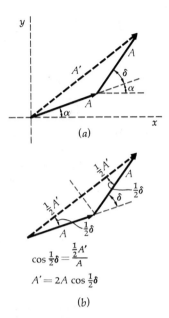

Figure 34-12
The vector addition for two waves having equal amplitude A and phase difference δ. (a) The vectors at a particular time at which $\alpha = \omega t$. (b) Construction for finding the amplitude of the resultant wave. The amplitude A' is found from $\cos \tfrac{1}{2}\delta = \tfrac{1}{2}A'/A$, giving $A' = 2A \cos \tfrac{1}{2}\delta$.

34-5 Interference Pattern of Three or Four Equally Spaced Sources

If we have three or more sources which are equally spaced and in phase with each other, the intensity pattern on a screen far away is similar to that due to two sources, but there are important differences. The position on the screen of the intensity maxima is the same no matter how many sources we have (assuming that they are equally spaced and in phase), but these maxima have much greater intensity and are much sharper if there are many sources. In this section we shall study the pattern produced by three or four sources. The extension of this study to the problem of finding the intensity pattern of a large number of equally spaced sources is not difficult and has important applications for diffraction and the diffraction grating, discussed in the next two sections.

We first consider the case of three sources, as shown in Figure 34-14. The geometry is the same as for two sources. At a great distance from the sources, the rays from the sources to a point P on the screen are approximately parallel. The path difference between the first and second source is then $d \sin \theta$, as before, and between the first and third source the path difference is $2d \sin \theta$. The wave at point P is the sum of three waves. Let $\alpha = kr_1 - \omega t_0$ be the phase of the first wave at point P, where t_0 is the time of observation and r_1 is the distance from the first

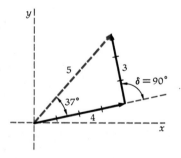

Figure 34-13
Vector model for the addition of the two waves in Example 34-3.

Figure 34-14
Geometry for calculating the intensity pattern far from three equally spaced sources in phase.

source to point P. We thus have the problem of addition of the three waves of the form

$$E_1 = A_0 \sin \alpha$$

$$E_2 = A_0 \sin (\alpha + \delta) \qquad \text{34-13}$$

$$E_3 = A_0 \sin (\alpha + 2\delta)$$

where

$$\delta = \frac{2\pi}{\lambda} d \sin \theta \approx \frac{2\pi}{\lambda} \frac{yd}{L} \qquad \text{34-14}$$

as in the two-slit problem.

It is easiest to analyze the resulting pattern in terms of the phase angle δ between the first and second sources or between the second and third sources instead of directly in terms of the space angle θ. If we know the resultant amplitude due to the three waves at some point P corresponding to a particular phase angle δ, we can relate this phase angle to θ by Equation 34-14.

The addition of these waves is easiest with the vector model of addition. We shall be most interested in the points of perfectly constructive interference and those of perfectly destructive interference, i.e., the interference maxima and minima.

At the central maximum point $\theta = 0$, the phase angle δ is zero, and the amplitude of the resultant wave is 3 times that of each individual wave. Since the intensity is proportional to the square of the amplitude, the intensity at this central maximum is 9 times that from each source acting separately. As we move away from this central maximum, the angle θ increases and the phase angle δ also increases.

Figure 34-15 shows the vector addition of three waves for a phase angle δ of about $30° = \pi/6$ rad. (This corresponds to a point P on the screen for which θ is given by $\sin \theta = \lambda\delta/2\pi d = \lambda/12d$.) The resultant amplitude is considerably less than 3 times that of each source. As the phase angle δ increases, the resultant amplitude decreases until the amplitude is zero at $\delta = 120°$. For this phase difference, the three vectors form an equilateral triangle (Figure 34-16). This first interference minimum occurs at a smaller phase angle (and therefore at a smaller angle θ) than the $180°$ for only two sources. As δ increases from $120°$, the resultant amplitude increases, reaching a secondary maximum when δ is $180°$. At the phase angle $\delta = 180°$ the amplitude is the same as that from

Figure 34-15
Vector addition to determine the resultant amplitude due to three waves, each of amplitude A_0, which have phase differences of δ and 2δ due to path differences of $d \sin \theta$ and $2d \sin \theta$. The angle $\alpha = kx - \omega t$ varies with time but does not affect the calculation of the resultant amplitude.

Figure 34-16
The resultant amplitude for the waves from three sources is zero when δ is $120°$. This interference minimum occurs at a smaller angle θ than the first minimum for two sources, which occurs when δ is $180°$.

a single source since the waves from the first two sources cancel each other, leaving only the third. The intensity of the secondary maximum is one-ninth that of the central maximum. As δ increases beyond 180°, the amplitude again decreases and is zero at $\delta = 180° + 60° = 240°$. For δ greater than 240° the amplitude increases and is again 3 times that of each source when $\delta = 360°$. This phase angle corresponds to a path difference of 1 wavelength for the waves from the first two sources and 2 wavelengths for the waves from the first and third sources. Hence the three waves are in phase at this point. Figure 34-17 shows the intensity pattern on a screen far from three equally spaced sources. The maxima are at the same positions as for just two sources, i.e., at points corresponding to the angles θ given by

$$d \sin \theta = m\lambda \qquad\qquad 34\text{-}15$$

where m, the *order number*, is 0, 1, 2, ... These maxima are stronger and narrower than those for two sources.

These results can be generalized to more sources. For example, if we have four equally spaced sources in phase, the interference maxima are again given by Equation 34-15 but the maxima are still narrower and there are two small secondary maxima between each pair of principal maxima. At $\theta = 0$, the intensity is 16 times that from a single source.

The first interference minimum occurs when δ is 90°, as can be seen by the vector diagram of Figure 34-18. The first secondary maximum is near $\delta = 120°$, where the waves from three of the sources cancel, leaving only the wave from the fourth source. The intensity of the secondary maximum is approximately one-sixteenth that of the central maximum. There is another minimum at $\delta = 180°$, a secondary maximum near $\delta = 240°$, and another minimum at $\delta = 270°$ before the next principal maximum at $\delta = 360°$. This discussion shows that as we increase the number of sources, the intensity becomes more and more concentrated in the maxima given by Equation 34-15 and the maxima become narrower. For N sources, the intensity at the maxima is N^2 times that of a single source, and the first minimum occurs at a phase angle of $\delta = 360°/N$. In that case the N vectors form a closed polygon of N sides. There is a secondary maximum between each two minimum points. These secondary maxima are very weak compared with the principal maxima.

Example 34-4 Four equally spaced slits are separated by a distance $d = 0.1$ mm and uniformly illuminated by light of wavelength 500 nm. The interference pattern is viewed on a screen at a distance of 1 m. The intensity on the screen due to each slit when the other three slits are covered is I_0. Find the positions of the interference maxima and compare their width with that for just two slits of the same spacing.

Figure 34-17
A plot of the relative-intensity pattern on a screen far from three equally spaced sources in phase. The major maxima occur when $d \sin \theta = m\lambda$ at the same position as for two sources, but for three sources the maxima are narrower and of greater intensity. Between successive major maxima there is a secondary maximum with one-ninth the intensity of the major maxima.

Figure 34-18
Vector diagram for the first minimum from four equally spaced sources in phase. The amplitude is zero when the phase difference of the waves from adjacent sources is 90°.

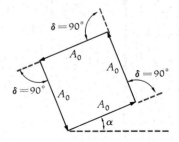

According to Equation 34-15, the maxima are at angles given by

$$\sin \theta = m \frac{\lambda}{d} = m \frac{5 \times 10^{-7} \text{ m}}{1 \times 10^{-4} \text{ m}} = 5 \times 10^{-3} m$$

where $m = 0, 1, 2, 3, \ldots$. Since θ is small, we can approximate $\sin \theta \approx \tan \theta \approx \theta$. The distance y measured along the screen from the central maximum is related to θ by

$$y = L \tan \theta \approx L\theta$$

The position of the mth maximum is thus

$$y_m = L\theta_m = m(1 \text{ m}) (5 \times 10^{-3}) = m(5 \text{ mm})$$

The maxima are thus separated by 5 mm on the screen. The first minimum occurs when the phase difference between two adjacent sources is $\delta = 90° = \pi/2$. This corresponds to a path difference of $\lambda/4$. The angle θ of this minimum is given by $d \sin \theta = \lambda/4$, or

$$\sin \theta = \frac{\lambda}{4d} = \frac{5 \times 10^{-7} \text{ m}}{4 \times 10^{-4} \text{ m}} = 1.25 \times 10^{-3}$$

The position y of this minimum is

$$y = L\theta = (1 \text{ m})(1.25 \times 10^{-3}) = 1.25 \text{ mm}$$

We shall take for the width of the maximum the distance between the first minima on either side of the first maximum, or $2y = 2.5$ mm. If we had only two sources with this same spacing, the maxima would be at the same points but the first minimum would be at an angle θ corresponding to a path difference of $\lambda/2$. The width of this maximum would be twice as great as with four sources (Figure 34-19).

Interference patterns for (a) two, (b) three, and (c) four coherent sources. In (b) there is a secondary maximum between each pair of principal maxima, and in (c) there are two such secondary maxima. (From M. Cagnet, M. Françon, and J. C. Thrierr, Atlas of Optical Phenomena, plate 19, Springer-Verlag, Berlin, 1962.)

Figure 34-19
Comparison of relative-intensity pattern for four sources with that for two sources when the spacing d is the same. The maxima for four sources are stronger and narrower than those for two sources.

Question

5. How many submaxima would there be between the main maxima in the interference pattern produced by eight equally spaced slits? Why would they be difficult to see?

34-6 Diffraction Pattern of a Single Slit

In our calculations of the interference patterns produced by two or more slits we have assumed the widths of the slits to be much less than the wavelength of the light, enabling us to consider the slits to be line sources of cylindrical waves, which in our two-dimensional diagrams are point sources of circular waves. (We can justify this assumption by our observation of circular waves in a ripple tank emanating from a small opening in a barrier, as shown in Figure 15-7.) We can also justify this assumption from Huygens' principle, according to which we can find the wave to the right of a single slit opening by assuming that each point along the line of the opening is a point source. If we replace the slit by a line of point sources separated by a distance d and the total width of the slit is much less than the wavelength, d will also be much less than the wavelength, and the path difference for waves from adjacent sources, $d \sin \theta$, will be much smaller than the wavelength. The waves at the screen from the different point sources will then have a negligible phase difference no matter what the angle θ; therefore we might as well treat the line of sources as a single point source. However, if the slit width is *not* small compared with the wavelength of light, we must take into account the different phases of the waves from the different portions of the slit. Our calculations of the interference pattern of a line of sources in the previous section will be very useful.

In Figure 34-20 we have a single slit of width a, which we have divided into N equal intervals. We assume that there is a point source of waves at the midpoint of each interval. If d is the distance between two adjacent sources and a the width of the opening, we have

$$d = \frac{a}{N}$$

For convenience we shall take the number of sources N to be an even number. Since the screen on which we are calculating the intensity is very far from the sources, the rays from the sources to a point P on the screen are approximately parallel. The path difference between any two adjacent sources is then $d \sin \theta$, and the phase difference is $\delta = (2\pi/\lambda)d \sin \theta$. If A_0 is the amplitude due to a single source, the amplitude at the central maximum point $\theta = 0$, where all the waves are in phase, is $A_{\max} = NA_0$ (Figure 34-21). We find the amplitude at some other point P at an angle θ by using the vector method of addition of waves. As in the addition of two, three, or four waves, the intensity is zero at a point such that the vectors representing the waves form a closed polygon. In this case the polygon has N sides (Figure 34-22). At point P the wave from the first source near the top of the opening and that from the source just below the middle of the opening are 180° out of phase. Consider, for example, 100 sources. These sources give zero intensity at a point P when the waves from the first source and the fifty-first source are out of phase by 180° and cancel (Figure 34-23). Then

Figure 34-20
Diagram for the calculation of the interference pattern far from a narrow slit. The slit of width a is assumed to contain a large number of in-phase point sources separated by a distance d. The rays from these sources to a point very far away are approximately parallel. The path difference for the waves from adjacent sources is then $d \sin \theta$.

Figure 34-21
A single slit is represented by N sources of amplitude A_0. At the central maximum point at $\theta = 0$ the waves from the sources add in phase, giving a resultant amplitude $A_{\max} = NA_0$.

Figure 34-22
Vector diagram for the first minimum in the single-slit diffraction pattern. When the waves from the N sources completely cancel, the N vectors form a closed polygon. The phase difference between waves from adjacent sources is then $\delta = 360°/N$. When N is very large, the waves from the first and last sources are approximately in phase.

the waves from the second and fifty-second are also out of phase by $180°$ and cancel, as do those from the third and fifty-third sources, etc. In this case the waves from the sources near the top and bottom of the opening differ in phase by nearly $360°$. (The phase difference is, in fact, $360° - 360°/N$.) Thus if the number of sources is very large, *we get complete cancellation when the waves from the first and last sources are out of phase by $360°$, corresponding to a path difference of 1 wavelength.* Since this path difference is $a \sin \theta$, where a is the width of the opening, the condition for the first minimum is

$$a \sin \theta = \lambda \qquad\qquad 34\text{-}16$$

Single-slit diffraction minimum

We get a second minimum when the path difference for the waves from the top and bottom of the opening is 2 wavelengths. This can be seen by dividing the group of N sources into two parts. The waves from the top half of the sources add to zero, as we have just noted. Similarly the waves from the bottom half of the sources also add to zero. Thus the condition for a minimum in the diffraction pattern is

$$a \sin \theta = m\lambda \qquad m = 1, 2, 3 \ldots \qquad\qquad 34\text{-}17$$

We now calculate the amplitude at a general point for which the waves from two adjacent sources differ in phase by δ. Figure 34-24 shows the vector diagram for the addition of N waves which differ in phase from the first wave by $\delta, 2\delta, \ldots, (N-1)\delta$. When N is very large and δ is very small, the vector diagram is approximately an arc of a circle. The resultant amplitude A is the length of the chord of this arc. We calculate this resultant amplitude in terms of the phase difference ϕ between the first and last waves. From Figure 34-24 we have

Figure 34-23
At the first diffraction minimum of a single slit, the waves from the source near the top and those from the source just below the middle of the slit are $180°$ out of phase and cancel.

$$\sin \tfrac{1}{2}\phi = \frac{\tfrac{1}{2}A}{r} \qquad \text{or} \qquad A = 2r \sin \tfrac{1}{2}\phi \qquad\qquad 34\text{-}18$$

where r is the radius of the arc. Since the length of the arc is $A_{\max} = NA_0$ and the angle subtended is ϕ, we have

$$\phi = \frac{A_{\max}}{r} \qquad\qquad 34\text{-}19$$

or

$$r = \frac{A_{\max}}{\phi}$$

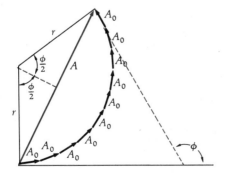

Figure 34-24
Vector model for the calculation of the amplitude of the waves from N sources in terms of the phase difference ϕ between the waves from the first source near the top of the slit and the last source near the bottom of the slit. For large N, the resultant amplitude A is the chord of a circular arc of length NA_0.

Substituting this into Equation 34-18 gives

$$A = \frac{2A_{\max}}{\phi} \sin \tfrac{1}{2}\phi = A_{\max} \frac{\sin \tfrac{1}{2}\phi}{\tfrac{1}{2}\phi} \qquad 34\text{-}20$$

Since the amplitude at the central maximum point ($\theta = 0$) is $A_{\max}$, the ratio of the intensity at any other point to that at the central maximum point is

$$\frac{I}{I_0} = \frac{A^2}{A_{\max}^2} = \left(\frac{\sin \tfrac{1}{2}\phi}{\tfrac{1}{2}\phi}\right)^2$$

or

$$I = I_0 \left(\frac{\sin \tfrac{1}{2}\phi}{\tfrac{1}{2}\phi}\right)^2 \qquad 34\text{-}21$$

Single-slit diffraction intensity

The phase difference ϕ between the first and last waves is $2\pi/\lambda$ times the path difference $a \sin \theta$ between the top and bottom of the opening:

$$\phi = \frac{2\pi}{\lambda} a \sin \theta \qquad 34\text{-}22$$

The function I/I_0 is plotted in Figure 34-25. The first minimum occurs when $\phi = 360°$, at the point where the waves from the top and bottom of the opening have a path difference of λ and are in phase. The second minimum occurs at $\phi = 720°$, where the waves from the top and bottom of the opening have a path difference of 2λ.

(a)

Figure 34-25
(a) Fraunhofer diffraction pattern of a single slit. (*From M. Cagnet, M. Françon, and J. C. Thierr,* Atlas of Optical Phenomena, *plate 18, Springer-Verlag, Berlin, 1962.*) (b) Plot of intensity versus phase difference for the pattern in (a).

Intensity

π 2π ϕ

(b)

Approximately midway between these minima is a secondary maximum. Figure 34-26 shows the vector diagram for determining the relative intensity of this secondary maximum. The phase difference between the first and last waves is approximately $360° + 180°$. The vectors thus complete $1\tfrac{1}{2}$ circles. The resultant amplitude is the diameter

A of a circle with a circumference C which is two-thirds the total length $NA_0 = A_{max}$. Thus

$$C = \tfrac{2}{3}A_{max} = \pi A \qquad A = \frac{2}{3\pi} A_{max}$$

and

$$A^2 = \frac{4}{9\pi^2} A_{max}^2 \qquad\qquad 34\text{-}23$$

The intensity at this point is

$$I = \frac{4}{9\pi^2} I_0 = \frac{1}{22.2} I_0 \qquad\qquad 34\text{-}24$$

The intensity pattern consists of a broad central maximum which has intensity I_0 at $\theta = 0$ and drops to zero intensity at the first diffraction minimum at an angle θ given by

$$a \sin \theta = \lambda$$

There is then a secondary maximum of intensity approximately $I_0/22.2$ at an angle θ given approximately by

$$a \sin \theta = \tfrac{3}{2}\lambda$$

The second minimum occurs at an angle θ given by $a \sin \theta = 2\lambda$.

The intensity pattern of Figure 34-25 is called a *Fraunhofer diffraction pattern* of a single slit. The assumptions made in deriving Equation 34-21 describing the pattern were (1) that plane waves are incident normally on the slit (we assumed that the amplitudes and phases of the many Huygens sources are equal) and (2) that the pattern is observed at a great distance from the slit compared with the size of the openings. (We assumed that the rays from the sources to a point on the screen are approximately parallel to simplify the geometry.) Another condition that is usually necessary is that the width of the slit be of the order of the wavelength. If the opening, or slit, is much wider than the wavelength, the angle of the first minimum will be very small. For example, if $a = 1000\lambda$, the first minimum will occur at an angle θ given by $\sin \theta = 1/1000 \approx \theta$. Unless the screen is extremely far away, this small angle is not much different from the angle made by the rays from the top and bottom of the slit to the central maximum, rays which we assumed to be parallel in our derivation.

When the diffraction pattern is observed near the opening, it is called a *Fresnel diffraction pattern*. Because of the geometry, this pattern is much more difficult to calculate. Figure 34-27 illustrates the difference between the Fresnel and Fraunhofer patterns.*

When there are two or more slits, the intensity pattern on a screen far away is a combination of the single-slit diffraction pattern and the multiple-slit interference pattern we have studied. Figure 34-28 shows the intensity pattern on a screen far from two slits whose separation d is 10 times the width a of each slit. The pattern is the same as the two-slit pattern for a line source (Figure 34-9) except that it is modulated by the Fraunhofer single-slit diffraction pattern. The intensity can be calculated from the product of the intensity for the two-slit pattern (Equation 34-11) and the intensity due to each slit given by Equation 34-21:

$$I = 4I_0 \left(\frac{\sin \tfrac{1}{2}\phi}{\tfrac{1}{2}\phi}\right)^2 \cos^2 \tfrac{1}{2}\delta \qquad\qquad 34\text{-}25$$

* See Richard E. Haskell, "A Simple Experiment on Fresnel Diffraction," *American Journal of Physics*, vol. 38, p. 1039, 1970.

Circumference $\times \tfrac{3}{2} = NA_0 = \pi A \times \tfrac{3}{2}$

$A = \tfrac{2}{3\pi}NA_0$

$A^2 = \tfrac{4}{9\pi^2}(NA_0)^2$

Figure 34-26
Vector diagram for calculating the amplitude of the first secondary maximum of the single-slit diffraction pattern. This secondary maximum occurs when the N vectors complete approximately $1\tfrac{1}{2}$ circles.

Assumptions for Fraunhofer diffraction

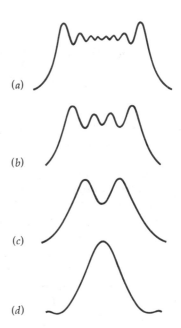

(a)

(b)

(c)

(d)

Figure 34-27
Diffraction pattern of a single slit for various screen distances. The Fresnel pattern observed near the slit (a) gradually merges into the Fraunhofer pattern (d) as the screen is moved to a great distance from the slit. (*Courtesy of Richard E. Haskell, Oakland University.*)

where ϕ is related to the width of each slit by

$$\phi = \frac{2\pi}{\lambda} a \sin \theta \qquad\qquad 34\text{-}26$$

and where δ is related to the slit separation by

$$\delta = \frac{2\pi}{\lambda} d \sin \theta \qquad\qquad 34\text{-}27$$

and I_0 is the intensity at $\theta = 0$ due to one slit alone. Note that in Figure 34-28 the central diffraction maximum contains 19 interference maxima, the central interference maximum and 9 maxima on either side. The tenth interference maximum on either side of the central one is at the angle θ given by $\sin \theta = 10\lambda/d = \lambda/a$, since $d = 10a$. This coincides with the position of the first diffraction minimum, and so this interference maximum is not seen. At this point the light from the two slits would be in phase and interfere constructively, but there is no light from either slit because the point is a diffraction minimum.

Questions

6. As the width of the slit producing a single-slit diffraction pattern is slowly and steadily reduced, how will the diffraction pattern change?

7. How many interference maxima will be contained in the central diffraction maximum in the diffraction-interference pattern of two slits if the separation d is 5 times the slit width a? How many if $d = Na$ for any N?

34-7 Diffraction and Resolution

Diffraction occurs whenever a portion of the wavefront is limited by an obstacle or aperture of some kind. The intensity of light at any point in space can be computed using Huygens' principle by taking each point on the wavefront to be a point source and computing the resulting interference pattern. Fraunhofer patterns can be observed at great distances from the obstacle or aperture so that the rays reaching any point are approximately parallel, or they can be observed using a lens to focus parallel rays on a viewing screen placed at the focal point of the lens. Fresnel patterns are observed at points close to the source. Diffraction of light is often difficult to observe because the wavelength is so small or because the light intensity is not great enough. Except for the Fraunhofer pattern of a long narrow slit, diffraction patterns are usually difficult to calculate.

Figure 34-29a shows the Fresnel diffraction pattern of a straight edge illuminated by light from a point source. A graph of the intensity versus distance measured along a line perpendicular to the edge is shown in Figure 34-29b. The light intensity does not fall abruptly to zero in the geometric shadow but decreases rapidly and is negligible within a few wavelengths of the edge. The Fresnel diffraction pattern of a circular aperture is shown in Figure 34-30. Note the similarity of this pattern and the diffraction pattern of an opaque disk shown in Figure 32-5. The Fresnel diffraction pattern of a rectangular opening is shown in Figure 34-31. These patterns cannot be seen with broad light sources like an ordinary light bulb because the dark fringes of the pattern produced by

Figure 34-28
Interference-diffraction pattern of two slits with separation d equal to 10 times their width a. The tenth interference maximum on either side of the central interference maximum is missing because it falls at the first diffraction minimum. (*From M. Cagnet, M. Françon, and J. C. Thrierr, Atlas of Optical Phenomena, plate 18, Springer-Verlag, Berlin, 1962.*)

Battelle-Northwest Photography

When waves are incident on a barrier having two slits with size and separation of the order of the wavelength, both diffraction and interference can be observed. Here plane ultrasonic waves in water are incident on slits of width 2λ and separation 8λ. The photograph was made using a laser and a Schlieren technique, which is sensitive to the small changes in the optical index of refraction of the water introduced by the compressions and rarefactions of the ultrasonic waves.

light from one point on the source overlap the bright fringes of the pattern produced by light from another point.

Figure 34-32 shows the Fraunhofer diffraction pattern of a circular aperture, which has important applications to the resolution of many optical instruments. The angle θ subtended by the first diffraction minimum is related to the wavelength and the diameter of the opening D by

$$\sin \theta = 1.22 \frac{\lambda}{D} \qquad 34\text{-}28$$

Diffraction by circular aperture

(a)

(b)

Intensity

Distance

Figure 34-29
(a) Fresnel diffraction of a straight edge. (*From M. Cagnet, M. Françon, and J. C. Thrierr,* Atlas of Optical Phenomena, *plate 32, Springer-Verlag, Berlin, 1962.*) (b) Intensity versus distance along a line perpendicular to the edge.

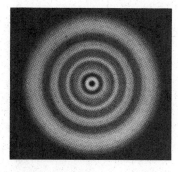

Figure 34-30
Fresnel diffraction of a circular aperture. Compare this pattern with Figure 32-5, the Fresnel diffraction pattern of an opaque circular disk. The patterns are complements of each other. (*From M. Cagnet, M. Françon, and J. C. Thrierr,* Atlas of Optical Phenomena, *plate 33, Springer-Verlag, Berlin, 1962.*)

Figure 34-31
Fresnel diffraction of a rectangular opening. (*From M. Cagnet, M. Françon, and J. C. Thrierr,* Atlas of Optical Phenomena, *plate 34, Springer-Verlag, Berlin, 1962.*)

Figure 34-32
Fraunhofer diffraction pattern of a circular aperture. (*From M. Cagnet, M. Françon, and J. C. Thrierr,* Atlas of Optical Phenomena, *plate 16, Springer-Verlag, Berlin, 1962.*)

Equation 34-28 is similar to Equation 34-16 except for the factor 1.22; it arises from the mathematical analysis, which is similar to that for a single slit but more complicated because of the cylindrical geometry. In many applications, the angle θ is small, so that $\sin \theta$ can be replaced by θ. Then

$$\theta \approx 1.22 \frac{\lambda}{D} \qquad\qquad 34\text{-}29$$

Figure 34-33 shows two incoherent point sources which subtend an angle α at a circular aperture far from the sources. The Fraunhofer diffraction intensity patterns are also sketched in this figure. If α is much greater than $1.22\ \lambda/D$, there will be little overlap in the diffraction patterns of the two sources and they will be seen as two sources. However, as α is decreased, the overlap of the diffraction patterns increases and it becomes difficult to distinguish the two sources from one source. At the critical-angular separation

$$\alpha_c = 1.22 \frac{\lambda}{D} \qquad\qquad 34\text{-}30$$

the first minimum of the diffraction pattern of one source falls at the central maximum of the other source. These objects are said to be just resolved by the Rayleigh criterion for resolution. Figure 34-34 shows the diffraction patterns for two sources for $\alpha > \alpha_c$ and $\alpha = \alpha_c$.

Rayleigh criterion for resolution

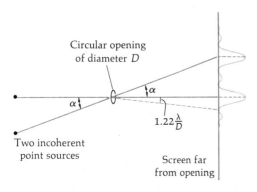

Figure 34-33
Two distant sources which subtend an angle α at a circular aperture are easily resolved if α is greater than $1.22\lambda/D$, where λ is the wavelength of light and D the diameter of the aperture. Then the diffraction patterns have little overlap. If α is not greater than $1.22\lambda/D$, the overlap of the diffraction patterns makes it difficult to distinguish two sources from one.

Figure 34-34
Diffraction patterns of a circular aperture and two incoherent point sources for (a) α greater than $1.22\lambda/D$ and (b) the limit of resolution $\alpha = 1.22\lambda/D$. (From M. Cagnet, M. Françon, and J. C. Thrierr, *Atlas of Optical Phenomena, plate 16, Springer-Verlag, Berlin, 1962.*)

(a)

(b)

Example 34-5 What is the minimum angular separation of two point objects if they are to be just resolved by the eye? How far apart must they be if they are 100 m away? Assume that the diameter of the pupil of the eye is 5 mm and the wavelength of the light is 600 nm.

Using Equation 34-30 with $D = 5$ mm and $\lambda = 600$ nm, we have for the minimum angular separation

$$\alpha_c = 1.22 \frac{6 \times 10^{-7}\ \text{m}}{5 \times 10^{-3}\ \text{m}} = 1.46 \times 10^{-4}\ \text{rad}$$

If the objects are separated by a distance y and are 100 m away, they will be just barely resolved if $\tan \alpha_c = y/(100\ \text{m})$. Then

$$\begin{aligned} y &= (100\ \text{m}) \tan \alpha_c \approx (100\ \text{m})\alpha_c \\ &= 1.46 \times 10^{-2}\ \text{m} \\ &= 1.46\ \text{cm} \end{aligned}$$

where we have used the small-angle approximation $\alpha_c \approx \tan \alpha_c$.

34-8 Diffraction Gratings

A useful tool for the analysis of light is the diffraction grating, which consists of a large number of equally spaced slits. Such a grating can be made by cutting parallel, equally spaced grooves on a glass or metal plate with a precision ruling machine. A grating with 10,000 slits per centimetre is not uncommon. The spacing of the slits for such a grating is $d = (1 \text{ cm})/10{,}000 = 10^{-4}$ cm. Consider a plane light wave incident normally on such a grating and assume that the width of each slit is very small, so that each slit produces a wide diffracted beam. Then the interference pattern produced on a screen a large distance from the grating is just that due to a large number of equally spaced line sources. The interference maxima are at angles θ given by

$$d \sin \theta = m\lambda \qquad\qquad 34\text{-}31$$

The position of an interference maximum does not depend on the number of sources, but the more sources there are, the sharper the maximum. We can see this by finding the angle $\theta_{\min}$ of the first minimum produced by the light from the N slits after the central maximum at $\theta = 0$. The calculation of the angle of the first minimum due to N sources in phase is the same as that for the first minimum of a single slit of width $a = Nd$. The minimum is therefore at the angle $\theta_{\min}$, given by

$$\sin \theta_{\min} = \frac{\lambda}{Nd}$$

For small angles we can replace $\sin \theta_{\min}$ by $\theta_{\min}$. Since the central maximum is at $\theta = 0$, the angular separation $\Delta\theta$ between the central maximum and the first minimum is approximately

$$\Delta\theta = \frac{1}{N}\frac{\lambda}{d} \qquad\qquad 34\text{-}32$$

We note that this "angular width" is approximately $1/N$ times the angular separation of the orders, which is λ/d for small θ.

Figure 34-35 shows a typical spectroscope, which uses a diffraction grating to analyze light from a source, usually a tube containing atoms of a gas, e.g., helium or sodium vapor. The atoms are excited because of bombardment by electrons accelerated by the high voltage across the tube. The light emitted by such a source does not consist of a continuous spectrum but contains only certain wavelengths characteristic of the atoms in the source. Light from the source passes through a narrow slit and is made parallel by a collimating lens. Parallel light from the lens is incident on the grating. Instead of falling on a screen a large distance

Figure 34-35
Typical student spectroscope. Light from a slit near a source is made parallel by a lens and falls on a grating. The diffracted light is viewed with a telescope at an angle which can be accurately measured. The wavelength of the light is found from Equation 34-31. (*Courtesy of Larry Langrill, Oakland University.*)

Spectrometer used in the late nineteenth century by Gustav Kirchhoff. In this drawing, a prism replaces the diffraction grating.

away, the parallel light from the grating is focused by a telescope and viewed by the eye. The telescope is mounted on a rotating platform, calibrated so that the angle θ can be measured. At angle $\theta = 0$ the central maximum for all wavelengths is seen. If light of a particular wavelength λ is emitted by the source, the first interference maximum is seen at angle θ, given by Equation 34-31 with $m = 1$. Each wavelength emitted by the source produces a separate image of the slit in the spectroscope and is called a *spectral line*. Thus by measuring the angle θ and knowing the spacing d of the slits on the grating we can measure the wavelengths emitted by the source.

An important feature of such a spectroscope is its ability to measure light of two nearly equal wavelengths λ_1 and λ_2. For example, the two prominent yellow lines in the spectrum of sodium have wavelengths 589.00 and 589.59 nm, which can be seen as two separate wavelengths if their interference maxima do not overlap. According to the Rayleigh criterion of resolution, these wavelengths are resolved if the angular separation of their interference maxima is greater than the angular separation between one interference maximum and the first interference minimum on either side of it. Let us consider small angles so that we can replace $\sin \theta$ by θ. The angular deviation of the mth-order maximum for light of wavelength λ (Equation 34-31 with θ replacing $\sin \theta$) is

$$\theta_1 = m\frac{\lambda_1}{d}$$

and that for light of wavelength λ_2 is

$$\theta_2 = m\frac{\lambda_2}{d}$$

The angular separation of these maxima is then

$$\Delta\theta = \theta_2 - \theta_1 = m\frac{\Delta\lambda}{d}$$

These wavelengths will just be resolved by the Rayleigh criterion if this angular separation equals that between the maximum and minimum of one of the wavelengths, which is given by Equation 34-32:

$$m\frac{\Delta\lambda}{d} = \frac{1}{N}\frac{\lambda}{d} \qquad\qquad 34\text{-}33$$

The resolving power of a diffraction grating is defined to be $\lambda/|\Delta\lambda|$, where $|\Delta\lambda|$ is the difference between two nearby wavelengths, each approximately equal to λ. From Equation 34-33 we see that the resolving power is

$$R = \frac{\lambda}{|\Delta\lambda|} = mN \qquad\qquad 34\text{-}34$$

Resolving power of a grating

This result, which we have obtained using the small-angle approximation, holds in general whether the angles are small or not. A general derivation is given below.

We see from Equation 34-34 that to resolve the two yellow lines in the sodium spectrum the resolving power must be

$$R = \frac{589.00\ \text{nm}}{589.59 - 589.00\ \text{nm}} = 998$$

Thus to resolve the two yellow sodium lines in the first order ($m = 1$) we need a grating containing about 1000 slits in the area illuminated by the light.

Optional

Derivation of Equation 34-34 for General Angles

The mth-order interference maximum corresponding to some wavelength λ is at an angle θ given by Equation 34-31:

$$\sin\theta = \frac{m\lambda}{d}$$

If we change the wavelength by a small amount $d\lambda$, the angular change $d\theta$ is found by differentiating this expression:

$$\cos\theta\ d\theta = \frac{m}{d}\ d\lambda$$

or

$$\Delta\theta \approx \frac{m\ \Delta\lambda}{d\cos\theta} \qquad\qquad 34\text{-}35$$

This is the angular separation of the mth-order interference maximum for two nearly equal wavelengths. The phase difference ϕ between the light from two adjacent slits is given by

$$\phi = \frac{2\pi d}{\lambda}\sin\theta \qquad\qquad 34\text{-}36$$

This phase difference is $m2\pi$ for the mth-order interference maximum. The first interference *minimum* occurs when the vectors representing the N sources form a closed polygon, as in Section 34-5. The phase difference between two adjacent sources is then $2\pi/N$. If we change the angle θ by a small amount $d\theta$, the change in phase is

$$d\phi = 2\pi d\cos\theta\ \frac{d\theta}{\lambda}$$

Setting this change in phase equal to $2\pi/N$, we obtain for the angular separation $d\theta$ between the interference maximum and the first minimum

$$2\pi d\cos\theta\ \frac{d\theta}{\lambda} = d\phi = \frac{2\pi}{N}$$

or

$$d\theta = \frac{\lambda}{Nd\cos\theta} \approx \Delta\theta \qquad\qquad 34\text{-}37$$

According to the Rayleigh criterion of resolution, these wavelengths are resolved if the angle $\Delta\theta$ between the interference maximum and minimum given by Equation 34-37 equals the angle $\Delta\theta$ between the interference maxima of the two wavelengths given by Equation 34-35. Thus the wavelengths are resolved if

$$\frac{\lambda}{Nd \cos \theta} = \frac{m \, \Delta\lambda}{d \cos \theta} \qquad \text{or} \qquad \frac{\lambda}{\Delta\lambda} = mN$$

Review

A. Define, explain, or otherwise identify:

Newton's rings, 903

Interferometer, 905

Lloyd's mirror, 908

Fraunhofer diffraction pattern, 917

Fresnel diffraction pattern, 917

Rayleigh's criterion for resolution, 920

Diffraction grating, 921

B. True or false:

1. When waves interfere destructively, the energy is converted into heat energy.

2. Interference is observed only for waves from coherent sources.

3. In single-slit diffraction, the narrower the slit, the wider the diffraction pattern.

4. A circular aperture can produce both a Fraunhofer and a Fresnel diffraction pattern.

5. The ability to resolve two point sources depends on the wavelength of the light.

Exercises

Section 34-1, Interference in Thin Films

1. Light of wavelength 500 nm is incident normally on a film of water 10^{-4} cm thick. The index of refraction of water is 1.33. (a) What is the wavelength of the light in the water? (b) How many wavelengths are contained in the distance $2t$, where t is the film thickness? (c) What is the phase difference between the wave reflected from the top of the film and the one reflected from the bottom after it has traveled this distance?

2. A loop of wire is dipped in soapy water and held so that the soap film is vertical. Viewed by reflection with white light, the top of the film appears black. Explain why. Below the black region are colored bands. Is the first band red or blue? Describe the appearance of the film when viewed by transmitted light.

3. A wedge-shaped film of air is made by placing a small slip of paper between the edges of two flat pieces of glass. Light of wavelength 700 nm is incident normally on the glass plates, and interference bands are observed by reflection. (a) Is the first band near the point of contact of the plates dark or bright? Why? (b) There are five dark bands per centimetre. What is the angle of the wedge?

4. A thin layer of a transparent material with index of refraction 1.30 is used as a nonreflective coating on the surface of glass with index of refraction 1.50. What should the thickness be for the film to be nonreflecting for light of wavelength 600 nm (in vacuum)?

5. The diameter of fine wires can be accurately measured by interference patterns. Two optically flat pieces of glass of length L are arranged with the wire as shown in Figure 34-36. The setup is illuminated by monochromatic light, and the resulting interference fringes are detected. Suppose $L = 20$ cm and yellow

sodium light ($\lambda \approx 590$ nm) is used for illumination. If 19 bright fringes are seen along this 20-cm distance, what are the limits on the diameter of the wire? *Hint:* The 19th fringe might not be right at the end, but you do not see 20 fringes.

Figure 34-36
Exercise 5.

Section 34-2, The Michelson Interferometer

6. A thin film of index of refraction $n = 1.5$ for light of wavelength 600 nm is inserted in one arm of a Michelson interferometer. (*a*) If a fringe shift of 12 fringes occurs, what is the thickness of this film? (*b*) If the illuminating light is changed to 400 nm, the fringe shift as this film is inserted becomes 16 fringes. What is the index of refraction of this film to light of wavelength 400 nm?

7. A hollow cell of length 5 cm with glass windows is inserted into one arm of a Michelson interferometer. The air is pumped out of the cell and the mirrors are adjusted to give a bright fringe at the center. As the air is gradually let back into the cell, there is a shift of 49.6 fringes when light of wavelength 589.29 nm is used. (*a*) How many waves are there in the 5.0-cm-long cell when it is evacuated? (*b*) How many waves are in the cell when it contains air? (*c*) What is the index of refraction of air as determined by this experiment?

Section 34-3, The Two-Slit Interference Pattern

8. Two narrow slits separated by a distance d are illuminated by light of wavelength λ, and the interference pattern is viewed on a screen a long distance L away ($L > d$). The mth interference maximum is a distance y_m from the central maximum measured along the screen. (*a*) Show that y_m is given approximately by $y_m = m\lambda L/d$. (*b*) For $\lambda = 600$ nm, $L = 2$ m, and $d = 1$ mm, find the number of bright fringes per centimetre on the screen.

9. A long narrow horizontal slit lies 1 mm above a plane mirror. The interference pattern produced by the slit and its image is viewed on a screen a distance 1 m from the slit. The wavelength of the light is 600 nm. (*a*) Find the distance above the mirror to the first maximum. (*b*) How many dark bands per centimetre are seen on the screen?

10. (See Exercise 8.) When light of wavelength 589 nm is used and the screen is 3 m from the slits, there are 28 bright fringes per centimetre on the screen. What is the slit separation?

11. In a lecture demonstration, laser light is used to illuminate two slits separated by 0.5 mm, and the interference pattern is observed on a screen 5 m away. The distance on the screen to the 37th bright fringe is 25.7 cm. What is the wavelength of the light? [See part (*a*) of Exercise 8.]

12. Light is incident at an angle ϕ with the normal to a plane containing two slits of separation d (Figure 34-37). Show that the interference maxima are now located at angles θ given by $\sin \theta + \sin \phi = m(\lambda/d)$.

Figure 34-37
Light incident on two slits at an angle ϕ with the normal for Exercise 12.

13. Light from a helium-neon laser of wavelength 633 nm is shone normally on a plane containing two slits. The first interference maximum is 82 cm from the central maximum on a screen 12.0 m away. (*a*) Find the separation of the slits. (*b*) How many interference maxima can be observed?

Section 34-4, The Vector Model for Addition of Harmonic Waves, and Section 34-5, Interference Pattern of Three or Four Equally Spaced Sources

14. With a compass, draw circular arcs representing wave crests for each of three point sources a distance d apart for $d = 6$ cm and $\lambda = 1$ cm. Connect the intersections corresponding to points of constant path difference and label the path difference for each line. How do the lines of constructive interference differ from those for just two sources of the same spacing?

15. Three equally spaced slits are separated by $d = 0.1$ mm and uniformly illuminated by light of wavelength 600 nm. The interference pattern is viewed on a screen 2 m away. Find the positions of the interference maxima and minima.

16. Five equally spaced slits are uniformly illuminated, and the interference pattern is viewed on a screen far away such that the approximation $\sin \theta \approx \theta$ is good for the first few maxima. (a) Find the angle θ_1 between the first interference maximum and the central maximum ($\theta = 0$) and compare this angle with the angle to the first minimum. (b) Sketch the interference pattern.

17. Do Exercise 16 for six equally spaced slits.

Section 34-6, Diffraction Pattern of a Single Slit

18. Equation 34-17, $a \sin \theta = m\lambda$, and Equation 34-5, $d \sin \theta = m\lambda$, are sometimes confused. For each equation define the symbols and explain its application.

19. Light of wavelength 600 nm is incident on a long narrow slit. Find the angle of the first diffraction minimum if the width of the slit is (a) 0.1 mm, (b) 1 mm, (c) 0.01 mm.

20. The single-slit diffraction pattern of light is observed on a screen a large distance L from the slit. (a) Show that if the angle θ is small, the distance y from the central maximum on the screen to the first minimum is given by $y = L\lambda/a$. Note that the width of the central maximum $2y$ varies inversely with the width of the slit a. Find this width $2y$ for $L = 2$ m, $\lambda = 500$ nm, and (b) $a = 0.1$ mm, (c) $a = 0.01$ mm, and (d) $a = 0.001$ mm.

21. In a lecture demonstration of diffraction, a laser beam of wavelength 700 nm passes through a vertical slit 0.5 mm wide and hits a screen 6 m away. Find the horizontal length of the principal diffraction maximum on the screen; i.e., find the distance between the first minimum on the left and the first minimum on the right of the central maximum.

22. Plane microwaves are incident on a long narrow slit of width 5 cm. The first diffraction minimum is observed at $\theta = 37°$. What is the wavelength of the microwaves?

23. A two-slit Fraunhofer diffraction-interference pattern is observed with light of wavelength 500 nm. The slits have a separation $d = 0.1$ mm and width a. Find the width a if the fifth interference maximum is at the same angle as the first diffraction minimum. For this case, how many bright fringes will be seen in the central diffraction maximum?

24. A two-slit Fraunhofer diffraction-interference pattern is observed with light of wavelength 700 nm. The slits have a width $a = 0.01$ mm and are separated by $d = 0.2$ mm. How many bright fringes will be seen in the central diffraction maximum?

Section 34-7, Diffraction and Resolution

25. Light of wavelength 700 nm is incident on a pinhole of diameter 0.1 mm. (a) What is the angle between the central maximum and the first diffraction minimum for Fraunhofer diffraction? (b) What is the distance between the central maximum and the first diffraction minimum on a screen 8 m away?

26. Two light sources ($\lambda = 700$ nm) are 10 m away from the pinhole in Exercise 25. How far apart must the sources be for their diffraction patterns to be resolved by the Rayleigh criterion?

27. (a) How far apart must two objects be on the moon to be resolved by the eye? Take the diameter of the pupil of the eye to be 5.0 mm, $\lambda = 600$ nm, and $L = 380,000$ km for the distance to the moon. (b) How far apart must the objects on the moon be to be resolved by a telescope which has a 5.00-m-diameter mirror?

28. Two sources of wavelength 700 nm are separated by a horizontal distance x. They are 5 m from a vertical slit of width 0.5 mm. What is the least value of x

permitting the diffraction pattern of the sources to be resolved by the Rayleigh criterion?

29. What aperture (in millimetres) is required for perfect opera glasses (binoculars) for the observer to be able to distinguish the soprano's individual eyelashes (separated by 0.5 mm) at 25 m? Assume the effective wavelength of light to be 550 nm.

30. The headlights on a small car are separated by 112 cm. At what maximum distance could you resolve them if the diameter of your pupils is 5.0 mm and the effective wavelength of the light is 550 nm?

31. You are told not to shoot until you see the whites of their eyes. If the eyes are separated by 6.5 cm and the diameter of your pupil is 5.0 mm, at what distance can you resolve the two eyes using light of wavelength 550 nm?

Section 34-8, Diffraction Gratings

32. A diffraction grating with 2000 slits per centimetre is used to measure the wavelengths emitted by hydrogen gas. At what angles θ would you expect to find the two blue lines of wavelength 434 and 410 nm?

33. With the grating used in Exercise 32, two other lines in the hydrogen spectrum are found in first order at angles $\theta_1 = 9.72 \times 10^{-2}$ rad and $\theta_2 = 1.32 \times 10^{-1}$ rad. Find the wavelengths of these lines.

34. Repeat Exercise 32 for a grating with 15,000 slits per centimetre.

35. A grating of 2000 slits per centimetre is used to analyze the spectrum of mercury. (a) Find the angular deviation in first order of the two lines of wavelength 579.0 and 577.0 nm. (b) How wide must the beam be on the grating for these lines to be resolved?

36. What is the longest wavelength that can be observed in fifth order using a grating with 4000 slits per centimetre?

Problems

1. Laser light falls normally on three evenly spaced slits. When one of the side slits is covered, the first-order maximum is at 0.60° from the normal. If the center slit is covered with the other two open, find (a) the angle of the first-order maximum and (b) the order number of the maximum that now occurs at the same angle the fourth-order maximum did before.

2. Two slits are separated by a distance d. Their interference pattern is to be observed on a screen a large distance L away. Calculate the spacing y of the maxima on the screen for light of wavelength 500 nm with $L = 1$ m and $d = 1$ cm. Would you expect to observe the interference of light on the screen for this situation? How close together should you place the slits for the maxima to be separated by 1 mm for this wavelength and screen distance?

3. Suppose that the central *diffraction* maximum for two slits contained 17 interference fringes for some wavelength of light. How many interference fringes would you expect in the first secondary diffraction maximum?

4. The ceiling of your lecture hall is probably covered with acoustic tile which has small holes separated by about 6.0 mm. (a) Using $\lambda = 500$ nm, how far could you be from this tile and still resolve these holes? The diameter of the pupil of your eye is about 5.0 mm. (b) Could you "see" these holes better with red or blue light?

5. The telescope on Mount Palomar has a diameter of about 5.0 m. Assuming "ideal" sky conditions, the resolution would be diffraction-limited. Suppose a double star were 4 light-years away. What would the stellar separation have to be for their images to be resolved?

6. A mica sheet 1.20 μm thick is suspended in air. In reflected light there are gaps in the visible spectrum at 421, 474, 542, and 632 nm. Find the index of refraction of mica.

7. A thin film having index of refraction 1.5 is surrounded by air. It is illuminated normally by white light and viewed by reflection. Analysis of the resulting reflected light shows that the wavelengths 360, 450, and 600 nm are the only missing wavelengths near the visible portion of the spectrum. That is, for these wavelengths there is destructive interference. (a) What is the thickness of the film? (b) What visible wavelengths would be extra bright in the reflected interference pattern? (c) If this film is supported on glass whose index of refraction is 1.6, what wavelengths in the visible spectrum will be missing from the reflected light?

8. For a ruby laser of wavelength 694 nm, the ends of the ruby crystal are the aperture that determines the diameter of the light beam emitted. If the diameter is 2.00 cm and the laser is aimed at the moon, 0.38 Gm away, find the approximate diameter of the light beam reaching the moon.

9. Sodium light of wavelength 589 nm falls normally on a 2-cm-square diffraction grating ruled with 4000 lines per centimetre. The Fraunhofer diffraction pattern is projected onto a screen at 1.5 m by a lens of focal length 1.5 m placed immediately in front of the grating. Find (a) the positions of the first two intensity maxima on one side of the central maximum, (b) the width of the central maximum, and (c) the resolution in the first order.

10. At the second secondary maximum of the diffraction pattern of a single slit, the phase difference between the waves from the top and bottom of the slit is approximately 5π. The vectors used to calculate the amplitude at this point complete $2\frac{1}{2}$ circles. If I_0 is the intensity at the central maximum, find the intensity I at this second secondary maximum.

11. White light falls at an angle of 30° to the normal of a plane containing a pair of slits separated by 2.5 μm. What visible wavelengths give a bright interference maximum in the transmitted light in the direction normal to the plane? (See Exercise 12.)

12. A Newton's-ring apparatus consists of a glass lens with radius of curvature R which rests upon a flat glass plate, as shown in Figure 34-38. Usually both pieces of glass have the same index of refraction n, and so the thin film is air of variable thickness. The pattern is viewed by reflected light. (a) Show that for a thickness t the condition for a bright (constructive) interference fringe is $t = \frac{1}{2}(m + \frac{1}{2})\lambda$, $m = 0, 1, 2, \ldots$ (b) Show that as long as $t/R \ll 1$, the radius r of a bright circular fringe is given by $r = \sqrt{(m + \frac{1}{2})\lambda R}$, $m = 0, 1, 2, \ldots$ (c) How would the transmitted pattern look in comparison with the reflected one? (d) Use $R = 10$ m and a diameter of 4.0 cm for the lens. How many bright fringes would you see if the apparatus were illuminated by yellow sodium light ($\lambda \approx$ 590 nm) and viewed by reflection? (e) What would be the diameter of the sixth bright fringe? (f) If the glass used in the apparatus has index of refraction $n = $ 1.5 and water is placed between the two pieces of glass, what change will take place in the bright fringes?

Figure 34-38
Newton's-rings setup for Problem 12.

13. A *Jamin refractometer* is a device for measuring or comparing the indexes of refraction of fluids. A beam of monochromatic light is split into two parts, each directed along the axis of a separate cylindrical tube before being combined into a single beam again and viewed through a telescope. Suppose that each tube is 0.40 m long and sodium light of wavelength 589 nm is used. Both tubes are initially evacuated, and constructive interference is observed in the center of the field of view. As air is slowly allowed to enter one of the tubes, the central field of view changes to dark and back to bright a total of 198 times. (a) What is the index of refraction of air? (b) If the fringes can be counted to ±0.25 fringe, where one fringe is equivalent to one complete cycle of intensity variation at the center of the field of view, to what accuracy can the index of refraction of air be determined by this experiment?

14. A camera lens is made of glass whose index of refraction is 1.6. This lens is coated with a magnesium fluoride film ($n = 1.38$) to enhance its light transmission. This film is to produce a zero reflection for light of wavelength 540 nm. Treat the lens surface as a flat plane and the film as a uniformly thick flat film. (a) How thick must the film be to accomplish its objective in first order? (b) Would there be destructive interference for any other visible wavelengths? (c) By what factor would the reflection be reduced by this film for 400 and 700 nm? Neglect the variation in the reflected light amplitudes from the two surfaces.

15. For a diffraction grating one is interested not only in resolving power R, which is the ability of the grating to separate two close wavelengths, but also in the dispersion D of the grating. This is defined by $D = \Delta\theta_m/\Delta\lambda$ in mth order. (a) Show that D can be written

$$D = \frac{m}{\sqrt{d^2 - m^2\lambda^2}}$$

where d is the slit spacing. (b) If a diffraction grating with 2000 slits per centimetre is to resolve the two sodium yellow lines (wavelengths 589.0 and 589.6 nm) in second order, how many slits must be illuminated by the beam? (c) What would the separation be between these resolved yellow lines if the pattern were viewed on a screen 4 m from the grating?

16. (a) Show that the positions of the interference minima on a screen a large distance L away from three equally spaced sources (spacing d with $d \gg \lambda$) are given approximately by

$$y = n\frac{\lambda L}{3d} \quad \text{where } n = 1, 2, 4, 5, 7, 8, 10, \ldots$$

that is, n is not a multiple of 3. (b) For $L = 1$ m, $\lambda = 5 \times 10^{-7}$ m, and $d = 0.1$ mm, calculate the width of the principal interference maxima (distance between successive minima) for three sources.

17. Show that the positions of the interference minima on a screen a large distance L from four equally spaced sources are given by

$$y = n\frac{\lambda L}{4d} \quad \text{where } n = \text{integer not a multiple of 4, and } d \geq \lambda$$

Compare the width of the principal interference maxima for four sources with that for two sources for $d = 0.1$ mm, $L = 2$ m, and $\lambda = 6 \times 10^{-7}$ m.

18. Light of wavelength λ is diffracted through a single slit of width a, and the resulting pattern is viewed on a screen a long distance L away from the slit. (a) Show that the width of the central maximum on the screen is approximately $2L\lambda/a$. (b) If a slit of width $2L\lambda/a$ is cut in the screen and illuminated, show that the width of its central diffraction maximum at the same distance L (that is, back on the plane of the slit) is a to the same approximation.

19. In a pinhole camera, the image is fuzzy because of the finite size of the pinhole and because of diffraction. As the pinhole is made smaller, the fuzziness due to its size, i.e., to rays arriving at the image point from different parts of the object, is reduced, but that due to diffraction is increased. The optimum size of the aperture for the sharpest possible image occurs when the spread due to diffraction equals that due to the aperture size. Estimate the optimum size of the aperture if the distance from the pinhole to the screen is 10.0 cm and the wavelength of the light is 550 nm.

20. Light of wavelength 480 nm falls normally on four slits; each is 2.0 μm wide and separated by 6.0 μm from the next. (a) Find the angle from the center to the first zero of the single-slit diffraction pattern. (b) Find the angles of any bright interference maxima that lie inside the central diffraction maximum. (c) Find the angular spread between the central interference maximum and the first interference minimum on either side of it. (d) Sketch the intensity as a function of angle on the distant screen.

CHAPTER 35   Relativity

Objectives After studying this chapter you should:

1. Be able to state the Einstein postulates of special relativity.

2. Be able to discuss the results and significance of the Michelson-Morley experiment.

3. Be able to define proper time and proper length and state the equations for time dilation and length contraction.

4. Be able to discuss the lack of synchronization of clocks in moving reference frames.

5. Be able to use the Lorentz transformation to work problems relating time or space intervals in different reference frames.

6. Be able to discuss the twin paradox.

7. Be able to state the definition of relativistic momentum and the equations relating the kinetic energy and total energy of a particle to its speed.

8. Be able to discuss the relation between mass and energy in special relativity and compute the binding energy of various systems from the known rest masses of their constituents.

9. Be able to state the principle of equivalence and discuss three predictions derived from it.

The theory of relativity consists of two rather different theories, the special theory and the general theory. The special theory, developed by Einstein and others in 1905, concerns the comparison of measurements made in different inertial reference frames moving with constant velocity relative to each other. Its consequences, which can be derived with a minimum of mathematics, are applicable in a wide variety of situations encountered in physics and engineering. On the other hand, the general theory, also developed by Einstein and others (around 1916), is concerned with accelerated reference frames and gravity. A thorough understanding of the general theory requires sophisticated mathematics, e.g., tensor analysis, and the applications of this theory are chiefly in the area of gravitation. It is of great importance in cosmology but is rarely encountered in other areas of physics or in engineering. We shall therefore concentrate on the special theory (often referred to as special relativity) and discuss the general theory only briefly in the last section of this chapter.*

The theory of special relativity can be derived from two postulates proposed by Einstein in a paper on the electrodynamics of moving bodies published when he was only 26 years old.** These two postulates, simply stated, are:

1. Absolute, uniform motion cannot be detected.

2. The speed of light is independent of the motion of the source.

Einstein postulates

Although each postulate seems quite reasonable, many of the implications of the two together are quite surprising and contradict what is often called common sense. For example, a direct consequence is that all observers measure the same number for the speed of light in vacuum independent of their relative motion. We shall derive this and other results later. First we shall look at an historically important experiment related to the theory of special relativity, the Michelson-Morley experiment.

35-1   The Michelson-Morley Experiment

From our study of wave motion we know that all mechanical waves require a medium for their propagation. The speed of such waves depends only on properties of the medium. With sound waves, for example, absolute motion, i.e., motion relative to the still air, can be detected. For example, the doppler effect for sound depends not only on the relative motion of the source and listener but on the absolute motion of each relative to the air. It was natural to expect that some kind of medium supports the propagation of light and other electromagnetic waves. Such a medium, called the *ether,* was proposed in the nineteenth century. The ether as proposed had to have unusual properties. Although it required great rigidity to support waves of such high velocity (recall that the velocity of waves on a string depends on the tension of the string), it could introduce no drag force on the planets, as their motion is fully accounted for by the law of gravitation.

It was of considerable interest to determine the velocity of the earth relative to the ether. Maxwell pointed out that in measurements of the speed of light (such as Fizeau's toothed-wheel method), the earth's speed v relative to the ether appears only in the second order v^2/c^2, an effect then considered too small to measure. In these measurements, the time for a light pulse to travel to and from a mirror is determined. Fig-

* There are several good books that the interested student may want to consult to pursue the subject of relativity at the elementary level: R. Resnick, *Introduction to Special Relativity*, Wiley, New York, 1968; A. P. French, *Special Relativity*, Norton, New York, 1968; and C. Kacser, *Introduction to the Special Theory of Relativity*, Prentice-Hall, Englewood Cliffs, N.J., 1967.

** *Annalen der Physik*, vol. 17, p. 841, 1905. For a translation from the original German, see W. Perrett and G. B. Jeffery (trans.), *The Principle of Relativity: A Collection of Original Memoirs on the Special and General Theory of Relativity* by H. A. Lorentz, A. Einstein, H. Minkowski, and H. Weyl, Dover, New York, 1923.

Figure 35-1
Light source and mirror moving with speed v relative to the "ether." According to classical theory, the speed of light relative to the source and mirror would be $c - v$ toward the mirror and $c + v$ away from the mirror.

ure 35-1 shows a light source and mirror a distance L apart. If we assume that both are moving with speed v through the ether, classical theory predicts that the light will travel toward the mirror with speed $c - v$ and back with speed $c + v$ (both speeds relative to the mirror and light source). The time for the total trip will be

$$t_1 = \frac{L}{c - v} + \frac{L}{c + v} = \frac{2cL}{c^2 - v^2} = \frac{2L}{c}\left(1 - \frac{v^2}{c^2}\right)^{-1} \qquad 35\text{-}1$$

For v much less than c, we can expand this result using the binomial expansion:

$$(1 + x)^n \approx 1 + nx + \cdots \qquad \text{for } x \ll 1$$

Then

$$t_1 \approx \frac{2L}{c}\left(1 + \frac{v^2}{c^2} + \cdots\right) \qquad 35\text{-}2$$

If we take the orbital speed of the earth about the sun for an estimate of v, we have $v \approx 3 \times 10^4$ m/s $= 10^{-4}c$ and $v^2/c^2 = 10^{-8}$. Thus the correction for the earth's motion is small indeed. Michelson realized that though this effect is too small to be measured directly, it should be possible to determine v^2/c^2 by a difference measurement. For this measurement he used the Michelson interferometer discussed in Section 34-2. Let us assume that the interferometer (see Figure 34-7) is oriented so that the beam transmitted by the beam splitter is in the direction of the assumed motion of the earth. Equation 35-2 then gives the classical result for the round-trip time t_1 for the transmitted beam. Since the reflected beam travels (relative to the earth) perpendicular to the earth's velocity, the velocity of this beam relative to earth (according to classical theory) is the vector difference $\mathbf{u} = \mathbf{c} - \mathbf{v}$ (Figure 35-2). The magnitude of $\mathbf{u}$ is $\sqrt{c^2 - v^2}$; so the round-trip time for this beam is

$$t_2 = \frac{2L}{\sqrt{c^2 - v^2}} = \frac{2L}{c}\left(1 - \frac{v^2}{c^2}\right)^{-1/2} \approx \frac{2L}{c}\left(1 + \frac{1}{2}\frac{v^2}{c^2} + \cdots\right) \qquad 35\text{-}3$$

where again the binomial expansion has been used. Hence there is a time difference:

$$\Delta t = t_1 - t_2 \approx \frac{2L}{c}\left(1 + \frac{v^2}{c^2}\right) - \frac{2L}{c}\left(1 + \frac{1}{2}\frac{v^2}{c^2}\right) = \frac{Lv^2}{c^3} \qquad 35\text{-}4$$

The time difference is to be detected by observing the interference of the two beams of light. Because of the difficulty of making the two paths of equal length to the precision required, the interference pattern of the two beams is observed and then the whole apparatus rotated 90°. The rotation produces a time difference given by Equation 35-4 for each beam. The total time difference of $2\,\Delta t$ is equivalent to a path difference

Figure 35-2
A light beam from the beam splitter A to the mirror M_2 in a Michelson interferometer moves with velocity $\mathbf{c}$ at an angle as shown in a reference frame in which the interferometer is moving to the right with speed v. Relative to the earth in which the interferometer is fixed, the light beam travels with velocity $\mathbf{u} = \mathbf{c} - \mathbf{v}$ according to classical theory. The speed of the light beam relative to the earth is then $u = (c^2 - v^2)^{1/2} = c(1 - v^2/c^2)^{1/2}$ according to classical theory.

of $2c\,\Delta t$. The interference fringes observed in the first orientation should thus shift by a number of fringes ΔN given by

$$\Delta N = \frac{2c\,\Delta t}{\lambda} = \frac{2L}{\lambda}\frac{v^2}{c^2} \qquad 35\text{-}5$$

where λ is the wavelength of the light. In Michelson's first attempt, in 1881, L was about 1.2 m and λ was 590 nm. For $v^2/c^2 = 10^{-8}$, ΔN was expected to be 0.04 fringe. Even though the experimental uncertainties were estimated to be about this same magnitude, when no shift was observed, Michelson reported this as evidence that the earth did not move relative to the ether. In 1887, when he repeated the experiment with Edward W. Morley, he used an improved system for rotating the apparatus without introducing a fringe shift because of mechanical strains, and he increased the effective path length L to about 11 m by a series of multiple reflections. Figure 35-3 shows the configuration of the Michelson-Morley apparatus. For this attempt, ΔN was expected to be about 0.4 fringe, about 20 to 40 times the minimum possible to observe. Once again, no shift was observed. The experiment has since been repeated under various conditions by a number of people, and no shift has ever been found.

The null result of the Michelson-Morley experiment is easily understood in terms of the Einstein postulates. According to postulate 1, absolute uniform motion cannot be detected. We can consider the whole apparatus and the earth to be at rest. No fringe shift is expected when the apparatus is rotated 90° since all directions are equivalent. It should be pointed out that Einstein did not set out to explain this experiment. His theory arose from his considerations of the theory of electricity and magnetism and the unusual property of electromagnetic waves, i.e., that they propagate in a vacuum. In his first paper, which contains the complete theory of special relativity, he made only a passing reference to the Michelson-Morley experiment, and in later years he could not recall whether he was aware of the details of this experiment before he published his theory.

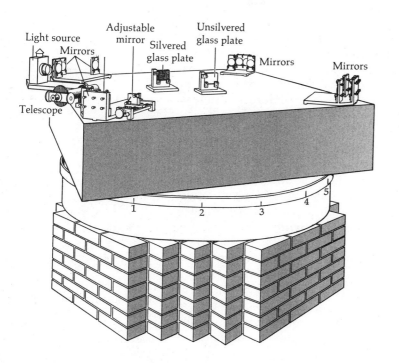

Figure 35-3
Drawing of Michelson-Morley apparatus used in their 1887 experiment. The optical parts were mounted on a sandstone slab 5 ft square, which was floated in mercury, thereby reducing the strains and vibrations that had affected the earlier experiments. Observations could be made in all directions by rotating the apparatus in the horizontal plane. (*From R. S. Shankland, "The Michelson-Morley Experiment," Copyright © November 1964 by Scientific American, Inc. All rights reserved.*)

35-2 Consequences of Einstein's Postulates

An immediate consequence of the two Einstein postulates is that

Every observer measures the same value for the speed of light independent of the relative motion of the sources and observers.

Consider a light source S and two observers, R_1 at rest relative to S and R_2 moving toward S with speed v, as shown in Figure 35-4a. The speed of light measured by R_1 is $c = 3 \times 10^8$ m/s. What is the speed measured by R_2? The answer is *not* $c + v$. By postulate 1, Figure 35-4a is equivalent to Figure 35-4b, in which R_2 is pictured at rest and the source S and R_1 are moving with speed v. That is, since absolute motion has no meaning, it is not possible to say which is really moving and which is at rest. By postulate 2, the speed of light from a moving source is independent of the motion of the source. Thus, looking at Figure 35-4b, we see that R_2 measures the speed of light to be c, just as R_1 does.

This result—that all observers measure the same value for the speed of light—contradicts our intuitive ideas about relative velocities. If a car moves at 50 km/h away from an observer and another car moves at 80 km/h in the same direction, the velocity of the second car relative to the first car is 30 km/h. This result is easily measured and conforms to our intuition. However, according to Einstein's postulates, if a light beam is moving in the direction of the cars, observers in both cars will measure the same speed for the light beam. We shall see later that our intuitive ideas about the combination of velocities are approximations which hold only when the speeds are very small compared with the speed of light. In practice, even in a plane moving with the speed of sound, it is not possible to measure the speed of light accurately enough to distinguish the difference between the result c and $c \pm v$, where v is the speed of the plane. In order to make such a distinction, we must either move with a very great velocity (much greater than that of sound) or make extremely accurate measurements, as in the Michelson-Morley experiment.

Einstein recognized that his postulates have important consequences for measuring time intervals and space intervals as well as relative velocities. He showed that the size of time and space intervals between two events depends on the reference frame in which the events are observed. The changes in such measurements from one reference frame to another are called *time dilation* and *length contraction*. Instead of developing the general formalism of relativity, known as the *Lorentz transformation*, we shall derive these famous relativistic effects directly from the Einstein postulates by considering some simple special cases of measurement of time and space intervals. We shall also study the problems of clock synchronization and simultaneity, which are of central importance in understanding special relativity. In these discussions we shall be comparing measurements made by observers who are moving relative to each other. We shall use a rectangular coordinate system xyz with origin O, called the S *reference frame*, and another system $x'y'z'$ with origin O', called the S' *frame*, which is moving with constant velocity relative to the S frame. For simplicity, we shall consider the S' frame to be moving with speed v along the x (or x') axis relative to S. In each frame we assume that there are as many observers as needed equipped with clocks, metresticks, etc. that are identical when compared at rest (see Figure 35-5).

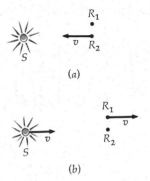

(a)

(b)

Figure 35-4
If absolute motion cannot be detected, the reference frame shown in (a), in which the receiver R_1 and source S are stationary and receiver R_2 is moving toward the source with speed v, is equivalent to the reference frame shown in (b), in which R_2 is stationary and the source S and receiver R_1 are moving with speed v. Since the speed of light does not depend on the source speed, receiver R_2 measures the same value for that speed as receiver R_1.

Figure 35-5
Coordinate reference frames S and S' moving with relative speed v. In each frame are observers with metresticks and clocks.

Albert Einstein (1879–1955)

Gerald Holton
Harvard University

Einstein at 17. (*From Carl Seelig*, Albert Einstein: Eine Dokumentarische Biographie, *Europa Verlag, Zurich, 1954.*)

A friend visiting Albert Einstein in his modest walkup apartment in Bern, Switzerland found him seated at a kitchen table, "pipe in his mouth, his left hand moving a children's carriage back and forth, in his right hand a shabby stub of a pencil." The year was 1905, and Einstein was writing his great paper on the special theory of relativity. Some of Einstein's personality traits can already be glimpsed here—the simplicity of his personal life and his ability to lift himself out of his surroundings by concentrating on the work before him.*

The paper of 1905 was the culmination of ideas that had started to preoccupy Einstein 10 years earlier, when he was about 16. At that age, he later wrote in his autobiography,** the following paradox occurred to him. As one knows from ample experience, galilean relativity holds in mechanics; if you throw a ball forward in a moving carriage, observing the motion of the ball cannot tell you how fast you and the carriage are moving. But when it comes to optics, it seems to be different. For example, if I move along "a beam of light with the velocity c [velocity of light in a vacuum], I should observe such a beam of light as a spatially oscillatory electromagnetic field." Looking back along the beam a distance of one whole wavelength, one should see that the local magnitudes of the electric and magnetic field vectors increase point by point from, say, zero to full strength, and then decrease again to zero, one wavelength away. Seeing such a curious field in free space would tell me that I am going at the speed of light with respect to absolute space, or the ether.

Einstein suspected that the imagined result of this *Gedanken experiment* (thought experiment) must somehow be in error. In any case, he said later, "from the very beginning it appeared to me intuitively clear that, judged from the standpoint of such an observer, everything would have to happen according to the same laws as for an observer who, relative to the earth, was at rest."

Einstein solved the apparent paradox in 1905 by showing that the expectation of what one would see in pursuing a light beam is false, being grounded in a wrong idea that "unrecognizedly was anchored in the unconscious" of all scientists of the time—namely the absolute character of time and of simultaneity. Instead, Einstein showed that a sound view of how physical nature operates can be gained by boldly postulating two principles.

The first principle of relativity for inertial systems generalizes galilean relativity to encompass not only mechanics but also optics and electromagnetism; the second principle says that light in a vacuum will always be found to move with the same speed c, regardless of the state of motion of the emitting body. The solutions to all paradoxes and experimentally puzzling observations at the time were derivable from these two principles.

At least as important to Einstein was the fact that this approach hugely simplified our view of nature in two ways: it broke down barriers between hitherto entirely separate notions, and it cleansed physics of unnecessary conceptions that had produced pseudo problems. Now electromagnetism was on the same footing as mechanics, instead of allowing "privileged" systems. Time and space were found to have interpenetrating meaning—what really existed was a space-time continuum. Electric fields and magnetic fields were at bottom the same reality perceived in different experimental conditions. Mass and energy were equivalent. Even the boldly intuitive approach to scientific

---

* See Einstein's essay "Motive of Research," pp. 224–227 in Sonja Bargmann (trans. and ed.), *Ideas and Opinions,* Crown, New York, 1954; look also at the other essays in this collection.

** Albert Einstein, *Autobiographical Notes,* in P. A. Schilpp (ed.), *Albert Einstein: Philosopher-Scientist,* Harper, New York, 1959.

discovery and the rigorously rational method were joined into one powerful approach in his paper. Moreover, the notion of the ether was at last declared to be "superfluous." It had long been embedded in physics but had required the assumption of ever more puzzling properties. While he was the first to give up the idea of an ether, to the end of his life Einstein was dedicated to the theme of the *field* (or, in general, the continuum) as the basic conceptual tool for the fundamental explanation of phenomena.

The preoccupation with the continuum, with explaining mysterious order-liness, with finding pleasing simplicity—these were conceptions in physics which characterized Einstein's work from the beginning. But as with other highly creative persons, the power of his work was derived not merely from good physical ideas. Instead it came from a fusion of his scientific interests and his characteristics as a human being, a synthesis of his life-style and his perception of the laws of nature. Much of what was most daring or novel in his great work in physics was present in Einstein's ways of everyday thinking and behaving, even as a child or a young student. Take, for example, the ability to come back, again and again, for years, to a difficult puzzle. In early life, this trait showed up as what may have seemed mere obstinacy. He was commonly reported to have been withdrawn as a child, preferring to play by himself, erecting complicated constructions. Before he was 10, with infinite patience he was making fantastic card houses that had as many as 14 floors. He was unable or unwilling to talk until the age of 3. In school, he was not an exceptional student, preferring to follow his own thoughts. Later, he stuck to his ideas with the same persistence when the experiments of others seemed to disconfirm his theories (in time, those experiments usually turned out to be wrong). And during the last 30 years of his life, he persisted in his skepticism concerning the fundamental explanatory power of quantum mechanics and in his dedication to the problems of field theory, unlike most physicists of the time.

With this single-mindedness and concentration on his own revolutionary ideas went his deep suspicion of established authority, in science no less than in daily life. The same young student who quietly challenged the established ideas in physics also abhorred the current political, religious, and social conventions. From the age of 12 years on, he confessed later, he had "suspicion against every kind of authority." One result was that his teachers were quite delighted to see him drop out of high school at the age of 15½, when he no longer could stand the regimented and militaristic way of life in his native Germany and in his school. Moving to the freer, more democratic atmosphere of Switzerland, he found at last a school to his liking, and there had a glorious year before entering the Polytechnic Institute of Zurich. It was there, too, that he had his first ideas on relativity.

Related to this trait of uncompromisingly sticking to his own identity was his search for what is really *necessary*. In his early work in physics, he said, the thought of having to explain electric and magnetic field effects as "two fundamentally different cases was for me unbearable." The highest aim was nothing less than finding the most economical, simple principles, the barest bones of nature's frame, cleansed of everything that is ad hoc, redundant, unsymmetrical. "What really interests me," he once said, "is whether God has any choice in the creation of the world." Nature does not like anything that is unnecessary. Nor did Einstein, in his personal life—in his clothing, in his manner of speech and writing, in his behavior, from his preference for the classical music of Bach and Mozart down to his preference for using the same bar of soap for washing and shaving instead of complicating life unnecessarily by facing two kinds of soap every morning.

The preference for an egalitarian democracy characterized Einstein's political life just as it did his physics. The man who declared every inertial system to be created equal before all the laws of physics was also, from youth on, fiercely opposed to every antidemocratic and narrowly nationalistic political or social system.

Einstein at 53. (*Courtesy of the Archives, California Institute of Technology.*)

Einstein kept an abiding interest in philosophy, which penetrated his work; his ideas, in turn, influenced the development of modern philosophy itself. Here, too, he was cutting across unnecessary barriers. This is not surprising. Genius discovers itself not in splendid solutions to little puzzles but in the struggle with deep and perhaps eternal problems at the point where science and philosophy join.*

* For further discussion on Einstein's relativity theory, its genesis and influence, see Gerald Holton, *Thematic Origins of Scientific Thought, Kepler to Einstein*, chaps. 5–10, Harvard University Press, Cambridge, Mass., 1973. Among more recent biographies, see Banesh Hoffmann, with the collaboration of Helen Dukas, *Albert Einstein: Creator and Rebel*, Viking, New York, 1972; Jeremy Bernstein, *Einstein*, Viking, New York, 1973; or the best of the older biographies: Philipp Frank, *Einstein: His Life and Times*, Knopf, New York, 1947. For a selection of Einstein's correspondence, see *Albert Einstein, the Human Side: New Glimpses from His Archives*, edited by Helen Dukas and Banesh Hoffman, Princeton University Press, Princeton, New Jersey, 1979.

35-3   Time Dilation and Length Contraction

Since the results of measurement of the time and space intervals between events do not depend on the kind of apparatus used for the measurements or on the events, we are free to choose any events and measuring apparatus which will help us understand the application of the Einstein postulates to the results of the measurement. Convenient events in relativity are those which produce light flashes. A convenient clock is a light clock, pictured schematically in Figure 35-6. A photocell detects the light pulse and sends a voltage pulse to an oscilloscope, giving a vertical deflection of the trace on the scope. The phosphorescent material on the face of the oscilloscope tube gives a persistent light that can be observed visually or photographed. The time between two light flashes is determined by measuring the distance between pulses on the scope and knowing the sweep speed of the scope. Such a clock, which can easily be calibrated and compared with other types of clocks, is often used in nuclear-physics experiments.

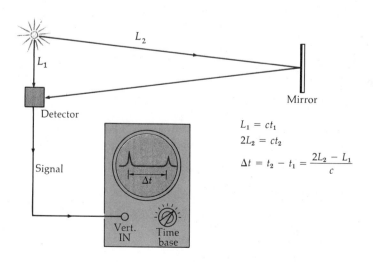

$$L_1 = ct_1$$
$$2L_2 = ct_2$$

$$\Delta t = t_2 - t_1 = \frac{2L_2 - L_1}{c}$$

Figure 35-6
Light clock for measuring time intervals. The time is measured by reading the distance between pulses on the oscilloscope after calibrating the sweep speed.

We first consider an observer A' at rest in frame S' a distance D from a mirror, as shown in Figure 35-7a. He explodes a flash gun and measures the time interval $\Delta t'$ between the original flash and the return flash from the mirror. Since light travels with speed c, this time is

$$\Delta t' = \frac{2D}{c} \qquad\qquad 35\text{-}6$$

We now consider these same two events, the original flash of light and the returning flash, as observed in reference frame S, where observer A' and the mirror are moving to the right with speed v. The events happen at two different places x_1 and x_2 in frame S because between the original flash and the return flash observer A' has moved a horizontal distance $v\,\Delta t$, where Δt is the time interval between the events measured in S. In Figure 35-7b we see that the path traveled by the light is longer in S than in S'. However, by Einstein's postulates, light travels with the same speed c in frame S as it does in frame S'.

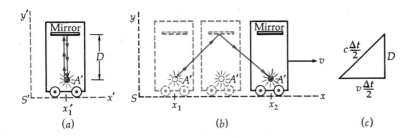

(a)   (b)   (c)

Figure 35-7
(a) A' and the mirror are at rest in S'. The time for a light pulse to reach the mirror and return is measured to be $2D/c$ by A'. (b) In frame S, A' and the mirror are moving with speed v. If the speed of light is the same in both reference frames, the time for the light to reach the mirror and return is longer than $2D/c$ in S because the distance traveled is greater than $2D$. (c) Right triangle for computing the time Δt in frame S.

Since it travels farther in S at the same speed, it takes longer in S to reach the mirror and return. The time interval in S is thus longer than it is in S'. We can easily calculate Δt in terms of $\Delta t'$. From the triangle in Figure 35-7c we have

$$\left(\frac{c\,\Delta t}{2}\right)^2 = D^2 + \left(\frac{v\,\Delta t}{2}\right)^2 \qquad\qquad 35\text{-}7$$

or

$$\Delta t = \frac{2D}{\sqrt{c^2 - v^2}} = \frac{2D}{c}\,\frac{1}{\sqrt{1 - v^2/c^2}}$$

Using $\Delta t' = 2D/c$, we have

$$\Delta t = \frac{\Delta t'}{\sqrt{1 - v^2/c^2}} = \gamma\,\Delta t' \quad \text{where} \quad \gamma = \frac{1}{\sqrt{1 - v^2/c^2}} \geq 1 \qquad 35\text{-}8$$ *Time dilation*

Observers in S would say that the clock held by A' runs slow since he claims a shorter time interval for these events.

A' measures the times of the light flash and return at the same point in S', while in S these events happen at two different places. A single clock can be used in S' to measure the time interval, but in S, two synchronized clocks are needed, one at x_1 and one at x_2. The time between events that happen at the *same place* in a reference frame (as with A' in S' in this case) is called *proper time*. The time interval measured in any other reference frame is always longer than the proper time. This expansion is called *time dilation*.

Time dilation is closely related to another phenomenon, *length contraction*. The length of an object measured in the reference frame in which the object is at rest is called its *proper length*. In a reference frame in which the object is moving, the measured length is shorter than its proper length. We can see this from our previous example using the light clocks. Suppose that x_1 and x_2 in that example are at the ends of a measuring rod of length $L_0 = x_2 - x_1$ measured in frame S, in which the rod is at rest. Since A' is moving relative to this frame with speed v, the distance moved in time Δt is $v\,\Delta t$. Since he moves from point x_1 to point x_2 in this time, this distance is $L_0 = x_2 - x_1 = v\,\Delta t$. According to observer A' in frame S' (Figure 35-8), the measuring rod moves with speed v and takes a time $\Delta t'$ to move past him. The length of the rod in his frame is $L' = v\,\Delta t'$. Since the time interval $\Delta t'$ is less than Δt, the length L' is less than L_0. These lengths are related by

$$L' = v\,\Delta t' = \frac{v\,\Delta t}{\gamma} = \frac{L_0}{\gamma} = \sqrt{1 - \frac{v^2}{c^2}}\,L_0 \qquad\qquad 35\text{-}9$$

Length contraction

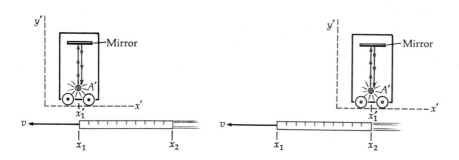

Figure 35-8
Measuring the length of a moving object. The rod has length $L_0 = x_2 - x_1 = v\,\Delta t$ as measured in frame S, in which it is at rest. In frame S' it moves past a fixed point in time $\Delta t' = \sqrt{1 - v^2/c^2}\,\Delta t$, so its length $L' = v\,\Delta t'$ is shorter in this frame.

(Both A and A' measure the same relative velocity. Otherwise there would be a lack of symmetry, and postulate 1 would be violated; i.e., we could choose the frame with the smaller or greater relative velocity to be a preferred frame.) Thus the length of the rod is smaller when measured in a frame in which it is moving. Since this contraction is just the amount proposed by Lorentz and FitzGerald to explain the Michelson-Morley experiment, it is often called the *Lorentz-FitzGerald contraction*.

An interesting example of the observation of these phenomena is afforded by the appearance of muons as secondary radiation from cosmic rays. Muons decay according to the statistical law of radioactivity:

$$N(t) = N_0 e^{-t/T} \qquad\qquad 35\text{-}10$$

where N_0 is the number at time $t = 0$, $N(t)$ is the number at time t, and T is the mean lifetime, which is about 2 μs for muons at rest. Since they are created (from the decay of π mesons) high in the atmosphere, usually several thousand metres above sea level, few muons should reach sea level. A typical muon moving with speed $0.998c$ would travel only about 600 m in 2 μs. However, the lifetime of the muon measured in the earth's reference frame is increased by the factor $1/\sqrt{1 - v^2/c^2}$, which is 15 for this particular speed. The mean lifetime measured in the earth's reference frame is therefore 30 μs, and a muon of this speed travels about 9000 m in this time. From the muon's point of view, it lives only 2 μs, but the atmosphere is rushing past it with a speed of $0.998c$. The distance of 9000 m in the earth's frame is thus contracted to only 500 m, as indicated in Figure 35-9.

It is easy to distinguish experimentally between the classical and relativistic predictions of the observation of muons at sea level. Suppose that we observe 10^8 muons at an altitude of 9000 m in some time interval with a muon detector. How many would we expect to observe at sea level in the same time interval? According to the nonrelativistic pre-

(a)

(b)

Figure 35-9
Although muons are created high above the earth and their mean lifetime is only about 2 μs when at rest, many appear at the earth's surface. (a) In the earth's reference frame a typical muon moving at $0.998c$ has a mean lifetime of 30 μs and travels 9000 m in this time. (b) In the reference frame of the muon, the distance traveled by the earth is only 600 m in the muon's lifetime of 2 μs.

diction, the time taken for these muons to travel 9000 m is $(9000 \text{ m})/0.998c \approx 30 \ \mu s$, which is 15 lifetimes. Putting $N_0 = 10^8$ and $t = 15T$ into Equation 35-10, we obtain

$$N = 10^8 e^{-15} = 30.6$$

We would thus expect all but about 31 of the original 100 million muons to decay before reaching sea level.

According to the relativistic prediction, the earth must travel only the contracted distance of 600 m in the rest frame of the muon. This takes only 2 $\mu s = T$. Thus the number expected at sea level is

$$N = 10^8 e^{-1} = 3.68 \times 10^7$$

Thus relativity predicts that we would observe 36.8 million muons in the same time interval. Experiments of this type have confirmed the relativistic predictions.

Question

1. You are standing on a corner and a friend is driving past in an automobile. Both of you note the times when the car passes two different intersections. You each determine from your watch readings the time that elapses between the two events. Which of you has determined the proper time interval?

35-4   Clock Synchronization and Simultaneity

At first glance, time dilation and length contraction seem contradictory not only to our intuition but to our ideas of self-consistency. If each reference frame can be considered at rest with the other moving, the clocks in the "other" frame should run slow. How can there be any self-consistency if each observer sees the clocks of the other run slow? The answer to this puzzle lies in the problems of clock synchronization and in the concept of simultaneity. We note that the time intervals Δt and $\Delta t'$ considered in Section 35-3 were measured in quite different ways. The events in frame S' happened at the same place, and the times could be measured on a single clock; but in S, the two events happened at different places. The time of each event was measured on a different clock and the interval found by subtraction. This procedure requires that the clocks be synchronized. We shall show in this section that:

Two clocks synchronized in one reference frame are not synchronized in any other frame moving relative to the first frame.

A corollary to this result is that:

Two events that are simultaneous in one reference frame are not simultaneous in another frame moving relative to the first.

(This is true unless the events and clocks are in the same plane perpendicular to the relative motion).

Comprehension of these facts usually resolves all relativity paradoxes. Unfortunately, the intuitive (and incorrect) belief that simultaneity is an absolute relation is difficult to correct.

How can we synchronize two clocks separated in space? The problem is not difficult but requires some thought. One obvious method is to bring the clocks together and set them to read the same time, then move them to their original positions. This method has the drawback that, as we have seen, each clock runs slowly during the time it is moved, according to the time-dilation result. Suppose the clocks are at rest at points A and B a distance L apart in frame S. If an observer at A looks at the clock at B and sets his clock to read the same time, the clocks will not be synchronized because of the time L/c it takes light to travel from one clock to another. To synchronize the clocks, the observer at A must set his clock ahead by the time L/c. Then he will see that the clock at B reads a time which is L/c behind the time on his clock, but he will calculate that the clocks are synchronized when he allows for the time L/c for the light to reach him. All observers except those midway between the clocks will see the clocks reading different times, but they will also compute that the clocks are synchronized when they correct for the time it takes the light to reach them. An equivalent method for the synchronization of two clocks would be for a third observer C at a point midway between the clocks to send a light signal and for observers at A and B to set their clocks to some prearranged time when they receive the signal.

We now examine the question of simultaneity. Suppose A and B agree to explode bombs at t_0 (having previously synchronized their clocks). Observer C will see the light from the two explosions at the same time, and since he is equidistant from A and B, he will conclude that the explosions were simultaneous. Other observers in S will see the light from A or B first, depending on their location, but after correcting for the time the light takes to reach them, they also will conclude that the explosions were simultaneous. *We shall thus define two events in a reference frame to be simultaneous if the light signals from the events reach an observer halfway between the events at the same time.* To show that two events which are simultaneous in frame S are not simultaneous in another frame S' moving relative to S we use an example introduced by Einstein. A train is moving with speed v past the station platform. We have observers A', B', and C' at the front, back, and middle of the train. (We shall consider the train to be at rest in S' and the platform in S.) We now suppose that the train and platform are struck by lightning at the front and back of the train and that the lightning bolts are simultaneous in the frame of the platform (S) (Figure 35-10). That is, an observer C halfway between the positions A and B, where the lightning strikes, observes the two flashes at the same time. It is convenient to suppose that the lightning scorches the train and platform so that the events can be easily located in each reference frame. Since C' is in the middle of the train, halfway between the places on the train which are scorched, the events can be simultaneous in S' only if C' sees the flashes at the same time. However, C' sees the flash from the front of the train before he

Simultaneity defined

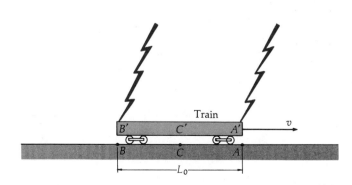

Figure 35-10
Simultaneous lightning bolts strike the ends of a train traveling with speed v in frame S attached to the platform. The light from these simultaneous events reaches observer C midway between the events at the same time. The distance between the bolts is L_0, which is also the length of the train measured in frame S.

sees the flash from the back. In frame S, when the light from the front flash reaches him, he has moved some distance toward it, so that the flash from the back has not yet reached him, as indicated in Figure 35-11. He must therefore conclude that the events are not simultaneous. The front of the train was struck before the back. As we have discussed above, all observers in S' on the train will agree with C' when they have corrected for the time it takes light to reach them.

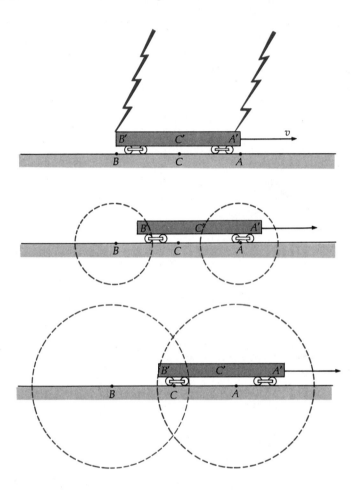

Figure 35-11
In frame S the light from the bolt at the front of the train reaches observer C' in the middle of the train before light from the bolt at the rear of the train because the train is moving. Since C' is midway between the events (which occur at the front and rear of the train), these events are not simultaneous for him.

Let L_0 be the distance between the scorch marks on the platform, which is also the length of the train measured in frame S. This distance is smaller than the proper length of the train L_T' because of length contraction. Figure 35-12 shows the situation in frame S', in which the train is at rest and the platform is moving. In this frame the distance between the burns on the platform is contracted and is related to L_0 by

$$L_P' = \frac{L_0}{\gamma} \qquad\qquad 35\text{-}11$$

Similarly the train length is the proper length related to L_0 by

$$L_T' = \gamma L_0 \qquad\qquad 35\text{-}12$$

(The drawing of this figure has been made for $\gamma = 1.5$.) The time interval in S' between these two events is the time it takes the platform to move the distance ΔL, where

$$\Delta L = L_T' - L_P' = \gamma L_0 - \frac{L_0}{\gamma} = \left(1 - \frac{1}{\gamma^2}\right)\gamma L_0 = \frac{v^2}{c^2}\gamma L_0$$

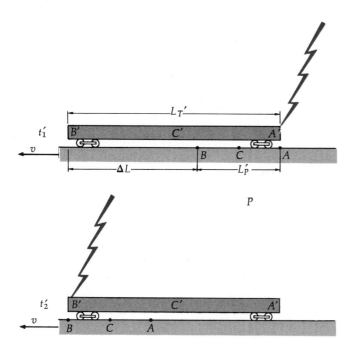

Figure 35-12
Lightning bolts of Figures
35-10 and 35-11 as seen in
reference frame S' fixed to the
train. In this frame the distance $L'_P = BA$ along the
moving platform is con-
tracted, and the train length
$L'_T = B'A'$ is its proper
length. At time t'_1 lightning
strikes the front of the train
when A' and A are coincident. At a later time t'_2 the
lightning strikes the rear of
the train when B' and B are
coincident. During the time
between flashes the platform
moves the distance $L'_T - L'_P$.

since

$$1 - \frac{1}{\gamma^2} = 1 - \left(1 - \frac{v^2}{c^2}\right) = \frac{v^2}{c^2}$$

Thus

$$t'_2 - t'_1 = \frac{\Delta L}{v} = \frac{\gamma L_0 v}{c^2}$$

During this time (according to observers in S'), the clocks at A and B
ticked off a time interval which is smaller because of time dilation

$$\Delta t_S = \frac{1}{\gamma}(t'_2 - t'_1) = \frac{L_0 v}{c^2} \qquad\qquad 35\text{-}13$$

This is the time interval by which the clocks in S are unsynchronized
according to observers in S'. That is, the clock at A is ahead of that at B
by the amount $L_0 v/c^2$. This result is worth remembering.

If two clocks are synchronized in the frame in which they are at rest, they
will be out of synchronization in another frame. In the frame in which
they are moving, the "chasing clock" leads by an amount $\Delta t_S = L_0 v/c^2$,
where L_0 is the proper distance between the clocks.

Lack of synchronization of
moving clocks

Example 35-1 A numerical example should help clarify time dilation,
clock synchronization, and the internal consistency of these results. Let
the light clock used in Section 35-3 be moving with speed $v = 0.8c$.
Then

$$1 - \frac{v^2}{c^2} = 1 - 0.64 = 0.36$$

and

$$\gamma = \frac{1}{\sqrt{1 - v^2/c^2}} = \frac{1}{\sqrt{0.36}} = \frac{1}{0.6} = \frac{5}{3}$$

Also, let the distance $x_2 - x_1$ be 40 light-minutes, that is, the distance light travels in 40 min (a convenient notation for this unit is $c \cdot$min):

$$x_2 - x_1 = 40 \ c \cdot \text{min}$$

The time for the light clock to travel this distance is Δt given by

$$\Delta t = \frac{x_2 - x_1}{v} = \frac{40 \ c \cdot \text{min}}{0.8c} = 50 \ \text{min}$$

(Note that the c in the unit $c \cdot$min cancels, giving the time in minutes.)
  The proper time interval in S' is shorter by the factor γ:

$$\Delta t' = \frac{\Delta t}{\gamma} = \frac{50 \ \text{min}}{5/3} = 30 \ \text{min}$$

(Thus we have taken $D = 15 \ c \cdot$min for the distance from the flash gun to the mirror in Figure 35-7.) The distance traveled is

$$\Delta x' = v \ \Delta t' = (0.8c)(30 \ \text{min}) = 24 \ c \cdot \text{min}$$

This is just the distance $x_2 - x_1$ that is contracted in S':

$$L' = \frac{L_0}{\gamma} = \frac{40 \ c \cdot \text{min}}{5/3} = 24 \ c \cdot \text{min}$$

Figure 35-13 shows the situation viewed in S'. We assume that the clock at x_1 reads noon at the time of the light flash. The clocks at x_1 and x_2 are synchronized in S but not in S'. In S', the clock at x_2, which is chasing the one at x_1, leads by

$$\frac{L_0 v}{c^2} = \frac{(40 \ c \cdot \text{min})(0.8c)}{c^2} = 32 \ \text{min}$$

When the light clock coincides with x_2, the clock there reads 50 min past noon. The time between events is therefore 50 min in S. Note that according to observers in S', this clock ticks off $50 - 32 = 18$ min for a trip which takes 30 min in S'. Thus this clock runs slow by the factor $30/18 = \frac{5}{3}$.

  Each observer sees clocks in the other frame run slow. According to observers in S who measure 50 min for the time interval, the time interval in S' is too small (30 min) because the single clock in S' runs too slow by the factor $\frac{5}{3}$. According to the observers in S', the observers in S measure a time which is too *long* despite the fact that their clocks run too slow, because they are out of synchronization. The clocks tick off only 18 min, but the second one leads the first by 32 min, so the time interval is 50 min.

Figure 35-13
Example 35-1. In frame S' the rod passes the light clock in a time $\Delta t' = 30$ min. During this time the clocks at each end of the rod tick off $30/\gamma$ min $= 18$ min. But the clocks are unsynchronized, with the chasing clock leading by $L_0 v/c^2 = 32$ min, where L_0 is the proper separation of the clocks. The time interval in S is therefore 32 min + 18 min $= 50$ min, which is longer than 30 min.

Questions

2. Two observers are in relative motion. In what circumstances can they agree on the simultaneity of two different events?

3. If event A occurs before event B in some frame, might it be possible for there to be a reference frame in which event B occurs before event A?

4. Two events are simultaneous in a frame in which they also occur at the same point in space. Are they simultaneous in other reference frames?

35-5 The Doppler Effect

In deriving the doppler effect for sound (Chapter 14) we found that the change in frequency for a given velocity v depends on whether it is the source or receiver that is moving with that speed. Such a distinction is possible for sound because there is a medium (the air) relative to which the motion takes place, and so it is not surprising that the motion of the source or the receiver relative to the still air can be distinguished. Such a distinction between motion of the source or receiver cannot be made for light or other electromagnetic waves in vacuum. The expressions we have derived for the doppler effect cannot be correct for light. We now derive the relativistic doppler-effect equations.

We shall consider a source moving toward a receiver with velocity v, and work in the frame of the receiver. Let the source emit N waves. If the source is moving toward the receiver, the first wave will travel a distance $c \, \Delta t_R$ and the source will travel $v \, \Delta t_R$ in the time Δt_R measured in the frame of the receiver. The wavelength will be $\lambda' = (c \, \Delta t_R - v \, \Delta t_R)/N$. The frequency f' observed by the receiver will therefore be

$$f' = \frac{c}{\lambda'} = \frac{c}{c - v}\frac{N}{\Delta t_R} = \frac{1}{1 - v/c}\frac{N}{\Delta t_R}$$

If the frequency of the source is f_0, it will emit $N = f_0 \, \Delta t_S$ waves in time Δt_S, measured by the source. Here Δt_S is the proper time interval (the first wave and the Nth wave are emitted at the same place in the source's reference frame). Times Δt_S and Δt_R are related by the usual time-dilation equation $\Delta t_S = \Delta t_R/\gamma$. Thus we obtain for the doppler effect for a moving source

$$f' = \frac{1}{1 - v/c}\frac{f_0 \, \Delta t_S}{\Delta t_R} = \frac{f_0}{1 - v/c}\frac{1}{\gamma} = \frac{\sqrt{1 - v^2/c^2}}{1 - v/c} f_0 \qquad 35\text{-}14a$$

which differs from our classical equation only in the time-dilation factor.

We now make the calculation in the reference frame of the source. That is, we assume that the source is at rest and the receiver moves with velocity v toward the source. In time Δt_S in the frame of the source, the receiver encounters all the waves in the distance $v \, \Delta t_S$ in addition to the waves in the distance $c \, \Delta t_S$, just as in the classical calculation. The number of waves encountered is thus

$$N = \frac{c \, \Delta t_S + v \, \Delta t_S}{\lambda} = \frac{(c + v) \, \Delta t_S}{c/f_0} = \left(1 + \frac{v}{c}\right) f_0 \, \Delta t_S$$

where we have used $\lambda = c/f_0$ for the wavelength. In the classical calculation, we need only divide N by the time interval Δt to obtain the frequency observed by the moving receiver, but here we must be careful to divide by the time interval in the receiver's frame to find the observed frequency. In this case the receiver's time interval Δt_R is proper time, i.e., the time interval between encountering the first wave and the Nth wave. Both events occur at the receiver. Thus $\Delta t_R = \Delta t_S/\gamma$, and the frequency observed is

$$f' = \frac{N}{\Delta t_R} = \left(1 + \frac{v}{c}\right) f_0 \frac{\Delta t_S}{\Delta t_R}$$

$$= \gamma \left(1 + \frac{v}{c}\right) f_0 = \frac{1 + v/c}{\sqrt{1 - v^2/c^2}} f_0 \qquad 35\text{-}14b$$

It is left to the student to show that this result is identical to Equation 35-14*a*.

When the source and receiver move away from each other with relative speed v, the frequency received is given by

$$f' = \frac{1 - v/c}{\sqrt{1 - v^2/c^2}} f_0 = \frac{\sqrt{1 - v^2/c^2}}{1 + v/c} f_0 \qquad 35\text{-}15$$

Question

5. How do the relativistic equations for the doppler effect for sound waves differ from those for light waves?

35-6   The Lorentz Transformation

We now consider the general relation between the coordinates x, y, z, and t of an event as seen in reference frame S and the coordinates x', y', z', and t' of the same event as seen in reference frame S', which is moving with uniform velocity relative to S. We shall consider only the simple special case in which the origins are coincident at time $t = t' = 0$ and S' is moving with speed v along the x (or x') axis. The classical relation, called the *galilean transformation,* is

$$x = x' + vt' \qquad y = y' \qquad z = z' \qquad t = t' \qquad 35\text{-}16$$

with the inverse transformation

$$x' = x - vt \qquad y' = y \qquad z' = z \qquad t' = t$$

(For the rest of this discussion we shall ignore the equations for y and z, which do not change for this special case of motion along the x and x' axes.) These equations are consistent with experiment as long as v is much less than c. They lead to the familiar classical addition law for velocities. If a particle has velocity $u_x = dx/dt$ in frame S, its velocity in frame S' is

$$u'_x = \frac{dx'}{dt'} = \frac{dx}{dt} - v = u_x - v \qquad 35\text{-}17$$

If we differentiate this equation again, we find that the acceleration of the particle is the same in both frames: $a_x = du_x/dt = du'_x/dt = a'_x$.

It should be clear that this transformation is not consistent with Einstein's postulates of special relativity. If light moves along the x axis with speed c in S, these equations imply that the speed in S' is $u'_x = c - v$ rather than $u'_x = c$, which is consistent with Einstein's postulates and with experiment. The classical transformation equations must therefore be modified to be consistent with Einstein's postulates and to reduce to the classical equations when v is much less than c. We shall give a brief outline of one method of obtaining the relativistic transformation, the *Lorentz transformation.* We assume that the equation for x is of the form

$$x = K(x' + vt')$$

where K is a constant which can depend on v and c but not on the coordinates. For this equation to reduce to the classical one, K must approach 1 as v/c approaches zero. The inverse transformation must look the same except for the sign of the velocity:

$$x' = K(x - vt)$$

Let us assume that a light pulse starts at the origin at $t = 0$. Since we have assumed that the origins are coincident at $t = t' = 0$, the pulse also starts out at the origin of S' at $t' = 0$. The equation for the wavefront of the light pulse is $x = ct$ in frame S and $x' = ct'$ in S'. Substituting these in our transformation equations gives

$$ct = K(ct' + vt') = K(c + v)t'$$

and

$$ct' = K(ct - vt) = K(c - v)t$$

We can eliminate either t' or t from these two equations and determine K. The result is $K^2 = (1 - v^2/c^2)^{-1}$, and K is the same as γ:

$$K = \gamma = \frac{1}{\sqrt{1 - v^2/c^2}}$$

The transformation is therefore $x = \gamma(x' + vt')$. We can obtain equations for t and t' by combining this transformation with its inverse:

$$x' = x - vt$$

Using this value for x' in the above equation, we get

$$x = \gamma[\gamma(x - vt) + vt']$$

which can be solved for t' in terms of x and t. The complete Lorentz transformation is

$$x = \gamma(x' + vt') \qquad y = y'$$
$$t = \gamma\left(t' + \frac{vx'}{c^2}\right) \qquad z = z'$$

35-18   *Lorentz transformation*

with the inverse

$$x' = \gamma(x - vt) \qquad y' = y$$
$$t' = \gamma\left(t - \frac{vx}{c^2}\right) \qquad z' = z$$

35-19

Example 35-2 Two events occur at the same place x_0 at times t_1 and t_2 in S. What is the time interval between them in S'?

$$t_1' = \gamma\left(t_1 - \frac{vx_0}{c^2}\right) \qquad t_2' = \gamma\left(t_2 - \frac{vx_0}{c^2}\right)$$

$$t_2' - t_1' = \gamma(t_2 - t_1)$$

This is our familiar result for time dilation since $t_2 - t_1$ is the proper time interval.

We can obtain the velocity transformation by differentiating the Lorentz transformation equations or merely by taking differences. We shall do the latter. Suppose a particle moves a distance Δx in time Δt in frame S. Its velocity is $u_x = \Delta x / \Delta t$. Using the transformation equations, we obtain

$$\Delta x' = \gamma(\Delta x - v\,\Delta t) \qquad \text{and} \qquad \Delta t' = \gamma\left(\Delta t - \frac{v\,\Delta x}{c^2}\right)$$

The velocity in S' is

$$u'_x = \frac{\Delta x'}{\Delta t'} = \frac{\gamma(\Delta x - v\,\Delta t)}{\gamma\left(\Delta t - \frac{v\,\Delta x}{c^2}\right)} = \frac{\dfrac{\Delta x}{\Delta t} - v}{1 - \dfrac{v}{c^2}\dfrac{\Delta x}{\Delta t}}$$

or

$$u'_x = \frac{u_x - v}{1 - vu_x/c^2} \qquad\qquad \text{35-20} \qquad \textit{Velocity addition}$$

Similarly the inverse velocity transformation is

$$u_x = \frac{u'_x + v}{1 + vu'_x/c^2} \qquad\qquad \text{35-21}$$

If a particle has components of velocity along y and z, it is not difficult to find the components in S'. In these cases, $\Delta y' = \Delta y$ and $\Delta z' = \Delta z$, with the same relation between $\Delta t'$ and Δt. Thus we obtain

$$u'_y = \frac{\Delta y'}{\Delta t'} = \frac{\Delta y}{\gamma\left(\Delta t - \frac{v\,\Delta x}{c^2}\right)} = \frac{\Delta y/\Delta t}{\gamma\left(1 - \frac{v}{c^2}\frac{\Delta x}{\Delta t}\right)}$$

or

$$u'_y = \frac{u_y}{\gamma(1 - vu_x/c^2)} \qquad\qquad \text{35-22}$$

with a similar result for u'_z. The inverse is

$$u_y = \frac{u'_y}{\gamma(1 + vu'_x/c^2)} \qquad\qquad \text{35-23}$$

Example 35-3 Light moves along the x axis with speed $u_x = c$. What is its speed in S'?

$$u'_x = \frac{c - v}{1 - vc/c^2} = \frac{c(1 - v/c)}{1 - v/c} = c$$

as required by the postulates.

Question

6. The Lorentz transformation for y and z is the same as the classical result: $y = y'$ and $z = z'$. Yet the relativistic velocity transformation does not give the classical result $u_y = u'_y$ and $u_z = u'_z$. Explain.

35-7 The Twin Paradox

Homer and Ulysses are identical twins. Ulysses travels at high speed to a planet beyond the solar system and returns while Homer remains at home. When they are together again, which twin is older, or are they the same age? The correct answer is that Homer, the twin who stays at home, is older. This problem, with variations, has been the subject of spirited debate for decades, though there are very few who disagree with the answer.*

The problem is a paradox because of the seemingly symmetric roles played by the twins with the asymmetric result in their aging. The paradox is resolved when the asymmetry of the twins' roles is noted. The relativistic result conflicts with common sense based on our strong but incorrect belief in absolute simultaneity. We shall consider a particular case with some numerical magnitudes which, though impractical, make the calculations easy.

Let planet P and Homer on earth be fixed in reference frame S a distance L_0 apart, as in Figure 35-14. We neglect the motion of the earth. Reference frames S' and S'' are moving with speed v toward and away from the planet, respectively. Ulysses quickly accelerates to speed v, then coasts in S' until he reaches the planet, when he stops and is momentarily at rest in S. To return he quickly accelerates to speed v toward earth and coasts in S'' until he reaches earth, where he stops. We can assume that the acceleration times are negligible compared with the coasting times. (Given the time needed to reach the speed v, we can formulate the problem with L_0 large enough to meet this condition.) We use the following values for illustration: $L_0 = 8$ light-years and $v = 0.8c$; then $\sqrt{1 - v^2/c^2} = \frac{3}{5}$.

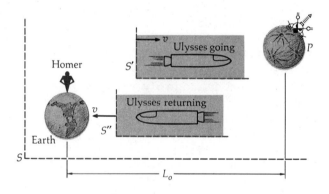

Figure 35-14
Twin paradox. The earth and a distant planet are fixed in frame S. Ulysses coasts in frame S' to the planet and then coasts in S'' back to earth. His twin Homer stays on earth. When Ulysses returns, he is younger than his twin. The roles played by the twins are not symmetric. Homer remains in one inertial reference frame, but Ulysses must accelerate if he is to return home.

It is easy to analyze the problem from Homer's point of view. According to his clock, Ulysses coasts in S' for a time $L_0/v = 10$ y and in S'' for an equal time. Thus Homer is 20 y older when Ulysses returns. The time interval in S' between leaving earth and arriving at the planet is shorter because it is proper time; the time to reach the planet by Ulysses' clock is

$$\Delta t' = \sqrt{1 - \frac{v^2}{c^2}} \frac{L_0}{v} = \tfrac{3}{5}(10) = 6 \text{ y}$$

Since the same time is required for the return trip, Ulysses will have recorded 12 y for the round trip and will be 8 y younger than Homer.

From Ulysses' point of view, the calculation of his trip time is not difficult. The distance from earth to planet is contracted and is only

$$L_0 \sqrt{1 - \frac{v^2}{c^2}} = \tfrac{3}{5}(8) = 4.8 \text{ light-years}$$

* A collection of some important papers concerning this paradox can be found in *Special Relativity Theory, Selected Reprints*, American Association of Physics Teachers, New York, 1963.

At $v = 0.8c$, it takes only 6 y each way. The real difficulty in this problem is for Ulysses to understand why his twin ages 20 y during his absence. If we consider Ulysses at rest and Homer moving away, doesn't Homer's clock measure proper time? When Ulysses measures 6 y, Homer should measure only $\frac{3}{5}(6) = 3.6$ y. Then why shouldn't Homer age only 7.2 y during the round trip? This, of course, is the paradox. The difficulty with the analysis from the point of view of Ulysses is that he does not remain in an inertial frame. What happens while Ulysses is stopping and starting? To investigate this problem in detail, we would need to treat accelerated reference frames, a subject dealt with in the study of general relativity but beyond the scope of this book. However, we can get some insight into the problem by considering the lack of synchronization of moving clocks.

Suppose that there is a clock at the planet P synchronized in S with Homer's clock at earth. In reference frame S' these clocks are unsynchronized by the amount $L_0 v/c^2$. For our example this is 6.4 y. Thus when Ulysses is coasting in S' near the planet, the clock at the planet leads that at the earth by 6.4 y. After he stops, he is in the frame S, in which these two clocks are synchronized. Thus in the negligible time (according to Ulysses) it takes him to stop, the clock at earth must gain 6.4 y. Accordingly, his twin on earth ages 6.4 y. This 6.4 y plus the 3.6 y that Homer aged during the coasting makes him 10 y older by the time Ulysses is stopped in frame S. When Ulysses is in frame S'' coasting home, the clock at earth leads that at the planet by 6.4 y, and it will run another 3.6 y before he arrives home. We do not need to know the detailed behavior of the clocks during the acceleration in order to know the cumulative effect; the special relativity theory is enough to show us that if the clocks on earth and at the planet are synchronized in S, the clock on earth lags that at P by $L_0/c^2 = 6.4$ y when viewed in S', and the clock on earth leads that at P by this amount when viewed in S''.

The difficulty in understanding the analysis of Ulysses lies in the difficulty of giving up the idea of absolute simultaneity. Suppose Ulysses sends a signal calculated to arrive at earth just as he arrives at P. If the arrival of this signal and the arrival of Ulysses at the planet are simultaneous in S', they are not simultaneous in S; in fact, the signal arrives at earth 6.4 y before Ulysses arrives at the planet, according to observers in S. The roles of the twins are not symmetric because Ulysses does not remain in an inertial reference frame but must accelerate.

It is instructive to have the twins send regular signals to each other so that they can record the other's age continuously. If they arrange to send a signal once a year, the age of the other can be determined merely by counting the signals received. The arrival frequency of the signals will not be 1 per year because of the doppler shift. The frequency observed will be given by Equation 35-14 or 35-15, which for our example is 3 per year or $\frac{1}{3}$ per year, depending on whether the source is approaching or receding.

Consider the situation first from the point of view of Ulysses. During the 6 y it takes him to reach the planet (remember that the distance is contracted in his frame), he receives signals at the rate of $\frac{1}{3}$ per year, and so he receives 2 signals. As soon as he turns around and starts back to earth, he receives 3 signals per year; in the 6 years it takes him to return he receives 18 signals, giving a total of 20 for the trip. He accordingly expects his twin to have aged 20 years.

We now consider the situation from Homer's point of view. He receives signals at the rate of $\frac{1}{3}$ per year not only for the 10 years it takes Ulysses to reach the planet but also for the 8 years it takes for the last signal sent by Ulysses from S' to get back to earth. (He cannot know that Ulysses has turned around until the signals reach him.) During the first 18 y Homer receives 6 signals. In the final 2 y before Ulysses arrives, Homer receives 6 signals, or 3 per year. (The first signal sent after Ulysses turns around takes 8 y to reach earth, whereas Ulysses, traveling at $0.8c$, takes 10 y to return and therefore arrives just 2 y after Homer begins to receive signals at the faster rate.) Thus Homer expects Ulysses to have aged 12 y. In this analysis, the asymmetry of the twins' roles is apparent. Both twins agree that when they are together again, the one who has been accelerated will be younger than the one who stayed home.

The predictions of the special theory of relativity concerning the twin paradox have been tested many times using small particles which can be accelerated to such large speeds that γ is appreciably greater than 1. Unstable particles can be accelerated and trapped in circular orbits in a magnetic field, for example, and their lifetimes compared with those of identical particles at rest. In all such experiments the accelerated particles live longer on the average than those at rest, as predicted. These predictions are also confirmed by the results of an experiment using high-precision atomic clocks flown around the world in commercial airplanes, but the analysis of this experiment is complicated by the necessity of including gravitational effects treated in the general theory of relativity.*

35-8 Relativistic Momentum

Our discussion of Newton's second law in Chapter 4 showed that if $\Sigma \mathbf{F} = m\mathbf{a}$ holds in one reference frame, it holds in any other reference frame moving with constant velocity relative to the first. According to the galilean transformation, the accelerations in the two frames are equal, $a'_x = a_x$, and forces, e.g., those due to stretching of springs, are also the same in the two frames. However, according to the Lorentz transformation, accelerations are not the same in two such reference frames. If a particle has acceleration a'_x and velocity u'_x in frame S', its acceleration in S obtained by computing du_x/dt from Equation 35-21 is

$$a_x = \frac{a'_x}{\gamma^3 (1 + u'_x v/c^2)^3}$$

Thus either the force transforms in a similar way or $\Sigma \mathbf{F} = m\mathbf{a}$ does not hold. It is reasonable to expect that $\Sigma \mathbf{F} = m\mathbf{a}$ does not hold at high speeds, for this equation implies that a constant force will accelerate a particle to unlimited velocity if it acts for a long time. However, if a particle's velocity were greater than c in some reference frame S', we could not transform from S' to the rest frame of the particle because γ becomes imaginary when $v > c$. We can see from the velocity transformation that if a particle's velocity is less than c in some frame S, it is less than c in all frames moving relative to S with $v < c$. Thus it seems reasonable to conclude that particles never have speeds greater than c.

We can avoid the question of how to transform forces by considering a problem in which the total force is zero, namely, a collision of two masses. In classical mechanics, the total momentum $\Sigma m_i \mathbf{u}_i$ is conserved.

* The details of these tests can be found in J. C. Hafele and Richard E. Keating, "Around-the-World Atomic Clocks: Predicted Relativistic Time Gains" and "Around-the-World Atomic Clocks: Observed Relativistic Time Gains," *Science*, July 14, 1972, p. 166.

We can see by a simple example that this quantity, the classical total momentum, is not conserved relativistically. That is, the conservation of the quantity $\Sigma m_i \mathbf{u}_i$ is an approximation which holds only at low speeds.

Consider an observer in frame S with a ball A, and one in S' with a ball B. The balls each have mass m and are identical when compared at rest. Each observer throws his ball along his y or y' axis with speed u_0 (measured in his own frame) so that the balls collide. Assuming the balls to be perfectly elastic, each observer will see his ball rebound with its original speed u_0. If the total momentum is to be conserved, the y component must be zero because the momentum of each ball is merely reversed by the collision. However, if we consider the relativistic velocity transformations, we can see that the quantity mu_y does not have the same magnitude for each ball as seen by either observer.

Let us consider the collision as seen in frame S (Figure 35-15). In this frame ball A moves along the y axis with velocity $u_{yA} = u_0$. Ball B has an x component of velocity $u_{xB} = v$ and a y component $u_{yB} = u'_{yB}/\gamma = -u_0\sqrt{1 - v^2/c^2}$. Here we have used the velocity-transformation equations 35-21 and 35-23 and the fact that $u'_{yB} = -u_0$ and $u'_{xB} = 0$. We see that the y component of velocity of ball B is smaller in magnitude than that of ball A. The factor $\sqrt{1 - v^2/c^2}$ comes from the time-dilation factor. The time taken for ball B to travel a given distance along the y axis in S is greater than the time measured in S' for the ball to travel this same distance. Thus in S, the total y component of classical momentum is not zero. Since the velocities are reversed in an elastic collision, momentum as defined by $\mathbf{P} = \Sigma m\mathbf{u}$ is not conserved in S. Analysis of this problem in S' leads to the same conclusion (Figure 35-15b). In the classical limit, $v \ll c$, momentum is conserved, of course, because $\gamma \approx 1$.

The reason for defining momentum to be $\Sigma m\mathbf{u}$ in classical mechanics was that this quantity is conserved when there are no external forces, as in collisions. We now see that this quantity is conserved only in the approximation $v \ll c$. We shall define *relativistic momentum* $\mathbf{p}$ of a particle to have the following properties:

1. $\mathbf{p}$ is conserved in collisions.

2. $\mathbf{p}$ approaches $m\mathbf{u}$ as u/c approaches zero.

We shall state without proof that the quantity meeting these conditions is

$$\mathbf{p} = \frac{m\mathbf{u}}{\sqrt{1 - u^2/c^2}} \qquad \text{35-24}$$

Relativistic momentum defined

where u is the speed of the particle. We thus take this equation for the definition of relativistic momentum of a particle. It is clear that this definition meets our second criterion because the denominator approaches 1 when u is much less than c. Proof that it also meets the first criterion involves much tedious algebra. From this definition, the momenta of the two balls A and B as seen in S are

$$p_{yA} = \frac{mu_0}{\sqrt{1 - u_0^2/c^2}} \qquad p_{yB} = \frac{mu_{yB}}{\sqrt{1 - (u_{xB}^2 + u_{yB}^2)/c^2}}$$

where $u_{yB} = -u_0\sqrt{1 - v^2/c^2}$ and $u_{xB} = v$. We shall leave the details of showing that $p_{yB} = -p_{yA}$ as an exercise. Because of the similarity of the

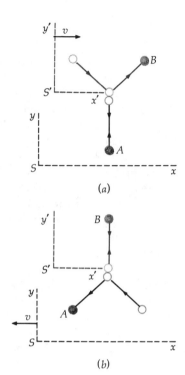

Figure 35-15
(a) Elastic collision of two identical balls as seen in frame S. The vertical component of the velocity of ball B is u_0/γ in S if it is u_0 in S'. (b) The same collision as seen in S'. In this frame ball A has vertical component of velocity u_0/γ.

factor $1/\sqrt{1 - u^2/c^2}$ and γ in the Lorentz transformation, Equation 35-24 is often written

$$\mathbf{p} = \gamma m\mathbf{u} \qquad \text{with } \gamma = \frac{1}{\sqrt{1 - u^2/c^2}} \qquad\qquad 35\text{-}25$$

This use of the symbol γ for two different quantities can cause some confusion. The notation is standard, however, and simplifies many of the equations. We shall use the notation except when considering transformations between reference frames. Then, to avoid confusion, we shall write out the factor $\sqrt{1 - u^2/c^2}$ and reserve γ for $1/\sqrt{1 - v^2/c^2}$, where v is the relative speed of the frames.

One interpretation of Equation 35-24 is that the mass of an object increases with speed. The quantity

$$\frac{m}{\sqrt{1 - u^2/c^2}} = \gamma m$$

is sometimes called the *relativistic mass,* written $m(u)$. The mass of the body in its rest frame, written m_0, is called the *rest mass.* Although this makes the expression $m(u)\mathbf{u}$ for relativistic momentum similar to the nonrelativistic expression, the use of relativistic mass often leads to mistakes. For example, the expression $\frac{1}{2}m(u)u^2$ is not the correct relativistic expression for kinetic energy. The measurement of relativistic mass involves measuring the force needed to produce a given change in momentum. The experimental evidence often cited as verification that mass depends on velocity can also be interpreted merely as evidence of the validity of the assumption that the force equals the time rate of change of relativistic momentum. We shall avoid using a symbol for relativistic mass. The symbol m in this book always refers to the rest mass.

35-9 Relativistic Energy

We have seen that the quantity $m\mathbf{u}$ is not conserved in collisions but that $\gamma m\mathbf{u}$ is, with $\gamma = 1/\sqrt{1 - u^2/c^2}$. Evidently Newton's law in the form $\Sigma\mathbf{F} = m\mathbf{a}$ cannot be correct relativistically, since it leads to the conservation of $m\mathbf{u}$. We can get a hint of the correct form of Newton's second law by writing it $\Sigma\mathbf{F} = d\mathbf{p}/dt$. Let us assume that this equation is correct if relativistic momentum $\mathbf{p}$ is used. The validity of this assumption can be determined only by examining its consequences, since an unbalanced force on a high-speed particle is measured by its effect on momentum and energy. We are essentially defining force by the equation $\Sigma\mathbf{F} = d\mathbf{p}/dt$. As in classical mechanics, we shall define kinetic energy as the work done by an unbalanced force in accelerating a particle from rest to some velocity. Considering one dimension only, we have

$$E_k = \int_{u=0}^{u} \Sigma F\, ds = \int_0^u \frac{d(\gamma m u)}{dt}\, ds = \int_0^u u\, d(\gamma m u) \qquad 35\text{-}26$$

using $u = ds/dt$. The computation of the integral in Equation 35-26 is not difficult but requires some messy algebra. It is left as a problem to show that

$$d(\gamma m u) = m\left(1 - \frac{u^2}{c^2}\right)^{-3/2} du$$

Substituting this into the integrand in Equation 35-26 gives

$$E_k = \int_0^u u\, d(\gamma m u) = \int_0^u m\left(1 - \frac{u^2}{c^2}\right)^{-3/2} u\, du$$

$$= mc^2\left(\frac{1}{\sqrt{1 - u^2/c^2}} - 1\right)$$

or

$$E_k = \gamma m c^2 - m c^2 \qquad\qquad\qquad 35\text{-}27$$

We can check this expression for low speeds by noting that for $u/c \ll 1$

$$\gamma = \left(1 - \frac{u^2}{c^2}\right)^{-1/2} \approx 1 + \frac{1}{2}\frac{u^2}{c^2} + \cdots$$

Thus

$$E_k \approx mc^2\left(1 + \frac{1}{2}\frac{u^2}{c^2} + \cdots -1\right) = \tfrac{1}{2}mu^2$$

The expression for kinetic energy consists of two terms. One, $\gamma m c^2$, depends on the speed of the particle (through the factor γ), and the other, mc^2, is independent of the speed. The quantity mc^2 is called the *rest energy* of the particle. The total energy E is then defined to be the sum of the kinetic energy and the rest energy:

Rest energy

$$E = E_k + mc^2 = \gamma m c^2 = \frac{mc^2}{\sqrt{1 - u^2/c^2}} \qquad 35\text{-}28$$

Total relativistic energy

Thus the work done by an unbalanced force increases the energy from the rest energy mc^2 to $\gamma m c^2$ (or increases the mass from m to γm).

The identification of the term mc^2 as rest energy is not merely a convenience. The rest energy of a system of particles is related to the internal energy of the system. For example, if two particles are bound together by attractive forces so that the potential energy U is negative (relative to zero potential energy at infinite separation) and the internal kinetic energy is less than U in magnitude (necessary if the system is to remain bound), the total internal energy $E_{\text{int}} = E_k + U$ will be negative. Then the rest mass of this system is less by the amount E_{int}/c^2 than the rest mass of the two particles when separated and at rest. On the other hand, if the rest mass of a system is greater than the rest mass of its individual parts, the system is not bound. This is true in radioactive decay, e.g., a radium nucleus decaying by emission of an alpha particle. In this case the sum of the masses of the alpha particle and the final nucleus is less than the mass of the original radium nucleus by an amount E_k/c^2, where E_k is the observed kinetic energy of the decay particles.

In practical applications the momentum or energy of a particle is often known rather than the speed. Equation 35-24 for the relativistic momentum and Equation 35-28 for the relativistic energy can be combined to eliminate the speed u. The result (see Exercise 24) is

$$E^2 = p^2 c^2 + (mc^2)^2 \qquad\qquad\qquad 35\text{-}29$$

If the energy of a particle is much greater than its rest energy mc^2, the second term on the right of Equation 35-29 can be neglected, giving the useful approximation

$$E \approx pc \quad \text{for} \quad E \gg mc^2 \qquad\qquad 35\text{-}30$$

Equation 35-30 is an exact relation between energy and momentum for particles with no rest mass, e.g., photons and neutrinos. If the kinetic energy of a particle is small compared with its rest energy, the term pc is small compared with mc^2. From Equation 35-29 we then have

$$E = [p^2c^2 + (mc^2)^2]^{1/2} = mc^2 \left(1 + \frac{p^2}{m^2c^2}\right)^{1/2}$$

$$\approx mc^2 \left(1 + \frac{1}{2}\frac{p^2}{m^2c^2} + \cdots\right)$$

$$= mc^2 + \frac{p^2}{2m} + \cdots \qquad\qquad 35\text{-}31$$

where we have used the approximation for small x, $(1 + x)^{1/2} \approx 1 + \frac{1}{2}x$. This result agrees with the classical result for the kinetic energy, $E_k = E - mc^2 = p^2/2m$. The velocity of a particle is often most easily calculated from its total energy and momentum. If we multiply Equation 35-28 by u and compare with Equation 35-24, we obtain the useful result

$$uE = pc^2$$

or

$$u = \frac{pc^2}{E} \qquad\qquad 35\text{-}32$$

The most convenient unit for expressing the energy of an electron or other subatomic particle is the electronvolt (eV) and its multiples, defined in Section 7-3. The electronvolt is related to the joule by

$$1 \text{ eV} = 1.602 \times 10^{-19} \text{ J}$$

Example 35-4 An electron with rest energy 0.511 MeV moves with speed $u = 0.8c$. Find its total energy, kinetic energy, and momentum.
From Equation 35-25

$$\gamma = \frac{1}{\sqrt{1 - 0.64}} = \frac{5}{3} = 1.67$$

The total energy is then

$$E = \gamma mc^2 = 1.67(0.511 \text{ MeV}) = 0.853 \text{ MeV}$$

The kinetic energy is the total energy minus the rest energy:

$$E_k = E - mc^2 = 0.853 \text{ MeV} - 0.511 \text{ MeV} = 0.342 \text{ MeV}$$

The momentum can be calculated from either Equation 35-25 or 35-32. From Equation 35-25 we have for its magnitude

$$p = \gamma mu = \gamma m(0.8c) = \frac{0.8\gamma mc^2}{c}$$

$$= \frac{(0.8)\,(1.67)\,(0.511 \text{ MeV})}{c} = 0.683 \text{ MeV}/c$$

The unit MeV/c is a convenient momentum unit.

Example 35-5 An electron has a total energy 5 times its rest energy. What is its momentum and speed?

From Equation 35-29

$$p^2c^2 = E^2 - (mc^2)^2 = (5mc^2)^2 - (mc^2)^2 = 24(mc^2)^2$$

$$pc = \sqrt{24}\, mc^2 = (4.90)\,(0.511\text{ MeV}) = 2.50\text{ MeV}$$

$$p = 2.50\text{ MeV}/c$$

We find its speed from Equation 35-32:

$$u = \frac{pc^2}{E} = \frac{\sqrt{24}\, mc^2}{5mc^2}\, c = 0.980c$$

35-10 Mass and Binding Energy

Some numerical examples from atomic and nuclear physics will illustrate changes in rest mass and rest energy. A unit of mass convenient for discussing atomic and nuclear masses is the *unified mass unit* u, defined as one-twelfth the mass of the neutral carbon atom consisting of the ^{12}C nucleus and six electrons. (This unit replaces the older atomic mass unit based on the oxygen atom.) Since 1 mol of carbon contains Avogadro's number of atoms and has a mass of 12 g, the relation between the unified mass unit and the gram is

$$1\text{ u} = \frac{1\text{ g}}{6.0220 \times 10^{23}} = 1.6606 \times 10^{-24}\text{ g}$$
$$= 1.6606 \times 10^{-27}\text{ kg} \qquad 35\text{-}33$$

In rough calculations we can write

$$1\text{ u} = 1.66 \times 10^{-24}\text{ g} = 1.66 \times 10^{-27}\text{ kg}$$

The rest energy of 1 g is

$$(1\text{ g})c^2 = (10^{-3}\text{ kg})\,(3 \times 10^8\text{ m/s})^2$$
$$= 9 \times 10^{13}\text{ J} = 5.61 \times 10^{32}\text{ eV} \qquad 35\text{-}34$$

The rest energy of a unified mass unit is

$$(1\text{ u})c^2 = 931.5\text{ MeV} \qquad 35\text{-}35$$

The rest masses and rest energies of some elementary particles and light nuclei are given in Table 35-1, from which we can see that the mass of a nucleus is not the same as the sum of the masses of its parts.

Example 35-6 The simplest example is that of the deuteron ^{2}H, consisting of a neutron and a proton bound together. Its rest energy is 1875.63 MeV. The sum of the rest energies of the proton and neutron is $938.28 + 939.57 = 1877.85$ MeV. Since this is greater than the rest energy of the deuteron, the deuteron cannot spontaneously break up into a neutron and a proton. The binding energy of the deuteron is $1877.85 - 1875.63 = 2.22$ MeV. In order to break up the deuteron into a proton and a neutron, at least 2.22 MeV must be added. This can be done by bombarding deuterons with energetic particles or electromagnetic radiation.

If a deuteron is formed by combination of a neutron and proton, energy must be released. When neutrons from a reactor are incident on protons, some neutrons are captured. The nuclear reaction is $n + p \rightarrow d + \gamma$. Most of these reactions occur for the low-energy neutrons (kinetic energy less than 1 eV). The energy of the photon plus the kinetic energy of the deuteron is 2.22 MeV.

Example 35-7 A free neutron decays into a proton plus an electron plus an antineutrino:

$$n \rightarrow p + e + \bar{\nu}$$

What is the kinetic energy of the decay products?

Here rest energy is converted into kinetic energy. Before decay, $(mc^2)_n = 939.57$ MeV. After decay, $(mc^2)_p + (mc^2)_e + (mc^2)_{\bar{\nu}} = 938.28 + 0.511 + 0 = 938.79$ MeV. Thus rest energy of $939.57 - 938.79 = 0.78$ MeV has been converted into kinetic energy of the decay products.

Example 35-8 The binding energy of the hydrogen atom (energy to remove the electron from the atom) is 13.6 eV. How much mass is lost when an electron and a proton form a hydrogen atom?

The mass of a proton plus that of an electron must be greater than that of the hydrogen atom by

$$\frac{13.6 \text{ eV}}{931.5 \text{ MeV/u}} = 1.46 \times 10^{-8} \text{ u}$$

This mass difference is so small that it is usually neglected.

Example 35-9 How much energy is needed to remove one proton from a ^{4}He nucleus?

Removal of one proton from ^{4}He leaves ^{3}H. From Table 35-1, the rest energy of ^{3}H plus that of a proton is $2808.94 + 938.23 = 3747.17$ MeV. This is about 19.8 MeV greater than the rest energy of ^{4}He.

These examples show that because atomic binding energies are so small (of the order of 1 eV to 1 keV), the mass changes are negligible in atomic (or chemical) reactions, but the nuclear binding energies are quite large and involve appreciable changes in mass.

Table 35-1

Rest energies of some elementary particles and light nuclei

Particle	Symbol	Rest energy, MeV
Photon	γ	0
Neutrino (antineutrino)	ν ($\bar{\nu}$)	0
Electron (positron)	e or e^- (e^+)	0.5110
Muon	μ	105.7
Pi meson	π^0	135
	$\pi^{\pm}$	139.6
Proton	p	938.280
Neutron	n	939.573
Deuteron	^{2}H or d	1875.628
Triton	^{3}H	2808.944
Alpha	^{4}He or α	3727.409

35-11 General Relativity

The generalization of relativity theory to noninertial reference frames by Einstein in 1916 is known as the general theory of relativity. It is much more difficult mathematically than the special theory of relativity, and there are fewer situations in which it can be tested. Nevertheless its importance calls for a brief qualitative discussion.

The basis of the general theory of relativity is the principle of equivalence:

A homogeneous gravitational field is completely equivalent to a uniformly accelerated reference frame.

Equivalence principle

This principle arises in newtonian mechanics because of the apparent identity of gravitational and inertial mass. In a uniform gravitational field, all objects fall with the same acceleration **g** independent of their mass because the gravitational force is proportional to the (gravitational) mass while the acceleration varies inversely with the (inertial) mass. Consider a compartment in space far from any matter and undergoing uniform acceleration **a**, as shown in Figure 35-16*a*. If objects are dropped in the compartment, they fall to the "floor" with acceleration **g** = −**a**. If people stand on a spring scale it will read their "weight" of magnitude *ma*. No mechanics experiment can be performed *inside* the compartment that will distinguish whether the compartment is actually accelerating in space or is at rest (or moving with uniform velocity) in the presence of a uniform gravitational field **g** = −**a**.

(a)   (b)

Figure 35-16
The results of experiments in a uniformly accelerated reference frame (*a*) cannot be distinguished from those in a uniform gravitational field (*b*) if the acceleration **a** and the gravitational field **g** have the same magnitude.

Einstein assumed that the principle of equivalence applied to all physics and not only to mechanics. In effect, he assumed that there is no experiment of any kind that can distinguish uniformly accelerated motion from the presence of a gravitational field. We shall look qualitatively at a few of the consequences of this assumption.

The first consequence of the principle of equivalence we shall discuss, the deflection of a light beam in a gravitational field, was one of

(a)

(b)

the first to be tested experimentally. Figure 35-17 shows a beam of light entering a compartment that is accelerating. Successive positions of the compartment are shown at equal time intervals. Because the compartment is accelerating, the distance it moves in each time interval increases with time. The path of the beam of light as observed from inside the compartment is therefore a parabola. But according to the equivalence principle, there is no way to distinguish between an accelerating compartment and one with uniform velocity in a uniform gravitational field. We conclude, therefore, that a beam of light, like massive objects, will accelerate in a gravitational field. For example, near the surface of the earth, light will fall with acceleration 9.81 m/s². This is difficult to observe because of the enormous speed of light. For example, in a distance of 3000 km, which takes about 0.01 s to traverse, a beam of light should fall about 0.5 mm. Einstein pointed out that the deflection of a light beam in a gravitational field might be observed when light from a distant star passes close to the sun, as illustrated in Figure 35-18. Because of the brightness of the sun, such a star cannot ordinarily be seen. Such a deflection was first observed in 1919 during an eclipse of the sun. This well-publicized observation brought instant worldwide fame to Einstein.

A second prediction from Einstein's theory of general relativity (which we shall not discuss in detail) is the excess precession of the

Figure 35-17
(a) Light beam moving in a straight line through a compartment that is undergoing uniform acceleration. The position of the beam is shown at equally spaced times t_1, t_2, t_3, and t_4. (b) In the reference frame of the compartment, the light travels in a parabolic path, as a ball would if it were projected horizontally. The vertical displacements are greatly exaggerated in both (a) and (b) for emphasis.

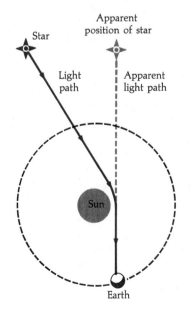

Figure 35-18
Deflection (greatly exaggerated) of a beam of light due to the gravitational attraction of the sun.

perihelion of the orbit of Mercury of about 0.01° per century. This effect had been known and unexplained for some time, so in some sense it represented an immediate success of the theory. There is, however, some difficulty in comparing the prediction of general relativity with experiment because of other effects, e.g., the perturbations due to other planets and the nonspherical shape of the sun.

A third prediction of general relativity concerns the change in time intervals and frequencies of light in a gravitational field. If Δt_1 is a time interval measured by a clock where the gravitational potential (potential energy per unit mass) is ϕ_1 and Δt_2 is the same interval measured by a clock where the gravitational potential is ϕ_2, general relativity predicts that the fractional difference will be approximately

$$\frac{\Delta t_2 - \Delta t_1}{\Delta t} = \frac{1}{c^2}\,(\phi_2 - \phi_1) \qquad\qquad 35\text{-}36$$

(Since this shift is usually very small, it does not matter by which interval we divide on the left side of the equation.) A clock in a region of low gravitational potential will therefore run slower than one in a region of high potential. Since a vibrating atom can be considered to be a clock, the frequency of vibration in a region of low potential (such as near the sun where the gravitational potential $-GM_s/r$ is negative) will be lower than that of the same atom on earth. This shift toward lower frequency and therefore longer wavelength is called the *gravitational red shift*.

Gravitational red shift

As our final example of the predictions of general relativity, we mention black holes, first predicted by Oppenheimer and Snyder in 1939. According to the general theory of relativity, if the density of an object such as a star is great enough, the gravitational attraction will be so great that nothing can escape, not even light or other electromagnetic radiation. A remarkable property of such an object is that nothing happening inside it can be communicated to the outside world. As is often true in physics, a simple but incorrect calculation gives the correct results for the relation between the mass and the critical radius of a black hole. In newtonian mechanics, the speed needed for escape of a particle from the surface of a planet or star is found by requiring the kinetic energy $\frac{1}{2}mv^2$ to be equal in magnitude to the potential energy $-GMm/r$ so that the total energy is zero. The resulting escape velocity is

$$v_e = \sqrt{\frac{2GM}{r}}$$

If we set the escape velocity equal to the speed of light and solve for the radius, we obtain the critical radius R_G, called the *Schwarzschild radius*:

$$R_G = \frac{2GM}{c^2} \qquad\qquad 35\text{-}37$$

For an object of mass equal to that of our sun to be a black hole, its radius must be about 3 km. Since no radiation is emitted from a black hole and its radius is expected to be small, the detection of such an object is not easy. The best chance of detection would occur if a black hole were a companion to a normal star in a binary star system. Measurements of the doppler shift of the light from the normal star might then allow a computation of the mass of the unseen companion to determine whether it is great enough to be a black hole. At present the binary x-ray source Cygnus X-1 (in the constellation Cygnus) appears to be an excellent candidate to be a black hole, but the evidence is not conclusive.

Black Holes

Alan P. Lightman
Cornell University

Imagine a region of space where attractive gravitational force is so intense that light rays venturing too near can be bent into circular orbit—a region from which no matter, radiation, or communication of any kind can ever escape. Such is a black hole, one of the most exciting creatures of theoretical physics and possibly the most bizarre object of space.

Although they were implicitly predicted by Einstein's 1915 theory of gravity, general relativity, black holes were first theoretically "discovered" by Oppenheimer and Snyder in 1939. Because of their highly nonintuitive properties, however, black holes were not taken seriously by most physicists and astronomers until the mid-1960s. We are now on the verge of confirming the discovery in space of the first black hole.

The concept of a black hole stretches our ideas of time and space to their limits. The surface of a black hole, called its *horizon,* is a closed boundary within which the escape speed exceeds the speed of light. The prediction of such a surface for sufficiently compact bodies can be made just on the basis of Newton's theory of gravity together with special relativity: the escape speed of a particle launched from the surface of a spherical mass M of radius R is $v_e = \sqrt{2GM/R}$ (see Chapter 7). When M/R satisfies $2GM/R > c^2$, v_e exceeds the speed of light and no particle or photon may escape, as required by special relativity. As a remarkable result, the interior of a black hole is causally disconnected from the rest of the universe; no physical process occurring inside the horizon can communicate its existence or effects to the outside. For a spherical black hole of mass M, the horizon is a sphere of circumference equal to 2π times the *Schwarzschild radius* of the hole R_G, where $R_G = 2GM/c^2$ (exact numerical coincidence of this radius with the newtonian analog above is accidental). A black hole with a mass equal to that of our sun would have a Schwarzschild radius of 2.95 km.

According to general relativity, time and space are warped by the gravitational field of massive bodies, with the warping most severe near a black hole. Gravity affects all physical systems in a universal way so that all clocks (whether transitions of an ammonia molecule or heartbeats of a human being) and all rulers would indicate that time is slowed down and space stretched out near a black hole (see Figure 1). Alternatively, one can describe the effects of the hole's gravitational field on a local measurement of time and space intervals as an acceleration of the local Lorentz frame (in which special relativity is valid) with respect to other local Lorentz frames at different locations.

Black holes are formed when massive stars undergo total gravitational collapse. While emitting heat and light into space, stars balance themselves against their own gravity with the outward thermal pressure force generated by heat from nuclear reactions in their deep interiors. But every star must die. When its nuclear fuel is exhausted, a star will contract. If its mass is less than about three times the mass of our sun, the shrinking star will stabilize at a smaller size when the inward pull of gravity cannot force the star's constituent particles any closer together. Such a star will spend eternity as a *white dwarf* or *neutron star.* But if the star's mass exceeds about 3 solar masses, theory indicates that no amount of outward pressure force can stave off the overwhelming crush of gravity and the star will plunge in upon itself, disappearing forever from sight and giving birth to a black hole.

Recent mathematical proofs, using Einstein's general relativity theory, demonstrate that a black hole is one of the simplest objects of nature and can be completely characterized by only three quantities: its mass, angular momentum, and net electrical charge. Other than these three quantities, all information of the progenitor star, e.g., whether it was composed of particles or antiparticles, whether it was pancake-shaped or spherical, is lost via gravita-

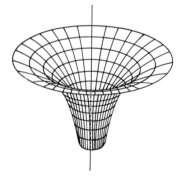

Figure 1
How the geometry of space is warped by a spherical (nonrotating) black hole. The radial distance from any point to the center of the black hole can be measured in two ways. One way is to measure the circumference of a circle passing through the point and centered on the hole, and divide by 2π. A second way is to measure the distance from the point to the center along a radial line on the curved surface. These two measures will be nearly equal far from the hole where space is "flat," but highly discrepant near the black-hole horizon where the funnel narrows to a vertical tube.

tional and electromagnetic waves shortly after formation of the hole. Rotating black holes, called *Kerr holes* (nonrotating holes are called *Schwarzschild holes*), surround themselves with a region called the *ergosphere* in which space and time are so strongly twisted that all particles, photons, and even local Lorentz frames are forced to rotate about the hole.

At great distances from a black hole, its gravitational field behaves as if generated by an ordinary star of mass M, obeying the newtonian inverse-square law. Close to a black hole, the gravitational field is far stronger than would be predicted by newtonian theory. A human being sucked into a solar-mass black hole would be ripped to shreds by the differential force of gravity across his body long before he reached the horizon.

Once the hole's horizon has formed we cannot receive any information about the ultimate fate of the collapsing star inside. Calculations of the collapse (idealized to be spherical) indicate that the star is crushed to zero volume and infinite density at the center of the black hole, forming there a point of infinite gravitational force called a *singularity*. For a collapsing star of a few solar masses, the final stellar death throes would be over in a few hundred-thousandths of a second (as measured locally). Quantum effects, omitted in the classical theory of general relativity, could possibly halt the stellar collapse at the unimaginable density $\rho \sim c^5/hG^2 \sim 5 \times 10^{93}$ g/cm³, where h is Planck's constant, and thus prevent creation of singularities; but such effects could not prevent the formation of black holes.

Current theories of stellar evolution suggest that there could be as many as 100 million black holes in our galaxy. The search for these black holes is not easy. We could never detect a tiny black speck a few miles across against the night sky. Instead, we must search for signs of black holes interacting with their astrophysical neighborhoods. Most vulnerable to discovery would be a black hole orbiting a normal star in a binary system. Using Kepler's laws, analysis of the magnitude and period of the doppler shift of the visible normal star allows us to compute whether the unseen companion is massive enough to be a black hole. In addition we might observe intense, flickering x-rays produced when gas from the normal star is sucked toward the black hole and heated to a billion degrees in its spiraling path to destruction in the hole. Such are the telltale signs of the binary x-ray source Cygnus X-1, an excellent candidate for a black hole about 8000 light-years from earth in the constellation Cygnus.

Black holes are a fundamental phenomenon of nature. Space may be littered with black holes, tempting us with their secrets of time and space behind a cloak of impenetrable darkness, a challenge and a prize for the persistence of astronomers and physicists.

Review

A. Define, explain, or otherwise identify:

Einstein postulates, 931
Ether, 931
Time dilation, 938
Proper time, 938
Proper length, 938
Length contraction, 939
Synchronized clocks, 941
Lorentz transformation, 947

Twin paradox, 948
Relativistic momentum, 952
Rest energy, 954
Unified mass unit, 956
Equivalence principle, 958
Gravitational red shift, 960
Black hole, 960
Schwarzschild radius, 960

B. True or false:

1. The speed of light is the same in all reference frames.

2. Proper time is the shortest time interval between two events.

3. Absolute motion can be determined by means of length contraction.

4. The light-year is a unit of distance.

5. A particle cannot be accelerated to a speed greater than the speed of light in vacuum.

6. Simultaneous events must occur at the same place.

7. If two events are not simultaneous in one frame, they cannot be simultaneous in any other frame.

8. $\Sigma \mathbf{F} = m\mathbf{a}$ holds at all speeds if the mass is the relativistic mass.

9. Rest mass can sometimes be converted into energy.

10. If two particles are tightly bound together by strong attractive forces, the mass of the system is less than the sum of the masses of the individual particles when separated.

Exercises

Section 35-1, The Michelson-Morley Experiment

1. In one series of measurements of the speed of light, Michelson used a path length L of 35.4 km (22 mi). (a) What is the time needed for light to make the round-trip distance of $2L$? (b) What is the classical correction term in seconds in Equation 35-1 assuming the earth's speed is $v = 10^{-4}c$? (c) From about 1600 measurements, Michelson quoted the result for the speed of light as 299,796 ± 4 km/s. Is this experiment accurate enough to be sensitive to the correction term in Equation 35-1?

2. An airplane flies with speed c relative to still air from point A to point B and returns. Compare the time required for the round trip when the wind blows from A to B with speed v with that when the wind blows perpendicularly to the line AB with speed v.

Section 35-2, Consequences of Einstein's Postulates

There are no exercises for this section.

Section 35-3, Time Dilation and Length Contraction

3. Derive the following results for v much less than c and use when applicable in the exercises and problems.

(a) $\gamma \approx 1 + \dfrac{1}{2}\dfrac{v^2}{c^2}$

(b) $\dfrac{1}{\gamma} \approx 1 - \dfrac{1}{2}\dfrac{v^2}{c^2}$

(c) $\gamma - 1 \approx 1 - \dfrac{1}{\gamma} \approx \dfrac{1}{2}\dfrac{v^2}{c^2}$

4. How great must the relative speed of two observers be for their time-interval measurements to differ by 1 percent (see Exercise 3)?

5. The proper mean life of π mesons is 2.6×10^{-8} s. If a beam of such particles has speed $0.9c$, (a) what would their mean life be as measured in the laboratory? (b) How far would they travel on the average before they decay? (c) What would your answer be to part (b) if you neglected time dilation?

6. (a) In the reference frame of the π meson in Exercise 5, how far does the laboratory travel in a typical lifetime of 2.6×10^{-8} s? (b) What is this distance in the laboratory frame?

7. A metrestick moves with speed $v = 0.6c$ relative to you in the direction parallel to the stick. (a) Find the length of the stick measured by you. (b) How long does it take for the stick to pass you?

8. Supersonic jets achieve maximum speeds of about $3 \times 10^{-6}c$. (a) By what percentage would you see such a jet contracted in length? (b) During a time of $1\,y = 3.15 \times 10^7$ s on your clock, how much time would elapse on the pilot's clock? How many minutes are lost by the pilot's clock in 1 y of your time?

Section 35-4, Clock Synchronization and Simultaneity

Exercises 9 to 13 refer to the following situation: an observer in S' lays out a distance $L' = 100$ c·min between points A' and B' and places a flashbulb at the midpoint C'. He arranges for the bulb to flash and for clocks at A' and B' to be started at 0 when the light from the flash reaches the clocks (see Figure 35-19). Frame S' is moving to the right with speed 0.6c relative to an observer C in S who is at the midpoint between A' and B' when the blub flashes and sets his clock to zero at that time.

Figure 35-19
Exercises 9 to 13.

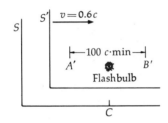

9. What is the separation distance between clocks A' and B' according to the observer in S?

10. As the light pulse from the flashbulb travels toward A' with speed c, A' travels toward C with speed 0.6c. Show that the clock in S reads 25 min when the flash reaches A'.

11. Show that the clock in S reads 100 min when the light flash reaches B', which is traveling away from C with speed 0.6c.

12. The time interval between reception of the flashes is 75 min according to the observer in S. How much time does he expect to have elapsed on the clock at A' during this 75 min?

13. The time interval calculated in Exercise 12 is the amount that the clock at A' leads that at B' according to observers in S. Compare this result with $L_0 v/c^2$.

Section 35-5, The Doppler Effect

14. (a) Show that Equations 35-14a and b are identical and can both be written

$$f' = f_0 \sqrt{\frac{1+\beta}{1-\beta}} \qquad \text{where } \beta = \frac{v}{c}$$

(b) Show that Equation 35-15 can be written

$$f' = f_0 \sqrt{\frac{1-\beta}{1+\beta}}$$

15. How fast must you move toward a red light ($\lambda = 650$ nm) for it to appear green ($\lambda = 525$ nm)?

16. A distant galaxy is moving away from us at speed 2×10^7 m/s. Calculate the fractional red shift $(\lambda' - \lambda_0)/\lambda_0$ in the light from this galaxy.

17. A distant galaxy is moving away from the earth with speed such that each wavelength is shifted by a factor of 2; that is, $\lambda' = 2\lambda_0$. What is the speed of the galaxy relative to us (see Exercise 14)?

Section 35-6, The Lorentz Transformation

18. Two events happen at the same point x'_0 in frame S' at times t'_1 and t'_2. (*a*) Use Equations 35-18 to show that in frame S the time interval between the events is greater than $t'_2 - t'_1$ by the factor γ. (*b*) Why are Equations 35-19 less convenient than Equations 35-18 for this problem?

19. The length of an object moving in your frame is measured by recording the positions of each end x_2 and x_1 at the same time t_0. (*a*) Use Equations 35-19 to show that the result is smaller than the proper length $x'_2 - x'_1$ measured in the frame in which the object is at rest. (*b*) Why are Equations 35-19 more convenient for this problem than Equations 35-18?

20. Two spaceships are approaching each other. (*a*) If the speed of each is $0.9c$ relative to the earth, what is the speed of one relative to the other? (*b*) If the speed of each relative to the earth is 30,000 m/s (about 100 times the speed of sound), what is the speed of one relative to the other?

21. A light beam moves along the y' axis with speed c in frame S' which is moving to the right with speed v relative to frame S. (*a*) Find u_x and u_y, the x and y components of the velocity of the light beam in frame S. (*b*) Show that the magnitude of the velocity of the light beam in S is c.

22. A particle moves with speed $0.9c$ along the x'' axis of frame S'', which moves with speed $0.9c$ along the x' axis relative to frame S'. Frame S' moves with speed $0.9c$ along the x axis relative to frame S. (*a*) Find the speed of the particle relative to frame S'. (*b*) Find the speed of the particle relative to frame S.

Section 35-7, The Twin Paradox

There are no exercises for this section.

Section 35-8, Relativistic Momentum, and Section 35-9, Relativistic Energy

23. Show that $p_{yA} = -p_{yB}$, where p_{yA} and p_{yB} are the relativistic momenta of the balls in Figure 35-15, given by

$$p_{yA} = \frac{mu_0}{\sqrt{1 - u_0^2/c^2}} \qquad p_{yB} = \frac{mu_{yB}}{\sqrt{1 - (u_{xB}^2 + u_{yB}^2)/c^2}}$$

$$u_{yB} = -u_0 \sqrt{1 - \frac{v^2}{c^2}} \qquad \text{and} \qquad u_{xB} = v$$

24. Combine Equations 35-24 and 35-28 to derive the equation $E^2 = p^2c^2 + m^2c^4$.

25. Make a sketch of the total energy E of an electron as a function of its momentum p. (See Equations 35-30 and 35-31 for the behavior of E at the limits of large and small values of p.)

26. Make a sketch of the kinetic energy of an electron E_k versus its speed u (see Equation 35-27).

27. An electron of rest energy 0.511 MeV has a total energy of 5 MeV. (*a*) Find its momentum in units of MeV/c from Equation 35-29. (*b*) Find the ratio of its speed u to the speed of light.

28. How much energy would be required to accelerate a particle of mass m from rest to speeds of (*a*) $0.5c$, (*b*) $0.9c$, (*c*) $0.99c$? Express your answers as multiples of the rest energy.

29. The rest energy of a proton is about 938 MeV. If its kinetic energy is also 938 MeV, find (*a*) its momentum and (*b*) its speed.

30. If the kinetic energy of a particle equals its rest energy, what error is made by using $p = mu$ for its momentum?

Section 35-10, Mass and Binding Energy

31. How much energy is required to remove one of the neutrons from ^{3}H to yield ^{2}H plus the neutron?

32. The rest mass of ^{3}He (two protons and one neutron) is 3.01440 u. (a) What is the rest energy of ^{3}He in millions of electronvolts? (b) How much energy is needed to remove a proton to make ^{2}H plus the proton?

33. The energy released when sodium and chlorine combine to form NaCl is 4.2 eV. (a) What is the increase in mass (in unified mass units) when a molecule of NaCl is dissociated into an atom of Na and an atom of Cl? (b) What percentage error is made in neglecting this mass difference? (The mass of Na is about 23 u, and that of Cl is about 35.5 u.)

34. In nuclear fusion, two ^{2}H atoms are combined to produce ^{4}He. (a) Calculate the decrease in rest mass in unified mass units. (b) How much energy is released in this reaction? (c) How many such reactions must take place per second to produce 1 W of power?

Section 35-11, General Relativity

There are no exercises for this section.

Problems

1. If a plane flies at a speed of 2000 km/h, how long must it fly before its clock loses 1 s because of time dilation?

2. (a) Show that the speed u of a particle of mass m and total energy E is given by $u/c = [1 - (mc^2/E)^2]^{1/2}$ and that if E is much greater than mc^2, the approximation $u/c \approx 1 - \frac{1}{2}(mc^2/E)^2$ holds. (b) Find the speed of an electron of kinetic energy 0.51 MeV and that of an electron of kinetic energy 10 MeV.

3. What percent error is made in using $\frac{1}{2}mu^2$ for the kinetic energy of a particle if its speed is (a) $u = 0.1c$, (b) $u = 0.9c$?

4. A stick has proper length L and makes an angle θ with the x axis in frame S. Show that the angle made with the x' axis in frame S' moving along the $+x$ axis with speed v is θ' given by $\tan \theta' = \gamma \tan \theta$ and that the length of the stick in S' is $L' = L[(\cos \theta/\gamma)^2 + \sin^2 \theta]^{1/2}$.

5. Two spaceships each 100 m long when measured at rest travel toward each other with speeds $0.8c$ relative to earth. (a) How long is each ship as measured by someone on earth? (b) How fast is each ship traveling as measured by the other? (c) How long is one ship when measured by the other? (d) At some time $t = 0$ (on earth clocks) the fronts of the ships are together as they begin to pass each other. At what time (on earth clocks) are their backs together? (e) Sketch diagrams in the frame of one of the ships showing the passing of the other ship.

6. Show that if a particle moves at an angle θ with the x axis with speed u in frame S, it moves at an angle θ' with the x' axis in S' given by

$$\tan \theta' = \frac{\sin \theta}{\gamma(\cos \theta - v/u)}$$

7. For the special case of a particle moving with speed u along the y axis in S, show that the momentum and energy in frame S' are related to the momentum and energy in S by the transformation equations

$$p'_x = \gamma\left(p_x - \frac{vE}{c^2}\right) \qquad p'_y = p_y$$

$$p'_z = p_z \qquad \frac{E'}{c} = \gamma\left(\frac{E}{c} - \frac{vp_x}{c}\right)$$

Compare these equations with the Lorentz transformation for x', y', z', and t'.

These show that the quantities p_x, p_y, p_z, and E/c transform in the same way as x, y, and ct.

8. The equation for a spherical wavefront of a light pulse which begins at the origin at time $t = 0$ is $x^2 + y^2 + z^2 - (ct)^2 = 0$. Using the Lorentz transformation equations, show that such a light pulse also has a spherical wavefront in frame S' by showing that $x'^2 + y'^2 + z'^2 - (ct')^2 = 0$ in S'.

9. In Problem 8 you showed that the quantity $x^2 + y^2 + z^2 - (ct)^2$ has the same value (0) in both S and S'. Such a quantity is called an *invariant*. From the results of Problem 7 the quantity $p_x^2 + p_y^2 + p_z^2 - (E/c)^2$ must also be invariant. Show that this quantity has the value $-m^2c^2$ in both the S and S' reference frames.

10. Show that if v is much less than c, the doppler frequency shift is approximately given by $\Delta f/f = \pm v/c$, both classically and relativistically. A radar transmitter-receiver bounces a signal off an aircraft and observes a fractional increase in the frequency of $\Delta f/f = 8 \times 10^{-7}$. What is the speed of the aircraft? (Assume the aircraft to be moving directly toward the transmitter.)

11. Two observers agree to test time dilation. They use identical clocks, and one observer in frame S' moves with speed $v = 0.6c$ relative to the other observer in frame S. When their origins coincide, they start their clocks. They agree to send a signal when their clocks read 60 min and to send a confirmation signal when each receives the other's signal. (a) When does the observer in S receive the first signal from the observer in S'? (b) When does he receive the confirmation signal? (c) Make a table showing the times in S when the observer sent the first signal, received the first signal, and received the confirmation signal. How does this table compare with one the observer in S' might have constructed?

12. A double star rotates about its center of mass such that one member is moving toward the earth and the other away from the earth. The center of mass is fixed relative to the earth. The hydrogen line of wavelength 656.3 nm in the laboratory has wavelength 660.0 nm from one of the stars and 650.0 nm from the other. What is the ratio of the masses of the stars? *Hint:* Show that the nonrelativistic doppler-effect equations have sufficient accuracy for this problem.

13. Two events in S are separated by a distance $D = x_2 - x_1$ and time $T = t_2 - t_1$. (a) Use the time-transformation equations to show that in frame S' moving with speed v relative to S the time separation is $t_2' - t_1' = \gamma(T - vD/c^2)$. (b) Show that the events can be simultaneous in frame S' only if D is greater than cT. (c) If one of the events is the *cause* of the other, the separation D must be less than cT since D/c is the smallest time a signal can take to travel from x_1 to x_2 in frame S. Show that if D is less than cT, t_2' is greater than t_1' in all reference frames. This shows that the cause must precede the effect in all reference frames (assuming that it does in one frame). (d) Suppose that a signal could be sent with speed c' greater than c so that in frame S the cause precedes the effect by the time $T = D/c'$, which is less than D/c. Show that there is then a reference frame moving at speed v less than c in which the effect precedes the cause.

14. Two identical particles each have rest mass m. In reference frame S which is the center-of-mass frame, the two particles are moving toward each other with speed u. According to the mass-energy relation, the rest mass of this two-particle system is now $M = \gamma 2m$, where $\gamma = (1 - u^2/c^2)^{-1/2}$ because of the internal (kinetic) energy of the system. The two particles collide inelastically with a spring which locks shut (Figure 35-20) and come to rest in S with their initial kinetic energy transformed into potential energy. In this problem you are going to show that $M = \gamma 2m$ is a necessary consequence of momentum conservation in a reference frame S' moving with speed $v = u$ relative to frame S so that one of the particles is initially at rest in S'. (a) Show that the speed of the particle not at rest in frame S' is u', where $u' = 2u/(1 + u^2/c^2)$. Use this result to show that

$$\sqrt{1 - \frac{u'^2}{c^2}} = \frac{1 - u^2/c^2}{1 + u^2/c^2}$$

(b) Show that the initial momentum in frame S' can be written $P_i' = 2mu/(1 - u^2/c^2)$. (c) After the collision, the two masses move with speed u in S' (since they are at rest in S). Write the total momentum in S' after the collision in terms of the mass of the system M and show that this equals the initial momentum only if $M = \gamma 2m$. (d) Find the total energy before and after the collision in each reference frame and show that the total energy is conserved in each frame.

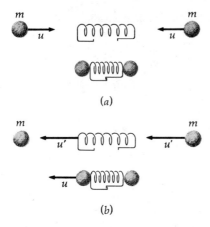

(a)

(b)

Figure 35-20
Problem 14. Inelastic collision of two equal masses in (a) center-of-mass frame S and (b) frame S' moving to right with speed u relative to frame S so that one of the particles is initially at rest. The spring is assumed to be massless.

15. A horizontal turntable rotates with angular speed ω. There is a clock at the center of the turntable and one at a distance r from the center. In an inertial reference frame the clock at distance r is moving with speed $v = r\omega$. (a) Show that, from time dilation in special relativity, time intervals Δt_0 for the clock at rest and Δt_r for the moving clock are related by

$$\frac{\Delta t_r - \Delta t_0}{\Delta t_0} \approx -\frac{r^2\omega^2}{2c^2} \quad \text{if} \quad r\omega \ll c$$

(b) In a reference frame attached to the turntable, both clocks are at rest. Show that the clock at distance r experiences a pseudo (centrifugal) force $F_r = mr\omega^2$ in this accelerated frame. Show that this is equivalent to a difference in gravitational potential between r and the origin of $\phi_r - \phi_0 = \frac{1}{2}r^2\omega^2$. Use this potential difference in Equation 35-36 to show that in this frame the difference in time intervals is the same as in the inertial frame.

CHAPTER 36   Quantization

Objectives   After studying this chapter you should:

1. Be able to sketch the spectral distribution curve for blackbody radiation and the curve predicted by the Rayleigh-Jeans law.

2. Be able to discuss which features of the photoelectric effect are in accord with the predictions of classical physics and which are not.

3. Be able to state the Einstein equation for the photoelectric effect.

4. Be able to discuss how the photon concept explains all the features of the photoelectric effect and Compton scattering of x rays.

5. Be able to state the Bohr postulates and use them to derive the energy levels of the hydrogen atom and the wavelengths of the spectral lines of hydrogen.

6. Be able to draw an energy-level diagram for hydrogen and indicate on it transitions involving the emission of a photon.

7. Be able to state the de Broglie relations for the frequency and wavelength of electron waves and use them and the standing-wave condition to derive the Bohr condition for the quantization of angular momentum in the hydrogen atom.

8. Be able to discuss the experimental evidence for the existence of electron waves.

9. Be able to discuss wave-particle duality.

10. Be able to discuss the uncertainty principle.

In Chapter 31 we showed that a wave equation for electromagnetic waves traveling at the speed of light can be derived from Maxwell's equations, which summarize the experimental laws of electricity and magnetism. This unification of the previously separate subjects of optics and electromagnetism through Maxwell's equations was one of the great triumphs of classical physics in the nineteenth century. Other predictions of electromagnetism, some of which we did not study, are also important for understanding the modifications of classical physics in the twentieth century.

1. Light and other electromagnetic waves carry momentum as well as energy. The momentum p and the energy E are related by $p = E/c$, where c is the speed of the waves.

2. If a charged particle has acceleration a, it radiates electromagnetic energy at a rate proportional to a^2.

3. If the acceleration of the charged particle is associated with periodic motion, e.g., simple harmonic oscillation or circular motion, the particle emits electromagnetic radiation with a frequency equal to that of the motion.

All these predictions have been verified in experiments involving macroscopic phenomena. Near the end of the nineteenth century it was generally believed that all natural phenomena could be described by Newton's laws, the laws of thermodynamics, and the laws of electromagnetism (as expressed in Maxwell's equations). There was nothing left for scientists to do but apply these laws to various phenomena and measure the next decimal in the fundamental constants (the speed of light, the electron charge, the gas constant, Avogadro's number, etc.). However, this optimism (or pessimism if you were a scientist hoping to

Table 36-1
Approximate dates of some important experiments and theories, 1881–1932

1881	Michelson obtains null result for absolute velocity of earth
1884	Balmer finds empirical formula for spectral lines of hydrogen
1887	Hertz produces electromagnetic waves, verifying Maxwell's theory and accidently discovering photoelectric effect
1887	Michelson repeats his experiment with Morley, again obtaining null result
1895	Röntgen discovers x rays
1896	Becquerel discovers nuclear radioactivity
1897	J. J. Thomson measures e/m for cathode rays, showing that electrons are fundamental constituents of atoms
1900	Planck explains blackbody radiation using energy quantization involving new constant h
1900	Lenard investigates photoelectric effect and finds energy of electrons independent of light intensity
1905	Einstein proposes special theory of relativity
1905	Einstein explains photoelectric effect by suggesting quantization of radiation
1907	Einstein applies energy quantization to explain temperature dependence of heat capacities of solids
1908	Rydberg and Ritz generalize Balmer's formula to fit spectra of many elements
1909	Millikan's oil-drop experiment shows quantization of electric charge
1911	Rutherford proposes nuclear model of atom based on alpha-particle scattering experiments of Geiger and Marsden
1912	Friedrich and Knipping and von Laue demonstrate diffraction of x rays by crystals showing that x rays are waves and crystals are regular arrays

make important discoveries) proved premature. We have already described how newtonian mechanics must be replaced with the special theory of relativity when the speed of a particle is not small compared with the speed of light, and we showed that when the particle's speed is small, the special theory of relativity reduces to Newton's laws. Many startling discoveries, both experimental and theoretical, in the last 20 years of the nineteenth century and first 30 years of the twentieth century demonstrated that the laws of classical physics also break down when applied to microscopic systems such as the particles within an atom. This failure is even more drastic than the failure of newtonian mechanics at high speeds. The interior of the atom can be described only in terms of quantum theory, which requires the modification of some of our fundamental ideas about the relationships between physical theory and the physical world. As with special relativity, quantum theory reduces to classical physics when applied to macroscopic systems. Table 36-1 lists the approximate dates of some of the important experiments and theories from 1881 to 1932.

A proper study of quantum theory is beyond the scope of this book. In this chapter we shall concentrate on the fundamental ideas of energy quantization and the wave-particle duality of nature.

Table 36-1
(continued)

1913	Bohr proposes model of hydrogen atom
1914	Moseley analyzes x-ray spectra using Bohr model to explain periodic table in terms of atomic number
1914	Franck and Hertz demonstrate atomic energy quantization
1915	Duane and Hunt show that the short-wavelength limit of x rays is determined from quantum theory
1916	Wilson and Sommerfeld propose rules for quantization of periodic systems
1916	Millikan verifies Einstein's photoelectric equation
1923	Compton explains x-ray scattering by electrons as collision of photon and electron and verifies results experimentally
1924	De Broglie proposes electron waves of wavelength h/p
1925	Schrödinger develops mathematics of electron wave mechanics
1925	Heisenberg invents matrix mechanics
1925	Pauli states exclusion principle
1927	Heisenberg formulates uncertainty principle
1927	Davisson and Germer observe electron wave diffraction by single crystal
1927	G. P. Thomson observes electron wave diffraction in metal foil
1928	Gamow and Condon and Gurney apply quantum mechanics to explain alpha-decay lifetimes
1928	Dirac develops relativistic quantum mechanics and predicts existence of positron
1932	Chadwick discovers neutron
1932	Anderson discovers positron

36-1   The Origin of the Quantum Constant: Blackbody Radiation

One of the most puzzling phenomena studied near the end of the nine-teenth century was the spectral distribution of blackbody radiation. A blackbody is an ideal system which absorbs all the radiation incident on it; it can be approximated by a cavity with a very small opening (see Section 18-4). The characteristics of the radiation in such a cavity of a body in thermal equilibrium depend only on the temperature of the walls. The fraction of the radiant energy density of wavelength λ in the interval $d\lambda$ is called the *spectral distribution* $f(\lambda, T)\, d\lambda$. The function $f(\lambda, T)$ was calculated from classical physics in a straightforward way and the result compared with experiment. The calculation involved finding the number of standing waves in a three-dimensional cavity in the wavelength interval $d\lambda$ and multiplying by the average energy per wave, which the equipartition theorem predicts to be kT. The result, known as the *Rayleigh-Jeans law*, is

$$f(\lambda, T) = 8\pi kT\lambda^{-4} \qquad\qquad 36\text{-}1 \qquad \textit{Rayleigh-Jeans law}$$

This result agrees with experiment in the region of long wavelengths but disagrees violently at short wavelengths. As λ approaches zero, the experimentally determined $f(\lambda, T)$ also approaches zero but the calcu-lated function becomes infinite as λ^{-4}. This result was known as the *ultraviolet catastrophe*. In 1900 the German physicist Max Planck an-nounced that by making a somewhat peculiar modification in the clas-sical calculation he could derive a function $f(\lambda, T)$ that agreed with the experimental data at all wavelengths. He first found an empirical function that fitted the data and then searched for a way to modify the usual calculation. His empirical function is

$$f(\lambda, T) = \frac{8\pi hc\lambda^{-5}}{e^{hc/\lambda kT} - 1} \qquad\qquad 36\text{-}2 \qquad \textit{Planck's law}$$

where h is an adjustable constant. Figure 36-1 compares this function with the Rayleigh-Jeans law and with experimental data. At large wave-lengths, the exponential in the denominator of Equation 36-2 is small, and we can use the approximation

$$e^{hc/\lambda kT} \approx 1 + \frac{hc}{\lambda kT}$$

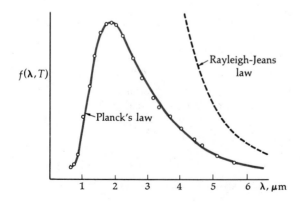

$f(\lambda, T)$

Rayleigh-Jeans law

Planck's law

1   2   3   4   5   6 λ, μm

Figure 36-1
Spectral distribution of black-body radiation versus wave-length at $T = 1600$ K. *(Adapted from F. K. Richtmyer, E. H. Kennard, and J. N. C. Lauritsen,* Introduction to Modern Physics, *5th ed., McGraw-Hill Book Company, New York, 1955, by permission.)*

Then

$$f(\lambda,\ T) \to \frac{8\pi hc\lambda^{-5}}{hc/\lambda kT} = 8\pi kT\lambda^{-4}$$

which is the Rayleigh-Jeans formula. For small wavelengths, the exponential is very large, and we can neglect the 1 in the denominator of Equation 36-2, giving

$$f(\lambda,\ T) \to 8\pi hc\lambda^{-5}e^{-hc/\lambda kT} \qquad \text{36-3}$$

This approaches zero as λ approaches zero. The constant h can be found by fitting this formula to the data points at small λ. Its presently accepted value is

$$h = 6.626 \times 10^{-34}\ \text{J·s} = 4.136 \times 10^{-15}\ \text{eV·s} \qquad \text{36-4}$$ *Planck's constant*

Planck found that he could derive Equation 36-2 by modifying the calculation of the average energy per wave in the cavity. Instead of dealing directly with the radiation in the cavity, Planck considered the emission and absorption of radiation by the cavity walls, which he represented by a set of oscillators of all frequencies. He reasoned that, in equilibrium, the average energy of an oscillator of frequency f would be associated with the average energy of the electromagnetic radiation of that frequency in the cavity. In the usual calculation of the average energy of an oscillator in thermal equilibrium with its surroundings the energy E is a continuous variable, and integration is used. The result for a one-dimensional oscillator is kT. Planck modified this calculation by assuming that the energy of an oscillator is discrete; i.e., it can take on only certain values E_n, given by

$$E_n = nhf \qquad \text{36-5}$$

where n is an integer, f is the frequency, and h is a constant. With this assumption Planck obtained for the average energy of an oscillator

$$E_{\text{av}} = \frac{hc/\lambda}{e^{hc/\lambda kT} - 1} \qquad \text{36-6}$$

Replacing kT in Equation 36-1 by this expression gives Equation 36-2. Although Planck tried to fit the constant h into the framework of classical physics, he was unable to do so. The fundamental importance of his assumption of energy quantization, implied by Equation 36-5, was not generally appreciated until Einstein applied similar ideas to explain the photoelectric effect and suggested that quantization is a fundamental property of electromagnetic radiation.

36-2 Quantization of Electromagnetic Radiation: Photons

As mentioned in Chapter 32, the photoelectric effect was discovered by Hertz in 1887 and studied by Lenard in 1900. Figure 36-2 shows a schematic diagram of the basic apparatus. When light is incident on a clean metal surface, the cathode C, electrons are emitted. If some of these electrons strike the anode A, there is a current in the external circuit. The number of the emitted electrons reaching the anode can be increased or decreased by making the anode positive or negative with respect to the cathode. Let V be the increase in potential from the cathode to anode. Figure 36-3 shows the current versus V for two values

Figure 36-2
Schematic drawing of photo-electric-effect apparatus. Light strikes the cathode C and ejects electrons. The number of electrons which reach the anode A is measured by the current in the ammeter. The anode can be made positive or negative with respect to the cathode to attract or repel the electrons.

of the intensity of light incident on the cathode. When V is positive, the electrons are attracted to the anode. At sufficiently large V all the emitted electrons reach the anode and the current reaches its maximum value. A further increase in V does not affect the current. Lenard observed that the maximum current is proportional to the light intensity, an expected result since doubling the energy per unit time incident on the cathode should double the number of electrons emitted. When V is negative, the electrons are repelled from the anode. Only electrons with initial kinetic energy $\frac{1}{2}mv^2$ greater than $|eV|$ can then reach the anode.

Figure 36-3
Photoelectric current versus voltage V for two values of light intensity. There is no current when V is less than $-V_0$. The saturation current observed for large V is proportional to the light intensity.

From Figure 36-3 we see that if V is less than $-V_0$, no electrons reach the anode. The potential V_0 is called the *stopping potential*. It is related to the maximum kinetic energy of the emitted electrons by

$$(\tfrac{1}{2}mv^2)_{\max} = eV_0$$

Stopping potential

The experimental result that V_0 is independent of the incident light intensity was surprising. Apparently increasing the rate of energy falling on the cathode does not increase the maximum kinetic energy of the electrons emitted. In 1905 Einstein demonstrated that this result can be understood if light energy is not distributed continuously in space but quantized in small bundles called *photons*. The energy of each photon is hf, where f is the frequency and h is Planck's constant. An electron emitted from a metal surface exposed to light receives its energy from a single photon. When the intensity of the light of a given frequency is increased, more photons fall on the surface in a unit time but the energy absorbed by each electron is unchanged. If ϕ is the energy necessary to remove an electron from the surface of a metal, the maximum kinetic energy of the electrons emitted will be

$$(\tfrac{1}{2}mv^2)_{\max} = eV_0 = hf - \phi \qquad\qquad 36\text{-}7$$

Einstein's photoelectric equation

The quantity ϕ, called the *work function*, is a characteristic of the metal. Some electrons will have kinetic energy less than $hf - \phi$ because of energy loss traversing the metal. Equation 36-7 is known as *Einstein's photoelectric equation*.

> If the derived formula is correct, then V_0 when represented in cartesian coordinates as a function of the frequency of the incident light must be a straight line whose slope is independent of the nature of the emitting substance.*

From Equation 36-7 we see that the slope of V_0 versus f should equal h/e. Einstein's equation was a bold prediction, for at that time there

* From Einstein's original paper translated by A. B. Arons and M. B. Peppard, *American Journal of Physics*, vol. 33, p. 367, 1965.

was no evidence that Planck's constant had any applicability outside of blackbody radiation, and there were no experimental data on the stopping potential V_0 as a function of frequency. The experimental verification of Einstein's theory was quite difficult. Careful experiments by R. C. Millikan reported in 1914 and in more detail in 1916 showed that the Einstein equation was correct, and measurements of h agreed with the value found by Planck. Figure 36-4 shows a plot of Millikan's data.

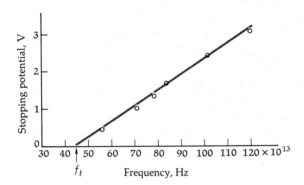

Figure 36-4
Millikan's data for the stopping potential versus frequency for the photoelectric effect. The data fall on a straight line which has slope h/e as predicted by Einstein a decade before the experiment. (*From R. C. Millikan,* Physical Review, *vol. 7, p. 362, 1916.*)

The threshold frequency f_t and the corresponding threshold wavelength λ_t are related to the work function ϕ by setting V_0 equal to zero in Equation 36-7. Then

$$\phi = hf_t = \frac{hc}{\lambda_t} \qquad\qquad 36\text{-}8$$

Threshold frequency and wavelength

Photons of frequency less than f_t (and therefore wavelengths greater than λ_t) do not have enough energy to eject an electron from the metal. Work functions for metals are typically a few electronvolts. Since wavelengths are usually given in nanometres and energies in electronvolts, it is useful to have the value of hc in electronvolt-nanometres. We have

$$hc = (4.14 \times 10^{-15} \text{ eV·s})(3 \times 10^8 \text{ m/s})$$
$$= 1.24 \times 10^{-6} \text{ eV·m}$$

or

$$hc = 1240 \text{ eV·nm} \qquad\qquad 36\text{-}9$$

Example 36-1 The threshold wavelength for potassium is 564 nm. What is the work function for potassium? What is the stopping potential when light of wavelength 400 nm is used?

$$\phi = hf_t = \frac{hc}{\lambda_t} = \frac{1240 \text{ eV·nm}}{564 \text{ nm}} = 2.20 \text{ eV}$$

The energy of a photon of wavelength 400 nm is

$$E = \frac{hc}{\lambda} = \frac{1240 \text{ eV·nm}}{400 \text{ nm}} = 3.10 \text{ eV}$$

The maximum kinetic energy of the emitted electrons is then

$$(\tfrac{1}{2}mv^2)_{\max} = hf - \phi = 3.10 \text{ eV} - 2.20 \text{ eV} = 0.90 \text{ eV}$$

The stopping potential is therefore 0.90 V.

Another interesting feature of the photoelectric effect is the absence of lag between the time the light is turned on and the time the electrons appear. In the classical theory, given the intensity (power per unit area), the time can be calculated for enough energy to fall on the area of an atom to eject an electron. However, even when the intensity is so small that such a calculation gives time lags of hours, essentially no time lag is observed. The explanation of this result is that although the number of photons hitting the metal per unit time is very small when the intensity is low, each photon has enough energy to eject an electron and there is a chance that one photon will be absorbed immediately. The classical calculation gives the correct average number of protons absorbed per unit time.

Further evidence of the correctness of the photon concept was furnished by Arthur H. Compton, who measured the scattering of x rays by free electrons. According to classical theory, when an electromagnetic wave of frequency f_1 is incident on material containing charges, the charges will oscillate with this frequency and reradiate electromagnetic waves of the same frequency. Compton pointed out that if the scattering process were considered to be a collision between a photon and electron, the electron would absorb energy due to recoil and the scattered photon would have less energy and therefore a lower frequency than the incident photon. According to classical theory, the energy and momentum of an electromagnetic wave are related by $E = pc$. This result is also consistent with the relativistic expression relating the energy and momentum of a particle (Equation 35-29):

$$E^2 = p^2c^2 + (mc^2)^2 \qquad \text{36-10}$$

if the mass of the photon is assumed to be zero. Figure 36-5 shows the geometry of a collision between a photon of wavelength λ_1 and an electron at rest. Compton related the scattering angle θ to the incident and scattered wavelengths λ_1 and λ_2 by treating the scattering as a relativistic-mechanics problem using conservation of energy and momentum. Let $\mathbf{p}_1$ be the momentum of the incident photon, $\mathbf{p}_2$ that of the scattered photon, and $\mathbf{p}_e$ that of the recoiling electron. Conservation of momentum gives

$$\mathbf{p}_1 = \mathbf{p}_2 + \mathbf{p}_e$$

or

$$p_e^2 + (p_1 - p_2)^2 = p_1^2 + p_2^2 - 2\mathbf{p}_1 \cdot \mathbf{p}_2$$
$$= p_1^2 + p_2^2 - 2p_1p_2 \cos\theta \qquad \text{36-11}$$

The energy before the collision is $p_1c + mc^2$, where mc^2 is the rest energy of the electron. After the collision the electron has energy $\sqrt{(mc^2)^2 + p_e^2c^2}$. Conservation of energy then gives

$$p_1c + mc^2 = p_2c + \sqrt{(mc^2)^2 + p_e^2c^2} \qquad \text{36-12}$$

Compton eliminated the electron momentum p_e from Equations 36-11 and 36-12 and expressed the photon momenta in terms of the wavelengths, to obtain an equation relating the incident and scattered wavelengths λ_1 and λ_2 and the angle θ. The algebraic details are left as a problem (Problem 4). Compton's result is

$$\lambda_2 - \lambda_1 = \frac{h}{mc}(1 - \cos\theta) \qquad \text{36-13}$$

The change in wavelength is independent of the original wavelength.

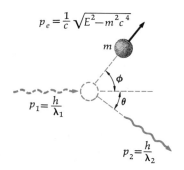

$$p_e = \frac{1}{c}\sqrt{E^2 - m^2c^4}$$

$$p_1 = \frac{h}{\lambda_1}$$

$$p_2 = \frac{h}{\lambda_2}$$

Figure 36-5
Compton scattering of an x ray by an electron. The scattered photon has less energy and therefore greater wavelength than the incident photon because of the recoil energy of the electron. The change in wavelength $\lambda_2 - \lambda_1$ is found from conservation of energy and momentum.

The quantity h/mc depends only on the mass of the electron. It has dimensions of length and is called the *Compton wavelength*. Its value is

$$\lambda_c = \frac{h}{mc} = \frac{hc}{mc^2} = \frac{1240 \text{ eV·nm}}{5.11 \times 10^5 \text{ eV}} = 2.43 \text{ pm}$$ 36-14   *Compton wavelength*

Because $\lambda_2 - \lambda_1$ is small, it is difficult to observe unless λ_1 is so small that the fractional change $(\lambda_2 - \lambda_1)/\lambda_1$ is appreciable. Compton used x rays of wavelength 71.1 pm. The energy of a photon of this wavelength is $E = hc/\lambda = (1240 \text{ eV·nm})/(0.0711 \text{ nm}) = 17.4 \text{ keV}$. Since this is much greater than the binding energy of the valence electrons in carbon, these electrons can be considered to be essentially free. Compton's experimental results for $\lambda_2 - \lambda_1$ as a function of scattering angle θ agreed with Equation 36-13, thereby confirming the correctness of the photon concept.

36-3 Quantization of Atomic Energies: The Bohr Model

In Planck's treatment of blackbody radiation he represented the walls of the body by a set of oscillators of frequency f whose energy is quantized in units of hf. In 1907 Einstein applied similar reasoning to explain the temperature dependence of the heat capacity of a solid. The most famous application of energy quantization to microscopic systems was that of Niels Bohr. In 1913 Bohr proposed a model of the hydrogen atom which had spectacular successes in calculating the wavelengths of lines in the known hydrogen spectrum and in predicting new lines (later found experimentally) in the infrared and ultraviolet spectra.

Many data were collected near the turn of the century on the emission of light by atoms in a gas when excited by an electric discharge. Viewed through a spectroscope with a narrow-slit aperture, this light appears as a discrete set of lines of different color or wavelength; the spacing and intensities of the lines are characteristic of the element. It was possible to determine the wavelengths of these lines accurately, and much effort went into finding regularities in the spectra. In 1884 a Swiss schoolteacher, Johann Balmer, found that the wavelengths of some of the lines in the spectrum of hydrogen can be represented by the formula

$$\lambda = 364.6 \frac{m^2}{m^2 - 4} \text{ nm}$$ 36-15   *Balmer formula*

where m is a variable integer which takes on the values $m = 3, 4, 5, \ldots$. Figure 36-6 shows the set of spectral lines of hydrogen, now known as the *Balmer series*, whose wavelengths are given by Equation 36-15. Balmer suggested that his formula might be a special case of a more general expression applicable to the spectra of other elements.

Figure 36-6
Balmer series in hydrogen. The wavelengths of these lines are given by Equation 36-15 for different values of the integer m. [From G. Herzberg, Annalen der Physik, vol. 84, p. 565 (1927).]

Such an expression, found by Johannes R. Rydberg and Walter Ritz, gives the reciprocal wavelength as

$$\frac{1}{\lambda} = R\left(\frac{1}{m^2} - \frac{1}{n^2}\right) \qquad n > m \qquad\qquad 36\text{-}16$$

Rydberg-Ritz formula

where R, called the *Rydberg constant* or the *Rydberg*, is the same for all series of the same element and varies only slightly in a regular way from element to element. For hydrogen, the value of R is $R_H = 10.96776\ \mu m^{-1}$. For very massive elements, R approaches the value $R_\infty = 10.97373\ \mu m^{-1}$. (The quantity $1/\lambda$ was used instead of the frequency $f = c/\lambda$ because λ can be measured much more accurately than the speed of light c.) Such empirical expressions were successful in predicting other spectra; e.g., other hydrogen lines outside the visible spectrum were predicted and found.

Many attempts were made to construct a model of the atom that would yield these formulas for its radiation spectrum. The most popular model, due to J. J. Thomson, considered various arrangements of electrons embedded in some kind of fluid that contained most of the mass of the atom and had enough positive charge to make the atom electrically neutral. Since classical electromagnetic theory predicted that a charge oscillating with frequency f would radiate light of that frequency, Thomson searched for configurations that were stable and had normal modes of vibration of frequencies equal to those of the spectrum of the atom. A difficulty of this model and all others was that electric forces alone cannot produce stable equilibrium. Thomson was unsuccessful in finding a model which predicted the observed frequencies for any atom.

The Thomson model was essentially ruled out by a set of experiments by H. W. Geiger and E. Marsden under the supervision of E. Rutherford about 1911, in which alpha particles from radioactive radium were scattered by atoms in a gold foil. Rutherford showed that the number of alpha particles scattered at large angles could not be accounted for by an atom in which the positive charge was distributed throughout the atomic size (known to be about 0.1 nm in diameter) but required that the positive charge and most of the mass of the atom be concentrated in a very small region, now called the nucleus, of diameter of the order of 10^{-6} nm = 1 fm. Niels Bohr, working in the Rutherford laboratory at the time, proposed a model of the hydrogen atom which combined the work of Planck, Einstein, and Rutherford and successfully predicted the observed spectra. Bohr assumed that the electron in the hydrogen atom moved under the influence of the Coulomb attraction to the positive nucleus according to classical mechanics, which predicts circular or elliptical orbits with the force center at one focus, as in the motion of the planets about the sun. For simplicity he chose a circular orbit. Although mechanical stability is achieved because the Coulomb attractive force provides the centripetal force necessary for the electron to remain in orbit, such an atom is unstable electrically according to classical theory because the electron must accelerate when moving in a circle and therefore radiate electromagnetic energy of frequency equal to that of its motion. According to classical electromagnetic theory, such an atom would quickly collapse, the electron spiraling into the nucleus as it radiates away its energy. Bohr "solved" this difficulty, modifying the laws of electromagnetism by *postulating* that the electron could move in certain orbits without radiating. He called these stable orbits *stationary states*.

Bohr's first postulate: stationary states

The atom radiates when the electron somehow makes a transition from one stationary state to another. The frequency of radiation is not the frequency of motion in either stable orbit but is related to the energies of the orbits by

$$f = \frac{W_i - W_f}{h} \qquad\qquad 36\text{-}17$$

Bohr's second postulate: photon frequency from energy conservation

where h is Planck's constant and W_i and W_f are the total energies in the initial and final orbits. This assumption, which is equivalent to that of energy conservation with the emission of a photon, is a key one in the Bohr theory because it deviates from the classical theory, which requires the frequency of radiation to be that of the motion of the charged particle. Planck made a similar assumption in his blackbody radiation theory, but in that case the energy of the oscillators was nhf and the energy difference between two adjacent states hf, so that when the oscillator emitted or absorbed radiation, the frequency of the radiation was the same as that of the oscillator.

If the nuclear charge is $+Ze$ and the electron charge $-e$, the potential energy at a distance r is

$$U = -\frac{kZe^2}{r}$$

where $k = 1/4\pi\epsilon_0$ is the Coulomb constant. (For hydrogen, $Z = 1$, but it is convenient to not specify Z at this time so that the results can be applied to other atoms.) The total energy of the electron moving in a circular orbit with speed v is then

$$W = \tfrac{1}{2}mv^2 + U = \tfrac{1}{2}mv^2 - \frac{kZe^2}{r}$$

The kinetic energy can be obtained as a function of r by using Newton's law $F = ma$. Setting the Coulomb attractive force equal to the mass times the centripetal acceleration, we have

$$\frac{kZe^2}{r^2} = m\frac{v^2}{r}$$

or

$$\tfrac{1}{2}mv^2 = \frac{1}{2}\frac{kZe^2}{r} \qquad\qquad 36\text{-}18$$

For circular orbits the kinetic energy equals half the magnitude of the potential energy, a result which holds for circular motion in any inverse-square-law force field. The total energy is then

$$W = -\frac{1}{2}\frac{kZe^2}{r} \qquad\qquad 36\text{-}19$$

Using Equation 36-17 for the frequency of radiation when the electron changes from orbit 1 of radius r_1 to orbit 2 of radius r_2, we obtain

$$f = \frac{W_1 - W_2}{h} = \frac{1}{2}\frac{kZe^2}{h}\left(\frac{1}{r_2} - \frac{1}{r_1}\right) \qquad\qquad 36\text{-}20$$

To obtain the Balmer-Ritz formula $f = c/\lambda = cR(1/m^2 - 1/n^2)$ it is evident that the radii of stable orbits must be proportional to the squares of integers. Bohr searched for a quantum condition for the radii of the stable orbits which would yield this result. After much trial and error, he found that he could obtain the correct results if he postulated that in a stable orbit the angular momentum of the electron equals an integer times Planck's constant divided by 2π. Since the angular momentum of a circular orbit is just mvr, this postulate is

$$mvr = \frac{nh}{2\pi} = n\hbar \qquad\qquad 36\text{-}21$$

Bohr's third postulate: quantized angular momentum

where $\hbar = h/2\pi$. (The constant $\hbar = h/2\pi$ is often more convenient than h itself, just as angular frequency $\omega = 2\pi f$ is often more convenient than the frequency f.) We can determine r by eliminating v between Equations 36-18 and 36-21:

$$v^2 = n^2\frac{\hbar^2}{m^2 r^2} = \frac{kZe^2}{mr}$$

or

$$r = n^2\frac{\hbar^2}{mkZe^2} = n^2\frac{r_0}{Z} \qquad\qquad 36\text{-}22$$

where

$$r_0 = \frac{\hbar^2}{mke^2} \approx 0.0529 \text{ nm} \qquad\qquad 36\text{-}23$$

is called the *first Bohr radius*. Combining Equations 36-22 and 36-20, we get

$$f = Z^2\frac{mk^2 e^4}{4\pi\hbar^3}\left(\frac{1}{n_2^2} - \frac{1}{n_1^2}\right) \qquad\qquad 36\text{-}24$$

When we compare this for $Z = 1$ with the empirical Balmer-Ritz formula (Equation 36-16), we have for the Rydberg constant

$$R = \frac{mk^2 e^4}{4\pi c\hbar^3} \qquad\qquad 36\text{-}25$$

Using the values of m, e, and $\hbar$ known in 1913, Bohr calculated R and found his result to agree (within the limits of the uncertainties of the constants) with the value obtained from spectroscopy.

The possible values of the energy of the hydrogen atom predicted by the Bohr model are given by Equation 36-19, with r given by Equation 36-22. They are

$$W_n = -\frac{k^2 e^4 m}{2\hbar^2}\frac{Z^2}{n^2} = -Z^2\frac{E_0}{n^2} \qquad\qquad 36\text{-}26$$

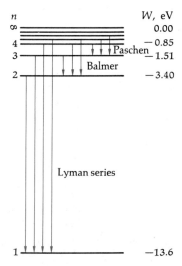

Figure 36-7
Energy-level diagram for hydrogen showing a few transitions in each of the Lyman, Balmer, and Paschen series. The energies of the levels are given by Equation 36-26.

where $E_0 = k^2 e^4 m/2\hbar^2 \approx 13.6$ eV. It is convenient to represent these energies in an energy-level diagram, as in Figure 36-7. Various series of transitions are indicated in this diagram by vertical arrows between the levels. The frequency of the light emitted in one of these transitions is the energy difference divided by h, according to Equation 36-17. At the time of Bohr's paper (1913) the Balmer series, corresponding to $n_2 = 2$, $n_1 = 3, 4, 5, \ldots$, and the Paschen series, corresponding to $n_2 = 3$, $n_1 = 4, 5, 6 \ldots$, were known. In 1916 T. Lyman found the series corresponding to $n_2 = 1$, and in 1922 and 1924 F. Brackett and H. A. Pfund, respectively, found series corresponding to $n_2 = 4$ and $n_2 = 5$. As can be determined by computing the wavelengths of these series, only the

Balmer series lies in the visible portion of the electromagnetic spectrum.

In our derivations we have assumed the electron to revolve around a stationary nucleus. This is equivalent to assuming the nucleus to have infinite mass. Since the mass of the hydrogen nucleus is not infinite but only about 2000 times that of the electron, a correction must be made for the motion of the nucleus. In the center-of-mass frame of the atom, the nucleus and electron have equal and opposite momenta, giving zero total momentum to the atom. If p is the magnitude of the momentum of either, the kinetic energy of the nucleus is $p^2/2M$, where M is its mass and the kinetic energy of the electron is $p^2/2m$. The total kinetic energy is

$$E_k = \frac{p^2}{2M} + \frac{p^2}{2m} = \frac{M+m}{mM}\frac{p^2}{2} = \frac{p^2}{2\mu}$$

where μ, called the *reduced mass*, is given by

$$\mu = \frac{mM}{m+M} = \frac{m}{1+m/M} \qquad \text{36-27} \qquad \textit{Reduced mass}$$

This is slightly different from the kinetic energy of just the electron because the reduced mass μ differs slightly from the electron mass m. We can correct any of our results for the motion of the nucleus by replacing the electron mass m with the reduced mass μ. (The validity of this procedure is shown in most intermediate and advanced mechanics books.) This correction leads to a very slight dependence of the Rydberg constant as given in Equation 36-25 on the nuclear mass, in precise agreement with the observed variation.

Example 36-2 Find the energy and wavelength of the longest-wavelength line in the Lyman series.

From Figure 36-7 we see that the Lyman series corresponds to transitions ending at the ground state of energy, $W_1 = -13.6$ eV. Since λ varies inversely with energy, the longest-wavelength transition is the lowest-energy transition from the first excited state $n = 2$ to the ground state. The first excited state has energy $W_2 = -13.6$ eV$/4 = -3.40$ eV. Since this is 10.2 eV above the ground-state energy, the energy of the photon emitted is 10.2 eV. The wavelength of this photon is

$$\lambda = \frac{hc}{E} = \frac{1240 \text{ eV·nm}}{10.2 \text{ eV}} = 121.6 \text{ nm}$$

This photon is outside the visible spectrum toward the ultraviolet. Since all the other lines in the Lyman series have even greater energies and shorter wavelengths, the Lyman series is completely in the ultraviolet region.

Questions

1. If an electron moves to a larger orbit, does its total energy increase or decrease? Does its kinetic energy increase or decrease?

2. How does the spacing of adjacent energy levels change as n increases?

3. What is the energy of the shortest-wavelength photon emitted by the hydrogen atom?

36-4   Electron Waves

In 1924 a French student, L. de Broglie, suggested in his dissertation that since light is known to have both wave and particle properties, perhaps matter—especially electrons—may also have wave as well as particle characteristics. This suggestion was highly speculative; there was no evidence at that time for any wave aspects of electrons. For the frequency and wavelength of electron waves de Broglie chose the equations

$$f = \frac{E}{h} \qquad\qquad 36\text{-}28$$

De Broglie relations

$$\lambda = \frac{h}{p} \qquad\qquad 36\text{-}29$$

where p is the momentum and E the energy of the electron. These equations also hold for photons, as we have seen. De Broglie pointed out that with these relations, the Bohr quantum condition (Equation 36-21) is equivalent to a standing-wave condition. We have

$$mvr = n\frac{h}{2\pi}$$

Substituting h/λ for the momentum mv gives

$$\frac{h}{\lambda}r = n\frac{h}{2\pi}$$

or

$$n\lambda = 2\pi r = C \qquad\qquad 36\text{-}30$$

where C is the circumference of the circular Bohr orbit. Thus Bohr's quantum condition is equivalent to saying that an integral number of electron waves must fit into the circumference of the circular orbit (Figure 36-8). The idea of explaining the discrete energy states of matter by standing waves seemed promising. In classical wave theory applied to waves on a string, for example, application of the boundary conditions that the string be fixed at both ends leads to a discrete set of allowed frequencies—those for which an integral number of half wavelengths fit into the length of the string. If energy is associated with the frequency of a standing wave, as in Equation 36-28, standing waves imply quantized energies. These ideas were developed into a detailed mathematical theory by Erwin Schrödinger in 1925. Schrödinger found a wave equation for electron waves and solved the standing-wave problem for the hydrogen atom, the simple harmonic oscillator, and other systems of interest. He found that the allowed frequencies, combined with the de Broglie relation $E = hf$, led to the same set of energy levels for the hydrogen atom as found by Bohr (Equation 36-26) and to almost the same set of energies for the simple harmonic oscillator as assumed by Planck and Einstein.* Schrödinger's *wave mechanics* therefore gave a general method of finding the quantization condition for a given system.

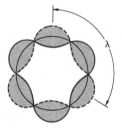

Figure 36-8
Standing waves around the circumference of a circle.

---

* Schrödinger found $E_n = (n + \frac{1}{2})hf$ for the energy levels of the simple harmonic oscillator, which differs slightly from $E_n = nhf$ assumed by Planck and Einstein. This difference amounts to a shift of each energy level by $\frac{1}{2}hf$ and has no effect on their results, which depend only on the distance between energy levels.

The first direct measurements of the wavelengths of electrons were made in 1927 by C. J. Davisson and L. H. Germer, who were studying electron scattering from a nickel target at Bell Telephone Laboratories. After heating the target to remove an oxide coating that had accumulated during an accidental break in the vacuum system, they found that the scattered-electron intensity as a function of the scattering angle showed maxima and minima. Their target had crystallized, and by accident they had observed electron diffraction. They then prepared a target consisting of a single crystal of nickel and investigated this phenomenon extensively. From measurements of the location of the observed diffraction maxima and minima, the wavelength of the electrons could be calculated, and agreement with the de Broglie relation (Equation 36-29) was found in all cases. In the same year G. P. Thomson (son of J. J. Thomson) also observed electron diffraction in transmission of electrons through thin metallic foils. Since then, diffraction has been observed for neutrons, protons, and other particles of small mass. Figure 36-9 shows a diffraction pattern produced by electrons incident on two narrow slits. This pattern is identical to that observed with photons of the same wavelength. Figures 36-10 and 36-11 compare the diffraction patterns of x rays, electrons, and neutrons of similar wavelength in transmission through thin foils.

Figure 36-9
Two-slit electron diffraction pattern. This pattern is the same as that usually obtained with photons. (*Courtesy of Claus Jönsson.*)

(a)

(b)

Figure 36-10
Diffraction pattern produced by (a) x rays and (b) electrons on an aluminum foil target. (*Courtesy of Film Studio, Education Development Center, Newton, Mass.*)

Figure 36-11
Diffraction pattern produced by neutrons on a target of polycrystalline copper. Note the similarity in the patterns produced by x rays, electrons, and neutrons. (*Courtesy of C. G. Shull.*)

We can understand why the wave properties of matter are not observed in macroscopic experiments by calculating the de Broglie wavelength for a macroscopic object.

Example 36-3 Find the de Broglie wavelength of a particle of mass 10^{-6} g moving with speed 10^{-6} m/s.

From Equation 36-29 we have

$$\lambda = \frac{h}{p} = \frac{h}{mv} = \frac{6.63 \times 10^{-34} \text{ J·s}}{(10^{-9} \text{ kg})(10^{-6} \text{ m/s})}$$

$$= 6.63 \times 10^{-19} \text{ m}$$

Since this wavelength is much smaller than any possible apertures or obstacles (the diameter of the nucleus of the atom is about 10^{-15} m, roughly 10,000 times this wavelength), diffraction or interference of such waves cannot be observed. As we have discussed, the propagation of waves of very small wavelength is indistinguishable from the propagation of particles. Note that we have chosen an extremely small value for the momentum in Example 36-3. Any other macroscopic particle with greater momentum will have an even smaller de Broglie wavelength. However, the situation is different for low-energy electrons. Consider an electron that has been accelerated through a potential difference V. Its energy is then (assuming the electron to be nonrelativistic)

$$E = \frac{p^2}{2m} = eV$$

and its wavelength is

$$\lambda = \frac{h}{p} = \frac{h}{\sqrt{2mE}} = \frac{h}{\sqrt{2meV}}$$

Putting in the values for m, e, and h, we obtain an expression for λ in nanometres when V is in volts:

$$\lambda = \frac{1.226}{\sqrt{V}} \text{ nm} \qquad\qquad 36\text{-}31$$

Example 36-4 Find the de Broglie wavelength of an electron whose kinetic energy is 13.6 eV.
From Equation 36-31 we have for $V = 13.6$ V,

$$\lambda = \frac{1.226 \text{ nm}}{\sqrt{13.6}} = 0.332 \text{ nm} = 2\pi(0.0529 \text{ nm})$$

which is the circumference of the first Bohr orbit in the hydrogen atom.

From Equation 36-31 we see that electrons with energies of the order of tens of electronvolts have de Broglie wavelengths of the order of nanometres, which is the order of magnitude of the size of the atom and of the spacing of atoms in a crystal.

36-5 Wave-Particle Duality

We have seen that light, which we ordinarily think of as a wave motion, exhibits particle properties when it interacts with matter, as in the photoelectric effect or in Compton scattering, whereas electrons, which we usually think of as particles, exhibit the wave properties of interference and diffraction. All phenomena—electrons, atoms, light, sound, etc.—have both particle and wave characteristics. It might be tempting to say that an electron, for example, is both a wave and a particle, but the meaning of such a statement is not clear. In classical physics the concepts of waves and particles are mutually exclusive. A *classical particle* behaves like a piece of shot: it can be localized and scattered; it exchanges energy suddenly in lumps; and it obeys the laws of conservation of energy and momentum in collisions. It does *not* exhibit interference or diffraction. A *classical wave*, on the other hand, behaves like a water wave; it exhibits diffraction and interference, and its energy is spread out continuously in space and time. Nothing can be both a classical particle and a classical wave at the same time.

Until the twentieth century, it was thought that light was a classical wave and electrons were classical particles. We now see that these concepts of classical waves and particles do not adequately describe the complete behavior of any phenomenon. Everything propagates like a classical wave and exchanges energy like a classical particle. Of course, there are times when the classical-particle and classical-wave concepts give the same results. When the wavelength is very small, the propagation of a classical wave cannot be distinguished from that of a classical particle. Similarly, the time averages of energy and momentum exchanges are the same for classical waves as for classical particles. For example, the wave theory of light correctly predicts that the total electron current in the photoelectric effect is proportional to the intensity of light.

36-6 The Uncertainty Principle

The wave-particle duality of nature has many important consequences. In newtonian mechanics the rate of change of momentum of a particle is related to the resultant force acting on the particle by Newton's second law. To determine the position of the particle at any time we must know the forces acting and the position and velocity of the particle at some time; i.e., we must know the initial conditions. Although there are always experimental uncertainties in any measurement of the initial position and velocity, it is assumed in classical mechanics that such uncertainties can in principle be made as small as desired. Because of the wave-particle duality of both radiation and matter, we now understand that it is impossible in principle to measure both the position and velocity of a particle simultaneously with infinite precision. This result is known as the *uncertainty principle*, first enunciated by Werner Heisenberg in 1927. For one-dimensional motion it is expressed in terms of the momentum p and the position x of a particle. Let Δx and Δp be the uncertainties in the position and momentum, respectively. According to the uncertainty principle, the product of Δx and Δp can never be less than $h/4\pi = \frac{1}{2}\hbar$:

$$\Delta x\, \Delta p \geq \tfrac{1}{2}\hbar \qquad\qquad\qquad 36\text{-}32$$

Uncertainty principle

Usually the uncertainty product is much greater than $\frac{1}{2}\hbar$. The equality holds only if the measurements of both x and p have ideal gaussian distributions and the experiments are ideal.

We can get a qualitative understanding of the uncertainty principle by considering the measurement of the position and momentum of a particle. If we know the mass of the particle, we can determine its momentum by measuring its position at two nearby times and computing its velocity. A common way to measure the position of an object is to look at it with light. When we do this, we scatter light from the object and determine the position by the direction of the scattered light. If we use light of wavelength λ, we can measure the position only to an uncertainty of the order of λ because of diffraction effects. To reduce the uncertainty in position we therefore use light of very short wavelength, perhaps even x rays. In principle, there is no limit to the accuracy of such a position measurement because there is no limit on how short a wavelength of electromagnetic radiation can be used. However, since all electromagnetic radiation carries momentum, the scattering of the radiation by the particle will deflect the radiation and change its original momentum in an uncontrollable way. By momentum conservation, the momentum of the particle also changes in an uncontrollable way. According to classical wave theory, this effect on the momentum of the particle could be reduced by reducing the intensity of the radiation. However, the energy and momentum of the radiation are quantized; each photon has momentum h/λ. When the wavelength of the radiation is small, the momentum of each photon will be large and the momentum measurement will have a large uncertainty. This uncertainty cannot be eliminated by reducing the intensity of light; such a reduction merely reduces the number of photons in the beam. To "see" the particle we must scatter at least one photon. Therefore, the uncertainty in the momentum measurement of the particle will be large if λ is small, and the uncertainty in the position measurement will be large if λ is large. A detailed analysis shows that the product of these uncertainties will always be at least of the order of Planck's constant h. Of course we could "look at" the particle by scattering electrons instead of photons, but we have the same difficulty. If we use low-momentum electrons to reduce the uncertainty in the momentum measurement, we have a large uncertainty in the position measurement because of diffraction of the electrons. The relation between the wavelength and momentum $\lambda = h/p$ is the same for electrons as for photons.

The Expanding Universe

Martin Rees
Institute of Astronomy, Cambridge University

Over 400 years has passed since Copernicus argued that the earth must be dethroned from the privileged central position accorded to it by Ptolemy's cosmology, and described the general layout of the solar system in the form accepted today. Only within the twentieth century, however, have we fully recognized the sun's status as just one of the 100 billion (10^{11}) stars in the Milky Way. Harlow Shapley, Jan Oort, and others showed that the Milky Way is a flat disk-shaped system of stars spinning around its axis about once every 2×10^8 years. The sun lies near the edge of the disk, roughly 30,000 light-years from its center. Furthermore, the Milky Way, our own galaxy, is merely one of the 100 billion galaxies on the cosmic scene. If our galaxy were viewed from a distance of several million light-years, it would look rather like the spiral galaxies illustrated (Figures 1 and 2).

Photographs of the sky taken with a large telescope may show many thousands of galaxies, some so far away that the light we receive has taken billions of years on its journey toward us. Although individual stars cannot be resolved, it is possible to obtain spectra of the integrated light from all the stars in a remote galaxy. Such spectra imply that the galaxies contain more or less the same mixture of chemical elements as the sun and that the basic physical constants (e.g., the masses and charges of electrons and protons and Planck's constant) whose values determine the properties of the radiation from stellar surfaces do indeed have uniform values throughout the observed part of the universe.

Such spectra do, however, reveal a significant systematic trend: the spectral lines are all shifted toward the red, i.e., to longer wavelengths. This effect was studied by Edwin Hubble, who interpreted the red shift as a doppler effect and inferred that distant galaxies are receding from us. Moreover, Hubble showed that the recession velocity of a given galaxy is proportional to its distance. The velocity-distance relationship is also found to be the same in all parts of the sky; in other words, the apparent "expansion of the universe" is *isotropic* (the same in all directions). At first sight, one might wonder whether this implies that the earth (or at least our galaxy) is in a privileged central position; but it is easy to convince oneself that this is not so and that a hypothetical astronomer on another galaxy could also have discovered Hubble's law: he would find all the other galaxies apparently receding isotropically from *him*.

If the galaxies had always had the same relative velocities, Hubble's law implies that at some definite time in the past (about 15 billion years ago) they would all have been "squashed together" at one point. Georges Lemaître and others proposed that the universe actually evolved from a primordial state of high density. The primary aim of observational cosmology since the 1930s has therefore been to test whether there was indeed a "big bang" or whether (as postulated by a rival theory) the universe has existed forever in a steady state, new material and new galaxies being continually created so that its average properties remain unaltered despite the Hubble expansion.

By observing distant objects, one can in principle probe the early history of the universe and perhaps learn whether galaxies were more crowded together or looked systematically different in the past. But ordinary galaxies become invisibly faint, even with the largest optical telescopes, before one reaches the distance where such effects might really be expected to show up. Fortunately, however, galaxies occasionally undergo mysterious violent explosions, in the course of which a small central region flares up until it outshines all the rest of the galaxy by a factor of 100. These rare bright sources of light, called *quasars*, permit optical astronomers to probe much deeper into space—and thus much farther into the past—than is possible using normal galaxies. These "galactic explosions" are sometimes strong sources of radio-frequency emission and are one of the important cosmic phenomena that can be investigated by the techniques of radio astronomy. Very little is understood about the physical nature of quasars. The problems they pose are among the most important and challenging that confront modern astrophysicists. However, by comparing the number of quasars observed at different distances one can infer that quasars (whatever they may actually be) were very much commoner in early epochs than they are now. This is quite incompatible with the steady-state model; in

Figure 1
The *Andromeda* galaxy, Messier 31. (*Courtesy of the Mount Wilson and Palomar Observatories.*)

an evolutionary cosmology, however, there are reasons for expecting that galaxies might indeed have been more prone to violent explosions when they were young.

In 1965, Arno Penzias and Robert Wilson made a discovery (more or less by accident) which is generally regarded as the most compelling evidence for a big bang. They found that the earth is apparently bathed in microwave radiation which comes from all directions but has no apparent source. Subsequent measurements have established that this microwave background seems to have the spectrum of blackbody radiation at a temperature of 2.7 K (see Section 18-4). Most cosmologists interpret this background as a relic of a hot, dense early phase when the universe was opaque and any radiation would have established the blackbody spectrum characteristic of thermal equilibrium. This possibility had in fact been predicted by George Gamow in 1948.

By synthesizing the disparate strands of evidence bearing on the problem, most cosmologists have reached a tentative consensus on how the universe might have evolved to its present state. About 15 billion years ago, according to this picture, all the material in the universe — all the stuff of which galaxies are now composed — constituted an exceedingly compressed and hot gas (hotter, in fact, than the center of the sun). The intense radiation in this fireball, though cooled and diluted by the expansion, would still be around, pervading the whole universe: this is the interpretation of the microwave background radiation discovered by Penzias and Wilson — an "echo," as it were, of the "explosion" which initiated the universal expansion. The early universe would not have been *completely* smooth and homogeneous (it may even have been turbulent), and the primordial irregularities eventually developed into galaxies. It may, however, have taken a billion years for these fluctuations to condense out into gravitationally bound systems. The most distant quasars are so far away that the radiation now reaching us has been traveling toward us for 80 or even 90 percent of the time elapsed since the initial big bang. That is, it set out toward us when the universe was only 10 or 20 percent of its present age. The universe would have appeared much more violent and active at these early epochs: the cosmic radio background would have been 100 times more intense; and an astronomer then would have found his nearest quasar perhaps 50 times closer to him than in these relatively quiescent times.

The microwave background is a "fossil" from even earlier stages in the expansion. There seems to be another important vestige of the primordial fireball in the present universe: the element helium, which constitutes about a quarter of the mass of most stars, including the sun. The other chemical elements could have been synthesized via the nuclear reactions which provide the power source in the cores of ordinary stars. It is believed that all the carbon, nitrogen, oxygen, and iron on the earth — and indeed in our own bodies — was manufactured in stars which exhausted their energy supply and exploded as *supernovae* before the sun formed. The solar system then condensed from gas contaminated by debris ejected from early generations of stars. One of the major achievements of theoretical astrophysics has been to understand some quantitative details of these processes of *cosmic nucleosynthesis*, e.g., the relative abundances of different elements. But it proved hard to account in this fashion for all the observed helium; it was therefore gratifying both to cosmologists and to experts on nucleosynthesis when the expected composition of material emerging from the big bang was calculated and found to be about 75 percent hydrogen and 25 percent helium. The synthesis of helium would have occurred within only a *few minutes* of the big bang. This epitomizes how the discovery of the microwave background has extended the scope of cosmology by bringing remote eras that were previously entirely speculative within the scope of quantitative scientific discussion. Such discussions do, however, entail extrapolating the locally determined physical laws into domains where one cannot be overwhelmingly sure of their continued validity.

The consistency of most of the available data (limited though it is) with this general picture encourages most cosmologists to adopt the hot big-bang theory as a working hypothesis, which should form the basis for interpreting new observations until some better theory emerges or until some glaring contradiction reveals itself.

Having drawn some tentative conclusions about how the universe has evolved from the primordial fireball to its present state, one is tempted to speculate about its future and its eventual fate. The conventional theories offer two alternative scenarios: either the universe will continue expanding forever, or else the expansion is slowing down to such an extent that it will eventually stop and be followed by a recontraction. This question can in principle be tackled observationally by extending Hubble's work out to very large distances: the velocity-distance relationship obviously refers to the velocity of the galaxies at the time when the light we now receive was emitted, so that any deceleration should be measurable if one observes sufficiently remote objects. In practice, however, this technique has yet to provide reliable results, because the observational difficulties are severe and also because of various uncertain corrections which must be incorporated into the analysis.

But there is another, more indirect, way of trying to determine how much the universal expansion is slowing down. Imagine that a big sphere is shattered by an explosion, the debris flying off in all directions. Each fragment feels the gravitational pull of all the others, and this causes the expansion to decelerate. If the explosion were sufficiently violent, the debris would fly apart forever; but if the fragments were not moving quite so fast, gravity might bind them together strongly enough to bring the expansion to a halt. The material would then collapse again. More or less the same argument probably holds for the universe. One might feel somewhat uneasy about applying a result based on Newton's theory of gravity to the whole universe. But even though one cannot describe the global properties of the universe properly (or the propagation of light) without using a more sophisticated theory such as Einstein's general relativity, the dynamics of the expansion are thought to be the same as in Newton's theory. One can therefore rephrase our earlier question: Does the universe have the escape velocity or not?

In the case of the galaxies (which, for the purposes of this argument, are regarded as fragments of the expanding universe) we know the expansion velocity from Hubble's law. What we do *not* know is the amount of gravitating matter that is causing the deceleration. It is a straightforward procedure to calculate how much material would be needed in order to halt the expansion: it works out at about one atom per cubic metre. If the average concentration of material were *below* this so-called critical density, we would expect the uni-

Figure 2
NGC 4594, spiral galaxy in
Virgo, seen edge on. (*Courtesy
of Hale Observatories.*)

verse to continue expanding forever; but if the mean density *exceeded* the critical density, the universe would seem destined eventually to recontract. There are various ways of estimating the masses of individual galaxies. Also, of course, one knows roughly how many galaxies there are in a typical volume of space. These estimates are bedeviled by many uncertainties, but it looks as though the material in galaxies, if spread uniformly through space, would fall short of the critical density by a factor of at least 30. At first sight, one might accept this as evidence that the universe will go on expanding forever; but this inference would really be unjustified, because there may be a lot *more* material embodied in some form *other than* ordinary galaxies. Galaxies are, admittedly, the most prominent features in the sky when we look with an optical telescope, but there is no reason to believe that everything in the universe shines. There may be many objects so cool that they radiate predominantly in the infrared and *absorb* light instead of emitting it—"dead" galaxies, for example, whose stars have all exhausted their nuclear energy, or objects shrouded by opaque clouds of dust. Alternatively, some objects may be so hot that their emission is concentrated in the ultraviolet and x-ray bands of the electromagnetic spectrum. This kind of astronomical work was not feasible until quite recently because since the air is very opaque in these wavebands, one must send equipment above the earth's atmosphere to make observations. Space astronomy is still in its pioneering stage, and one would not be surprised if it were to disclose many unsuspected sorts of objects. So our present inventory of the contents of the universe may well prove exceedingly biased and incomplete. For instance, there may be a large amount of diffuse gas *between* the galaxies. (There is, after all, no reason to expect all, or even most, of the primordial hydrogen and helium to have condensed into galaxies.) This intergalactic gas may in fact be responsible for emitting most of the cosmic x-rays reaching us from beyond our own galaxy.

At present, therefore, there is no definite reason for believing that there is enough gravitating material in the universe to bring the expansion to a halt. But it remains quite conceivable that the universe contains a great deal of stuff even more elusive than intergalactic gas. For instance, the critical density could be provided by neutrinos, gravitational waves, or black holes without there being the slightest chance of detection by present techniques.

We therefore cannot rule out the possibility that eventually the expansion will stop and turn into a contraction. Distant galaxies, displaying *blue* shifts instead of red shifts, would eventually collide and merge with one another. As the contraction proceeded further, the sky would become brighter and brighter; and eventually all the stars would explode (because the sky would be hotter than the fuel in their interiors!), everything in the universe being finally engulfed in a fireball like that from which, according to most cosmologists, it emerged: the ultimate, universal, gravitational collapse. But there is no immediate cause for concern; our breathing space before this cataclysm should be billions of years, at the very least. If the universe did *not* have the critical density, it would continue to expand forever. Each galaxy would fade to a dull glow as its constituent stars exhausted their available energy and the supply of gas from which new bright stars can condense was inexorably depleted.

An important aim of current cosmological research is to decide between these two contrasting fates for the universe. Also, we hope to eventually fill in more details of the big-bang picture and to understand how the primordial material aggregates into galaxies; why galaxies have the sizes and shapes that are observed; how they evolve; and why they sometimes explode. But one should remain aware that present data in cosmology are still limited, ambiguous, and fragmentary; and they all depend on complex instruments stretched right to the limits of their sensitivity and performance. Therefore, even if this general picture seems consistent with what is known at the moment, it would be rash to bet *too* heavily on its being correct. Moreover, self-consistency is, of course, no guarantee of truth in itself. Many more observations are essential before we can be sure whether we really know, even in outline, the basic overall structure of the physical universe or whether our current ideas will eventually be discarded and superseded as surely as Ptolemy's epicycles were.

Review

A. Define, explain, or otherwise identify:

Blackbody radiation, 972
Ultraviolet catastrophe, 972
Photoelectric effect, 973
Work function, 974
Einstein's photoelectric equation, 974
Compton wavelength, 977
Balmer series, 977
The Rydberg, 978
Stationary states, 978

Bohr quantization condition, 980
Energy-level diagram, 980
Reduced mass, 981
De Broglie wavelength, 982
Classical particle, 984
Classical wave, 984
Wave-particle duality, 984
Uncertainty principle, 985

B. True or false:

1. The spectral distribution of radiation in a blackbody depends only on the temperature of the body.

2. In the photoelectric effect, the maximum current is proportional to the intensity of the incident light.

3. The work function of a metal depends on the frequency of the incident light.

4. The maximum kinetic energy of electrons emitted in the photoelectric effect varies linearly with the frequency of the incident light.

5. In the Bohr model of the hydrogen atom the electron moves in circular orbits.

6. One of Bohr's assumptions is that atoms never radiate light.

7. In a stationary state in the Bohr model the total energy is negative and has the same magnitude as the kinetic energy.

8. In the ground state of the hydrogen atom the potential energy is −27.2 eV.

9. The de Broglie wavelength of an electron varies inversely with its momentum.

10. Electrons can be diffracted.

11. Neutrons can be diffracted.

12. Classical particles can be diffracted.

13. The position and momentum of an object cannot be measured simultaneously.

14. The uncertainty principle could be violated in principle if the experimental measurements were ideal.

Exercises

Section 36-1, The Origin of the Quantum Constant: Blackbody Radiation

There are no exercises for this section.

Section 36-2, The Quantization of Electromagnetic Radiation: Photons

1. Find the range of photon energies in the visible spectrum from about 400 to 700 nm wavelength.

2. Find the photon energy if the wavelength is (a) 0.1 nm (about 1 atomic diameter) and (b) 1 fm (10^{-15} m, about 1 nuclear diameter).

3. Find the photon energy for an electromagnetic wave in the FM band of frequency 100 MHz.

4. The work function for tungsten is 4.58 eV. (a) Find the threshold frequency and wavelength for the photoelectric effect. Find the stopping potential if the wavelength of the incident light is (b) 200 nm; (c) 250 nm.

5. When light of wavelength 300 nm is incident on potassium, the emitted electrons have maximum kinetic energy of 2.03 eV. (a) What is the energy of the incident photon? (b) What is the work function for potassium? (c) What would be the stopping potential if the incident light had a wavelength of 400 nm?

6. The threshold wavelength for the photoelectric effect for silver is 262 nm. (a) Find the work function for silver. (b) Find the stopping potential if the incident radiation has a wavelength of 200 nm.

7. The work function for cesium is 1.9 eV. (a) Find the threshold frequency and wavelength for the photoelectric effect. Find the stopping potential if the wavelength of the incident light is (b) 300 nm; (c) 400 nm.

8. The wavelength of Compton-scattered photons is measured at $\theta = 90°$. If $\Delta\lambda/\lambda$ is to be 1 percent, what should the wavelength of the incident photons be?

9. Compton used photons of wavelength 0.0711 nm. (a) What is the energy of these photons? (b) What is the wavelength of the photon scattered at $\theta = 180°$? (c) What is the energy of the photon scattered at $\theta = 180°$? (d) What is the recoil energy of the electron for $\theta = 180°$?

Section 36-3, Quantization of Atomic Energies: The Bohr Model

10. Use the known values of the constants in Equation 36-23 to show that r_0 is approximately 0.0529 nm.

11. The wavelength of the longest wavelength in the Lyman series was calculated in Example 36-2. Find the wavelengths for the transitions (a) $n = 3$ to $n = 1$ and (b) $n = 4$ to $n = 1$. (c) Find the shortest wavelength in the Lyman series.

12. (a) Use the Balmer formula (Equation 36-15) to calculate the three longest wavelengths in the Balmer series. (b) Find the photon energies corresponding to each of these wavelengths.

13. Calculate the longest three wavelengths and the series limit (shortest wavelength) for the Paschen series ($n_2 = 3$) and indicate their positions on a horizontal linear scale.

14. Repeat Exercise 13 for the Brackett series ($n_2 = 4$).

Section 36-4, Electron Waves

15. Use Equation 36-31 to calculate the de Broglie wavelength for an electron of kinetic energy (a) 1 eV, (b) 100 eV, (c) 1 keV, and (d) 10 keV.

16. When the kinetic energy of an electron is much greater than its rest energy, the relativistic approximation $E \approx pc$ is good. (a) Show that in this case photons and electrons of the same energy have the same wavelength. (b) What is the de Broglie wavelength of an electron of energy 100 MeV?

17. A thermal neutron in a reactor has kinetic energy of about 0.02 eV. Calculate the de Broglie wavelength of this neutron.

18. Find the de Broglie wavelength of a proton of energy 1 MeV.

Section 36-5, Wave-Particle Duality

There are no exercises for this section.

Section 36-6, The Uncertainty Principle

19. A particle of mass 10^{-6} g moves with speed 1 cm/s. If its speed is uncertain by 0.01 percent, what is the minimum uncertainty in its position?

20. The uncertainty in the position of an electron in an atom cannot be greater than the diameter of the atom. (a) Calculate the minimum uncertainty in momentum associated with an uncertainty in position of 0.1 nm. (b) If an electron has momentum p equal in magnitude to the uncertainty Δp found in (a), what is its kinetic energy? Express your answer in electronvolts.

21. An electron has kinetic energy 25 eV. If its momentum is uncertain by 10 percent, what is the minimum uncertainty in its position?

Problems

1. Data for stopping potential versus wavelength for the photoelectric effect using sodium are

λ, nm	200	300	400	500	600
V_0, V	4.20	2.06	1.05	0.41	0.03

Plot these data so as to obtain a straight line and from your plot find (a) the work function, (b) the threshold frequency, and (c) the ratio h/e.

2. A light bulb radiates 100 W uniformly in all directions. (a) Find the intensity at a distance of 1 m. (b) Find the number of photons per second that strike a 1-cm² area oriented so that its normal is along the line to the bulb if the wavelength of the light is 600 nm.

3. This problem is one of estimating the time lag (expected classically but not observed) in the photoelectric effect. Let the intensity of the incident radiation be 0.01 W/m². (a) Assuming that the area of an atom is 0.01 nm² find the energy per second falling on an atom. (b) If the work function is 2 eV, how long would it take classically for this much energy to fall on one atom?

4. (a) Solve Equation 36-12 for p_e^2 to obtain $p_e^2 = p_1^2 + p_2^2 - 2p_1p_2 + 2mc \times (p_1 - p_2)$. (b) Eliminate p_e^2 from your result in part (a) and Equation 36-11 to obtain $mc(p_1 - p_2) = p_1p_2(1 - \cos\theta)$. (c) Multiply both sides of your result in part (b) by h/mcp_1p_2 and use the de Broglie relation $h/p = \lambda$ to obtain the Compton formula (Equation 36-13).

5. The photoelectric effect can occur only with electrons that are bound (in atoms); i.e., a photon cannot be absorbed by a single free electron. Prove this by considering the problem of conservation of energy and momentum in the reference frame in which the total momentum of the initial photon and electron is zero.

6. The total energy density of radiation in a blackbody is given by $\eta = \int_0^\infty f(\lambda, T) \, d\lambda$, where $f(\lambda, T)$ is given by the Planck formula (Equation 36-2). Change the variable to $x = hc/\lambda kT$ and show that the total energy density can be written

$$\eta = \left(\frac{kT}{hc}\right)^4 8\pi hc \int_0^\infty \frac{x^3}{e^x - 1} \, dx = \alpha T^4$$

where α is a constant independent of T. This result, that the total energy density is proportional to T^4, is the *Stefan-Boltzmann law*.

7. Find the ratio of the reduced mass for the hydrogen atom to the electron mass μ/m. Show that this ratio equals the ratio of the Rydberg constant for hydrogen to that for an infinitely massive nucleus R_H/R_∞, and compare your results with the values of R_H and R_∞ given in the text.

8. The frequency of revolution of an electron in a circular orbit of radius r is $f_{rev} = v/2\pi r$, where v is the speed. (a) Show that in the nth stationary state

$$f_{rev} = \frac{k^2 Z^2 e^4 m}{2\pi \hbar^3} \frac{1}{n^3}$$

(b) Show that when $n_1 = n$, $n_2 = n - 1$, and n is much greater than 1,

$$\frac{1}{n_2^2} - \frac{1}{n_1^2} \approx \frac{2}{n^3}$$

(c) Use your result of part (b) in Equation 36-24 to show that in this case the frequency of radiation emitted equals the frequency of motion. This result is an example of Bohr's correspondence principle: when n is large, so that the energy difference between adjacent states is a small fraction of the total energy, classical and quantum physics must give the same result.

9. The kinetic energy of rotation of a diatomic molecule can be written $E_k = L^2/2I$, where L is its angular momentum and I the moment of inertia. (a) Assuming that the angular momentum is quantized, as in the Bohr model of the hydrogen atom, show that the energy is given by $E_n = n^2 E_1$, where $E_1 = \hbar^2/2I$, and make an energy-level diagram for such a molecule. (b) Estimate E_1 for the hydrogen molecule assuming the separation of the atoms to be 0.1 nm and considering rotation about an axis through the center of mass and perpendicular to the line joining the atoms. Express your estimate in electronvolts. (c) When E_1 is greater than kT, molecular collisions do not excite rotation and rotational energy does not contribute to the internal energy of the gas. Use your result of part (b) to find the critical temperature T_c, defined by $kT_c = E_1$. (d) Estimate E_1 for rotation about the line joining the atoms. In this case, the moment of inertia I is of the order $m_e r_a^2$, where m_e is the electron mass and r_a is the radius of the atom. Find T_c for rotation about this axis.

10. A particle moves about in a one-dimensional box of length L. Its energy is quantized by the condition $n(\lambda/2) = L$, where λ is the de Broglie wavelength of the particle. (a) Show that the allowed energies are given by $E_n = n^2 E_1$, where $E_1 = h^2/8mL^2$ and m is the mass of the particle. Evaluate E_1 for (b) an electron in a box of size $L = 0.1$ nm and (c) a proton in a box of size $L = 10^{-15}$ m. (Take the potential energy of the particle in the box to be zero so that its total energy is its kinetic energy $p^2/2m$.)

11. An important consequence of the uncertainty principle is that a particle confined to some region of space has a minimum kinetic energy greater than zero. If Δp is taken to be the standard deviation, it is given by $(\Delta p)^2 = [(p - p_{av})^2]_{av}$. (a) Show that this is equivalent to $(\Delta p)^2 = (p^2)_{av} - (p_{av})^2$. (b) Consider a particle of mass m moving in a box of length L such that $p_{av} = 0$. Its average kinetic energy is then $E_{k,av} = (p^2)_{av}/2m = (\Delta p)^2/2m$. Show that its average kinetic energy must be at least as large as $\hbar^2/8mL^2$ by using the fact that Δx cannot be larger than L.

CHAPTER 37   Nuclear Physics

Objectives   After studying this chapter you should:

1. Be able to give the order of magnitude of the radius of an atom and of a nucleus.

2. Be able to sketch the N-versus-Z curve for stable nuclei.

3. Be able to sketch the binding energy per nucleon versus A and discuss the significance of this curve for fission and fusion.

4. Know the exponential law of radioactive decay and be able to work problems using it.

5. Be able to describe the nuclear-fission chain reaction and discuss the advantages and disadvantages of fission reactors.

6. Be able to state the Lawson criterion for nuclear-fusion reactors.

7. Be able to discuss the chief mechanisms of energy loss of particles in matter and explain why some particles have well-defined ranges and others do not.

8. Be able to discuss the radiation dosage units of roentgen, rad, and rem.

The first information about the atomic nucleus came from the discovery of radioactivity by A. H. Becquerel in 1896. The rays emitted by radioactive nuclei were studied by many physicists in the early decades of the twentieth century. They were first classified by E. Rutherford as α, β, and γ rays, according to their ability to penetrate matter and ionize air: α radiation penetrates the least and produces the most ionization and γ penetrates the most with the least ionization. It was later found that α rays are helium nuclei, β rays are electrons or positrons, and γ rays are high-energy photons, i.e., electromagnetic radiation of very short wavelength. (A positron is an antielectron, a particle identical to the electron except that its charge is positive.) The α-particle scattering experiments of H. W. Geiger and E. Marsden in 1911 and the successes of the Bohr model of the atom led to the modern picture of an atom as consisting of a tiny massive nucleus with a radius of the order of 1 to 10 fm (1 fm = 10^{-15} m) surrounded by a cloud of electrons at a relatively great distance, of the order of 0.1 nm = 100,000 fm, from the nucleus. In 1919 Rutherford bombarded nitrogen with α particles and observed scintillations on a zinc sulfide screen due to protons, which have a much longer range in air than α particles. This was the first observation of artificial nuclear disintegration. Such experiments were extended to many other elements in the next few years.

In 1932 the neutron was discovered by Chadwick and the positron by Anderson, and the first nuclear reaction using artificially accelerated particles was observed by Cockcroft and Walton. It is quite reasonable to mark this year as the beginning of modern nuclear physics. With the discovery of the neutron, it became possible to understand some of the properties of nuclear structure; and the advent of nuclear accelerators made many experimental studies possible without the severe limitations on particle type and energy imposed by naturally radioactive sources.

In this chapter we first discuss some of the general properties of the atomic nucleus and then consider the important features of radioactivity. We then look at some nuclear reactions, including the important reactions of fission and fusion. Finally, we look at the interactions of nuclear particles with matter, a subject important for understanding the detection of nuclear particles, the shielding of reactors, and the effects of radiation on the human body. Our discussions will be descriptive and phenomenological, with the aim of presenting general information rather than a theoretical understanding of nuclear physics.

37-1 Atoms

Slightly more than 100 different elements have been discovered, 92 of which are found in nature. Each is characterized by an atom which contains a number Z of electrons, an equal number of protons, and a number N of neutrons. The number Z of electrons or protons is called the *atomic number*. The lightest atom, hydrogen (H), has $Z = 1$; the next lightest, helium (He), has $Z = 2$; the next lightest, lithium (Li), has $Z = 3$, etc. The structure of an atom is somewhat like that of our solar system. Nearly all the mass of the atom is in a tiny nucleus, which contains the protons and neutrons. The nuclear radius is typically about 1 to 10 fm. The distance between the nucleus and the electrons in the atom is about 0.1 nm = 100,000 fm. This distance determines the "size" of the atom. The chemical and physical properties of an element are determined by the number and arrangement of the electrons in the atom. Because each proton has a positive charge $+e$, the nucleus has a total positive charge $+Ze$. The electrons are negatively charged ($-e$) and so are attracted to the nucleus and repelled by each other. The electrons are arranged in shells: the first shell has up to 2 electrons; the second shell, about 4 times farther out, can contain up to 8 electrons; etc. This shell structure accounts for the periodic nature of the properties of the elements as shown in the periodic table. Elements with a single electron in an outer shell (hydrogen, lithium, sodium, etc.) or those with a single vacancy in an outer shell (fluorine, chlorine, bromine, etc.) are very active chemically; those with completely filled outer shells (helium, neon, argon, etc.) are chemically inert. The calculation of the electron configurations of the atoms and the resulting chemical properties was one of the great triumphs of quantum mechanics in the 1920s.

Since electrons and protons have equal but opposite charges and there are an equal number of electrons and protons in an atom, atoms are electrically neutral. Atoms that lose or gain one electron (or more) are then electrically charged and are called *ions*. Atoms with just 1 electron in an outer shell (such as sodium which has 11 electrons) tend to lose it readily and become positive ions; those lacking just one electron to have a complete shell tend to gain an electron to become negative ions, e.g., chlorine. Atoms bond together to form molecules, e.g., H_2O, or solids. This bonding involves only the outer electrons. In a molecule or solid, the separation of the atomic nuclei is about 1 atomic diameter, i.e., of the order of 0.1 nm.

37-2 Properties of Nuclei

The nucleus of an atom contains just two kinds of particles, protons and neutrons (aside from hydrogen, which contains a single proton). These particles are similar except that the proton has a charge $+e$ and the neutron is uncharged. They have approximately equal masses, the neutron being about 0.2 percent more massive; the mass of the electron is smaller by a factor of about 2000. A convenient unit of mass for discussing nuclei is the unified mass unit u (formerly the atomic mass unit amu), which is defined so that Avogadro's number of unified mass units equals 1 g and Avogadro's number of ^{12}C atoms has a mass of 12 g. The value of Avogadro's number is

Unified mass unit

$$N_A = 6.02204 \times 10^{23} \qquad 37\text{-}1$$

Avogadro's number

Another unit of mass convenient for nuclear physics is the MeV/c^2, which comes from the relativistic expression for rest energy $E = mc^2$. It is related to the unified mass unit by

$$1\ u = 931.50\ MeV/c^2 \qquad 37\text{-}2$$

The masses of the proton, neutron, and electron are

$$m_p = 1.6726 \times 10^{-27}\ kg = 1.007276\ u = 938.28\ MeV/c^2$$

$$m_n = 1.6750 \times 10^{-27}\ kg = 1.008665\ u = 939.573\ MeV/c^2 \qquad 37\text{-}3$$

$$m_e = 9.109 \times 10^{-31}\ kg = 5.486 \times 10^{-4}\ u = 0.511\ MeV/c^2$$

The lightest element, hydrogen, has in its most common form a nucleus consisting of just a single proton. Another less common form of hydrogen, called deuterium, has a nucleus with 1 proton and 1 neutron in it. Although the deuterium atom is about twice as massive as the ordinary hydrogen atom, its chemical behavior is the same because they each have 1 electron. A particular nuclear species is called a *nuclide*. Two or more different nuclides corresponding to the same atom are called *isotopes*. Isotopes have the same Z value (because they correspond to the same atom) but different N values and therefore different atomic mass numbers A. The two isotopes of hydrogen just mentioned are written ^1H and ^2H with the A number as a presuperscript. The Z number is sometimes written as a presubscript as in ^1_1H or ^2_1H, but this is not necessary because the chemical symbol H tells you that $Z = 1$. Another even less common isotope of hydrogen is tritium, ^3H, which has a nucleus consisting of 1 proton and 2 neutrons. On the average, there are about three stable isotopes for each atom (some have only one; others have five or six) so there are about 300 stable nuclei for the 100 different atoms. The most common isotope of the second lightest atom, helium, is ^4He (which is also known as an α particle), but ^3He also exists.

Isotopes

Inside the nucleus, the nucleons exert strong attractive forces on their nearby neighbors. This force, called the *strong nuclear force* or the *hadronic force,* is much stronger than the electric force of repulsion between the protons and very much stronger than the gravitational forces between the nucleons. (Gravity is so weak it can always be neglected in nuclear physics.) The force between two neutrons is roughly the same as that between two protons or between a neutron and proton, except that between two protons there is also the electric force due to their charges, which tends to weaken the attractions somewhat. The hadronic force decreases rapidly with distance, so that when two nucleons are more than a few femtometres apart, the force is negligible.

Size and Shape

The size and shape of the nucleus can be determined by bombarding it with high-energy particles and observing the scattering, similar to the experiments of Geiger and Marsden; or in some cases from measurements of radioactivity. The results depend somewhat on the kind of experiment. For example, a scattering experiment with electrons measures the charge distribution of the nucleus, whereas an experiment done with neutrons determines the region of influence of the hadronic force. Despite these differences, a wide variety of experiments suggest that most nuclei are approximately spherical, with radii given by

$$R = R_0 A^{1/3} \qquad\qquad 37\text{-}4 \qquad\qquad \textit{Nuclear radius}$$

where R_0 is about 1.5 fm. (Some nuclei in the rare-earth region of the periodic table are ellipsoidal, the major and minor axes differing by about 20 percent or less.) The fact that the radius of a spherical nucleus is proportional to $A^{1/3}$ implies that the volume of the nucleus is proportional to A, which in turn implies that the density is the same for all nuclei since the mass is approximately proportional to the number of nucleons A. The fact that a liquid drop has a constant density independent of its size has led to the analogy of the nucleus with a liquid drop, a model that has proved quite successful in understanding nuclear behavior, especially the fission of heavy nuclei.

N and Z Numbers

One characteristic of the nuclear force is that for light nuclei greatest stability is achieved with approximately equal proton and neutron numbers, $N = Z$. For heavier nuclei, the electric repulsion of the protons leads to greater stability with more neutrons than protons. You can see this by looking at N and Z numbers for the most abundant isotopes of various elements, for example, $^{40}_{20}\text{Ca}$, $^{16}_{8}\text{O}$, $^{56}_{26}\text{Fe}$, $^{207}_{82}\text{Pb}$, and $^{238}_{92}\text{U}$. Figure 37-1 shows a plot of N versus Z for the known stable nuclei. The straight line $N = Z$ is followed for small N and Z.

Mass and Binding Energy

The mass of a nucleus is not equal to the sum of the masses of the nucleons making up the nucleus. When two or more nucleons fuse together to form a nucleus, the total mass decreases and energy is given off. Conversely, to break up a nucleus into its parts, energy must be put into the system to be changed into the increase in rest mass. The energy

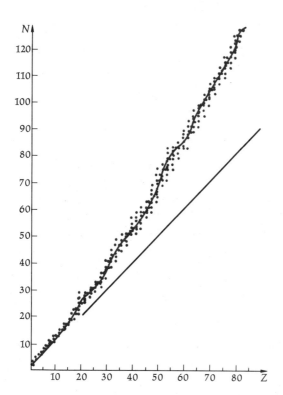

Figure 37-1
Plot of neutron number N versus proton number Z for the stable nuclides. The straight line is $N = Z$.

involved is c^2 times the change in mass, where c is the speed of light in vacuum. The difference between the rest energy of the parts of a nucleus and the rest energy of the nucleus is called the total *binding energy* of the nucleus. Consider ^{4}He, for example, which consists of two protons and two neutrons. The mass of an atom can be accurately measured in a mass spectrometer, which measures q/M for ions by bending them in a magnetic field. The mass of the ^{4}He atom is 4.00263 u. This includes the masses of the two electrons in the atom. The mass of the ^{1}H atom is 1.007825 u, and that of the neutron is 1.008665 u. The sum of the masses of two ^{1}H atoms plus two neutrons is 2(1.007825 u) + 2(1.008665 u) = 4.03298 u, which is greater than the mass of the ^{4}He atom by 0.030377 u. Note that by using the masses of two ^{1}H atoms rather than two protons, the masses of the electrons in the atom cancel out. We do this because it is atomic masses that are measured directly and listed in mass tables.*

This mass difference of 0.030377 u can be converted into the binding energy of the He atom by using the mass conversion factor 1 u·c^2 = 931.50 MeV from Equation 37-2. Then 0.030377 u·c^2 = 0.030377(931.5) = 28.20 MeV, which is the total binding energy of ^{4}He. In general, the binding energy of a nucleus of atomic mass M_A containing Z protons and N neutrons is found from

$$E_b = ZM_H c^2 + Nm_n c^2 - M_A c^2 \qquad\qquad 37\text{-}5$$

Binding energy

where M_H is the mass of the ^{1}H atom and m_n that of the neutron. The atomic masses for some selected nuclei are listed in Table 37-1. Once the

---

* A slight error is made because the mass of the ^{1}H atom is slightly less than the sum of the mass of an electron plus a proton because of the binding energy of the atom. However, since atomic binding energies are very small compared with nuclear binding energies, this error is negligible.

Table 37-1
Atomic masses of selected isotopes

Element	Symbol	Z	Atomic Mass, u
Hydrogen	^{1}H	1	1.007 825
	^{2}H	1	2.014 102
	^{3}H	1	3.016 050
Helium	^{3}He	2	3.016 030
	^{4}He	2	4.002 603
Lithium	^{6}Li	3	6.015 125
Boron	^{10}B	5	10.012 939
Carbon	^{12}C	6	12.000 000
	^{14}C	6	14.003 242
Oxygen	^{16}O	8	15.994 915
Sodium	^{23}Na	11	22.989 771
Potassium	^{39}K	19	38.963 710
Iron	^{56}Fe	26	55.939 395
Copper	^{63}Cu	29	62.929 592
Silver	^{107}Ag	47	106.905 094
Gold	^{197}Au	79	196.966 541
Lead	^{208}Pb	82	207.976 650
Polonium	^{212}Po	84	211.989 629
Radon	^{222}Rn	86	222.017 531
Radium	^{226}Ra	88	226.025 360
Uranium	^{238}U	92	238.048 608
Plutonium	^{242}Pu	94	242.058 725

atomic mass has been determined, the binding energy can be computed from Equation 37-5. It is found that the total binding energy of a nucleus is approximately proportional to the total number of nucleons A in the nucleus. Figure 37-2 shows the binding energy per nucleon E_b/A versus A. The mean value is about 8.3 MeV. The fact that this curve is approximately constant (meaning that E_b is approximately proportional to A) indicates that there is saturation of nuclear forces in the nucleus; i.e., each nucleon bonds to only a certain number of other nucleons independent of the total number of nucleons in the nucleus. If, for example, there were no saturation and each nucleon bonded to each other nucleon, there would be $A - 1$ bonds for each nucleon and a total of $A(A - 1)$ bonds altogether. The total binding energy, which is a measure of the energy needed to break all these bonds, would then be proportional to $A(A - 1)$ and E_b/A would not be approximately constant. Figure 37-2 indicates that, instead, there is a fixed number of bonds per nucleon, as would be the case if each nucleon were attracted only to its nearest neighbors. Such a situation also leads to a constant nuclear density consistent with the measurements of radius. If the binding energy per nucleon were proportional to the number of nucleons, the radius would be approximately constant, as it is for atoms.

Figure 37-2
The binding energy per particle versus mass number A. (*From R. Leighton*, Principles of Modern Physics, *McGraw-Hill Book Company, New York, 1959, by permission.*)

Questions

1. How does the nuclear force differ from the electromagnetic force?

2. What property of the nuclear force is indicated by the fact that all nuclei have about the same density?

3. The mass of ^{12}C which contains 6 protons and 6 neutrons is exactly 12.000 u, by the definition of the unified mass unit. Why isn't the mass of ^{16}O which contains 8 protons and 8 neutrons exactly 16.0000 u?

37-3 Radioactivity

Nuclei that are not stable are radioactive. The three kinds of radioactivity, alpha (α) decay, beta (β) decay, and gamma (γ) decay, were named before it was known that α particles are ^{4}He nuclei, β particles are electrons (e^-) or positrons (e^+), and γ rays are photons (γ). All very heavy nuclei ($Z > 83$) are theoretically unstable to α decay; i.e., for heavy nuclei the mass of the original radioactive nucleus is greater than the sum of the masses of the decay products consisting of an α particle and what is called the daughter nucleus. An example of α decay is ^{232}Th ($Z = 90$), which decays into ^{228}Ra ($Z = 88$) plus an α particle, written

$$^{232}\text{Th} \rightarrow {}^{228}\text{Ra} + {}^4\text{He} \qquad\qquad 37\text{-}6$$

The mass of the ^{232}Th atom is 232.038124 u. The mass of the daughter atom ^{228}Ra is 228.031139 u. Adding this to the mass of ^{4}He, we get 232.03374 for the total mass on the right side of the reaction. This is less than that of ^{232}Th by 0.00438 u, which multiplied by 931.50 MeV/c^2 gives 4.08 MeV/c^2 for the excess rest mass of ^{232}Th over the decay products. ^{232}Th is therefore theoretically unstable to α decay, and this decay does in fact occur in nature with the emission of an α particle of energy 4.08 MeV. (The energy is actually somewhat less than 4.08 MeV because some of the decay energy is shared by the recoiling ^{228}Ra nucleus.) Examination of the possible α decays of other heavy nuclei shows that all nuclei heavier than bismuth ($Z = 83$) are theoretically unstable to α decay whereas all lighter nuclei are stable; i.e., the rest mass of the decay products is greater than that of the original nucleus for Z less than 83.

Beta decay occurs for light or intermediate nuclei that have too many or too few neutrons for stability, as indicated on the N-versus-Z curve of Figure 37-1. The energy released in β decay can also be determined by computing the rest mass of the original nucleus and that of the decay products. Note that in α decay the daughter nucleus has a mass number A that is less by 4 than that of the original nucleus and an atomic number Z that is less by 2, whereas in β decay A remains the same while Z either increases by 1 (e^- decay) or decreases by 1 (e^+ decay). γ decay usually happens very quickly and is observed only because it follows either α or β decay. In γ decay, the radioactive nucleus remains the same nucleus as it decays from an excited state to the ground state or an excited state of lower energy than that of the original nucleus. The question of stability of a nucleus to α or β decay can always be answered by looking at the rest masses of the nuclei and particles involved.

In 1900 Rutherford discovered that the rate of emission of radioactive particles from a substance is not constant in time but decreases exponentially. This exponential time dependence is characteristic of all radioactivity and indicates that the decay is a statistical process. Because each nucleus is well shielded from others by the atomic electrons, pressure and temperature changes have no effect on nuclear properties. For a statistical decay, in which the decay of any individual nucleus is a random event, the number of nuclei decaying in a time interval dt is proportional to dt and to the number of radioactive nuclei. If $N(t)$ is the number of radioactive nuclei at time t and $-dN$ is the number that decay in time dt (the negative sign is necessary because N decreases), we have

$$-dN = \lambda N \, dt \qquad\qquad 37\text{-}7$$

where the constant of proportionality λ is called the *decay constant*. It is the probability per unit time of the decay of any given radioactive nucleus. The solution of this equation is

$$N = N_0 e^{-\lambda t} \qquad\qquad 37\text{-}8$$

where N_0 is the number of nuclei at $t = 0$. The decay rate is

$$R = \frac{dN}{dt} = \lambda N_0 e^{-\lambda t} = R_0 e^{-\lambda t} \qquad\qquad 37\text{-}9 \qquad \textit{Decay rate}$$

where $R_0 = \lambda N_0$ is the initial decay rate. Note that both the number of nuclei and the rate of decay decrease exponentially. It is the decrease in the rate of decay that is determined experimentally. Figure 37-3 shows N versus t. If we multiply the numbers on the N axis by λ, this figure becomes a graph of R versus t.

An important parameter of radioactive decay related to the decay constant is the half-life $t_{1/2}$, defined as the average time for half of a given number of radioactive nuclei to decay. It is also the time for which the decay rate decreases to half its original value. If we set N equal to $\frac{1}{2}N_0$ and $t = t_{1/2}$ in Equation 37-8, we obtain

$$\tfrac{1}{2}N_0 = N_0 e^{-\lambda t_{1/2}}$$

$$e^{+\lambda t_{1/2}} = 2$$

$$\lambda t_{1/2} = \ln 2 = 0.693$$

or

$$t_{1/2} = \frac{\ln 2}{\lambda} = \frac{0.693}{\lambda} \qquad\qquad 37\text{-}10 \qquad \textit{Half-life}$$

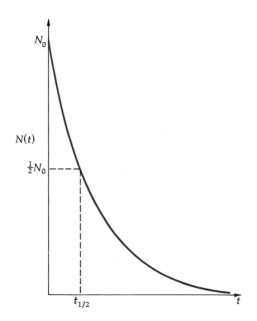

N_0

$N(t)$

$\frac{1}{2}N_0$

$t_{1/2}$

t

Figure 37-3
Exponential radioactive-decay law. The number of nuclei remaining at time t decreases exponentially with time t. The half-life $t_{1/2}$ is indicated. The decay rate $R(t) = \lambda N(t)$ has the same time dependence.

The decay of any one nucleus is completely random; all we can say is that, on the average, half of some original number of radioactive nuclei will decay in one half-life, then half of those left will decay in the next half-life, etc. The half-lives of various radioactive nuclei vary from very small times (less than a microsecond) to very large times, for example 10^{16} y. Some nuclei that are theoretically unstable to α decay as calculated from the mass-energy relations have a half-life that is long compared with the age of the universe; for all practical purposes such nuclei can be considered to be stable.

Example 37-1 A radioactive source has a half-life of 1 min. At time $t = 0$ it is placed near a detector, and the counting rate is observed to be 2000 counts/s. Find the decay constant and the counting rate at times $t = 1, 2, 3,$ and 10 min.

From Equation 37-10 we find for the decay constant:

$$\lambda = \frac{0.693}{t_{1/2}} = \frac{0.693}{1 \text{ min}} = 1.16 \times 10^{-2} \text{ s}^{-1}$$

Since the half-life is 1 min, the counting rate will be half as great at $t = 1$ min as at $t = 0$, so $R_1 = 1000$ counts/s at 1 min, $R_2 = 500$ counts/s at $t = 2$ min, and $R_3 = 250$ counts/s at 3 min. At $t = n$ min $= nt_{1/2}$ the rate will be

$$R = (\tfrac{1}{2})^n R_0 \qquad\qquad\qquad 37\text{-}11$$

so that after 10 min $= 10t_{1/2}$ the rate will be $R_{10} = (\tfrac{1}{2})^{10}(2000) = 1.95$ counts/s.

Example 37-2 If the detection efficiency in Example 37-1 is 20 percent, how many radioactive nuclei are there at time $t = 0$? At time $t = 1$ min? How many decay in the first minute?

The detection efficiency depends on the distance from the source to the detector and the chance that a radioactive decay particle entering the detector will produce a count. If the source is smaller than the detector and placed very close, about half the emitted particles will enter the detector. If the counting rate at $t = 0$ is 2000 counts/s and the efficiency is 20 percent, the decay rate at $t = 0$ must be 10,000 s^{-1}. The number of radioactive nuclei can be found from Equation 37-9 using $\lambda = 0.693$ min^{-1} from Example 37-1:

$$N_0 = \frac{R_0}{\lambda} = \frac{10{,}000 \text{ s}^{-1}}{0.693 \text{ min}^{-1}} = 8.66 \times 10^5$$

At time $t = 1$ min $= t_{1/2}$ there are half as many nuclei as at $t = 0$; so $N_1 = \frac{1}{2}(8.66 \times 10^5) = 4.33 \times 10^5$. The number that decay in the first minute is therefore 4.33×10^5. Note that this is not equal to the initial decay rate of 10,000 s^{-1} times 60 s because the decay rate is not constant during the 1-min interval but decreased by half during each minute. Note also that the average decay rate is not the mean of 10,000 and 5000 s^{-1} because the decrease is exponential, not linear in time.

The SI unit of radioactive decay is the becquerel (Bq), which is defined as one decay per second:

$$1 \text{ Bq} = 1 \text{ decay/s} \qquad \qquad 37\text{-}12 \qquad \textit{Becquerel}$$

An historical unit that applies to all types of radioactivity is the Curie (Ci), defined as

$$1 \text{ Ci} = 3.7 \times 10^{10} \text{ decays/s} = 3.7 \times 10^{10} \text{ Bq} \qquad 37\text{-}13 \qquad \textit{Curie}$$

The curie is the amount of radiation emitted by 1 g of radium. Since this is a very large unit, the millicurie (mCi) or microcurie (μCi) is often used.

Alpha Decay

When a nucleus emits an α particle, both N and Z decrease by 2 and A decreases by 4. The daughter of a radioactive nucleus is often itself radioactive, decaying either by α or β decay or both. If the original decaying nucleus has a mass number A that is 4 times an integer, the daughter nucleus and all those in the chain will also have mass numbers equal to 4 times an integer. Similarly, if the mass number of the original nucleus is $4n + 1$, where n is an integer, all the nuclei in the decay chain will have mass numbers given by $4n + 1$. We see therefore that there are four possible α-decay chains, depending on whether A equals $4n$, $4n + 1$, $4n + 2$, or $4n + 3$, where n is an integer. All but one of these are found in nature. The $4n + 1$ series is not, because its longest-lived member (other than the stable end product ^{209}Bi), ^{237}Np, has a half-life of only 2×10^6 y, which is much smaller than the age of the earth.

Figure 37-4 illustrates the thorium series, which has $A = 4n$ and begins with an α decay from ^{232}Th to ^{228}Ra. The daughter nuclide of an α decay is on the left or neutron-rich side of the stability curve (dashed line), so that it often decays by β^- decay, in which one neutron changes to a proton by emitting an electron. In Figure 37-4, ^{228}Ra decays by β decay to ^{228}Ac, which in turn decays to ^{228}Th. There are then four α decays to ^{212}Pb, which decays by β^- to ^{212}Bi. There is a branch point at ^{212}Bi which decays either by α decay to ^{208}Tl or by β decay to ^{212}Po. The branches meet at the stable lead isotope ^{208}Pb.

The energy released in α decay is determined by the difference in mass of the parent nucleus and the decay products, which include the daughter nucleus and the α particle. If all the α decays proceeded from the ground state of the parent nucleus to the ground state of the daugh-

Figure 37-4
The thorium (4n) α-decay series.

ter nucleus, the emitted α particles would all have the same energy (slightly less than the energy released because of the small recoil energy of the daughter nucleus needed to conserve momentum). When the energies of the emitted α particles are measured with high resolution, a spectrum of energies is observed, as shown in Figure 37-5. The peak in the spectrum labeled α_0 corresponds to α decays to the ground state of the daughter nucleus with a total energy of 6.04 MeV, as determined from the masses of the particles. The peak labeled α_{30} indicates α particles with energy 30 keV less than those of maximum energy, indicating that the decay is to an excited state of the daughter nucleus at 30 keV above the ground state. The energy spectrum of α particles then indicates the energy levels in the daughter nucleus, as in the case shown in Figure 37-5. This interpretation of the particle energy spectrum is confirmed by the observation of a 30-keV γ ray emitted as the daughter nucleus decays to its ground state. Because the half-life for γ decay is very short, this decay follows essentially immediately after the α decay. Figure 37-6 shows an energy-level diagram for ^{223}Ra obtained from the measurement of the α-particle energies in the decay of ^{227}Th

Figure 37-5
The α-particle spectrum from ^{227}Th. The highest-energy α particles correspond to decay to the ground state of ^{223}Ra with a transition energy of $Q = 6.04$ MeV. The next-highest-energy particles, α_{30}, result from transitions to the first excited state of ^{223}Ra, 30 keV above the ground state. The energy levels of the daughter nucleus, ^{223}Ra, can be determined by measurement of the α-particle energies. (*Data taken from R. C. Pilger, Ph.D. thesis, University of California Radiation Laboratory Report UCRL-3877, 1957.*)

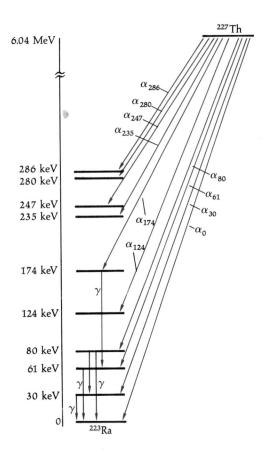

6.04 MeV

^{227}Th

α_{286}
α_{280}
α_{247}
α_{235}

286 keV
280 keV

α_{80}
α_{61}

247 keV
235 keV

α_{30}

α_{174}

α_0

α_{124}

174 keV

γ

124 keV

80 keV
61 keV

γ γ

30 keV

γ

0

^{223}Ra

Figure 37-6
Energy levels of ^{223}Ra determined by measurement of α-particle energies from ^{227}Th, as shown in Figure 37-5. Only the lowest-lying levels and some of the γ-ray transitions are shown.

and from the observed γ-decay energies. Only the lowest-lying levels and some of the γ-ray transitions are indicated. This nuclear energy-level diagram is similar to the atomic energy-level diagram for hydrogen in our discussion of the Bohr model (Chapter 36).

Beta Decay

The simplest example of β decay is that of the free neutron, which decays into a proton plus an electron with a half-life of about 10.8 min. The energy of decay is 0.78 MeV, which is the difference between the rest energy of the neutron (939.57 MeV) and that of the proton plus electron (938.28 + 0.511 MeV). More generally, in β^- decay, a nucleus of mass number A and atomic number Z decays into one with mass number A and atomic number $Z' = Z + 1$ with the emission of an electron. If the decay energy is shared by the daughter nucleus and the emitted electron, the energy of the electron is uniquely determined by conservation of energy and momentum, just as in α decay. Experimentally, however, the energies of the electrons emitted in decay are observed to vary from zero to the maximum energy available. A typical energy spectrum is shown in Figure 37-7; compare it with the discrete spectrum of particle energies of Figure 37-5.

To explain the apparent nonconservation of energy in β decay, W. Pauli in 1930 suggested that a third particle, which he called the *neutrino,* is emitted. The mass of the neutrino was assumed to be very much less than that of the electron because the maximum energy of the electrons emitted is nearly equal to the total available for the decay. The neutrino was observed experimentally in 1957. It is now known that there are at least two kinds of neutrinos, one (ν_e) associated with elec-

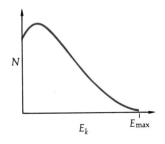

Figure 37-7
Energy spectrum of electrons emitted in β decay. The number of electrons per unit energy interval is plotted versus energy. The fact that all the electrons do not have the same energy E_{max} suggests that there is another particle emitted which shares the energy available for decay.

Neutrino

trons and one (ν_μ) associated with muons, and possibly one associated with a new particle, the heavy lepton, recently discovered. Moreover, each neutrino has an antiparticle, written $\bar{\nu}_e$ and $\bar{\nu}_u$. It is the antineutrino that is emitted in the decay of the neutron, which is written

$$n \rightarrow p + \beta^- + \bar{\nu}_e \qquad \text{37-14}$$

In β^+ decay a proton changes into a neutron with the emission of a positron (and a neutrino). A free proton cannot decay by positron emission because of conservation of energy (the rest mass of the neutron plus positron is greater than that of the proton), but because of binding-energy effects, a proton inside a nucleus can. A typical decay is

$$^{13}_7\text{N} \rightarrow {}^{13}_6\text{C} + \beta^+ + \nu_e \qquad \text{37-15}$$

The electrons or positrons emitted in decay do not exist inside the nucleus. They are created in the process of decay, just as photons are created when an atom makes a transition from a higher to a lower energy state.

An important example of decay is that of ^{14}C, which is used in radioactive dating:

$$^{14}\text{C} \rightarrow {}^{14}\text{N} + \beta^- + \bar{\nu}_e \qquad \text{37-16}$$

Carbon dating

The half-life for this decay is 5730 y. The radioactive isotope ^{14}C is produced in the upper atmosphere from nuclear reactions caused by cosmic rays. The chemical behavior of carbon atoms with ^{14}C nuclei is the same as those with ordinary ^{12}C nuclei; e.g., atoms with these nuclei combine with oxygen to form CO_2 molecules. Since living organisms continually exchange CO_2 with the atmosphere, the ratio of ^{14}C to ^{12}C in a living organism is the same as the equilibrium ratio in the atmosphere, which is about 1.3×10^{-12}. When an organism dies, it no longer absorbs ^{14}C from the atmosphere, so that the ratio $^{14}\text{C}/^{12}\text{C}$ continually decreases due to the radioactive decay of ^{14}C. A measurement of the decay rate per gram of carbon makes it possible to calculate when the organism died.

Example 37-3 Calculate the decay rate of a living organism per gram of carbon, assuming that the ratio $^{14}\text{C}/^{12}\text{C} = 1.3 \times 10^{-12}$.

From Equation 37-9 the decay rate is related to the number of radioactive nuclei as follows:

$$R = -\frac{dN}{dt} = \lambda N = \frac{0.693}{t_{1/2}} N$$

The number of ^{12}C nuclei in 1 g of carbon is

$$N_{12\text{C}} = \frac{6.02 \times 10^{23} \text{ nuclei/mol}}{12 \text{ g/mol}} = 5.02 \times 10^{22} \text{ nuclei/g}$$

The number of radioactive ^{14}C nuclei is 1.3×10^{-12} times this number. The decay rate is therefore

$$R = \frac{0.693}{5730 \text{ y}} (1.3 \times 10^{-12}) \frac{5.02 \times 10^{22}}{\text{g}} \frac{1 \text{ y}}{3.16 \times 10^7 \text{ s}} \frac{60 \text{ s}}{\text{min}}$$

$$= 15 \text{ decays/min·g}$$

The decay rate for a living organism is therefore 15 decays per minute per gram of carbon.

Example 37-4 A bone containing 200 g of carbon has a β-decay rate of 400 decays/min. How old is the bone?

We first obtain a rough estimate. If the bone were from a living organism, we would expect the decay rate to be (15 decays/min·g) (200 g) = 3000 decays/min. Since 400/3000 is roughly $\frac{1}{8}$ (actually 1/7.5), the sample must be about 3 half-lives old, or about 3(5730) y. To find the age more accurately, we note that after n half-lives, the decay rate decreases by a factor of $(\frac{1}{2})^n$. We therefore find n from

$$(\tfrac{1}{2})^n = \frac{400}{3000}$$

or

$$2^n = \frac{3000}{400} = 7.5$$

$$n \ln 2 = \ln 7.5$$

$$n = \frac{\ln 7.5}{\ln 2} = 2.91$$

The age is therefore $t = nt_{1/2} = 2.91(5730 \text{ y}) = 16{,}700$ y.

Gamma Decay

In γ decay a nucleus in an excited state decays to a lower-energy state by the emission of a photon. This decay is the nuclear analog of the emission of light by atoms. Since the spacing of the nuclear energy levels is of the order of millions of electronvolts (compared with electronvolts in atoms), the wavelengths of the emitted photons are of the order of

$$\lambda = \frac{hc}{E} \sim \frac{1240 \text{ MeV·fm}}{1 \text{ MeV}} = 1240 \text{ fm}$$

Emission of γ rays usually follows β or α decay. For example, if a radioactive parent nucleus decays by β decay to an excited state of the daughter nucleus, the daughter nucleus often decays down to its ground state by γ emission. The mean life for γ decay is often very short. Direct measurements of mean lives down to about 1 μs are possible, and sophisticated coincidence techniques can be used to measure mean lives as short as 10^{-11} s. Measurements of lifetimes smaller than 10^{-11} s are difficult but can sometimes be accomplished by determining the natural line width Γ and using the uncertainty relation $\tau = \hbar/\Gamma$.

Questions

4. Why do extreme changes in temperature and pressure of a radioactive sample have little or no effect on the radioactivity?

5. Why is the decay series $A = 4n + 1$ not found in nature?

6. A decay by α emission is often followed by β decay. When this occurs, it is by β^- and not β^+ decay. Why?

7. The half-life of ^{14}C is much less than the age of the universe, yet ^{14}C is found in nature. Why?

8. What effect would a long-term variation in the cosmic-ray activity have on the accuracy of ^{14}C dating?

37-4   Nuclear Reactions

Most of the information about nuclei is obtained by bombarding them with various particles and observing the results. Although the first experiments were limited by the need to use radiation from naturally occurring sources, they produced many important discoveries. In 1932 J. D. Cockcroft and E. T. S. Walton succeeded in producing the reaction

$$p + {}^7\text{Li} \rightarrow {}^8\text{Be} \rightarrow {}^4\text{He} + {}^4\text{He}$$

using artificially accelerated protons. At about the same time, the Van de Graaff electrostatic generator was built (by R. Van de Graaff in 1931), as was the first cyclotron (by E. O. Lawrence and M. S. Livingston in 1932). Since then, an enormous technology has been developed for accelerating and detecting particles, and many nuclear reactions have been studied.

When a particle is incident on a nucleus, several different things can happen. The particle may be scattered elastically or inelastically (in which case the nucleus is left in an excited state and decays by emitting photons or other particles), or the original particle may be absorbed and another particle or particles emitted.

Consider a general reaction of particle x on nucleus X, resulting in nucleus Y and particle y. The reaction is usually written

$$x + X \rightarrow Y + y + Q$$

or, for short, $X(x, y)\, Y$. The quantity Q, defined by

$$Q = (m_x + m_X - m_y - m_Y)c^2 \qquad \text{37-17}$$

Q value

is the energy released in the reaction. The Q value for the reaction $n + {}^1\text{H} \rightarrow {}^2\text{H} + \gamma + Q$ is +2.22 MeV, whereas for the inverse reaction, $\gamma + {}^2\text{H} \rightarrow n + {}^1\text{H} + Q$, it is −2.22 MeV. If Q is positive, the reaction is said to be *exothermic*. It is energetically possible even if the particles are at rest. If Q is negative, the reaction is *endothermic* and cannot occur below a threshold energy. In the reference frame in which the total momentum is zero (the center-of-mass frame) the threshold energy for the two particles is just $|Q|$. However, most reactions occur with one particle X at rest. In this frame, called the *laboratory frame*, the incident particle must have energy greater than $|Q|$ because, by conservation of momentum, the kinetic energy of y and Y cannot be zero. Consider the nonrelativistic case of x, of mass m, incident on X, of mass M. We showed in Equation 8-44 that if $|Q|$ is the threshold in the center-of-mass reference frame, the threshold in the laboratory frame is

$$E_{\text{th}} = \frac{m + M}{m}\,|Q| \qquad \text{37-18}$$

A measure of the effective size of a nucleus for a particular reaction is the *cross section* σ, defined as *the number of reactions per unit time per nucleus divided by the incident intensity* (number of incident particles per unit time per unit area). Consider, for example, the bombardment of ${}^{13}\text{C}$ by protons. A number of reactions might occur. Elastic scattering is written ${}^{13}\text{C}(p, p)\,{}^{13}\text{C}$: the first p indicates an incident proton; the second indicates that the particle that leaves is also a proton. If the scattering is inelastic, the outgoing proton is indicated by p' and the nucleus in an excited state by ${}^{13}\text{C}^*$; one then writes ${}^{13}\text{C}(p, p')\,{}^{13}\text{C}^*$. Some other pos-

Cross section

The Cockcroft-Walton accelerator. Walton is sitting in the foreground. J. D. Cockcroft and E. T. S. Walton produced the first transmutation of nuclei with artificially accelerated particles in 1932, for which they received the Nobel prize (1951). (*Courtesy of Cavendish Laboratory.*)

sible reactions are

$$(p, n):\ ^{13}\text{C}(p, n)\ ^{13}\text{N}$$

$$\text{Capture}:\ ^{13}\text{C}(p, \gamma)\ ^{14}\text{N}$$

$$(p, \alpha):\ ^{13}\text{C}(p, \alpha)\ ^{10}\text{B}$$

The total cross section is the sum of the partial cross sections

$$\sigma = \sigma_{p,p} + \sigma_{p,p'} + \sigma_{p,n} + \sigma_{p,\gamma} + \sigma_{p,\alpha} + \cdots$$

Cross sections have the dimensions of area. Since nuclear cross sections are of the order of the square of the nuclear radius, a convenient unit for them is the *barn*, defined by

$$1 \text{ barn} = 10^{-28} \text{ m}^2 \qquad\qquad 37\text{-}19$$

The cross section for a particular reaction is a function of energy. For an endothermic reaction, it is zero for energies below the threshold.

Reactions with Neutrons

Nuclear reactions involving neutrons are important for understanding nuclear reactors. The most likely reaction with a nucleus for a neutron of energy of more than about 1 MeV is scattering. However, even if the scattering is elastic, the neutron loses some energy to the nucleus. If a neutron is scattered many times in a material, its energy decreases until it is of the order of kT. The neutron is then equally likely to gain or lose energy from a nucleus when it is elastically scattered. A neutron with energy of the order of kT is called a *thermal neutron*. At low energies, a neutron is likely to be captured, with the emission of a γ ray from the excited nucleus:

$$n + {}^{A}_{Z}M \rightarrow {}^{A+1}_{Z}M + \gamma \qquad\qquad 37\text{-}20$$

Since the binding energy of a neutron is of the order of 6 to 10 MeV and the kinetic energy of the neutron is negligible in comparison, the excitation energy of the nucleus after capturing the neutron is from 6 to 10 MeV and γ rays of this energy are emitted. If there are no resonances, the cross section $\sigma(n, \gamma)$ varies smoothly with energy, decreasing with increasing energy roughly as $E^{-1/2} \propto 1/v$, where v is the speed of the neutron. This energy dependence is easily understood because the time spent by a neutron near a nucleus is inversely proportional to the speed of the neutron and the capture cross section is therefore proportional to this time. Superimposed on the $1/v$ dependence are large fluctuations in the capture cross section due to resonances. Figure 37-8 shows the neutron-capture cross section for silver as a function of energy. The dashed line indicates the $1/v$ dependence. At the maximum of the resonance, the value of the cross section is very large (> 5000 barns) compared with the value of about 10 barns just past the resonance. Many elements show similar resonances in the neutron-capture cross section.

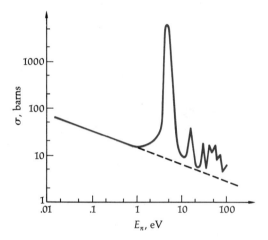

Figure 37-8
Neutron-capture cross section for Ag versus energy. The dashed-line extension would be expected if there were no resonances and the cross section were merely proportional to the time spent near the nucleus, i.e., proportional to $1/v$. (*From R. Evans,* The Atomic Nucleus, *McGraw-Hill Book Company, New York, 1955, by permission.*)

Since the maximum cross section for ^{113}Cd is about 57,000 barns, this material is useful for shielding against low-energy neutrons.

An important nuclear reaction that involves neutrons is fission. Very heavy nuclei (atomic numbers greater than 100) are subject to spontaneous fission; i.e., they break apart into two nuclei even if left to themselves with no outside disturbance. We can understand this by considering the analogy of a charged liquid drop. If the drop is not too large, surface tension can overcome the repulsive forces of the charges and hold the drop together, but there is a certain maximum size beyond which the drop will be unstable and spontaneously break apart. Spontaneous fission puts an upper limit on the size of a nucleus and therefore on the number of elements that are possible. Although the very heavy elements of uranium and polonium are stable as far as spontaneous fission is concerned, they can be induced to fission by the capture of a neutron. We shall study fission and another important nuclear reaction, fusion, in the next section.

George Gerster, Photo Researchers

Three-mile-long linear accelerator at Stanford University, used for accelerating electrons to very high energies.

Questions

9. What is meant by the cross section for a nuclear reaction?

10. Why is the neutron-capture cross section (excluding resonances) proportional to $E^{-1/2}$?

11. What is meant by the Q value of a reaction? Why is the reaction threshold not equal to Q?

12. Why isn't there an element with $Z = 130$?

37-5   Fission, Fusion, and Nuclear Reactors

Two nuclear reactions, fission and fusion, are of particular importance. In the fission of ^{235}U, for example, the uranium nucleus is excited by the capture of a neutron and splits into two nuclei, each with about half the total mass. The Coulomb force of repulsion drives the fission fragments apart, the energy eventually showing up as thermal energy. In fusion, two light nuclei such as those of deuterium and tritium (^2H and ^3H) fuse together to form a heavy nucleus (in this case ^4He plus a neutron). A typical fusion reaction is

$$^2H + {}^3H \rightarrow {}^4He + n + 17.6 \text{ MeV} \qquad 37\text{-}21$$

A plot of the mass difference per nucleon $(M - Zm_p - Nm_n)/A$ versus A, in units of MeV/c^2, is shown in Figure 37-9. This is just the negative of the binding-energy curve of Figure 37-2. From this figure we see that the rest mass per particle for both very heavy ($A \approx 200$) and very light ($A \approx 20$) nuclides is more than that for nuclides of intermediate mass.

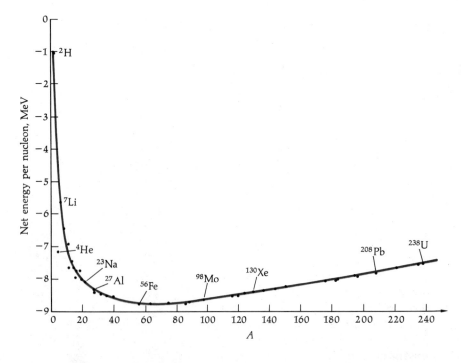

Figure 37-9
Plot of the mass difference $(M - Zm_p - Nm_n)/A$ versus A in units of MeV/c^2. The rest mass per particle is less for intermediate-mass nuclei ($A \approx 80$) than for either very light or very heavy nuclei.

Thus in both fission and fusion the total rest mass decreases and energy is released. Since for $A = 200$ the rest energy is about 1 MeV per nucleon greater than for $A = 100$, about 200 MeV is released in the fission of a heavy nucleus. This is much more energy than is released in atomic or chemical reactions such as combustion (about 4 eV). The energy release in fusion depends on the particular reaction. For the $^2H + {}^3H$ reaction above, 17.6 MeV is released. Although this is less than the energy released in fission, it is a greater amount of energy per unit mass.

Example 37-5 Calculate the total energy released in kilowatt-hours in the fission of 1 g of ^{235}U, assuming that 200 MeV is released per fission.

The number of ^{235}U nuclei in 1 g is

$$N = \frac{6.02 \times 10^{23}}{235} = 2.56 \times 10^{21}$$

The energy released per gram is then

$$\frac{200 \text{ MeV}}{\text{Nucleus}} \frac{2.56 \times 10^{21} \text{ nuclei}}{\text{g}} \frac{1.6 \times 10^{-19} \text{ J}}{\text{eV}} \frac{1 \text{ h}}{3600 \text{ s}} \frac{1 \text{ kW}}{1000 \text{ J/s}}$$

$$= 2.28 \times 10^4 \text{ kW·h}$$

Example 37-6 Compare the energy released per gram in the fusion of deuterium and tritium with that of fission.

We can compare the energy per gram by first comparing the energy per nucleon. In fission this is about 1 MeV per nucleon. In the fusion of ^{2}H + ^{3}H it is (17.6 MeV)/5 = 3.52, or about $3\frac{1}{2}$ times as great. The energy released per gram in this fusion reaction is therefore about $3\frac{1}{2}$ times that in fission.

The application of both fission and fusion to the development of nuclear weapons has had a profound effect on our lives in the past 40 years. The peaceful application of these reactions to the development of our energy resources may have an even greater effect on the future. In this section we shall look at some of the features of fission and fusion that are important for their application in reactors to generate power.

The fission of uranium was discovered in 1939 by Hahn and Strassmann, who found, by careful chemical analysis, that medium-mass elements such as barium and lanthanum were produced in the bombardment of uranium with neutrons. The discovery that several neutrons are emitted in the fission process led to speculation concerning the possibility of using these neutrons to cause further fissions, thereby producing a chain reaction. Just 3 years later, in 1942, a group led by Fermi at the University of Chicago constructed the first nuclear reactor, using such a chain reaction. When ^{235}U captures a neutron, the nucleus emits γ rays as it deexcites to the ground state about 15 percent of the time and fissions about 85 percent of the time. The fission process is analogous to the oscillation of a liquid drop. If the oscillations are violent enough, the drop splits in two, as shown in Figure 37-10. Using a liquid-drop model, Bohr and Wheeler calculated the critical energy E_c needed by the compound nucleus ^{236}U to fission. For this nucleus the energy is 5.3 MeV, which is less than the binding energy of the last neutron in ^{236}U, 6.4 MeV. The addition of a neutron to ^{235}U therefore

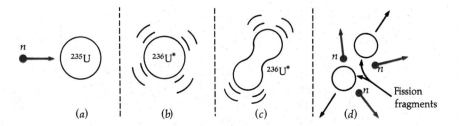

(a)   (b)   (c)   (d)

Figure 37-10
Schematic description of nuclear fission. (a) Absorption of a neutron by ^{235}U leads to ^{236}U* in excited state (b). In (c), the oscillation of ^{236}U* has become unstable. (d) The nucleus splits apart, emitting several neutrons which can produce fission in other nuclei.

produces an excited state of the ^{236}U nucleus with more than enough energy to break apart. On the other hand, the critical energy for the ^{239}U nucleus is 5.9 MeV, while the binding energy of the last neutron is only 5.2 MeV, so that when a neutron is captured by ^{238}U to form ^{239}U, the excitation energy is not great enough for fission and the nucleus deexcites by γ emission.

A fissioning nucleus can break into two medium-mass fragments in many different ways. A typical breakup is

$$n + {}^{235}\text{U} \rightarrow ({}^{236}\text{U}^*) \rightarrow {}^{141}_{56}\text{Ba} + {}^{92}_{36}\text{Kr} + 3n + Q \qquad \text{37-22}$$

For this reaction Q is 175 MeV. Other fission reactions induced by neutron capture of ^{235}U lead to other end products.

Figure 37-11 shows the distribution of fission products versus mass number. The average number of neutrons emitted in the fission of ^{235}U is about 2.5. In order to sustain a chain reaction, one of these neutrons (on the average) must be captured by another ^{235}U nucleus and cause it to fission. The *reproduction constant k* of a reactor is defined as the average number of neutrons from each fission that cause a further fission. The maximum possible value of k is 2.5, but it is less than this for two important reasons: (1) neutrons may escape from the region containing fissionable nuclei, and (2) neutrons may be captured by other nuclei in the reactor not leading to fission. If k is exactly 1, the reaction will be self-sustaining. If k is less than 1 it will die out. If k is significantly greater than 1, the reaction rate will increase rapidly and "run away." In the design of nuclear bombs, such a runaway reaction is desired; *in power reactors, the value of k must be kept very nearly equal to 1.*

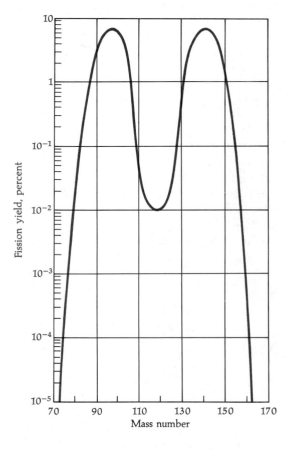

Figure 37-11
Distribution of fission fragments of ^{235}U. Symmetric fission, in which the uranium nucleus splits into two nuclei of equal mass, is less likely than asymmetric fission, in which the fragments have unequal masses.

Since the neutrons emitted in fission have energies of the order of 1 MeV whereas the cross section for capture leading to fission in ^{235}U is largest at small energies, the chain reaction is most easily sustained if the neutrons are slowed down before they escape from the reactor. At high energies (1 to 2 MeV) neutrons lose energy rapidly by inelastic scattering from ^{238}U, leaving the nucleus in an excited state that decays by γ emission. Once the neutron energy is below the excitation energies of the reactor elements (about 1 MeV), the main process of energy loss is by elastic scattering, in which a fast neutron collides with a nucleus essentially at rest and transfers some of its kinetic energy. Such energy transfers by elastic scattering are efficient only if the masses of the two bodies are comparable, e.g., a neutron will not transfer much energy in an elastic collision with a heavy ^{238}U nucleus. (Natural uranium contains 99.3 percent ^{238}U and only 0.7 percent fissionable ^{235}U.) A *moderator* consisting of material, such as water or carbon, containing light nuclei is therefore placed around the fissionable material in the core of a reactor to slow the neutrons down. Neutrons are slowed down by elastic collisions with the nuclei of the moderator until the neutrons are in thermal equilibrium with the moderator. Because of the relatively large neutron-capture cross section of the hydrogen in water, reactors using ordinary water as a moderator must use enriched uranium, in which the ^{235}U content is increased from 0.7 to 1 to 4 percent. Natural uranium can be used if heavy water (D_2O) replaces ordinary (light) water (H_2O) as the moderator. Although heavy water is expensive, most Canadian reactors use it for a moderator to avoid the cost of constructing uranium-enrichment facilities.

The ability to control the reproduction factor k is important if a power reactor is to operate with any degree of safety. There are both natural negative-feedback mechanisms and mechanical methods of control. If k is greater than 1 and the reaction rate increases, the temperature of the reactor increases. If water is used as a moderator, its density decreases with increasing temperature and it becomes a less effective moderator. A second important method is the use of control rods made of a material such as cadmium, which has a very large neutron-capture cross section. When a reactor is started up, the control rods are inserted, so that k is less than 1. As they are gradually withdrawn from the reactor the neutron capture decreases and k increases to 1. If k becomes greater than 1, the rods are inserted again.

Control of the reaction rate of a nuclear reactor with mechanical control rods is possible only because some of the neutrons emitted in the fission process are *delayed*. The time needed to slow down a neutron from 1 or 2 MeV to thermal energy is only of the order of a millisecond. If all the neutrons emitted in fission were prompt neutrons, i.e., emitted immediately in the fission process, mechanical control would not be possible because the reactor would run away before the rods could be inserted. However, about 0.65 percent of the neutrons emitted are delayed by an average time of about 14 s. These neutrons are emitted not in the fission process itself but in the decay of the fission fragments. A typical decay is

$$^{87}Br \rightarrow {}^{87}Kr^* + \beta^-$$
$$\downarrow$$
$$^{86}Kr + n$$

In this decay, which has a 56-s half-life, the $^{87}Kr^*$ nucleus has enough excitation energy to emit a neutron. This neutron is thus delayed by 56 s on the average. The effect of the delayed neutrons can be seen in the following examples.

Example 37-7 If the average time between fission generations (i.e., the time for a neutron emitted in one fission to cause another) is 1 ms and the reproduction factor is 1.001, how long would it take for the reaction rate to double?

If $k = 1.001$, the rate after N generations is 1.001^N. Setting this rate equal to 2 and solving for N, we obtain $N = 700$ generations for the rate to double. The time is then $700(1 \text{ ms}) = 0.70$ s. This is not enough time for the mechanical control of such an excursion.

Example 37-8 Assuming that 0.65 percent of the neutrons emitted are delayed by 14 s, find the average generation time and the doubling time if $k = 1.001$.

Since 99.35 percent of the generations are 10^{-3} s and 0.65 percent are 14 s, the average time is

$$t_{av} = 0.9935 \times 10^{-3} + (0.0065)(14) = 0.092 \text{ s}$$

The time for 700 generations is then

$$700(0.092 \text{ s}) = 64.4 \text{ s}$$

Because of the limited supply of the small fraction of ^{235}U in natural uranium and the limited capacity of enrichment facilities, reactors using the fission of ^{235}U cannot meet our energy needs for very long. Two possibilities hold much promise for the future: breeder reactors and controlled nuclear-fusion reactors.

The breeder reactor makes use of the fact that when the relatively plentiful but nonfissionable ^{238}U nucleus captures a neutron, it decays by β decay (half-life 20 min) to ^{239}Np, which in turn decays by β decay (half-life 2 days) to the fissionable nuclide ^{239}Pu. Since plutonium fissions with fast neutrons, no moderator is needed. A reactor with a mixture of ^{238}U and ^{239}Pu will breed as much fuel as it uses or more if one or more of the neutrons emitted in the fission of ^{239}Pu is captured by ^{238}U. Practical studies indicate that a typical breeder reactor can be expected to double its fuel supply in 7 to 10 years.

There are several safety problems inherent with breeder reactors. Since fast neutrons are used, the time between generations is essentially determined by the fraction of delayed neutrons, which is only 0.3 percent for the fission of ^{239}Pu. Mechanical control is therefore much more difficult. Also, since the operating temperature of a breeder reactor is relatively high and a moderator is not desired, a heat exchanger such as liquid sodium metal is used rather than water (which is the moderator as well as the heat-transfer material in an ordinary reactor). When the temperature of the reactor increases, the decreased density of the heat exchanger now leads to positive feedback since it now absorbs fewer neutrons than before. There is also the general safety problem with the large-scale production of plutonium, which is extremely poisonous; this is the material used in nuclear bombs. In addition, there is the problem of storage of the radioactive waste products produced in any reactor. For example, a single 1-GW ordinary reactor produces in 1 y about 3 million Ci of the radioactive ^{90}Sr (half-life 28.8 y), enough to contaminate Lake Michigan above the legal limit if it were mixed uniformly in the lake. Despite elaborate storage methods of this and other long-lived radioactive waste products, the long-term safety of large-scale production of these wastes is always open to question.

The production of power from the fusion of light nuclei holds great promise because of the relative abundance of the fuel and the lack of some of the dangers presented in fission reactors. Unfortunately, the technology has not yet been developed sufficiently for use of this plentiful energy source. We shall consider the ^{2}H + ^{3}H reaction (Equation 37-21); other reactions present similar problems.

Because of the Coulomb repulsion between the ^{2}H and ^{3}H nuclei, kinetic energies of the order of 10 keV or greater are needed to get the nuclei close enough for the attractive nuclear forces to become effective in causing fusion. Such energies can be obtained in an accelerator, but since the scattering of one nucleus by the other is much more probable than fusion, the bombardment of one nucleus by another from an accelerator requires the input of more energy than is recovered. To obtain energy from fusion, the particles must be heated to a temperature great enough for the fusion reaction to occur in random thermal collisions. The temperature corresponding to $kT = 10$ keV is of the order of 10^8 K. (This is roughly the temperature in the interiors of stars, where such reactions actually take place.) At such temperatures a gas consists of positive ions and negative electrons; such a gas is called a *plasma*. One of the problems arising in the attempt to produce controlled fusion reactions is that of confining such a plasma long enough for the reactions to take place. The energy required to heat a plasma is proportional to the density of the ions n, whereas the collision rate is proportional to the square of the density n^2. If τ is the confinement time, the output energy is proportional to $n^2\tau$. If the output energy is to exceed the input energy, we have

$$C_1 n^2 \tau > C_2 n$$

where C_1 and C_2 are constants. In 1957 the British physicist J. D. Lawson evaluated these constants from estimates of efficiencies of various hypothetical fusion reactors and derived the following relation between density and confinement time, known as *Lawson's criterion:*

$$n\tau > 10^{14} \text{ s·particles/cm}^3 \qquad 37\text{-}23$$

Lawson's criterion

Two schemes for achieving the Lawson criterion are currently under investigation. (At present, the product of density and confinement time attainable by either scheme is several orders of magnitude too small.) In one scheme a magnetic field is used to confine a hot plasma. Densities of 3×10^{13} particles/cm^3 at temperatures corresponding to $kT \approx 1$ keV have been achieved for times up to about 0.03 s, each of which is about a factor of 10 too low. In a second scheme, called *inertial confinement*, a pellet of solid deuterium and tritium is bombarded from all sides by intense pulsed laser beams of energy of the order of 10^4 J lashing about 10^{-8} s. Computer simulation studies indicate the pellet should be compressed to about 10^4 times its normal density and heated to a temperature greater than 10^8 K, producing about 10^6 J of fusion energy in 10^{-11} s, which is so brief that confinement is achieved by inertia alone.

Questions

13. Why is a moderator needed in an ordinary nuclear-fission reactor?

14. What happens to neutrons produced in fission that do not produce another fission?

15. What is the advantage of using ^{238}U as a fuel to breed ^{239}Pu rather than using ^{235}U? What are the disadvantages?

16. Why does fusion occur spontaneously in the sun but not on earth?

37-6   The Interaction of Particles with Matter

In this section we shall discuss briefly the main interactions of charged particles, neutrons, and photons with matter. Understanding these interactions is important for the study of nuclear detectors, shielding, and the effects of radiation. We shall not attempt to give a detailed theory, but shall indicate instead the principal factors involved in stopping or attenuating a beam of particles.

Charged Particles

When a charged particle traverses matter, it loses energy mainly by excitation and ionization of electrons in the matter. If the particle energy is large compared with the ionization energies of the atoms, the energy loss in each encounter with an electron will be only a small fraction of the particle energy. (A heavy particle cannot lose a large fraction of its energy to a free electron because of conservation of momentum.) Since the number of electrons in matter is so large, we can treat the problem as that of a continuous loss of energy. After a fairly well-defined distance, called the *range*, the particle has lost all its kinetic energy and stops. Near the end of the range, the continuous picture of energy loss is not valid because individual encounters are important. The statistical variation of the path length for monoenergetic particles is called *straggling*. For electrons, this can be quite important; however, for heavy particles of several million electronvolts or more, the path lengths vary by only a few percent or less.

Range

Figure 37-12 shows the rate of energy loss per unit path length $-dE/dx$ versus the energy of the ionizing particle. We see from this figure that the rate of energy loss is maximum at low energies and that for high energies the rate is approximately independent of energy. Particles with kinetic energy greater than their rest energy are called *minimum ionizing particles*. Their energy loss per unit path length or range is

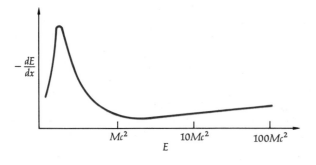

Figure 37-12
Energy loss $-dE/dx$ versus energy for a charged particle.

roughly proportional to the energy. Figure 37-13 shows the range-versus-energy curve for protons in air.

Since the energy loss of a charged particle is due to interactions with the electrons in the material, the greater the number of electrons, the greater the rate of energy loss. The energy loss rate $-dE/dx$ is approximately proportional to the density of the material. For example, the range of a 6-MeV proton is about 40 cm in air. In water, which is about 800 times denser, the range is only 0.5 mm.

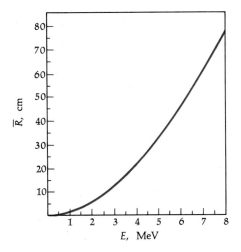

Figure 37-13
Mean range versus energy for protons in dry air. Except at low energies, the range is approximately linear with energy.

If the energy of the charged particle is large compared with its rest energy, the energy loss due to radiation as the particle slows down is important. This radiation is called *Bremsstrahlung*. The ratio of the energy lost by radiation and that lost through ionization is proportional to the energy of the particle and to the atomic number Z of the stopping material. This ratio equals 1 for electrons of about 10 MeV in lead.

The fact that the rate of energy loss for heavy charged particles is very great at very low energies (as seen from the low energy peak in Figure 37-12) has important applications in nuclear radiation therapy. Figure 37-14 shows the energy loss versus penetration distance of charged particles in water. Most of the energy is deposited near the end of the range. The peak in this curve is called the *Bragg peak*. A beam of heavy charged particles can be used to destroy cancer cells at a given depth in the body without destroying other, healthy cells if the energy is carefully chosen so that most of the energy loss occurs at the proper depth.

Figure 37-14
Energy loss of helium ions and neon ions in water versus depth of penetration. Most of the energy loss occurs near the end of the path in the Bragg peak. In general, the heavier the ion, the narrower the peak.

Neutrons

Since neutrons are uncharged, they do not interact with electrons in matter. *Neutrons lose energy by nuclear scattering, or they may be captured.* For energies large compared with kT, the most important processes are elastic and inelastic scattering. If we have a collimated beam of intensity I, any scattering or absorption will remove neutrons from the beam. If the sum of the cross sections for all such processes is σ, the number removed from the beam is σI per nucleus. If n is the number of nuclei per cubic centimetre, the number of neutrons removed in a distance dx is

$$-dN = \sigma InA\ dx$$

where A is the area of the neutron beam and $N = IA$ is the number of neutrons. Thus

$$-dN = \sigma nN\ dx$$

which gives

$$N = N_0 e^{-\sigma n x} \hspace{4cm} \text{37-24}$$

or

$$I = I_0 e^{-\sigma n x} \hspace{4cm} \text{37-25}$$

The number of neutrons in the beam decreases exponentially with distance, and there is no range.

The main source of energy loss for a neutron is usually elastic scattering. (In materials of intermediate weight, such as iron and silicon, inelastic scattering is also important. We shall neglect inelastic scattering here.) The maximum energy loss possible in one elastic collision occurs when the collision is head on. This can be calculated by considering a neutron of mass m with speed v_L making a head-on collision with a nucleus of mass M at rest in the laboratory frame (see Problem 8). The result is that the fractional energy lost by a neutron in one such collision is

$$-\frac{\Delta E}{E} = \frac{4mM}{(M + m)^2} = \frac{4\ m/M}{(1 + m/M)^2} \hspace{2cm} \text{37-26}$$

This fraction has a maximum value of 1 when $M = m$ and approaches $4\ m/M$ for $M \gg m$.

Photons

The intensity of a photon beam, like that of a neutron beam, decreases exponentially with distance through an absorbing material. The intensity is given by Equation 37-25, where σ is the cross section for absorption per atom. *The important processes that remove photons from a beam are the photoelectric effect at low energies, Compton scattering at intermediate energies, and pair production at high energies.* The total cross section for absorption is the sum of the partial cross sections σ_{pe}, σ_{cs}, σ_{pp} for these three processes. The photonuclear cross sections, such as $\sigma(\gamma, n)$, are very small compared with these atomic cross sections and can usually be neglected. The cross section for the photoelectric effect dominates at very low energy but decreases rapidly with increasing energy. The photoelectric effect cannot occur unless the electron is bound in an atom, which can recoil to conserve momentum.

If the photon energy is large compared with the binding energy of the electrons, the electrons can be considered to be free and Compton scattering is the principal mechanism for the removal of photons from the beam. If the photon energy is greater than $2m_e c^2 = 1.02$ MeV, the photon can disappear, with the creation of an electron-positron pair, a process called *pair production*.

The cross section for pair production increases rapidly with the photon energy and is the dominant term in σ at high energies. Like the photoelectric effect, pair production cannot occur in free space. If we consider the reaction $\gamma \to e^+ + e^-$, there is some reference frame in which the total momentum of the electron-positron pair is zero but there is no reference frame in which the photon's momentum is zero. Thus a nucleus is needed nearby to absorb the momentum by recoil. The cross section for pair production is proportional to Z^2 of the absorbing matter. The three partial cross sections σ_{pe}, σ_{cs}, and σ_{pp} are shown with the total cross section as functions of energy in Figure 37-15.

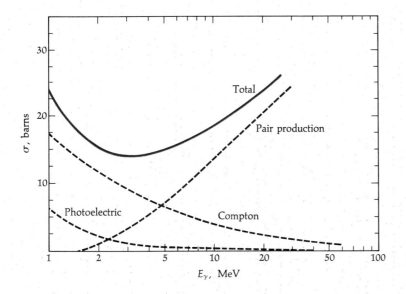

Figure 37-15
Photon attenuation cross section versus energy for lead. The cross sections for the photoelectric effect, Compton scattering, and pair production are shown by the dashed curves. The total cross section (solid curve) is the sum of these. (*From C. Davisson and R. Evans,* Reviews of Modern Physics, *vol. 24, p. 79, 1952.*)

Dosage

The biological effects of radiation are principally due to the ionization produced. Three different units are used to measure these effects. The roentgen (R) is defined as the amount of radiation that produces $\frac{1}{3} \times 10^{-9}$ C of electric charge (either positive ions or electrons) in 1 cm^3 of dry air at standard conditions. The roentgen has been largely replaced by the rad (*radiation absorbed dose*), which is defined in terms of the energy absorbed. One rad is the amount of radiation that deposits 10^{-2} J/kg of energy in any material.* Since 1 R is equivalent to the deposit of about 8.7×10^{-3} J/kg, these units are roughly equal. The amount of biological damage depends not only on the energy absorbed, which is equivalent to dependence on the number of ion pairs formed, but also on the spacing of the ion pairs; if the ion pairs are closely spaced, as in ionization by α particles, the biological effect is enhanced. The unit rem (*roentgen equivalent in man*) is the dose that has the same biological effect as 1 rad of β or γ radiation:

Rad

$$1 \text{ rem} = 1 \text{ rad} \times \text{RBE} \qquad\qquad 37\text{-}27$$

Rem

where the factor RBE (*relative biological effectiveness factor*) is tabulated for α particles, protons, and neutrons of various energies. For example, RBE is 1 for γ and β rays, about 4 or 5 for slow neutrons, about 10 for fast neutrons, and between 10 and 20 for α particles of energies from 5 to 10 MeV. Table 37-2 compares the various radiation units we have discussed. Some typical human radiation exposures are listed in Table 37-3, and some dose-limit recommendations are listed in Table 37-4.

* The SI unit joules per kilogram is now called a gray (Gy); 1 rad = 10^{-2} Gy.

Table 37-2
Radiation and dose units*

Quantity	Customary unit		SI unit		Conversion
	Name	Symbol	Name	Symbol	
Energy	Electronvolt	eV	Joule	J	$1 \text{ MeV} = 1.602 \times 10^{-13} \text{ J}$
Exposure	Roentgen	R	Coulomb per kilogram	C/kg	$1 \text{ R} = 2.58 \times 10^{-4} \text{ C/kg}$
Absorbed dose	Rad	rad or rd	Gray	Gy = J/kg	$1 \text{ rad} = 10^{-2} \text{ J/kg} = 10^{-2} \text{ Gy}$
Dose equivalent	Rem	rem	Rem	rem	
Activity	Curie	Ci	Becquerel	Bq = 1/s	$1 \text{ Ci} = 3.7 \times 10^{10} \text{ decays/s}$ $= 3.7 \times 10^{10} \text{ Bq}$

* From S. C. Bushong, *The Physics Teacher*, vol. 15, no. 3, p. 135, 1977.

Table 37-3
Sources and average intensities of human radiation exposure*

Source	mrad/y	
Natural		
Cosmic rays	45	
Terrestrial, external exposure	60	
Terrestrial, internally deposited radionuclides	25	
		130
Artificial		
Diagnostic x rays	70	
Weapons testing	<1	
Power generation	<1	
Occupational	≤1	
		70
Total		200

* From S. C. Bushong, *The Physics Teacher*, vol. 15, no. 3, p. 135, 1977.

Review

A. Define, explain, or otherwise identify:

Isotopes, 997
Binding energy, 999
α decay, 1001
β decay, 1002
γ decay, 1002
Decay constant, 1002
Half-life, 1002
Becquerel, 1004
Curie, 1004
Neutrino, 1006
Carbon dating, 1007
Q value of reaction, 1009
Cross section, 1009
Thermal neutron, 1010

Fission, 1012
Fusion, 1012
Moderator, 1015
Delayed neutron, 1015
Breeder reactor, 1016
Plasma, 1017
Lawson's criterion, 1017
Inertial confinement, 1017
Range, 1018
Bremsstrahlung, 1019
Bragg peak, 1019
Pair production, 1020
Rad, 1021
Rem, 1021

Table 37-4
Recommended dose limits*

	Maximum permissible dose equivalent for occupational exposure
Combined whole-body occupational exposure	
Prospective annual limit	5 rems in any one year
Retrospective annual limit	10–15 rems in any one year
Long-term accumulation to age N years	$(N - 18) \times 5$ rems
Skin	15 rems in any one year
Hands	75 rems in any one year (25 per quarter)
Forearms	30 rems in any one year (10 per quarter)
Other organs, tissue, and organ systems	15 rems in any one year (5 per quarter)
Fertile women (with respect to fetus)	0.5 rem in gestation period
Dose limits for nonoccupationally exposed	
Population average	0.17 rem in any one year
An individual in the population	0.5 rem in any one year
Students	0.1 rem in any one year

* Adapted from *NCRP Rep.* 39, 1971, as given in S. C. Bushong, *The Physics Teacher*, vol. 15, no. 3, p. 135, 1977.

B. True or false:

1. The atomic nucleus contains protons, neutrons, and electrons.

2. The mass of ^{2}H is less than the mass of a proton plus a neutron.

3. All nuclei are radioactive.

4. After two half-lives, all the radioactive nuclei in a given sample have decayed.

5. Exothermic reactions have no threshold energy.

6. In a breeder reactor, fuel can be produced as fast as it is consumed.

7. The number of protons in a beam decreases exponentially with distance in matter.

8. The number of photons in a beam decreases exponentially with distance in matter.

9. The rad is a measure of the energy deposited per unit volume in matter.

Exercises

Section 37-1, Atoms

There are no exercises for this section.

Section 37-2, Properties of Nuclei

1. Give the symbols for two other isotopes of (a) ^{14}N, (b) ^{56}Fe, (c) ^{118}Sn.

2. Use Table 37-1 to calculate the binding energy and the binding energy per nucleon for (a) ^{12}C, (b) ^{56}Fe, (c) ^{208}Pb.

3. Repeat Exercise 2 for (a) ^{6}Li, (b) ^{39}K, and (c) ^{238}U.

4. Given that the mass of a nucleus of mass number A is approximately $m = CA$, where C is a constant, find an expression for the nuclear density in terms of C and the constant R_0 in Equation 37-4. Compute the value of this density in grams per centimetre using the fact that C has the approximate value of 1 g per N_A nucleon.

5. Find the energy needed to remove a neutron from ^{4}He.

6. Use Equation 37-4 to compute the radii of the following nuclei: (a) ^{16}O, (b) ^{56}Fe, (c) ^{108}Ag, (d) ^{197}Au.

Section 37-3, Radioactivity

7. The stable isotope of sodium is ^{23}Na. What kind of radioactivity would you expect of (a) ^{22}Na, (b) ^{24}Na?

8. The half-life of radium is 1620 y. Calculate the number of disintegrations per second of 1 g of radium and show that the disintegration rate is approximately 1 Ci.

9. The counting rate from a radioactive source is 8000 counts/s at time $t = 0$, and 10 min later the rate is 1000 counts/s. (a) What is the half-life? (b) What is the decay constant? (c) What is the counting rate after 1 min?

10. A radioactive silver foil ($t_{1/2} = 2.4$ min) is placed near a Geiger counter, and 1000 counts/s are observed at time $t = 0$. (a) What is the counting rate at $t = 2.4$ min and $t = 4.8$ min? (b) If the counting efficiency is 20 percent, how many radioactive nuclei are there at time $t = 0$? At time $t = 2.4$ min? (c) At what time will the counting rate be about 30 counts/s?

11. Use Table 37-1 to calculate the energy of decay for the alpha decay of (a) ^{226}Ra and (b) ^{242}Pu.

12. The counting rate from a radioactive source is measured every minute. The resulting counts per second are 1000, 820, 673, 552, 453, 371, 305, 250, . . . Plot the counting rate versus time and use your graph to find the half-life.

13. A sample of wood contains 10 g of carbon and shows a ^{14}C decay rate of 100 counts/min. What is the age of the sample?

14. A bone claimed to be 10,000 y old contains 15 g of carbon. What should the decay rate of ^{14}C be from this bone?

Section 37-4, Nuclear Reactions

15. (a) Find the Q value for the reaction ^{2}H + ^{1}H → ^{3}He + n + Q using Table 37-1. (b) Find the threshold energy for this reaction if stationary ^{1}H nuclei are bombarded with ^{3}H nuclei from an accelerator. (c) Find the threshold energy for this reaction if stationary ^{3}H nuclei are bombarded with ^{1}H nuclei from an accelerator.

16. Find the Q values for the following fusion reactions:

(a) ^{2}H + ^{2}H → ^{3}He + n + Q
(b) ^{2}H + ^{2}H → ^{3}H + ^{1}H + Q
(c) ^{2}H + ^{3}He → ^{4}He + ^{1}H + Q
(d) ^{6}Li + n → ^{3}H + ^{4}He + Q

Section 37-5, Fission, Fusion, and Nuclear Reactors

17. Assuming an average energy of 200 MeV per fission, calculate the number of fissions per second needed for a 500-MW reactor.

18. If the reproduction factor in a reactor is $k = 1.1$, find the number of generations needed for the power level to (a) double, (b) increase by a factor of 10, and (c) increase by a factor of 100. Find the time needed in each case if (d) there were no delayed neutrons so that the time between generations is 1 ms and (e) if the delayed neutrons make the average time between generations 100 ms.

19. Compute the temperature T corresponding to $kT = 10$ keV, where k is Boltzmann's constant.

Section 37-6, The Interaction of Particles with Matter

20. The range of 4-MeV α particles in air is 2.5 cm ($\rho = 1.29$ mg/cm^3). Assuming the range to be inversely proportional to the density of matter, find the range of 4-MeV α particles in (a) water and (b) lead (density 11.2 g/cm^3).

21. The range of 6-MeV protons in air is approximately 45 cm. Find the approximate range of 6-MeV protons in (a) water and (b) lead (see Exercise 20).

22. If the absorption cross section for neutrons in iron is 2.5 barns, find the thickness of iron that will reduce the intensity of a neutron beam by a factor of 2 through absorption.

23. The total absorption cross section for γ rays in lead is about 20 barns at 15 MeV. (a) What thickness of lead will reduce the intensity of 15-MeV γ rays by $1/e$? (b) What thickness will reduce the intensity by a factor of 100?

Problems

1. Show that

$$ke^2 = \frac{1}{4\pi\epsilon_0} e^2 = 1.44 \text{ MeV·fm}$$

where k is the Coulomb constant.

2. Show that if the decay rate is R_0 at $t = 0$ and R_1 at some later time t_1, the decay constant λ and the half-life $t_{1/2}$ are given by

$$\lambda = t_1^{-1} \ln \frac{R_0}{R_1} \quad \text{and} \quad t_{1/2} = \frac{0.693 t_1}{\ln (R_0/R_1)}$$

3. (a) Calculate the radius of $^{141}_{56}$Ba and of $^{92}_{36}$Kr from Equation 37-4. (b) Assume that after fission of ^{235}U into ^{141}Ba and ^{92}Kr, the two nuclei are momentarily separated by a distance d equal to the sum of the radii found in (a). Calculate the electrostatic potential energy $U = k(Z_1 e)(Z_2 e)/d$ for these two nuclei at this separation and compare with the fission energy of 175 MeV.

4. Calculate the radii of ^{2}H and ^{3}H from Equation 37-4 and find the electrostatic energy when these two nuclei are "just touching"; i.e., when their centers are separated by d equal to the sum of their radii.

5. The intensity of a neutrino beam decreases exponentially with distance according to Equation 37-25, just like that of a beam of neutrons or photons. The absorption cross section for neutrinos is of the order of 10^{-20} barn. Find the thickness of iron needed to reduce the intensity of a neutrino beam by $1/e$. Compare this thickness with the distance from the earth to the sun (about 150 Gm).

6. Radioactive nuclei with decay constant λ are produced in an accelerator at a constant rate R_p. The number of radioactive nuclei N then obeys the equation $dN/dt = R_p - \lambda N$. (a) If the number N is zero at $t = 0$, sketch N versus t for this situation. (b) ^{62}Cu is produced at a rate of 100 per second by placing ordinary copper (^{63}Cu) in a beam of high-energy photons (the reaction is $\gamma + {}^{63}\text{Cu} \rightarrow {}^{62}\text{Cu} + n$). ^{62}Cu decays by β decay with a half-life of 10 min. After time long enough so that $dN/dt \approx 0$, how many ^{62}Cu nuclei are there?

7. Use the results of Problem 2 to find the decay constant and half-life if the counting rate is 1000 counts/s at time $t = 0$ and 800 counts/s 1 min later.

8. Consider a neutron of mass m moving with speed v_L and colliding head on with a nucleus of mass M. (a) Show that the speed of the center of mass in the lab frame is $V = mv_L/(m + M)$. (b) What is the speed of the nucleus in the center-of-mass frame before the collision? After the collision? (c) What is the speed of the nucleus in the original lab frame after the collision? (d) Show that the energy of the nucleus after the collision is $\frac{1}{2}M(2V)^2 = [4mM/(m + M)^2] \times \frac{1}{2}mv_L^2$ and use it to obtain Equation 37-26.

9. (a) Show that after N head-on collisions of a neutron with a carbon nucleus at rest, the energy of a neutron is approximately $0.72^N E_0$, where E_0 is its original energy. (b) How many head-on collisions are required to reduce the energy of a neutron from 2 to 0.02 MeV, assuming stationary carbon nuclei?

10. Energy is generated in the sun and other stars by fusion. One of the fusion cycles, the proton-proton cycle, consists of the following reactions:

$$^1\text{H} + {}^1\text{H} \rightarrow {}^2\text{H} + \beta^+ + \nu$$

$$^1\text{H} + {}^2\text{H} \rightarrow {}^3\text{He} + \gamma$$

followed by either

$$^3\text{He} + {}^3\text{He} \rightarrow {}^4\text{He} + {}^1\text{H} + {}^1\text{H}$$

or

$$^1\text{H} + {}^3\text{He} \rightarrow {}^4\text{He} + \beta^+ + \nu$$

(a) Show that the net effect of these reactions is

$$4\ {}^1\text{H} \rightarrow {}^4\text{He} + 2\beta^+ + 2\nu + \gamma$$

(b) Show that the rest-mass energy of 24.7 MeV is released in this cycle, not counting the 2×0.511 MeV released when each positron meets an electron and is annihilated according to $e^+ + e^- \rightarrow 2\gamma$. (c) The sun radiates energy at the rate of about 4×10^{26} W. Assuming that this is due to the conversion of four protons into helium plus γ rays and neutrinos, which releases 26.7 MeV, what is the rate of proton consumption in the sun? How long will the sun last if it continues to radiate at its present level? (Assume that protons constitute about half the total mass of the sun, which is about 2×10^{30} kg.)

APPENDIX A   SI Units

Basic units

Length	The *metre* (m) is the length equal to 1,650,763.73 wavelengths in vacuum of the radiation corresponding to the transition between the $2p_{10}$ and $5d_5$ levels of the ^{86}Kr atom
Time	The *second* (s) is the duration of 9,192,631,770 periods of the radiation corresponding to the transition between the two hyperfine levels of the ground state of the ^{133}Cs atom
Mass	The *kilogram* (kg) is the mass of the international standard body preserved at Sèvres, France
Current	The *ampere* (A) is that current in two very long parallel wires 1 m apart that gives rise to a magnetic force per unit length of 2×10^{-7} N/m
Temperature	The *kelvin* (K) is 1/273.16 of the thermodynamic temperature of the triple point of water
Luminous intensity	The *candela* (cd) is the luminous intensity, in the perpendicular direction, of a surface of area 1/600,000 m² of a blackbody at the temperature of freezing platinum at pressure of 1 atm

Derived units

Force	newton (N)	$1\ N = 1\ kg \cdot m/s^2$
Work, energy	joule (J)	$1\ J = 1\ N \cdot m$
Power	watt (W)	$1\ W = 1\ J/s$
Frequency	hertz (Hz)	$1\ Hz = s^{-1}$
Charge	coulomb (C)	$1\ C = 1\ A \cdot s$
Potential	volt (V)	$1\ V = 1\ J/C$
Resistance	ohm (Ω)	$1\ \Omega = 1\ V/A$
Capacitance	farad (F)	$1\ F = 1\ C/V$
Magnetic induction	tesla (T)	$1\ T = 1\ N/A \cdot m$
Magnetic flux	weber (Wb)	$1\ Wb = 1\ T \cdot m^2$
Inductance	henry (H)	$1\ H = 1\ J/A^2$

APPENDIX B   Numerical Data

For additional data see the front and back endpapers and the following tables in the text:

Physical constants

Gravitational constant	G	6.672×10^{-11} N·m^2/kg^2
Speed of light	c	2.997925×10^8 m/s
Electron charge	e	1.60219×10^{-19} C
Avogadro's number	N_A	6.0220×10^{23} particles/mol
Gas constant	R	8.314 J/mol·K
		1.9872 cal/mol·K
		8.206×10^{-2} L·atm/mol·K
Boltzmann's constant	$k = R/N_A$	1.3807×10^{-23} J/K
		8.617×10^{-5} eV/K
Unified mass unit	$u = 1/N_A$ g	1.6606×10^{-24} g
Coulomb constant	$k = 1/4\pi\epsilon_0$	8.98755×10^9 N·m^2/C^2
Permittivity of free space	ϵ_0	8.85419×10^{-12} C^2/N·m^2
Magnetic constant	$k_m = \mu_0/4\pi$	10^{-7} N/A^2 (exact)
Permeability of free space	μ_0	$4\pi \times 10^{-7}$ N/A^2
		1.256637×10^{-6} N/A^2
Planck's constant	h	6.6262×10^{-34} J·s
		4.1357×10^{-15} eV·s
	$\hbar = h/2\pi$	1.05459×10^{-34} J·s
		6.5822×10^{-16} eV·s
Mass of electron	m_e	9.1095×10^{-31} kg
		511.0 keV/c^2
Mass of proton	m_p	1.67265×10^{-27} kg
		938.28 MeV/c^2
Mass of neutron	m_n	1.67495×10^{-27} kg
		939.57 MeV/c^2
Bohr magneton	$m_B = e\hbar/2m_e$	9.2741×10^{-24} A-m^2
		5.7884×10^{-5} eV/T

Terrestrial data

Acceleration of gravity g	9.80665 m/s^2 standard value
	32.1740 ft/s^2
At sea level, at equator*	9.7804 m/s^2
At sea level, at poles*	9.8322 m/s^2
Mass of earth M_E	5.98×10^{24} kg
Radius of earth R_E, mean	6.37×10^6 m
	3960 mi
Escape speed $\sqrt{2R_Eg}$	1.12×10^4 m/s
	6.95 mi/s
Solar constant**	1.35 kW/m^2
Standard temperature and pressure (STP):	
T	273.15 K
P	101.325 kPa
	1.00 atm
Molecular mass of air	28.97 g/mol
Density of air (STP), ρ_{air}	1.293 kg/m^3
Speed of sound (STP)	331 m/s
Heat of fusion of H_2O (0°C, 1 atm)	333.5 kJ/kg
Heat of vaporization of H_2O (100°C, 1 atm)	2.257 MJ/kg

* Measured relative to the earth's surface.
** Average power incident normally on 1 m^2 outside the earth's atmosphere at the mean distance from the earth to the sun.

Astronomical data

Earth, distance to moon*	3.844×10^8 m
	2.389×10^5 mi
Distance (mean) to sun*	1.496×10^{11} m
	9.30×10^7 mi
	1.00 AU
Orbital speed (mean)	2.98×10^4 m/s
Moon, mass	7.35×10^{22} kg
Radius	1.738×10^6 m
Period	27.32 d
Acceleration of gravity at surface	1.62 m/s^2
Sun, mass	1.99×10^{30} kg
Radius	6.96×10^8 m

* Center to center.

APPENDIX C   Conversion Factors

Conversion factors are written as equations for simplicity; relations marked with an asterisk are exact.

Length

1 km = 0.6215 mi

1 mi = 1.609 km

1 m = 1.0936 yd = 3.281 ft = 39.37 in

*1 in = 2.54 cm

*1 ft = 12 in = 30.48 cm

*1 yd = 3 ft = 91.44 cm

1 light-year = 1 $c \cdot y$ = 9.461 × 10^{15} m

*1 Å = 0.1 nm

Area

*1 m^2 = 10^4 cm^2

1 km^2 = 0.3861 mi^2 = 247.1 acres

*1 in^2 = 6.4516 cm^2

1 ft^2 = 9.29 × 10^{-2} m^2

1 m^2 = 10.76 ft^2

*1 acre = 43,560 ft^2

1 mi^2 = 640 acres = 2.590 km^2

Volume

*1 m^3 = 10^6 cm^3

*1 L = 1000 cm^3 = 10^{-3} m^3

1 gal = 3.786 L

1 gal = 4 qt = 8 pt = 128 oz = 231 in^3

1 in^3 = 16.39 cm^3

1 ft^3 = 1728 in^3 = 28.32 L = 2.832 × 10^4 cm^3

Time

*1 h = 60 min = 3.6 ks

*1 d = 24 h = 1440 min = 86.4 ks

1 y = 365.24 d = 31.56 Ms

Speed

1 km/h = 0.2778 m/s = 0.6215 mi/h

1 mi/h = 0.4470 m/s = 1.609 km/h

1 mi/h = 1.467 ft/s

Angle and Angular Speed

*π rad = 180°

1 rad = 57.30°

1° = 1.745 × 10^{-2} rad

1 rev/min = 0.1047 rad/s

1 rad/s = 9.549 rev/min

Mass

*1 kg = 1000 g

*1 tonne = 1000 kg = 1 Mg

1 u = 1.6606 × 10^{-27} kg

1 kg = 6.022 × 10^{23} u

1 slug = 14.59 kg

1 kg = 6.852 × 10^{-2} slugs

1 u = 931.50 MeV/c^2

Density

*1 g/cm^3 = 1000 kg/m^3 = 1 kg/L

(1 g/cm^3)g = 62.4 lb/ft^3

Force

1 N = 0.2248 lb = 10^5 dyn

1 lb = 4.4482 N

(1 kg)g = 2.2046 lb

Pressure

*1 Pa = 1 N/m^2

*1 atm = 101.325 kPa = 1.01325 bars

1 atm = 14.7 lb/in^2 = 760 mmHg

     = 29.9 inHg = 33.8 ftH_2O

1 lb/in^2 = 6.895 kPa

1 torr = 1 mmHg = 133.32 Pa

1 bar = 100 kPa

Energy

*1 kW·h = 3.6 MJ

*1 cal = 4.1840 J

1 ft·lb = 1.356 J = 1.286 × 10^{-3} Btu

*1 L·atm = 101.325 J

1 L·atm = 24.217 cal

1 Btu = 778 ft·lb = 252 cal = 1054.35 J

1 eV = 1.602 × 10^{-19} J

1 u·c^2 = 931.50 MeV

*1 erg = 10^{-7} J

Power

1 horsepower = 550 ft·lb/s = 745.7 W

1 Btu/min = 17.58 W

1 W = 1.341 × 10^{-3} horsepower

     = 0.7376 ft·lb/s

Magnetic Induction

*1 G = 10^{-4} T

*1 T = 10^4 G

Thermal Conductivity

1 W/m·K = 6.938 Btu·in/h·ft^2·°F

1 Btu·in/h·ft^2·°F = 0.1441 W/m·K

APPENDIX D   # Mathematical Symbols and Formulas

Mathematical Symbols and Abbreviations

$=$	is equal to
$\neq$	is not equal to
$\approx$	is approximately equal to
$\sim$	is of the order of
$\propto$	is proportional to
$>$	is greater than
$\gg$	is much greater than
$<$	is less than
$\ll$	is much less than
Δx	change in x
$n!$	$n(n-1)(n-2)\,\cdots\,1$
Σ	sum
lim	limit
$\Delta t \to 0$	Δt approaches zero
$\int$	integral
$\dfrac{dx}{dt}$	derivative of x with respect to t
$\dfrac{\partial y}{\partial x}$	partial derivative of y with respect to x

Roots of the Quadratic Equation

If
$$ax^2 + bx + c = 0$$
then
$$x = \frac{-b}{2a} \pm \frac{1}{2a}\sqrt{b^2 - 4ac}$$

Binomial Expansion

$$(1 + x)^n = 1 + nx + \frac{n(n-1)}{2!}x^2 + \frac{n(n-1)(n-2)}{3!}x^3 + \cdots$$

To compute $(a + b)^n$ write
$$(a + b)^n = a^n(1 + x)^n \qquad \text{with } x = \frac{b}{a}$$
or
$$(a + b)^n = b^n(1 + y)^n \qquad \text{with } y = \frac{a}{b}$$

Trigonometric Formulas

See also Sec. 1-8.

$$\sin^2 \theta + \cos^2 \theta = 1 \qquad \sec^2 \theta - \tan^2 \theta = 1 \qquad \csc^2 \theta - \operatorname{ctn}^2 \theta = 1$$

$$\sin 2\theta = 2 \sin \theta \cos \theta$$

$$\cos 2\theta = \cos^2 \theta - \sin^2 \theta = 2 \cos^2 \theta - 1 = 1 - 2 \sin^2 \theta$$

$$\tan 2\theta = \frac{2 \tan \theta}{1 - \tan^2 \theta}$$

$$\sin \tfrac{1}{2}\theta = \sqrt{\frac{1 - \cos \theta}{2}} \qquad \cos \tfrac{1}{2}\theta = \sqrt{\frac{1 + \cos \theta}{2}} \qquad \tan \tfrac{1}{2}\theta = \sqrt{\frac{1 - \cos \theta}{1 + \cos \theta}}$$

$$\sin (A \pm B) = \sin A \cos B \pm \cos A \sin B$$

$$\cos (A \pm B) = \cos A \cos B \mp \sin A \sin B$$

$$\tan (A \pm B) = \frac{\tan A \pm \tan B}{1 \mp \tan A \tan B}$$

$$\sin A \pm \sin B = 2 \sin \left[\tfrac{1}{2}(A \pm B)\right] \cos \left[\tfrac{1}{2}(A \mp B)\right]$$

$$\cos A + \cos B = 2 \cos \left[\tfrac{1}{2}(A + B)\right] \cos \left[\tfrac{1}{2}(A - B)\right]$$

$$\cos A - \cos B = 2 \sin \left[\tfrac{1}{2}(A + B)\right] \sin \left[\tfrac{1}{2}(B - A)\right]$$

$$\tan A \pm \tan B = \frac{\sin (A \pm B)}{\cos A \cos B}$$

Exponential and Logarithmic Functions

$$e = 2.71828 \qquad e^0 = 1$$

If $y = e^x$, then $x = \ln y$.

$$e^{\ln x} = x$$

$$e^x e^y = e^{(x+y)}$$

$$(e^x)^y = e^{xy} = (e^y)^x$$

$$\ln e = 1 \qquad \ln 1 = 0$$

$$\ln xy = \ln x + \ln y$$

$$\ln \frac{x}{y} = \ln x - \ln y$$

$$\ln e^x = x \qquad \ln a^x = x \ln a$$

$$\ln x = (\ln 10) \log x = 2.3026 \log x$$

$$\log x = \log e \ln x = 0.43429 \ln x$$

$$e^x = 1 + x + \frac{x^2}{2!} + \frac{x^3}{3!} + \cdots$$

$$\ln (1 + x) = x - \frac{x^2}{2} + \frac{x^3}{3} - \frac{x^4}{4} + \cdots$$

APPENDIX E   Derivatives and Integrals

Table E-1 lists some important properties of derivatives and the derivatives of some particular functions which often occur in physics. It is followed by comments aimed at making these properties and rules plausible. More detailed and rigorous discussions can be found in most calculus books. Table E-2 lists some important integration formulas. More extensive lists of differentiation and integration formulas can be found in handbooks, e.g., Herbert Dwight, *Tables of Integrals and Other Mathematical Data*, 4th ed., Macmillan Publishing Co., Inc., New York, 1961.

Table E-1
Properties of derivatives and derivatives of particular functions

---

Linearity

---

1. The derivative of a constant times a function equals the constant times the derivative of the function:

$$\frac{d}{dt}[Cf(t)] = C\frac{df(t)}{dt}$$

2. The derivative of a sum of functions equals the sum of the derivatives of the functions:

$$\frac{d}{dt}[f(t) + g(t)] = \frac{df(t)}{dt} + \frac{dg(t)}{dt}$$

Chain rule

---

3. If f is a function of x and x is in turn a function of t, the derivative of f with respect to t equals the product of the derivative of f with respect to x and the derivative of x with respect to t:

$$\frac{d}{dt}f(x) = \frac{df}{dx}\frac{dx}{dt}$$

Derivative of a product

---

4. The derivative of a product of functions $f(t)g(t)$ equals the first function times the derivative of the second plus the second function times the derivative of the first:

$$\frac{d}{dt}[f(t)g(t)] = f(t)\frac{dg(t)}{dt} + \frac{df(t)}{dt}g(t)$$

Reciprocal derivative

---

5. The derivative of t with respect to x is the reciprocal of the derivative of x with respect to t, assuming that neither derivative is zero:

$$\frac{dx}{dt} = \left(\frac{dt}{dx}\right)^{-1} \quad \text{if} \quad \frac{dt}{dx} \neq 0$$

Derivatives of particular functions

6. $\dfrac{dC}{dt} = 0$ where C is a constant

7. $\dfrac{d(t^n)}{dt} = nt^{n-1}$

8. $\dfrac{d}{dt} \sin \omega t = \omega \cos \omega t$

9. $\dfrac{d}{dt} \cos \omega t = -\omega \sin \omega t$

10. $\dfrac{d}{dt} e^{bt} = be^{bt}$

11. $\dfrac{d}{dt} \ln bt = \dfrac{1}{t}$

Comment on Rules 1 to 5

Rules 1 and 2 follow from the fact that the limiting process is linear. We can understand rule 3, the chain rule, by multiplying $\Delta f/\Delta t$ by $\Delta x/\Delta x$ and noting that since x is a function of t, both Δx and Δf approach zero as Δt approaches zero. Since the limit of a product of two functions equals the product of their limits, we have

$$\lim_{\Delta t \to 0} \frac{\Delta f}{\Delta t} = \lim_{\Delta t \to 0} \frac{\Delta f}{\Delta x}\frac{\Delta x}{\Delta t} = \left(\lim_{\Delta x \to 0} \frac{\Delta f}{\Delta x}\right)\left(\lim_{\Delta t \to 0} \frac{\Delta x}{\Delta t}\right) = \frac{df}{dx}\frac{dx}{dt}$$

Rule 4 is not immediately apparent. The derivative of a product of functions is the limit of the ratio

$$\frac{f(t + \Delta t)g(t + \Delta t) - f(t)g(t)}{\Delta t}$$

If we add and subtract the quantity $f(t + \Delta t)g(t)$ in the numerator, we can write this ratio as

$$\frac{f(t + \Delta t)g(t + \Delta t) - f(t + \Delta t)g(t) + f(t + \Delta t)g(t) - f(t)g(t)}{\Delta t}$$

$$= f(t + \Delta t)\frac{g(t + \Delta t) - g(t)}{\Delta t} + g(t)\frac{f(t + \Delta t) - f(t)}{\Delta t}$$

As Δt approaches zero, the terms in brackets become $dg(t)/dt$ and $df(t)/dt$, respectively, and the limit of the expression is

$$f(t)\frac{dg(t)}{dt} + g(t)\frac{df(t)}{dt}$$

Rule 5 follows directly from the definition:

$$\frac{dx}{dt} = \lim_{\Delta t \to 0} \frac{\Delta x}{\Delta t} = \lim_{\Delta t \to 0} \left(\frac{\Delta t}{\Delta x}\right)^{-1} = \lim_{\Delta x \to 0} \left(\frac{\Delta t}{\Delta x}\right)^{-1} = \left(\lim_{\Delta x \to 0} \frac{\Delta t}{\Delta x}\right)^{-1}$$

Comment on Rule 7

We can obtain this important result using the binomial expansion (Appendix D). We have

$$f(t) = t^n$$

$$f(t + \Delta t) = (t + \Delta t)^n = t^n \left(1 + \frac{\Delta t}{t}\right)^n$$

$$= t^n \left[1 + n\frac{\Delta t}{t} + \frac{n(n-1)}{2!}\left(\frac{\Delta t}{t}\right)^2 + \frac{n(n-1)(n-2)}{3!}\left(\frac{\Delta t}{t}\right)^3 + \cdots\right]$$

Then

$$f(t - \Delta t) - f(t) = t^n \left[n\frac{\Delta t}{t} + \frac{n(n-1)}{2!}\left(\frac{\Delta t}{t}\right)^2 + \cdots\right]$$

and

$$\frac{f(t - \Delta t) - f(t)}{\Delta t} = nt^{n-1} + \frac{n(n-1)}{2!}t^{n-2}\Delta t + \cdots$$

The next term omitted from the last sum is proportional to $(\Delta t)^2$, the following to $(\Delta t)^3$, and so on. Each term except the first approaches zero as Δt approaches zero. Thus

$$\frac{df}{dt} = \lim_{\Delta x \to 0} \frac{f(t + \Delta t) + f(t)}{\Delta t} = nt^{n-1}$$

Comment on Rules 8 and 9

We first write $\sin \omega t = \sin \theta$ with $\theta = \omega t$ and use the chain rule,

$$\frac{d \sin \theta}{dt} = \frac{d \sin \theta}{d\theta}\frac{d\theta}{dt} = \omega\frac{d \sin \theta}{d\theta}$$

We then use the trigonometric formula for the sine of the sum of two angles θ and $\Delta\theta$:

$$\sin (\theta + \Delta\theta) = \sin \Delta\theta \cos \theta + \cos \Delta\theta \sin \theta$$

Since $\Delta\theta$ is to approach zero, we can use the small-angle approximations

$$\sin \Delta\theta \approx \Delta\theta \qquad \text{and} \qquad \cos \Delta\theta \approx 1$$

Then

$$\sin (\theta + \Delta\theta) \approx \Delta\theta \cos \theta + \sin \theta$$

and

$$\frac{\sin (\theta + \Delta\theta) - \sin \theta}{\Delta\theta} \approx \cos \theta$$

Similar reasoning can be applied to the cosine function to obtain rule 9.

Comment on Rule 10

Again we use the chain rule,

$$\frac{de^\theta}{dt} = b\frac{de^\theta}{d\theta} \qquad \text{with } \theta = bt$$

and use the series expansion for the exponential function (Appendix D):

$$e^{\theta + \Delta\theta} = e^\theta e^{\Delta\theta} = e^\theta \left[1 + \Delta\theta + \frac{(\Delta\theta)^2}{2!} + \frac{(\Delta\theta)^3}{3!} + \cdots\right]$$

Then

$$\frac{e^{\theta+\Delta\theta} - e^{\theta}}{\Delta\theta} = e^{\theta} + e^{\theta}\frac{\Delta\theta}{2!} + e^{\theta}\frac{(\Delta\theta)^2}{3!} + \cdots$$

As $\Delta\theta$ approaches zero, the right side of the equation above approaches e^{θ}.

Comment on Rule 11

Let

$$y = \ln bt$$

Then

$$e^y = bt \qquad \text{and} \qquad \frac{dt}{dy} = \frac{1}{b}e^y = t$$

Then using rule 5, we obtain

$$\frac{dy}{dt} = \left(\frac{dt}{dy}\right)^{-1} = \frac{1}{t}$$

Table E-2
Integration formulas*

1. $\int A\, dt = At$	5. $\int e^{bt}\, dt = \frac{1}{b}e^{bt}$
2. $\int At\, dt = \frac{1}{2}At^2$	6. $\int \cos \omega t\, dt = \frac{1}{\omega}\sin \omega t$
3. $\int At^n\, dt = A\frac{t^{n+1}}{n+1} \qquad n \neq -1$	7. $\int \sin \omega t\, dt = -\frac{1}{\omega}\cos \omega t$
4. $\int At^{-1}\, dt = A \ln t$	

* In these formulas A, b, and ω are constants. An arbitrary constant C can be added to the right side of each equation.

APPENDIX F Trigonometric Tables

Angle θ		$\sin \theta$	$\cos \theta$	$\tan \theta$	Angle θ		$\sin \theta$	$\cos \theta$	$\tan \theta$
Degree	Radian				Degree	Radian			
0	0.0000	0.0000	1.0000	0.0000					
1	0.0175	0.0175	0.9998	0.0175	46	0.8029	0.7193	0.6947	1.0355
2	0.0349	0.0349	0.9994	0.0349	47	0.8203	0.7314	0.6820	1.0724
3	0.0524	0.0523	0.9986	0.0524	48	0.8378	0.7431	0.6691	1.1106
4	0.0698	0.0698	0.9976	0.0699	49	0.8552	0.7547	0.6561	1.1504
5	0.0873	0.0872	0.9962	0.0875	50	0.8727	0.7660	0.6428	1.1918
6	0.1047	0.1045	0.9945	0.1051	51	0.8901	0.7771	0.6293	1.2349
7	0.1222	0.1219	0.9925	0.1228	52	0.9076	0.7880	0.6157	1.2799
8	0.1396	0.1392	0.9903	0.1405	53	0.9250	0.7986	0.6018	1.3270
9	0.1571	0.1564	0.9877	0.1584	54	0.9425	0.8090	0.5878	1.3764
10	0.1745	0.1736	0.9848	0.1763	55	0.9599	0.8192	0.5736	1.4281
11	0.1920	0.1908	0.9816	0.1944	56	0.9774	0.8290	0.5592	1.4826
12	0.2094	0.2079	0.9781	0.2126	57	0.9948	0.8387	0.5446	1.5399
13	0.2269	0.2250	0.9744	0.2309	58	1.0123	0.8480	0.5299	1.6003
14	0.2443	0.2419	0.9703	0.2493	59	1.0297	0.8572	0.5150	1.6643
15	0.2618	0.2588	0.9659	0.2679	60	1.0472	0.8660	0.5000	1.7321
16	0.2793	0.2756	0.9613	0.2867	61	1.0647	0.8746	0.4848	1.8040
17	0.2967	0.2924	0.9563	0.3057	62	1.0821	0.8829	0.4695	1.8807
18	0.3142	0.3090	0.9511	0.3249	63	1.0996	0.8910	0.4540	1.9626
19	0.3316	0.3256	0.9455	0.3443	64	1.1170	0.8988	0.4384	2.0503
20	0.3491	0.3420	0.9397	0.3640	65	1.1345	0.9063	0.4226	2.1445
21	0.3665	0.3584	0.9336	0.3839	66	1.1519	0.9135	0.4067	2.2460
22	0.3840	0.3746	0.9272	0.4040	67	1.1694	0.9205	0.3907	2.3559
23	0.4014	0.3907	0.9205	0.4245	68	1.1868	0.9272	0.3746	2.4751
24	0.4189	0.4067	0.9135	0.4452	69	1.2043	0.9336	0.3584	2.6051
25	0.4363	0.4226	0.9063	0.4663	70	1.2217	0.9397	0.3420	2.7475
26	0.4538	0.4384	0.8988	0.4877	71	1.2392	0.9455	0.3256	2.9042
27	0.4712	0.4540	0.8910	0.5095	72	1.2566	0.9511	0.3090	3.0777
28	0.4887	0.4695	0.8829	0.5317	73	1.2741	0.9563	0.2924	3.2709
29	0.5061	0.4848	0.8746	0.5543	74	1.2915	0.9613	0.2756	3.4874
30	0.5236	0.5000	0.8660	0.5774	75	1.3090	0.9659	0.2588	3.7321
31	0.5411	0.5150	0.8572	0.6009	76	1.3265	0.9703	0.2419	4.0108
32	0.5585	0.5299	0.8480	0.6249	77	1.3439	0.9744	0.2250	4.3315
33	0.5760	0.5446	0.8387	0.6494	78	1.3614	0.9781	0.2079	4.7046
34	0.5934	0.5592	0.8290	0.6745	79	1.3788	0.9816	0.1908	5.1446
35	0.6109	0.5736	0.8192	0.7002	80	1.3963	0.9848	0.1736	5.6713
36	0.6283	0.5878	0.8090	0.7265	81	1.4137	0.9877	0.1564	6.314
37	0.6458	0.6018	0.7986	0.7536	82	1.4312	0.9903	0.1392	7.115
38	0.6632	0.6157	0.7880	0.7813	83	1.4486	0.9925	0.1219	8.144
39	0.6807	0.6293	0.7771	0.8098	84	1.4661	0.9945	0.1045	9.514
40	0.6981	0.6428	0.7660	0.8391	85	1.4835	0.9962	0.0872	11.430
41	0.7156	0.6561	0.7547	0.8693	86	1.5010	0.9976	0.0698	14.301
42	0.7330	0.6691	0.7431	0.9004	87	1.5184	0.9986	0.0523	19.081
43	0.7505	0.6820	0.7314	0.9325	88	1.5359	0.9994	0.0349	28.636
44	0.7679	0.6947	0.7193	0.9657	89	1.5533	0.9998	0.0175	57.290
45	0.7854	0.7071	0.7071	1.0000	90	1.5708	1.0000	0.0000	∞

APPENDIX G   Periodic Table of the Elements

The values listed are based on $^{12}_{6}C = 12$ u exactly. For artificially produced elements, the approximate atomic mass of the most stable isotope is given in brackets.

PERIOD	SERIES	GROUP I	II	III	IV	V	VI	VII	VIII			0
1	1	1 H 1.00797										2 He 4.003
2	2	3 Li 6.942	4 Be 9.012	5 B 10.81	6 C 12.011	7 N 14.007	8 O 15.9994	9 F 19.00				10 Ne 20.183
3	3	11 Na 22.990	12 Mg 24.31	13 Al 26.98	14 Si 28.09	15 P 30.974	16 S 32.064	17 Cl 35.453				18 Ar 39.948
4	4	19 K 39.102	20 Ca 40.08	21 Sc 44.96	22 Ti 47.90	23 V 50.94	24 Cr 52.00	25 Mn 54.94	26 Fe 55.85	27 Co 58.93	28 Ni 58.71	
4	5	29 Cu 63.54	30 Zn 65.37	31 Ga 69.72	32 Ge 72.59	33 As 74.92	34 Se 78.96	35 Br 70.909				36 Kr 83.80
5	6	37 Rb 85.47	38 Sr 87.62	39 Y 88.905	40 Zr 91.22	41 Nb 92.91	42 Mo 95.94	43 Tc [98]	44 Ru 101.1	45 Rh 102.905	46 Pd 106.4	
5	7	47 Ag 107.870	48 Cd 112.40	49 In 114.82	50 Sn 118.69	51 Sb 121.75	52 Te 127.60	53 I 126.90				54 Xe 131.30
6	8	55 Cs 132.905	56 Ba 137.34	57-71 Lanthanide series*	72 Hf 178.49	73 Ta 180.95	74 W 183.85	75 Re 186.2	76 Os 190.2	77 Ir 192.2	78 Pt 195.09	
6	9	79 Au 196.97	80 Hg 200.59	81 Tl 204.37	82 Pb 207.19	83 Bi 208.98	84 Po [210]	85 At [210]				86 Rn [222]
7	10	87 Fr [223]	88 Ra [226]	89-103 Actinide series**								

*Lanthanide Series	57 La 138.91	58 Ce 140.12	59 Pr 140.91	60 Nd 144.24	61 Pm [147]	62 Sm 150.35	63 Eu 152.0	64 Gd 157.25	65 Tb 158.92	66 Dy 162.50	67 Ho 164.93	68 Er 167.26	69 Tm 168.93	70 Yb 173.04	71 Lu 174.97
**Actinide Series	89 Ac [227]	90 Th 232.04	91 Pa [231]	92 U 238.03	93 Np [237]	94 Pu [242]	95 Am [243]	96 Cm [247]	97 Bk [247]	98 Cf [251]	99 Es [254]	100 Fm [253]	101 Md [256]	102 No [254]	103 Lw [257]

Answers

These answers are calculated using $g = 9.81$ m/s² unless otherwise specified in the exercise or problem. The results are usually rounded to three significant figures. Differences in the last figure can easily result from differences in rounding the input data and are not important.

Chapter 20

True or False

1. False; only if the charge is positive 2. True 3. False; they diverge from a positive point charge. 4. True

Exercises

1. (a) 6.24×10^{12} (b) 6.24×10^6

3. 230 N

5. $1.5 \times 10^{-2}\mathbf{i}$ N

7. $0.914 kq^2/L^2$ away from negative charge

9. (a) $8.00\mathbf{i}$ kN/C at $x = 2$ m; $9.36\mathbf{i}$ kN/C at $x = 10$ m (b) $x = 4$ m (c) **E** points in the positive x direction just to the right of the origin and in the negative x direction just to the left of the origin.

11. (a) $-25.9\mathbf{j}$ kN/C (b) $-51.8\mathbf{j}$ μN

13. (a) $|q_{left}| = 4|q_{right}|$ (b) left positive; right negative (c) The field is strong close to charges and between them where the lines are most dense. The field is weak away from charges, particularly to the right of the negative charge, where the lines are less dense.

17. (a) $1.42\mathbf{i}$ MN/C (b) $1.92\mathbf{i}$ MN/C (c) $1.77\mathbf{i}$ MN/C (d) $125\mathbf{i}$ N/C

19. (a) $\dfrac{k\lambda}{y}\left(\dfrac{b}{\sqrt{y^2 + b^2}} + \dfrac{a}{\sqrt{y^2 + a^2}}\right)$

21. 1.8 MN/C, assuming an infinite line charge; 30 N/C, assuming a point charge

23. (a) 1.69 MN/C (b) 1.69 MN/C (c) 1.69 MN/C (d) 122 N/C

25. (a) 0 (b) 0 (c) 1.79 MN/C (d) 449 kN/C (e) 112 kN/C

27. (a) 176 GN/kg (b) 17.6 Tm/s² opposite the direction of **E** (c) 0.17 μs (d) 25.6 cm

29. (a) $-70.4\mathbf{j}$ Tm/s² (b) 50 ns (c) 8.8 cm in the negative y direction

31. The electric force is about 2.7×10^{12} times mg.

33. $8 \times 10^{-18}\mathbf{i}$ C·m

Problems

1. (a) $0.914\ kq^2/L^2$ toward other positive charge

3. (a) 4 and 2 μC (b) $3 + \sqrt{17}$ μC $= +7.12$ μC and $3 - \sqrt{17}$ μC $= -1.12$ μC

5. (c) For $x \gg a$, the separation of charges is not important, and so you expect the field to be like that of a point charge $+2q$.

7. (a) stable for displacements of q_0 along y axis and unstable for displacements along x axis, assuming other charges held fixed (b) stable for displacements of q_0 along x axis and unstable for displacements along y axis, assuming other charges held fixed (c) $q_0 = -\frac{1}{4}q$. If none of the charges is held fixed, the equilibrium is unstable for displacement of any of the charges. A stable equilibrium distribution of electric charges is not possible under electric forces only.

9. (a) $dE_x/dx = kQ/a^3$ (b) $E_x = E_x(0) + x(dE_x/dx)$ for small x, so $F_x = qE_x = -qx(kQ/a^3) = ma_x$ for small x. This is of the form $a_x = -\omega^2 x$ with $\omega = (kqQ/ma^3)^{1/2}$, the angular frequency of simple harmonic motion.

11. $E_x = kQx(x^2 + \frac{1}{4}L^2)^{-1}(x^2 + \frac{1}{2}L^2)^{-1/2}$, where $Q = 4L\lambda$. For a ring of radius $r = \frac{1}{2}L$ but carrying the same charge, $E_x = kQx(x^2 + \frac{1}{4}L^2)^{-3/2}$. *Hint:* Use Equation 20-16 for the magnitude of the field of each segment, replacing y with $r = (x^2 + \frac{1}{4}L^2)^{1/2}$. Then take 4 times the x component of this field.

13. (a) left: $-\sigma/\epsilon_0$; right: $+\sigma/\epsilon_0$; between: zero (the + direction is to the right)
(b) left: 0; right: 0; between: $+\sigma/\epsilon_0$

17. (a) 46° replacing tan θ with sin θ, or 36° replacing sin θ with tan θ (b) 40.7°

Chapter 21

True or False

1. False 2. False 3. True 4. False; it is zero in electrostatics. 5. False 6. False 7. False 8. True

Exercises

1. (a) 20.0 N·m²/C (b) 17.3 N·m²/C

3. (a) 3.14 m² (b) 71.9 kN/C (c) 226 kN·m²/C (d) No (e) 226 kN·m²/C

5. (a) N (b) N/6 (c) q/ϵ_0 (d) $q/6\epsilon_0$ (e) (b) and (d)

7. (a) 1.57 N·m²/C (b) 0 (c) 3.14 N·m²/C (d) 27.8 pC

9. (a) $E_r = kQr/R^3$ (b) $E_r = kQ/r^2$ where $Q = \frac{4}{3}\pi R^3 \rho$

11. (a) 2.85×10^{22} electrons (b) 4.56 kC

13. 11.3 kN/C

15. (a) 7.07 C/m² (b) 800 kN/C outward (c) 0 (d) 400 kN/C outward both just inside and just outside the hole. If the portion with its charge were replaced, its field would cancel the field just inside the shell and add to the field just outside.

Problems

1. (a) $Q = \pi AR^4$ (b) inside: $E_r = Ar^2/4\epsilon_0$; outside: $E_r = AR^4/4\epsilon_0 r^2 = kQ/r^2$

3. (a) $Q = 4\pi CR$ (b) inside: $E_r = C/\epsilon_0 r$; outside: $E_r = CR/\epsilon_0 r^2$

5. (a) for $r < R_1$, $E_r = 0$; for $R_1 < r < R_2$, $E_r = kq_1/r^2$; for $r > R_2$, $E_r = k(q_1 + q_2)/r^2$ (b) $q_2 = -q_1$

7. $Q = \rho \frac{4}{3}\pi(r_2^3 - r_1^3)$; for $r < r_1$, $E_r = 0$; for $r_1 < r < r_2$, $E_r = \rho(r^3 - r_1^3)/3\epsilon_0 r^2$; for $r > r_2$, $E_r = \rho(r_2^3 - r_1^3)/3\epsilon_0 r^2$

9. for $r < R$, $g_r = -GMr/R^3$; for $r > R$, $g_r = -GM/r^2$

11. (a) for $r < a$, $E_r = kq/r^2$; for $a < r < b$, $E_r = 0$; for $r > b$, $E_r = 0$ (c) uniform; the charge density would not change

15. for $r < a$, $E_r = 0$; for $a < r < b$, $E_r = (\rho/2\epsilon_0)(r^2 - a^2)/r$; for $r > b$, $E_r = (\rho/2\epsilon_0)(b^2 - a^2)/r$

Chapter 22

True or False

1. False; V must be constant in the region but need not be zero. 2. True; but if $V = 0$ at a *point*, $\mathbf{E}$ need *not* be zero at that point. If V is zero or some other constant in a *region* of space, it cannot be changing from point to point, so $\mathbf{E}$ must also be zero. 3. False 4. True 5. True

Exercises

1. (a) 24 mJ (b) -24 mJ (c) -8 kV (d) $V(x) = -2000x$ (e) $V(x) = 4000 - 2000x$ (f) $V(x) = -2000(x - 1)$ (V is in volts and x is in metres)

3. (a) positive (b) 25 kV/m

5. (a) V/m² (b) $\frac{1}{2}ax^2q_0$ (c) $V(x) = -\frac{1}{2}ax^2$

7. (a) 25.4 kV (b) 12.7 kV (c) 0

9. (a) 48.7 mJ (b) 0 (c) -12.7 mJ if like charges are adjacent, -23.3 mJ if like charges are diagonally opposite

11. (a) 12.0 kV (b) 60 mJ (c) 0; 0

13. (a) 0 (b) $E_x = 6000$ V/m (c) $V(3.01) = -60$ V; $\Delta V/\Delta x = 6000$ V/m, the same as E_x to this accuracy

15. (a) -3 kV/m (b) -3 kV/m (c) $+3$ kV/m (d) 0

17. 1.77 mm

21. $V(x)$ is maximum at $x = 0$, at which point $E_x = 0$

23. (a) 9 kV/m just outside; 0 just inside (b) 900 V both just inside and just outside (c) $V = 900$ V, $E = 0$ at center

25. 200 W

27. 300 kV

Problems

1. (a) $E_k = ke^2/2r$ (b) $\frac{1}{2}mv^2 = 2.18 \times 10^{-18}$ J $= 13.6$ eV; $E = \frac{1}{2}mv^2 + U = -13.6$ eV (c) 13.6 eV

3. (a) $V = \begin{cases} \sigma x/\epsilon_0 & \text{for } x < 0 \\ 0 & \text{for } 0 \leq x \leq a \\ -\sigma(x - a)/\epsilon_0 & \text{for } x > a \end{cases}$ (b) $V = \begin{cases} 0 & \text{for } x < 0 \\ -\sigma x/\epsilon_0 & \text{for } 0 \leq x \leq a \\ -\sigma a/\epsilon_0 & \text{for } x > a \end{cases}$

5. 44.6 fm

7. $-4.82kq^2/a$

9. (a) $\sigma d/\epsilon_0$ (b) $\sigma(d - a)/\epsilon_0$

13. $kq(b - a)/ab$

15. for $r < a$, $V = kq/r + kq(b - a)/ab$; for $a < r < b$, $V = kq/b$; for $r > b$, $V = kq/r$

17. (a) $V = \pm\dfrac{ke}{a}\left(1 - \dfrac{1}{2} + \dfrac{1}{3} - \dfrac{1}{4} + \cdots \dfrac{1}{n}\right)$; $+$ if the nth charge is $+e$, $-$ if it is $-e$ (b) $(ke^2/a)2 \ln 2 = 20.0$ eV

Chapter 23

True or False

1. False 2. False 3. False 4. True 5. True 6. False 7. True

Exercises

1. 1.13×10^7 m^2; 3.36 km

3. 466 μF

5. (a) 2 pF (b) 200 pF (c) 0.2 μF

7. (a) 9000 Mm (b) 1410

9. (a) 6.67 μF (b) 40 μC (c) 4 V across 10-μF capacitor and 2 V across 20-μF capacitor

11. (a) 14 μF (b) $\frac{8}{7}$ μF

13. (a) 100 (b) 10 V (c) $q = 10$ μC; $V = 1$ kV

15. 10 μF

17. 5×10^{-5} J

19. (a) 0.625 J (b) 1.875 J

21. (a) $\epsilon_0 A/2d$ (b) $2V$ (c) $\epsilon_0 A V^2/d$ (d) $\epsilon_0 A V^2/2d$

23. (a) 10^5 V/m (b) 4.42×10^{-2} J/m^3 (c) 8.85×10^{-5} J (d) 8.85×10^{-5} J, the same

25. (a) $E = 2.5 \times 10^4$ V/m; $\sigma = 2.21 \times 10^{-7}$ C/m^2; $U = 6.64 \times 10^{-7}$ J (b) 6.25 kV/m; 25 V (c) 1.66×10^{-7} J (d) 1.66×10^{-7} C/m^2

Problems

1. (a) 0.242 μF (b) 2.42 μC on 0.30-μF capacitor; 1.94 μC on 1.0-μF capacitor; 0.484 μC on 0.25-μF capacitor (c) 12.1 μJ

3. (a) 15.2 μF (b) Q_{12} = 2400 μC; Q_{15} = Q_4 = 632 μC (c) 0.303 J

5. (a) Q_{50} = 42.9 nC; Q_{20} = 17.1 nC (b) U_i = 90 μJ; U_f = 25.7 μJ; energy is lost

7. Each capacitor must have C = 5 μF. If all are in series, C_{eff} = 1.67 μF; if one is in series with the other two in parallel, C_{eff} = 3.33 μF; if one is in parallel with the other two in series, C_{eff} = 7.50 μF.

9. (a) 3.50 V (b) 984 μJ

11. (a) 1.2 kV (b) 640 μJ

13. (a) The potential difference between the plates is Ed, where E can be the field either in the dielectric or in the free space.

15. (a) 5.0 μF (b) 133 V

17. (a) 2.28 nF   *Note:* The approximation that this is a parallel-plate capacitor with area $2\pi rL$ and separation 2 mm gives 2.22 nF. (b) about 67 μC

19. Use 16 capacitors with 4 parallel sets of 4 in series.

21. (a) for $r < R_1$, $E = 0$, $\eta = 0$; for $R_1 < r < R_2$, $E = kQ/r^2$, $\eta = kQ^2/8\pi r^4$; for $r > R_2$, $E = 0$, $\eta = 0$ (b) $dU = (Q^2/8\pi\epsilon_0 r^2)\, dr$ (c) $\dfrac{Q^2}{8\pi\epsilon_0}\left(\dfrac{1}{R_1} - \dfrac{1}{R_2}\right) = \tfrac{1}{2}QV$

23. (a) for $r < R$, $\eta = kQ^2r^2/8\pi R^6$; for $r > R$, $\eta = kQ^2/8\pi r^4$ (b) for $r < R$, $dU = (kQ^2r^4/2R^6)\, dr$; for $r > R$, $dU = (kQ^2/2r^2)\, dr$ (c) $3kQ^2/5R$. For a conducting sphere, E is the same and therefore the energy is the same for $r > R$; but $E = 0$ for $r < R$, so there is no field energy inside the sphere.

Chapter 24   **True or False**   1. False 2. True 3. False; this is the *definition* of resistance. 4. False; true only for conductors obeying Ohm's law 5. True 6. True

Exercises

1. (a) 600 C (b) 3.75 $\times$ 10^{21} electrons

3. (a) 0.40 A (since ions and electrons flow in opposite directions, their currents add) (b) 566 A/m^2

5. (a) $f = v/2\pi r$ (b) $I = qv/2\pi r$

7. (a) 3.2 $\times$ 10^{13} protons per cubic metre (the proton speed is 6.2 $\times$ 10^7 m/s, obtained from $E_k = \tfrac{1}{2}mv^2$) (b) 3.75 $\times$ 10^{17} protons (c) $Q = 1.0 \times 10^{-3}t$ C, since current is 1.0 mA

9. (a) 1.0 V (b) 0.10 V/m

11. (a) V_{Cu}/V_{Fe} = 0.17 (b) E greater in iron

13. (a) 2.75 $\times$ 10^{-2} Ω (b) 3.0 $\times$ 10^{-2} Ω

15. 1.2 Ω

17. 46°C

19. 0.031 Ω

21. (a) 2.42 kW (b) 1.21 kW

23. (a) 48.4 Ω; 4.55 A (b) 250 W

25. (a) 0.364 cents (b) 96 cents

27. 0.03 Ω

29. (a) 30 A (b) 4 V

31. (a) 5000 A/m^2 (b) 370 nm/s (c) 5.0 $\times$ 10^{-14} s (d) about 5 nm

Problems

1. (a) 3.12 V (b) J = 4.59 $\times$ 10^6 A/m^2; E = 7.8 $\times$ 10^{-2} V/m (c) 18.7 W

3. (*a*) 10^{12} electrons (*b*) 1.6×10^{-4} A (*c*) 64 kW (*d*) 640 MW (*e*) 1/10,000

5. (*a*) 2.55×10^6 A/m² (*b*) $E_{Cu} = 0.0433$ V/m; $E_{Fe} = 0.255$ V/m (*c*) $V_{Cu} = 3.46$ V; $V_{Fe} = 12.48$ V (*d*) $7.97\ \Omega = R_{Cu} + R_{Fe}$, where $R_{Cu} = 1.73\ \Omega$ and $R_{Fe} = 6.24\ \Omega$

7. (*c*) $l_{Cu}/l_C = 264$

| Chapter 25 | True or False | 1. True 2. True 3. True 4. True 5. False 6. False |

Exercises

1. (*a*) $1.0\ \Omega$ (*b*) $I_{2\Omega} = 6$ A; $I_{3\Omega} = 4$ A; $I_{6\Omega} = 2$ A

3. (*a*) $3.6\ \Omega$ (*b*) $I_{2\Omega} = 1.33$ A; $I_{7\Omega} = 1.33$ A; $I_{6\Omega} = 2$ A

5. (*a*) 18 V (*b*) 2 A

7. (*a*) $5\ \Omega$ (*b*) upper branch: $I_{4\Omega} = 1.2$ A with 0.72 A in upper subbranch (10-Ω resistor) and 0.48 A in lower subbranch (5- and 10-Ω resistors); lower branch: $I_{5\Omega} = 1.2$ A with 0.6 A in each 10-Ω resistor

9. (*a*) 2 A through 1.5-Ω resistor; 0.75 A through each 4-Ω resistor; 0.5 A through 6-Ω resistor (*b*) 12 W

11. (*a*) $5\ \Omega$ (*b*) 30 V

13. (*a*) 4 A (*b*) 2 V (*c*) $1\ \Omega$

15. (*c*) $3\ \Omega$; 27.0 W

17. (*a*) $I_{4\Omega} = 0.667$ A; $I_{6\Omega} = 1.56$ A; $I_{3\Omega} = 0.889$ A (*b*) 9.33 V (*c*) 8 W by left battery; 10.7 W by right battery

19. (*a*) 600 μC (*b*) 0.2 A (*c*) 3 ms (*d*) 81.2 μC

21. 57.7 MΩ

23. (*a*) 3.79 μC (*b*) 2.21 μA (*c*) 2.21 μA (*d*) 13.2 μW (*e*) 4.87 μW (*f*) 8.37 μW

25. (*a*) 0.050 Ω (*b*) 4.9 kΩ

27. (*a*) 0.100 Ω (*b*) 0.100 Ω (*c*) 3.00 MΩ

29. (*a*) 1.49 kΩ (*b*) 1.50 kΩ (*c*) 13.5 kΩ

31. $R_1 = 9.90$ kΩ; $R_2 = 90.0$ kΩ; $R_3 = 900$ kΩ

33. (*a*) 43.9 Ω (*b*) 300 Ω (*c*) 3800 Ω

Problems

3. 63.0 A in second battery; 57.0 A in sick battery; 5.00 A in load. The emf of the second battery delivers 793.7 W, of which 39.7 W goes into heat in its own internal resistance, 32.5 W goes into heat in the internal resistance of the sick battery, 71.6 W goes into heat in the load resistance, and 650 W goes into internal energy of the sick battery.

5. (*a*) 2 A to the right through the 1-Ω and first 2-Ω resistor; 1 A down through the second 2-Ω resistor; 1 A down through the 6-Ω resistor (*b*) 16 W by the 8-V battery; 8 W by the upper 4-V battery and 4 W *to* the lower 4-V battery (*c*) 4 W in the 1-Ω resistor; 8 W in the horizontal 2-Ω resistor; 2 W in the vertical 2-Ω resistor; 6 W in the 6-Ω resistor

7. $V_a - V_b = 2.4$ V

9. (*a*) 1.974 A (*b*) 1.3% (*c*) 1.48 V (*d*) 0.001%

11. 600 Ω

13. (*b*) method *a* gives $R_c = 0.498\ \Omega$, 2.91 Ω, 44.4 Ω; method *b* gives 0.6 Ω, 3.1 Ω, 80.1 Ω

17. (*a*) $I = \dfrac{R_1 + R_2}{R_1 R_2}\ \varepsilon$ (*b*) $I = \varepsilon/R_2$ (*c*) $I(t) = \varepsilon\left(\dfrac{1}{R_2} + \dfrac{1}{R_1}e^{-t/R_1 C}\right)$ (*d*) 2.3×10^{-2} s

Chapter 26   **True or False**   1. True 2. True 3. True 4. False; the period is independent of the radius. 5. True

Exercises

1. $-1.28 \times 10^{-12}\mathbf{j}$ N

3. (a) $-7.2 \times 10^{-13}\mathbf{j}$ N (b) $4.8 \times 10^{-13}\mathbf{i}$ N (c) 0 (d) $9.6 \times 10^{-13}\mathbf{i} - 7.2 \times 10^{-13}\mathbf{j}$ N

5. 1.0 N

7. $20\mathbf{k}$ N/m

9. (a) 2 A·m² (b) 50 A·m

11. (a) 1 dyn·cm/G $= 10^{-3}$ A·m² (b) 1 eV/G $= 1.6 \times 10^{-15}$ A·m²

13. (a) 0.30 A·m² (b) 0.13 N·m

15. (a) 37°; $\hat{\mathbf{n}} = 0.8\mathbf{i} - 0.6\mathbf{j}$ (b) $0.48\hat{\mathbf{n}}$ A·m² $= 0.38\mathbf{i} - 0.29\mathbf{j}$ A·m² (c) $0.58\mathbf{k}$ N·m

17. 3.93×10^{-6} N·m

19. (a) 0.131 μs (b) 3.83×10^7 m/s (c) 7.66 MeV

21. (a) 2.7 mm (b) 3.5×10^{10} rad/s; 0.18 ns

23. (a) 0.13 μs (b) 2.41×10^7 m/s (c) 12.0 MeV

25. (a) $-10^4\mathbf{k}$ V/m (b) no

27. (a) 6.9 mm (b) 75 μT

29. (a) 1.44×10^8 rad/s (b) 27.1 MeV (c) assuming $m_d = 2m_p$, both answers are halved

31. (a) 1.07×10^{-2} cm/s (b) 5.84×10^{22} electrons/cm³

33. (a) 3.68×10^{-3} cm/s (b) 1.47 μV

Problems

1. 0.98 A

3. (a) $B = (Mg/IL) \tan \theta$ (b) $g \sin \theta$ up the rails

5. (a) 190 km/s (b) 9.9 mm

Chapter 27   **True or False**   1. True 2. False 3. False 4. True 5. True

Exercises

1. (a) $4.44 \times 10^{-11}\mathbf{j}$ T (b) $-1.11 \times 10^{-11}\mathbf{j}$ T (c) 0 (d) $-4.44 \times 10^{-11}\mathbf{i}$ T

3. (a) 2×10^{-11} T in direction of the unit vector $\hat{\mathbf{n}} = -(2/\sqrt{5})\mathbf{i} + (1/\sqrt{5})\mathbf{j}$ (b) $2/\sqrt{5} \times 10^{-11}\mathbf{j}$ T

5. (a) $-88.9\mathbf{k}$ μT (b) 0 (c) $88.9\mathbf{k}$ μT (d) $-160\mathbf{k}$ μT

7. (a) $-178\mathbf{k}$ μT (b) $-133\mathbf{k}$ μT (c) $-178\mathbf{k}$ μT (d) $+107\mathbf{k}$ μT

9. (a) $64\mathbf{j}$ μT (b) $-48\mathbf{k}$ μT

11. 0 due to segments along line through P; 56.6 μT inward due to each vertical segment; 113 μT inward due to horizontal segment above P; resultant $\mathbf{B}$ is 226 μT inward

13. 23.6 μT

15. (a) $32\sqrt{2}\ k_m I/L = 45.3 k_m I/L$ (b) $4\pi^2 k_m I/L = 39.5 k_m I/L$ (c) the square

17. (a) antiparallel (b) 44.7 mA

19. 28 A

21. (a) 4.5×10^{-4} N/m to right (b) 30 μT down

23. (a) $+10\mu_0$ for C_1; 0 for C_2; $-10\mu_0$ for C_3 (b) None of these paths can be used because of lack of symmetry.

25. (a) 8 G = 800 μT (b) 40 G (c) 28.6 G

27. (a) 6.03 mT (b) 0.377 A·m^2 (c) 1.51 A·m (d) 7.9 nT using the point-pole approximation

29. (a) 0.40 A·m (b) 7.76 nN

31. 4800 A/m

33. (a) 1.54 MA/m (b) 1.93 T

35. (a) 0 (b) 1.37×10^{-5} T·m^2 = 1.37×10^{-5} Wb (c) 0 (d) 1.19×10^{-5} Wb

37. 6.74×10^{-3} Wb

39. (a) 1.80×10^{15} N/C·s (b) $I_d = \epsilon_0 A \, dE/dt = 5$ A (c) 1.60×10^4 A/m^2

Problems

1. (a) $B_x = (\mu_0/4\pi L)(I_2 + 2I_1)$; $B_y = -(\mu_0/4\pi L)(2I_1 + I_2)$
   (b) $B_x = (\mu_0/4\pi L)(2I_3 - I_2)$; $B_y = (\mu_0/4\pi L)(I_2 - 2I_1)$
   (c) $B_x + (\mu_0/4\pi L)(I_2 - 2I_3)$; $B_y = -(\mu_0/4\pi L)(2I_1 + I_2)$

3. At the earth's surface the magnetic field due to the cable is 0.05 G, which is about 7% of the magnitude of the earth's field.

5. (a) 2×10^{-16} N directed opposite current (b) 3.2×10^{-16} N directed radially away from wire (c) 0

7. $v = (2\pi m/\mu_0 q I)v_t^2$

9. where $J = I/(\pi R^2 - \pi a^2)$, (a) $B_x = 0$; $B_y = \dfrac{\mu_0 J}{2}\left(\dfrac{a^2}{2R - b} - \dfrac{R}{2}\right)$

(b) $B_x = \dfrac{\mu_0 J}{2}\left(\dfrac{R}{2} - \dfrac{2Ra^2}{4R^2 - b^2}\right)$; $B_y = -\dfrac{\mu_0 J}{2}\left(\dfrac{a^2 b}{4R^2 + b^2}\right)$

13. (a) 5.40×10^{-2} T at $x = 5$ cm; 5.39×10^{-2} T at $x = 7$ cm; 5.26×10^{-2} T at $x = 9$ cm; 4.86×10^{-2} T at $x = 11$ cm

15. (a) 5.01×10^{-7} Wb (b) 7.14×10^{-5} N away from wire

17. 1.97×10^{-6} N·m

19. $k_m I$

Chapter 28

True or False

1. False; it is proportional to the rate of change of the flux. 2. True 3. False 4. False; it is an independent law. 5. True

Exercises

1. (a) 6.4×10^{-20} N (b) 0.40 V/m (c) 0.12 V

3. (a) 1.6 V (b) 0.8 A counterclockwise (c) 0.128 N (d) 1.28 W; 1.28 W

5. (a) clockwise; repel (b) counterclockwise; attract

7. (a) counterclockwise (b) clockwise

9. 199 T/s

11. (a) $\varepsilon = 0.4 - 0.2t$ V (b) at $t = 0$, $\phi_m = 0$, $\varepsilon = 0.4$ V; at $t = 2$ s, $\phi_m = -0.4$ T·m^2, $\varepsilon = 0$; at $t = 4$ s, $\phi_m = 0$, $\varepsilon = -0.4$ V; at $t = 6$ s, $\phi_m = 1.2$ T·m^2, $\varepsilon = -0.8$ V

13. (a) 1.26×10^{-3} C (b) 12.6 mA (c) 0.628 V

15. 0.80 G

17. (a) 24 Wb (b) $(-)1.60$ kV

19. (a) 60.3 G (b) 7.58×10^{-4} Wb (c) 253 μH (d) 37.9 mV

21. (b) $M = \mu_0 N_1 N_2 A_1/\ell_2$ (c) The flux through the solenoid depends on B due to the coil, and B is not uniform, especially outside the coil.

23. (a) $I = 0$; $dI/dt = 25$ A/s (b) $I = 2.26$ A; $dI/dt = 20.5$ A/s (c) $I = 7.90$ A; $dI/dt = 9.20$ A/s (d) $I = 10.8$ A; $dI/dt = 3.38$ A/s

25. (a) 13.5 mA (b) 7.44×10^{-44} A

27. 4.4 ms

29. (a) 44.06 W (b) 40.44 W (c) 3.62 W

31. (a) 2.0 A (b) 4.0 J

33. (a) $W_m = 3.98 \times 10^5$ J (b) $W_e = 4.43 \times 10^{-4}$ J (c) 3.98×10^5 J

35. 1.26 ms

37. (b) in the third circuit, since $I_{max} = \omega Q_0 = \omega CV$ and ω and V are the same for all three circuits

Problems

3. (a) for $0 \le t \le 5$ s, $\phi_m = 10^{-3}t$ Wb; for 5 s $\le t \le 10$ s, $\phi_m = 5 \times 10^{-3}$ Wb; for 10 s $\le t \le 15$ s, $\phi_m = (15 - t) \times 10^{-3}$ Wb; for 15 s $\le t$, $\phi_m = 0$ (b) for $0 \le t < 5$ s, $\mathcal{E} = -1$ mV; for 5 s $< t < 10$ s, $\mathcal{E} = 0$; for 10 s $< t < 15$ s, $\mathcal{E} = +1$ mV; for 15 s $< t$, $\mathcal{E} = 0$

5. $\mu_0 n_1 n_2 \pi r_1^2 \ell$

7. $R = 10 \ \Omega$; $L = 4$ H

9. (a) 3.53 J (b) 1.92 J

11. (b) 275 rad/s

13. (a) $I^2 R = (2400 \ W)e^{-6000t}$ (b) 0.40 J (c) 0.40 J

Chapter 29

True or False

1. True; but it is often masked by paramagnetism. 2. True 3. False 4. True 5. True 6. False

Exercises

1. (a) $\mathbf{H} = 8.00 \times 10^3 \mathbf{k}$ A/m; $\mathbf{B} = 101\mathbf{k}$ G (b) $\mathbf{H} = 8.00 \times 10^3 \mathbf{k}$ A/m; $\mathbf{B} = 1.52\mathbf{k}$ T

3. (a) 75 A·m (b) $\mathbf{H}_p = -1.19 \times 10^3 \mathbf{k}$ A/m (c) $\mathbf{H} = +6.81 \times 10^3 \mathbf{k}$ A/m (d) $\mathbf{B} = 714\mathbf{k}$ G

5. diamagnetic: H_2, CO_2, and N_2; paramagnetic: O_2

7. $\mathbf{H} = 8.00 \times 10^3 \mathbf{k}$ A/m; $\mathbf{M} = 0.544\mathbf{k}$ A/m; $\mathbf{B} = 101\mathbf{k}$ G

9. $1.69 m_B$

11. (a) $M_s = 5.6 \times 10^5$ A/m; $\mu_0 M_s = 0.70$ T (b) 5.2×10^{-4} (c) The measured magnetization includes that due to induced moments (diamagnetism), which reduces the net magnetization.

13. $\mu = 1.48 \times 10^{-5}$ H/m; $K_m = 11.7$

15. 11 A

17. (a) 1000 A/m (b) 1.26×10^6 A/m (c) 1.26×10^3

Problem

1. (b) $H = NI/2\pi R$, the same as without core since there are no end effects; $B = \mu_0 NI/2\pi R + \mu_0 M$

Chapter 30

True or False

1. True 2. True 3. True 4. False 5. True 6. True

Exercises

1. (a) 250 rad/s (b) 4.8π V = 15.1 V

3. (b) 170 V

25. (a) 8 G = 800 μT (b) 40 G (c) 28.6 G

27. (a) 6.03 mT (b) 0.377 A·m² (c) 1.51 A·m (d) 7.9 nT using the point-pole approximation

29. (a) 0.40 A·m (b) 7.76 nN

31. 4800 A/m

33. (a) 1.54 MA/m (b) 1.93 T

35. (a) 0 (b) 1.37 × 10⁻⁵ T·m² = 1.37 × 10⁻⁵ Wb (c) 0 (d) 1.19 × 10⁻⁵ Wb

37. 6.74 × 10⁻³ Wb

39. (a) 1.80 × 10¹⁵ N/C·s (b) $I_d = \epsilon_0 A\, dE/dt = 5$ A (c) 1.60 × 10⁴ A/m²

Problems

1. (a) $B_x = (\mu_0/4\pi L)(I_2 + 2I_1)$; $B_y = -(\mu_0/4\pi L)(2I_1 + I_2)$
(b) $B_x = (\mu_0/4\pi L)(2I_3 - I_2)$; $B_y = (\mu_0/4\pi L)(I_2 - 2I_1)$
(c) $B_x + (\mu_0/4\pi L)(I_2 - 2I_3)$; $B_y = -(\mu_0/4\pi L)(2I_1 + I_2)$

3. At the earth's surface the magnetic field due to the cable is 0.05 G, which is about 7% of the magnitude of the earth's field.

5. (a) 2 × 10⁻¹⁶ N directed opposite current (b) 3.2 × 10⁻¹⁶ N directed radially away from wire (c) 0

7. $v = (2\pi m/\mu_0 qI)v_t^2$

9. where $J = I/(\pi R^2 - \pi a^2)$, (a) $B_x = 0$; $B_y = \dfrac{\mu_0 J}{2}\left(\dfrac{a^2}{2R - b} - \dfrac{R}{2}\right)$

(b) $B_x = \dfrac{\mu_0 J}{2}\left(\dfrac{R}{2} - \dfrac{2Ra^2}{4R^2 - b^2}\right)$; $B_y = -\dfrac{\mu_0 J}{2}\left(\dfrac{a^2 b}{4R^2 + b^2}\right)$

13. (a) 5.40 × 10⁻² T at x = 5 cm; 5.39 × 10⁻² T at x = 7 cm; 5.26 × 10⁻² T at x = 9 cm; 4.86 × 10⁻² T at x = 11 cm

15. (a) 5.01 × 10⁻⁷ Wb (b) 7.14 × 10⁻⁵ N away from wire

17. 1.97 × 10⁻⁶ N·m

19. $k_m I$

Chapter 28

True or False

1. False; it is proportional to the rate of change of the flux. 2. True 3. False 4. False; it is an independent law. 5. True

Exercises

1. (a) 6.4 × 10⁻²⁰ N (b) 0.40 V/m (c) 0.12 V

3. (a) 1.6 V (b) 0.8 A counterclockwise (c) 0.128 N (d) 1.28 W; 1.28 W

5. (a) clockwise; repel (b) counterclockwise; attract

7. (a) counterclockwise (b) clockwise

9. 199 T/s

11. (a) $\varepsilon = 0.4 - 0.2t$ V (b) at t = 0, $\phi_m = 0$, $\varepsilon = 0.4$ V; at t = 2 s, $\phi_m = -0.4$ T·m², $\varepsilon = 0$; at t = 4 s, $\phi_m = 0$, $\varepsilon = -0.4$ V; at t = 6 s, $\phi_m = 1.2$ T·m², $\varepsilon = -0.8$ V

13. (a) 1.26 × 10⁻³ C (b) 12.6 mA (c) 0.628 V

15. 0.80 G

17. (a) 24 Wb (b) (−)1.60 kV

19. (a) 60.3 G (b) 7.58 × 10⁻⁴ Wb (c) 253 μH (d) 37.9 mV

21. (b) $M = \mu_0 N_1 N_2 A_1/\ell_2$ (c) The flux through the solenoid depends on B due to the coil, and B is not uniform, especially outside the coil.

23. (a) $I = 0$; $dI/dt = 25$ A/s (b) $I = 2.26$ A; $dI/dt = 20.5$ A/s (c) $I = 7.90$ A; $dI/dt = 9.20$ A/s (d) $I = 10.8$ A; $dI/dt = 3.38$ A/s

25. (a) 13.5 mA (b) 7.44×10^{-13} A

27. 4.4 ms

29. (a) 44.06 W (b) 40.44 W (c) 3.62 W

31. (a) 2.0 A (b) 4.0 J

33. (a) $W_m = 3.98 \times 10^2$ J (b) $W_c = 4.43 \times 10^{-1}$ J (c) 3.98×10^2 J

35. 1.26 ms

37. (b) in the third circuit, since $I_{max} = \omega Q_0 = \omega CV$ and ω and V are the same for all three circuits

Problems

3. (a) for $0 < t < 5$ s, $\phi_m = 10^{-3}t$ Wb; for 5 s $< t \le 10$ s, $\phi_m = 5 \times 10^{-3}$ Wb; for 10 s $< t < 15$ s, $\phi_m = (15 - t) \times 10^{-3}$ Wb; for 15 s $\le t$, $\phi_m = 0$ (b) for $0 \le t < 5$ s, $\mathcal{E} = -1$ mV; for 5 s $< t < 10$ s, $\mathcal{E} = 0$; for 10 s $< t < 15$ s, $\mathcal{E} = +1$ mV; for 15 s $< t$, $\mathcal{E} = 0$

5. $\mu_0 n_1 n_2 \pi r_1^2 \ell$

7. $R = 10\ \Omega$; $L = 4$ H

9. (a) 3.53 J (b) 1.92 J

11. (b) 275 rad/s

13. (a) $I^2R = (2400\ W)e^{-600t}$ (b) 0.40 J (c) 0.40 J

Chapter 29

True or False   1. True; but it is often masked by paramagnetism. 2. True 3. False 4. True 5. True 6. False

Exercises

1. (a) $\mathbf{H} = 8.00 \times 10^3\mathbf{k}$ A/m; $\mathbf{B} = 101\mathbf{k}$ G (b) $\mathbf{H} = 8.00 \times 10^3\mathbf{k}$ A/m; $\mathbf{B} = 1.52\mathbf{k}$ T

3. (a) 75 A·m (b) $\mathbf{H}_p = -1.19 \times 10^3\mathbf{k}$ A/m (c) $\mathbf{H} = +6.81 \times 10^3\mathbf{k}$ A/m (d) $\mathbf{B} = 714\mathbf{k}$ G

5. diamagnetic: H_2, CO_2, and N_2; paramagnetic: O_2

7. $\mathbf{H} = 8.00 \times 10^3\mathbf{k}$ A/m; $\mathbf{M} = 0.544\mathbf{k}$ A/m; $\mathbf{B} = 101\mathbf{k}$ G

9. $1.69 m_B$

11. (a) $M_s = 5.6 \times 10^5$ A/m; $\mu_0 M_s = 0.70$ T (b) 5.2×10^{-4} (c) The measured magnetization includes that due to induced moments (diamagnetism), which reduces the net magnetization.

13. $\mu = 1.48 \times 10^{-5}$ H/m; $K_m = 11.7$

15. 11 A

17. (a) 1000 A/m (b) 1.26×10^6 A/m (c) 1.26×10^3

Problem

1. (b) $H = NI/2\pi R$, the same as without core since there are no end effects; $B = \mu_0 NI/2\pi R + \mu_0 M$

Chapter 30

True or False   1. True 2. True 3. True 4. False 5. True 6. True

Exercises

1. (a) 250 rad/s (b) 4.8π V $= 15.1$ V

3. (b) 170 V

5. (a) 120π rad/s = 377 rad/s (b) I_{max} = 4.0 A; I_{rms} = $I_{max}/\sqrt{2}$ = 2.83 A (c) 48 W (d) 0 (e) P_{av} = $I_{rms}^2 R$ = $P_{max}/2$ = 24 W

7. (a) 2.65 MΩ (b) 26.5 kΩ (c) 26.5 Ω

11. (a) 15.9 kHz; angular frequency, 10^5 rad/s (b) 159 Hz; angular frequency, 10^3 rad/s (c) 1.59 MHz; angular frequency, 10^7 rad/s

13. (a) 0.377 Ω (b) 3.77 Ω (c) 37.7 Ω

15. 1.59 kHz

17. (a) 7.07×10^3 rad/s (b) 14.1 A (c) X_C = 62.5 Ω; X_L = 80 Ω (d) Z = 18.2 Ω; I_{rms} = 3.89 A (e) ϕ = 74.1°

19. (a) 1.13 kHz (b) X_C = 79.6 Ω; X_L = 62.8 Ω (c) Z = 17.5 Ω; I_{rms} = 0.404 A (d) $-73.4°$

21. 2000

23. (a) 14.1 (b) 500 rad/s (c) 0.275

25. (a) 0.539 (b) 95.3 mA (c) 0.726 W

27. (a) 0.553 (b) 608 mA (c) 37.0 W

29. (a) step-down (b) 2.2 V (c) $(-)$5 A

31. 10.4 turns for 2.5 V; 31.3 turns for 7.5 V; 37.5 turns for 9 V

Problems

1. (a) 0.333 (b) 26.7 Ω (c) 200 mH (d) lags; 70.5°

3. (a) L = 0.8 mH; C = 12.5 μF (b) 1.6 (c) for the given values of X_L and X_C, I_{max} = 2.0 A; at resonance, I_{max} = 5.2 A

5. (a) 3.00 A (b) 8.34 to 31.7 mH

7. (a) 933 W (b) 7.71 Ω (c) 99.9 μF (d) increase capacitance by 40.8 μF

11. (a) 12 Ω (b) R = 7.2 Ω; X = 9.6 Ω (c) capacitive

13. L = 4.00 mH; I_{max} = 100 mA

15. $\cos \phi$ = Z/R

Chapter 31 **True or False** 1. False 2. True 3. True 4. True 5. False; B = E/c 6. True

Exercises

1. (a) Equation 31-4 (b) Equation 31-5 (c) Equation 31-2 (d) Equation 31-3

3. (a) 333 nPa (b) 19.4 V/m (c) 647 nT

5. (a) 283 V/m (b) 943 nT (c) 212 W/m² (d) 708 nPa

7. (a) 40 nN (b) 80 nN

Problems

1. I = 1.91 kW/m²; E_{rms} = 849 V/m; B_{rms} = 2.83 μT; p_r = 6.37 μPa

3. (a) I = 553 mW/m²; p_r = 1.84 nPa (b) E_{rms} = 14.4 V/m; B_{rms} = 48.1 nT

5. (a) E = $\rho I/\pi a^2$ parallel to current (b) B = $\mu_0 I/2\pi a$ encircling current (c) S = $\rho I^2/2\pi^2 a^3$ radially in towards wire (d) SA = $\rho I^2 L/\pi a^2$ = $I^2 R$

Chapter 32 **Exercises** 3. 12.4 keV

5. 10 GHz

7. AM: 300 m for f = 1000 kHz; FM: 3.00 m for f = 100 MHz

9. (a) $I_1/9$ (b) $I_1/2$ (c) $I_1/16$

11. (a) 9.46×10^{12} km (b) 2.11×10^6 light years

13. 10.8 min

15. 92%

17. image is 5 cm below mirror surface

19. (*a*) 27.1° (*b*) 41.7° (*c*) 70.1° (*d*) no refracted ray; total internal reflection

21. 2.25×10^8 m/s in water; 2.00×10^8 m/s in glass

23. 2.46

25. blue: 40.8°; red: 41.8°

27. (*a*) $I_0/8$ (*b*) $3I_0/32$

29. (*a*) 30° (*b*) 1.73

31. 56.3°

Problems

5. (*a*) 62.7° (*b*) No, the critical angle for glass-air refraction is 41.8°. For angles greater than 41.8° but less than 62.7°, the light rays will enter the water, but they cannot get into the air.

7. (*a*) 37.0°

9. (*a*) 15 m (*b*) 15.4 m (*c*) 13.86 s, which is 0.19 s less than the time for the straight-line path

Chapter 33

True or False

1. True 2. True 3. False 4. True 5. False 6. True 7. False 8. True 9. False; it is smaller than the object. 10. True

Exercises

7. (*a*) $s' = 33.3$ cm; real; inverted; reduced, $y' = -y/3$
    (*b*) $s' = 50$ cm; real; inverted; same size
    (*c*) $s' = \infty$; no image
    (*d*) $s' = -16.7$ cm; virtual; erect; enlarged, $y' = 5y/3$

9. (*a*) $s' = -20$ cm; virtual; erect; reduced, $y' = y/5$
    (*b*) $s' = -16.7$ cm; virtual; erect; reduced, $y' = y/3$
    (*c*) $s' = -12.5$ cm; virtual; erect; reduced, $y' = y/2$
    (*d*) $s' = -7.14$ cm; virtual; erect; reduced, $y' = 0.714y$

13. 1.33 cm

15. $s = 10$ cm

17. (*a*) $s' = -46.15$ cm; virtual (*b*) $s' = -6.47$ cm; virtual (*c*) $s' = 44.1$ cm; real

19. (*a*) 13.6 cm (*b*) 20 cm (*c*) -10 cm (*d*) -40 cm

21. (*a*) -40 cm (*b*) $+40$ cm (*c*) -40 cm

23. A negative object distance occurs when a surface interrupts light that was converging to a point beyond the surface. The image point is then a virtual object, and the object distance is negative. (*a*) $s' = +10$ cm; real; erect (*b*) $s' = +15$ cm; real; erect

25. $s' = +10$ cm; $y' = 1$ cm

27. There are two solutions: (1) $s = 15$ cm produces a real, inverted image at $s' = 30$ cm; (2) $s = 5$ cm produces a virtual, erect image at $s' = -10$ cm.

31. The image is 30 cm to the right of the second lens. It is erect and twice as large as the object.

33. The computed distances from the vertex of the mirror are (to four significant figures): for 0.5 cm, $s' = 3.000$ cm; for 1.0 cm, $s' = 2.998$ cm; for 2.0 cm, $s' = 2.975$ cm; and for 4.0 cm, $s' = 2.674$ cm.

35. The lens should move 2.78 mm forward.

37. $\theta = 8 \times 10^{-5}$ rad; $\Delta x = 1.6$ mm

39. (a) 5.0 (b) 6.0

41. (a) 20.0 cm (b) -4 (c) -20 (d) 6.25 cm

43. (a) 0.9 cm (b) 0.180 rad

Problems

1. 0.129 m

3. concave; 90 cm

5. (a) $s' = -128$ cm (b) $s' = 14.7$ cm (c) real

7. 19.3 cm

9. The final image is the one formed by the mirror; it is real, erect, and at the center of the lens.

11. 1.6

13. The final image is virtual and 15 cm in front of the second lens.

15. (b) 2.55 m

17. $f_{\text{blue}} = 17.4$ cm

Chapter 34   **True or False**   1. False 2. True 3. True 4. True 5. True

Exercises

1. (a) 376 nm (b) $5\frac{1}{3}$ waves (c) 300°

3. (a) dark; the ray reflecting from air-glass surface suffers a phase change of 180° (b) 1.75 $\times 10^{-4}$ rad

5. 5.46 μm to 5.75 μm

7. (a) 84,848 (b) 84,873 (c) 1.0003

9. (a) 0.15 mm (b) 33.3 bands per centimetre

11. 695 nm

13. (a) 9.29 μm (b) 14

15. If $m = 0, 1, 2, \ldots$, then the maxima is $y_m = 1.2m$ cm, and the minima is $y_m = (m + \frac{1}{3})1.2$ cm or $y_m = (m + \frac{2}{3})1.2$ cm.

17. (a) $\theta_1 = \lambda/d$; $\theta_{\min} = \theta_1/6$

19. (a) 6.00×10^{-3} rad (b) 6.00×10^{-4} rad (c) 6.00×10^{-2} rad

21. 1.68 cm

23. 20.0 μm; 9

25. (a) 8.54×10^{-3} rad (b) 6.83 cm

27. (a) 55.6 km (b) 55.6 m

29. 33.6 mm

31. 484 m

Problems

1. (a) 0.300° (b) 8th order

3. 8

5. about 6.5×10^6 km using light with $\lambda = 700$ nm

7. (a) minimum thickness: 600 nm (b) 400, 514, and 720 nm (c) 400, 514, and 720 nm

9. (a) $y_1 = 35.3$ cm; $y_2 = 70.7$ cm (b) 88.4 μm (c) 8000

11. 625 and 417 nm

13. (a) 1.000292 (b) 3.7 × 10⁻⁵% (n − 1 is accurate to 0.126%)

15. (b) 491 (c) 0.99 mm

17. 12 mm for 2 slits; 6 mm for 4 slits

19. 0.37 mm

| Chapter 35 | True or False | 1. True 2. True 3. False 4. True 5. True 6. False 7. False 8. False 9. True 10. True |

Exercises

1. (a) 236 μs (b) 2.36 × 10⁻¹² s (c) no

3. *Hint:* Use the binomial expansion.

5. (a) 5.96 × 10⁻⁸ s (b) 16.1 m (c) 7.02 m

7. (a) 0.8 m (b) 4.44 ns

9. 80 c-min

13. $L_0 v/c^2 = 60$ min, which is the result of Exercise 12

15. about 0.21c = 63,000 km/s

17. 0.6c

19. (a) $x_2' - x_1' = \gamma(x_2 - x_1)$ (b) Times t_1 and t_2 in Equations 35-19 are equal to t_0, but times t_1' and t_2' in Equations 35-18 are not equal and are not known.

21. (a) $u_x = -v$; $u_y = c/\gamma$

27. (a) 4.974 MeV/c (b) 0.9948

29. (a) $938\sqrt{3}$ MeV/c = 1625 MeV/c (b) $v/c = \frac{1}{2}\sqrt{3} = 0.866$; $v = 2.60 × 10^8$ m/s

31. 6.26 MeV

33. (a) 4.5 × 10⁻⁹ u (b) about 8 × 10⁻⁹%, a very small error

Problems

1. 18,500 y

3. (a) 0.76% [$E_k = (\gamma - 1)mc^2 = 5.038 × 10^{-3}mc^2$; $\frac{1}{2}mu^2 = 5.00 × 10^{-3}mc^2$] (b) 68% ($E_k = 1.29mc^2$; $\frac{1}{2}mu^2 = 0.402mc^2$)

5. (a) 60 m (b) 0.9756c (c) 21.95 m (d) 0.25 μs

11. (a) 120 min (b) 240 min (c) 60 min; 120 min; 240 min; observer in S' makes the same table

| Chapter 36 | True or False | 1. True 2. True 3. False; it is a property of the metal. 4. True 5. True 6. False 7. True 8. True 9. True 10. True 11. True 12. False 13. True 14. False |

Exercises

1. 1.77 to 3.10 eV

3. 4.14 × 10⁻⁷ eV

5. (a) 4.13 eV (b) 2.10 eV (c) 1.00 V

7. (a) $f_t = 4.59 × 10^{14}$ Hz; $\lambda_t = 653$ nm (b) 2.23 V (c) 1.20 V

9. (a) 17.44 keV (b) 0.0760 nm (c) 16.32 keV (d) 1.12 keV

11. (a) 103 nm (b) 97.2 nm (c) 91.2 nm

13. $\lambda_{4\to3} = 1876$ nm; $\lambda_{5\to3} = 1282$ nm; $\lambda_{6\to3} = 1094$ nm; $\lambda_{\infty\to3} = 820.5$ nm

15. (a) 1.23 nm (b) 0.123 nm (c) 0.0388 nm (d) 0.0123 nm

17. 0.202 nm

19. about 5×10^{-20} m $= 5 \times 10^{-5}$ fm, where 1 fm $= 10^{-15}$ m is of the order of the diameter of the nucleus

21. 0.195 nm

Problems

1. Plot c/λ versus V_0. (a) 2.05 eV (b) about 5×10^{14} Hz (c) 4.14×10^{-15} V·s

3. (a) 10^{-22} W (b) about 53 min

7. $\mu/m = 0.99946 = R_H/R_\infty$

9. (b) $E_1 = h^2/m_p r^2 = 6.25 \times 10^{-3}$ eV (c) $T_c \approx 73$ K (d) about 13 eV using $r_a = 0.053$ nm; $T_c \approx 1.5 \times 10^5$ K

Chapter 37

True or False   1. True 2. True 3. False 4. False 5. True 6. True 7. False 8. True 9. True

Exercises

1. (a) ^{13}N, ^{15}N (b) ^{55}Fe, ^{58}Fe (or ^{57}Fe, ^{59}Fe, etc.) (c) ^{116}Sn, ^{120}Sn

3. (a) 31.99 MeV; 5.33 MeV (b) 333.7 MeV; 8.56 MeV (c) 1803.7 MeV; 7.58 MeV

5. 20.6 MeV

7. (a) β^+ (b) β^-

9. (a) 3.33 min (b) 0.208/min (c) 6500 counts per second

11. (a) 4.87 MeV (b) 7.00 MeV

13. 3350 y

15. (a) -0.764 (b) 3.05 MeV (c) 1.02 MeV

17. 1.56×10^{19} fissions per second

19. 1.2×10^8 K

21. (a) 0.58 mm (b) 51.4 μm

23. (a) 1.54 cm (b) 7.1 cm

Problems

3. (a) $R_{Ba} = 7.8$ fm; $R_{Kr} = 6.8$ fm (b) 199 MeV, slightly greater than 175 MeV

5. 1.18×10^{19} m, about 7.9×10^7 times distance from earth to sun

7. $\lambda = 3.72 \times 10^{-3}$/s; $t_{1/2} = 3.11$ min

9. (b) 55

Index

1 m = 39.37 in = 3.281 ft = 1.094 yd

1 m = 10^{15} fm = 10^{10} Å = 10^9 nm

1 km = 0.6215 mi

1 mi = 5280 ft = 1.609 km

1 in = 2.540 cm

1 L = 10^3 cm^3 = 10^{-3} m^3 = 1.057 qt

1 y = 365.24 d = 3.156×10^7 s

1 km/h = 0.278 m/s = 0.6215 mi/h

1 ft/s = 0.3048 m/s = 0.6818 mi/h

1 rev = 2π rad = 360°

1 rad = 57.30°

1 rev/min = 0.1047 rad/s

1 slug = 14.59 kg

1 tonne = 10^3 kg = 1 Mg

1 atm = 101.3 kPa = 1.013 bar = 76.00 cmHg = 14.70 lb/in^2

1 N = 10^5 dyn = 0.2248 lb

1 lb = 4.448 N

1 Pa·s = 10 poise

1 J = 10^7 erg = 0.7373 ft·lb = 9.869×10^{-3} L·atm

1 kW·h = 3.6 MJ

1 cal = 4.184 J = 4.129×10^{-2} L·atm

1 L·atm = 101.3 J = 24.22 cal

1 eV = 1.602×10^{-19} J

1 Btu = 778 ft·lb = 252 cal = 1054 J

1 horsepower = 550 ft·lb/s = 746 W

1 W/m·K = 6.938 Btu·in/h·ft^2·°F

1 T = 10^4 G

1 kg weighs about 2.205 lb